*(continued on back endpapers)*

# INTRODUCTORY ALGEBRA

## A JUST-IN-TIME APPROACH

**Alice Kaseberg**

*Lane Community College*
*Eugene, Oregon*

BROOKS/COLE PUBLISHING COMPANY

I(T)P • An International Thomson Publishing Company

Pacific Grove • Albany • Belmont • Bonn • Boston • Cincinnati • Detroit • Johannesburg • London • Madrid
Melbourne • Mexico City • New York • Paris • Singapore • Tokyo • Toronto • Washington

I(T)P ™
International Thomson Publishing
The trademark ITP is used under license.

**Library of Congress Cataloging-in-Publication Data**

Kaseberg, Alice.
    Introductory algebra : a just-in-time approach/Alice Kaseberg.
        p.    cm.
    Includes index.
    ISBN 0-534-94392-6
    1. Algebra.   I. Title.
    QA152.2.K37  1995                                95-5039
    512.9--dc20                                      CIP

Printed and bound in the United States of America
    97  98  99  —  10  9  8  7  6  5  4

**For more information, contact:**
**Brooks/Cole Publishing Company**
**511 Forest Lodge Road**
**Pacific Grove, CA 93950**

International Thomson Publishing Europe
Berkshire House I68-I73
High Holborn
London WC1V 7AA
England

Thomas Nelson Australia
102 Dodds Street
South Melbourne, 3205
Victoria, Australia

Nelson Canada
1120 Birchmont Road
Scarborough, Ontario
Canada M1K 5G4

Sponsoring Editor    *David Dietz*
Developmental Editor    *Maureen Brooks*
Production Coordinator    *Elise S. Kaiser*
Marketing Manager    *Marianne C. P. Rutter*
Manufacturing Coordinator    *Marcia A. Locke*
Production    *Lifland et al., Bookmakers*
Interior/Cover Designer    *Elise S. Kaiser*
Interior Illustrator    *Scientific Illustrators*
Cover Art    *Don Baker*
Compositor    *Beacon Graphics Corporation*
Cover Printer    *New England Book Components, Inc.*
Text Printer and Binder    *Courier/Westford*

**Credits**
Page 1—courtesy of Cleora Adkins; pages 91, 373—courtesy of Margilee Morse Kaseberg; page 161—Kevork Djansezian, AP/Wide World Photos; page 215—adapted with permission from Michelle Hymen, "Partying for Profit," *The Register Guard*, May 10, 1994; page 261—reprinted with permission of Cornell University.

 *This book is printed on recycled, acid-free paper.*

International Thomson Editores
Seneca 53
Col. Polanco
11560 México, D.F., México

International Thomson Publishing GmbH
Königswinterer Strasse 418
53227 Bonn, Germany

International Thomson Publishing Asia
221 Henderson Road
#05-10 Henderson Building
Singapore 0315

International Thomson Publishing Japan
Hirakawacho Kyowa Building, 31
2-2-1 Hirakawacho
Chiyoda-ku, Tokyo 102
Japan

Dedicated to the memory of
Oscar Schaaf, mentor, and
Scott McFadden, colleague and friend

# Contents

## SEVEN

## SQUARES AND SQUARE ROOTS: EXPRESSIONS AND EQUATIONS 317

## EIGHT

## RATIONAL EXPRESSIONS 373

INTRODUCTORY ALGEBRA: A JUST-IN-TIME APPROACH is a non-traditional approach to algebra, based on my long-time involvement in mathematics education, extensive teaching experience at the community college, high school, and junior high levels, and degrees in business administration, mathematics, and engineering science. Although somewhat less formal than many books, this text does not make for an easy course. Students are asked to integrate ideas from algebra and geometry, to make connections among tabular, graphical, and symbolic information, and to apply mathematics to real-world settings. By learning general principles and procedures as needed in the context of specific, concrete problems, students increase their understanding and retention. Stockpiling a large inventory of abstract vocabulary, definitions, and symbolic notation squanders students' valuable time and energy. Because students have their highest level of optimism and energy at the beginning of the term, I focus on introducing important "big picture" algebraic concepts first.

The college developmental algebra population comprises at least four different types of students: those who have never had algebra, those who have been out of school for a while and need to relearn forgotten math skills, those who are fresh out of high school but need to review, and those who failed algebra in high school. Each of these groups may benefit from an approach to algebra characterized by the following concerns:

♦ What purpose does algebra serve?

♦ How much can students reasonably retain?

♦ Why do students forget?

♦ Why do students fail algebra in the first place?

Developments such as the establishment of the AMATYC and NCTM standards, the calculus reform movement, and widespread acceptance of graphing technology represent the culmination of decades of research into problem solving, student learning styles, and teaching strategies. The AMATYC and NCTM standards are intended to address the questions above. Many curriculum projects at various levels have underscored the importance of a trifold presentation: numerical, visual, and symbolic. Such an approach serves many learning styles. The use of graphing technology—specifically, calculators that may be carried with the textbook—enhances algebra at all levels. *Introductory Algebra: A Just-in-Time Approach* reflects my agreement with the philosophy advocated by the standards and many new curriculum projects, as well as the use of graphing technology.

*Introductory Algebra: A Just-in-Time Approach* is intended for college students in any of the four categories above who have passed a test covering arithmetic skills through percent concepts. For a one-quarter course, some familiarity with algebra, even in the distant past, is helpful. For a two-quarter or semester course, no prior algebra work is needed. The text is sufficiently different in its approach to provide new learning and thinking experiences for the student with some algebra background and sufficiently complete in its presentation and practice to provide basic skills for the student with no prior algebra.

"**B**Y LEARNING GENERAL PRINCIPLES AND PROCEDURES AS NEEDED IN THE CONTEXT OF SPECIFIC, CONCRETE PROBLEMS, STUDENTS INCREASE THEIR UNDERSTANDING AND RETENTION."

"**T**HE COLLEGE DEVELOPMENTAL ALGEBRA POPULATION COMPRISES AT LEAST FOUR DIFFERENT TYPES OF STUDENTS."

## PEDAGOGY

### Objectives

The learning outcomes for each section are listed at the beginning of the section. They serve as summaries for students and instructors. If students get distracted by the complexities of the applications, the objectives return focus to what is important.

### Warm-ups

The Warm-up at the beginning of each section is designed to serve as a class opener, reviewing important concepts and linking prior and current topics. Warm-ups tend to be skill-oriented; the problems generally connect to the algebra needed to solve text examples. I am indebted to Steve Givant, of Mills College, who introduced me to the technique of using key parts of the lecture as class warm-ups. In the classroom setting, warm-ups permit examples to flow more evenly, and in the textbook setting, they make reading easier for the student. For more information, see "Before Class" in the Developmental Algebra Instruction section of the *Annotated Instructor's Edition*.

### Applications

Nontrivial applications are drawn from life experiences, as well as from agriculture, art, aviation, business, economics, engineering, environmental science, health, household math, science, sports, statistics, transportation, travel, etc. Examples are selected to emphasize not only the connection between math and the real world but also that between math and the student's world. In Section 1.2, students are introduced to tables through a college tuition example. In most cases students will have paid tuition recently. They will be aware of the cost per credit, and the instructor might use the school's own tuition as a basis for discussion. In several sections, grade calculations are invoked to illustrate equation solving or building equations from word problems. Algebra is used to show the impact on a grade of not doing some portion of the course work (such as homework). Several applications, such as the Chevron credit card payment schedule, are repeated several times throughout the text so that students may observe continuity and connections among topics.

To remind students to pick up a pencil and try to solve the worked examples, the solution to each in-text example is preceded by a pencil icon:

### Tables and Graphs

The text makes extensive use of data in tabular form. Tables encourage organization of information and promote pattern observation. They also prepare students for spreadsheet technology. Where appropriate, a graph related to a table underscores the connections among algebra, geometry, statistics, and the real world. Numbers and their corresponding equations and graphs are employed to emphasize the fact that algebra is the transition language between arithmetic and analysis.

### Calculators

At the very least a scientific calculator is required for this text. General keystrokes for scientific calculators are supplied where appropriate.

The use of graphing calculator technology—even if it is only by the instructor—enhances learning, as it gives students an understanding of basic concepts. Graphing calculator keystrokes are provided in a few Graphing Calculator Technique boxes, identified by the graphing calculator icon:

For more on calculators, see the Developmental Algebra Instruction section in the *Annotated Instructor's Edition*.

"CURRICULUM PROJECTS HAVE UNDERSCORED THE IMPORTANCE OF A TRIFOLD PRESENTATION: NUMERICAL, VISUAL, AND SYMBOLIC. SUCH AN APPROACH SERVES MANY LEARNING STYLES."

"*INTRODUCTORY ALGEBRA: A JUST-IN-TIME APPROACH* REFLECTS MY AGREEMENT WITH THE PHILOSOPHY ADVOCATED BY THE STANDARDS AND MANY NEW CURRICULUM PROJECTS."

"TABLES ENCOURAGE ORGANIZATION OF INFORMATION AND PROMOTE PATTERN OBSERVATION."

### Exercise Sets

A spiral curriculum is built into the exercises. Exercises in many sections use expressions that will be studied at a later time. Early on, students evaluate expressions that investigate the relationship between $-x^2$ and $(-x)^2$ and between $a^2 + b^2$ and $(a + b)^2$ and that lead to simplifications of $a - b$ divided by $b - a$. Some order-of-operations exercises require simplification of quadratic-formula expressions. Practice with exponents includes evaluation of expressions resulting from the binomial theorem. This repetition builds familiarity with material and makes connections among seemingly isolated topics.

Within many exercise sets, students are directed back to examples in the text and asked to try the same examples with different numbers or to extend the ideas presented. This encourages students to read the text and find patterns and underlying relationships.

### Small-Group Work

Some sections contain introductory material presented in question or activity form. These are intended to be done in class in small groups. In Section 1.1, Exercise 11 on Numerical Prefixes calls attention to the fact that students have different backgrounds and experiences. The exercise immediately demonstrates how each student may contribute to the class and, in turn, learn from others. It is important to emphasize how students improve their own understanding by helping others. This type of activity is vital to getting the course off to a good start.

### Projects

Starting in Section 1.4, most exercise sets contain Projects, which are intended for group work or individual effort. These may be more complicated problems related to the topic at hand, activity-based problems involving manipulatives, or real-world applications that require research outside class. Projects suited to in-class group work are highlighted with this icon:

### Mid-Chapter Test

To keep students engaged and build their confidence, a Mid-Chapter Test is included in each chapter. This test gives students the opportunity to check their progress.

### Chapter Summary, Review Exercises, and Test

Every chapter ends with a Chapter Summary, Chapter Review Exercises, and a Chapter Test.

### Glossary/Index

For student and instructor convenience, in the combined Glossary/Index at the very end of the text, essential vocabulary is defined and referenced by page.

## ANCILLARIES TO ACCOMPANY THE TEXT

### For Instructors

***Annotated Instructor's Edition***   For the instructor's convenience, the answer to each exercise appears adjacent to that exercise, space allowing (longer answers are included in the Additional Answers section of the AIE). Annotations in the margin offer teaching hints and strategies. The *Annotated Instructor's Edition* starts out with a section entitled "How to Use *Introductory Algebra: A Just-in-Time Approach*," which includes information on course planning and teaching techniques, as well as detailed section-by-section comments. This information is included to provide fresh ideas for the experienced instructor and extensive guidelines for successful mathematics instruction for the novice instructor. With the proliferation of part-time instructors, in-service support now needs to be part of the textbook material.

"THE USE OF GRAPHING TECHNOLOGY—EVEN IF IT IS ONLY BY THE INSTRUCTOR—ENHANCES LEARNING, AS IT GIVES STUDENTS AN UNDERSTANDING OF BASIC CONCEPTS."

"THE EXERCISE IMMEDIATELY DEMONSTRATES HOW EACH STUDENT MAY CONTRIBUTE TO THE CLASS AND, IN TURN, LEARN FROM OTHERS."

***Assessment Materials***    Test items, sample tests, and a project are included for each chapter.

***Computer Testing Software***    Computerized versions of the *Assessment Materials* are available for IBM and compatibles, Windows, and Macintosh.

### For Students

***Student's Solutions Manual***    Complete worked-out solutions for all odd-numbered exercises are provided in this separate manual. For Mid-Chapter Tests and Chapter Tests, complete worked-out solutions are given for all exercises.

## ACKNOWLEDGMENTS

I would like to thank the following reviewers and class-testers for their helpful comments and significant contributions:

Carol Achs
*Mesa Community College*

Rick Armstrong
*Florissant Valley Community College*

Paula Castagna
*Fresno City College*

Deann Christianson
*University of the Pacific*

Dennis C. Ebersole
*Northampton Community College*

Grace P. Foster
*Beaufort County Community College*

Donna L. Huck
*North Central High School, Spokane*

Charlotte Hutt
*Southwest Oregon Community College*

Marveen J. McCready
*Chemeketa Community College*

Alice A. Mullaly
*Southern Oregon State College*

Susan D. Poston
*Chemeketa Community College*

Douglas Robertson
*University of Minnesota*

John Thickett
*Southern Oregon State College*

Susan M. White
*DeKalb College*

My deepest gratitude goes to my husband, Rob Bowie, to Cosmo, and to our parents for their support and understanding. A special thank-you is due to Anne for getting me started, to Tamy for keeping me going, to Cindy Rubash for answers and support, and to everyone at PWS Publishing Company and Lifland et al., Bookmakers.

*Alice Kaseberg*

"JUST IN TIME" is an industrial engineering term that describes a modern inventory management scheme in a manufacturing plant. Materials used in the manufacturing process are scheduled for purchase and delivery at the precise moment at which they are needed. The old inventory method of accumulating huge stockpiles of all manufacturing materials was costly, and during the recessions of recent decades many manufacturers either changed to the just-in-time inventory method or went out of business.

*Introductory Algebra: A Just-in-Time Approach* presents material for an introductory algebra course. "Just in time" describes this algebra in two ways. First, it refers to the book's novel approach to the study of algebra, in which you work on real-world problems, learning algebraic principles and procedures just in time as you need them. Second, it describes the book's new curriculum, which allows you to concentrate on what you need to know in this age of modern technology, just in time for the twenty-first century.

*Introductory Algebra: A Just-in-Time Approach* will help you to understand algebraic concepts, not memorize manipulation skills. Tables and graphs are used from the beginning to give numerical and visual meaning to algebra. The text permits you to use calculator technology regularly in algebra and shows you how the technology will be used in future courses.

## BEGINNING THIS COURSE

A number of different items for you to think about as you begin this course follow.

### Learning Styles

Because you and your classmates have different cultural backgrounds, with a wide variety of past and present life experiences, no single example or presentation will appeal to all of you. Consider how you best learn directions to a destination—in words over the phone, from a map, or from having been there with someone else. As a student of algebra, you may prefer words (a verbal approach), drawings, pictures, and graphs (a visual approach), getting up and moving around (a kinesthetic approach), or a combination of these approaches. To be successful in mathematics you need to know your learning style and focus on those ways of learning information that best fit your style. It is advantageous, in the long run, if you also begin to learn in the other styles. To help you achieve success with algebra, *Introductory Algebra: A Just-in-Time Approach* presents concepts in as many ways as space permits.

### Independent Thinking

Although you will find the examples helpful, this text is designed to encourage your independent thinking. Look for patterns and relationships; discover concepts for yourself; seek out applications that are meaningful to you. The more involvement you have with the material, the more useful it will be and the longer you will remember it.

### Groups

You are encouraged to work with others throughout this course. One of the most important benefits of working on mathematics with other people is that you clarify your own understanding when you explain an idea to someone else. This is especially true for the kinesthetic learner.

### Alternative Approaches

Those of you who have had algebra before may find many familiar concepts in the text. Some concepts will appear just the way you learned them the first time; others will be presented quite differently. You are being asked to learn alternative approaches, not to discard your former skills.

Alternative approaches are important for several reasons. Your old way may not work in all situations. The new way may help introduce a later concept; it may be more efficient or give more useful results. Acknowledgment of alternative approaches validates your own discoveries. New and often better ways to do mathematics are being discovered all the time.

### Are You in the Right Course?

Each mathematics course has one or more prerequisite courses. Having passed a placement test does not ensure that you remember everything you need to be successful. If you have studied the background material recently, then usually with time, effort, confidence, and patience you will be able to learn the new material. If you have had a semester or a quarter or a summer break since your last math course, it is necessary to review. If the review provided in the text is not sufficient for you to remember, say, operations with fractions, you should immediately seek advice from your instructor or outside help. Use your prior book as a reference. If you took the prerequisite course more than a year ago, you may want to retake it before going on.

### Getting a Good Start

To succeed you need to attend class, read the book before class, and do the homework in a timely manner. Plan your study time. Some students set up their schedules to have the hour after math class free, to review notes and start the assignment.

Success also depends on being prepared with the proper equipment: an appropriate calculator, a six-inch ruler also marked in centimeters, and graph paper. Good six-inch rulers have protractors printed on them.

Keep in mind that your first homework paper is a "grade application," just as a cover letter and resume are part of a job application. First impressions count. Neatness and completeness make a lasting impression on the instructor. So does having homework ready to turn in as you walk into class.

Now, get on with the course. Come back to the following suggestions if you run into difficulty at a later time.

## KEEPING UP THE GOOD WORK

### Stuck on the Homework?

Sometimes we get too close to a problem and overlook the obvious. Think about the last time someone watched you play a game of solitaire and pointed out a move that you had not seen. A fresh point of view may help. Work on assignments as early as possible, right after class or early in the weekend. This gives you the option of going back later and spending more time on a difficult problem. Make notes to yourself on homework papers. Highlight problems that were difficult and that you want to review again later. Summarize the solution method in your own words so you won't forget how to do the problem later. Highlight formulas or key steps.

Obtain help from your teacher or resource center as you need it. Don't wait until just before a test.

### Falling Behind?

If you find yourself falling behind, let your instructor know that you are trying to catch up. Set up a plan that allows two to three days for each missed assignment. Most important, do current assignments first, even if you have to skip a few problems because you missed material. Work immediately after the class session. By doing the current assignment first you will stay with the class. If you gradually complete missed work, within a reasonable amount of time you will be completely caught up. Do not skip class because you are behind or confused.

### Forgetting Material?

Many students select one or two exercises from each section and write them on $3'' \times 5''$ cards, with complete solutions on the back. These "flash cards" may then be shuffled and practiced at any time for review. Cards provide an excellent way to study for tests and the final exam. Include vocabulary words in your card set.

## STRATEGIES FOR TAKING TESTS

### Prepare Yourself Academically

1. Attend class, and do the homework completely and regularly. If there are exercises on the homework that you do not know how to do, get help—from a classmate, the teacher, or another appropriate source.

2. Work under time pressure on a regular basis. Set yourself a limited amount of time to do portions of the homework. Use a time limit when doing review exercises or the practice tests at the middle and end of each chapter. Working in one- or two-hour blocks of time is usually more productive than spending all afternoon and evening on math one day a week.

### Prepare Yourself Physically

3. Get a good night's sleep. Being rested helps you think clearly, even if you know less material.

### Prepare Yourself Mentally

Psych yourself up! This is especially important if you have your test later in the day.

4. If you have a test at 8:00 A.M., use the last few minutes before bedtime to get everything ready for the next day. Make your lunch, set your books or pack on a chair by the door, set out your umbrella or appropriate weather gear, and make sure you have change for the bus or train or that your car's tires, battery, and gasoline level are okay.

5. Plan 10 or 15 minutes of quiet time before the test. Try to arrive early, if possible.

6. Imagine yourself taking the test.

   a. Imagine writing your name on the test.

   b. Imagine reading through the test completely to see where the instructor put various types of questions.

   c. Imagine writing notes, formulas, or reminders to yourself on the test.

   d. Imagine working your favorite type of problem first.

### Take the Test Right

7. Arrive early. Be ready—pencil sharpened and homework papers ready to turn in.

8. Concentrate on doing steps a–d in item 6 above.

9. Work quickly and carefully through those problems you know how to solve. Don't spend excessive time on one problem until you have tried every problem.

# Introduction to Expressions and the Coordinate Graph

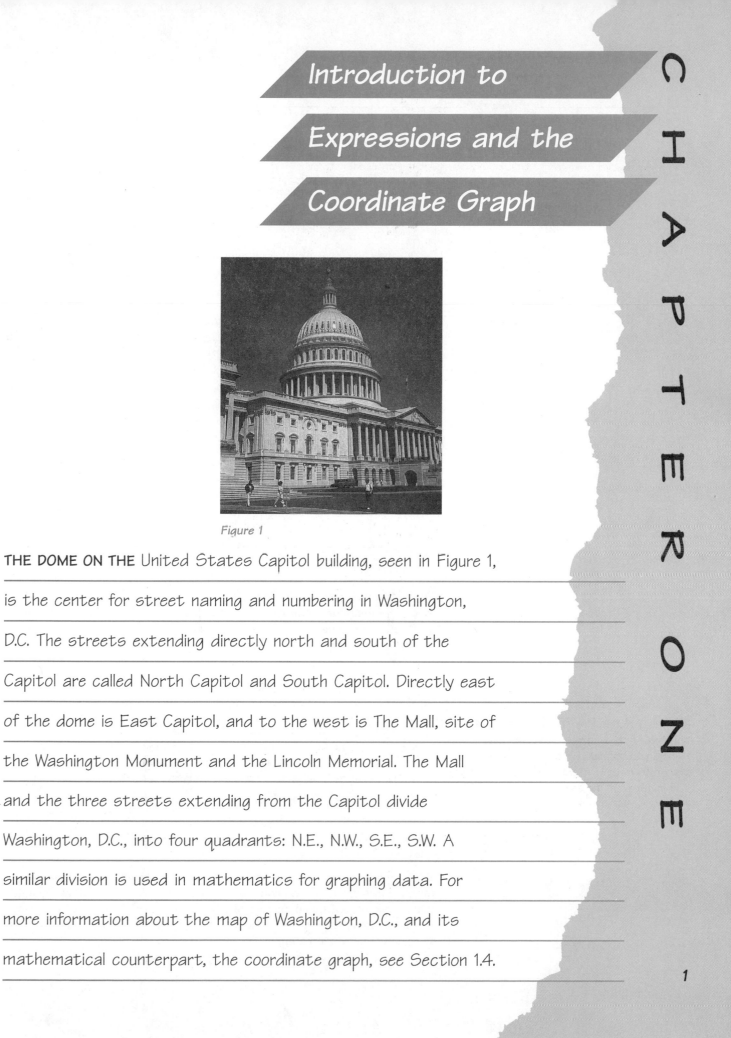

Figure 1

**THE DOME ON THE** United States Capitol building, seen in Figure 1, is the center for street naming and numbering in Washington, D.C. The streets extending directly north and south of the Capitol are called North Capitol and South Capitol. Directly east of the dome is East Capitol, and to the west is The Mall, site of the Washington Monument and the Lincoln Memorial. The Mall and the three streets extending from the Capitol divide Washington, D.C., into four quadrants: N.E., N.W., S.E., S.W. A similar division is used in mathematics for graphing data. For more information about the map of Washington, D.C., and its mathematical counterpart, the coordinate graph, see Section 1.4.

**1.1**    PROBLEM SOLVING

OBJECTIVES    Apply Polya's four strategies to problem-solving situations.  ♦  Identify assumptions in problem situations.  ♦  Determine number patterns from geometric figures.

> ### TIPS TO STUDENTS
>
> Use the Objectives to help you consider these questions:
>
> What is it that I need to learn from this section?
>
> Do I know it?
>
> How does this section fit in with prior concepts?

WARM-UP    Jake is setting up the agriculture events center for a pet show. He needs 20 pens for the pot-bellied pigs. He has a large number of straight, movable partitions that fasten together at the ends. He begins to set up partitions for the pens. How many partitions will he need for the 20 pens?    ♦

> ### STUDENT NOTE
>
> Warm-ups are intended to refresh your thinking about important concepts and to encourage your active participation in reading the text. The symbol ♦ designates the end of a Warm-up or of a solution to an example.

Mathematics—especially algebra—was developed to describe situations and to solve problems. In this section we look at four strategies for solving problems. We will examine problem-solving techniques throughout the course. In the next section we will study the language of algebra in order to describe problem situations.

### PROBLEM-SOLVING STRATEGY 1: UNDERSTAND THE PROBLEM

In the Warm-up, Jake is given the task of building pens from partitions. If he is new to the job and is given no instructions, he must first decide how to arrange the partitions. Like us, he must first understand the problem.

Sometimes the conditions or assumptions in a problem are clearly stated; sometimes they are not.

READING NOTE: Use the pencil symbol at the end of the example statement to remind you to pick up your pencil and try the problem before you read the solution. You might want to cover the solution with paper or a bookmark.

EXAMPLE    1    *Building pens: understanding the problem*    What conditions or assumptions must Jake consider?    ✎

SOLUTION    Two important conditions are access to the pens by the owners and an arrangement that allows the public to view each pen. Jake might have assumed that he had an unlimited number of partitions. He might also have assumed that he could build the pens in one long line. Are there other conditions or assumptions that might be important?

♦  ♦  ♦  ♦  ♦  ♦  ♦  ♦  ♦  ♦  ♦  ♦  ♦  ♦  ♦  ♦  ♦  ♦  ♦  ♦  ♦  ♦  ♦  ♦  ♦  ♦  ♦  ♦  ♦  ♦  ♦  ♦  ♦

In mathematical problem solving, our first step is to understand the problem. The meaning of each word in the problem is important, but understanding means more than just the words. We must also *consider conditions and assumptions.* There are times when we need to read the problem two or three times to understand it clearly and then again to gather the details.

## PROBLEM-SOLVING STRATEGY 2: MAKE A PLAN

E X A M P L E  **2**

Building pens: making a plan  Devise a plan to find the number of partitions needed to build the 20 pens. State any assumptions.

SOLUTION  Our plan might be to *draw a picture* of all 20 pens and count the partitions. In drawing a picture we make assumptions about the arrangement and the shape of the pens. For this plan we assume the 20 pens are square and form one long line. When viewed from above, 3 pens appear as in Figure 2. We assume there is no space restriction in the building and that this long arrangement will be compatible with other exhibits at the pet show.

What other pen arrangement might be appropriate?

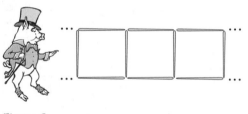

*Figure 2*

♦ ♦ ♦ ♦ ♦ ♦ ♦ ♦ ♦ ♦ ♦ ♦ ♦ ♦ ♦ ♦ ♦ ♦ ♦ ♦ ♦ ♦ ♦ ♦ ♦ ♦ ♦ ♦ ♦ ♦ ♦ ♦ ♦ ♦ ♦ ♦

In stating our assumptions we may create new and interesting problems. We may find that different assumptions lead to different solutions to the problem.

Another part of our planning is *thinking about whether we have seen similar ideas before.* We do this every time we look back through a textbook for an example similar to an exercise. Examples are useful in developing skills. Once we have mastered the skills, we use them in new and creative ways.

## PROBLEM-SOLVING STRATEGY 3: CARRY OUT THE PLAN—DO IT!

As we carry out our plan, we need to confirm that we are following our assumptions. We should check each step and *look for patterns.*

E X A M P L E  **3**

Building pens: drawing a picture  Suppose we assume that the pens are square and in one long line. Show the number of partitions needed for 20 pens.

SOLUTION  We draw a picture, such as Figure 3, showing 20 pens. How many partitions are there?

*Figure 3*

♦ ♦ ♦ ♦ ♦ ♦ ♦ ♦ ♦ ♦ ♦ ♦ ♦ ♦ ♦ ♦ ♦ ♦ ♦ ♦ ♦ ♦ ♦ ♦ ♦ ♦ ♦ ♦ ♦ ♦ ♦ ♦ ♦ ♦ ♦ ♦

Counting is tedious and may lead to error. We might look for patterns in the number of partitions by *considering a simpler problem.*

EXAMPLE 4

Building pens: making a table    If Jake first builds 1 pen and then attaches a second and a third, as shown in Figure 4, what pattern of partitions is formed? *Make a table* showing the number of pens and the number of partitions required.

SOLUTION

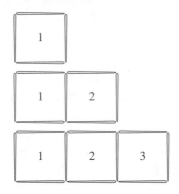

Figure 4

| Pens | Partitions |
|------|-----------|
| 1 | 4 |
| 2 | 7 |
| 3 | 10 |
| 4 | |

Table 1

We use 4 partitions in the first pen. We add 3 partitions to that pen to form a second pen. The third pen requires 3 more partitions. How do we extend Table 1 to predict the total number of partitions for 20 pens?

♦ ♦ ♦ ♦ ♦ ♦ ♦ ♦ ♦ ♦ ♦ ♦ ♦ ♦ ♦ ♦ ♦ ♦ ♦ ♦ ♦ ♦ ♦ ♦ ♦ ♦ ♦ ♦ ♦ ♦ ♦ ♦ ♦

## PROBLEM-SOLVING STRATEGY 4: CHECK THE SOLUTION AND EXTEND IT TO OTHER SITUATIONS

Go back to the Warm-up problem. Does the answer make sense in the problem? Are there patterns that confirm the result? What other assumptions and solutions are possible?

If we solve the problem with the same assumptions in two different ways and the results are the same, we may be reasonably sure the answer is correct. Do your results in Examples 3 and 4 agree?

## THE FOUR-STRATEGIES CYCLE

These four problem-solving steps—understand, plan, carry out, and check—were first published by George Polya in *How To Solve It* in 1945. Recent editions are still available in new and used bookstores. The book elaborates on the four strategies and contains many suggestions on how to draw a picture, look for a pattern, make a table, consider a simpler problem, and examine assumptions and conditions.

Sometimes we cycle through the four problem-solving steps repeatedly. For example, one approach to solving crossword puzzles uses such a cyclic method: plan, carry out, check. First, we plan to fill in all known words, do so, and then go back and check that there are no conflicting results. Second, we use letter clues to fill in more words and go back and check our results. Third, we use the newest letters to fill in still more blanks and then check. Each time through the crossword, a little more information is gained.

## THE FOUR STRATEGIES AND GROUP WORK

Problem solving as a repeated cycle works well with small groups of students. One group strategy is to work through the entire problem individually, answering everything possible. Next go through it again and find more answers. Answering one part may make another part clearer. Next *explain your thinking to others in the group and share information*. Together you may think of answers that none recalled individually.

You will find this strategy works well with homework exercises.

**Exercises 1 and 2 give traditional examples of problems in which we must think carefully about our assumptions.**

1. A man and his son were in an auto accident. The father was killed, the son critically injured. The son arrives at the hospital and is wheeled immediately into surgery. The surgeon declares, "I cannot operate on this boy; he is my son." How is this possible?

2. A researcher is following a bear. The bear travels due south 1 mile, due east another mile, and due north 1 mile. The bear has now returned to its original position. What color is the bear?

**Exercises 3 to 8 offer practice in using shapes to find number patterns and to solve problems.**

3. Cassandra is facilities manager for the agriculture events center featured in the Warm-up. When Jake comes to ask for more money to buy additional partitions, she suggests that he build the pens in a double row, as shown in the figure. How many partitions will he now need for the 20 pens?

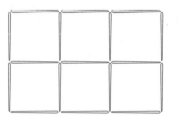

4. Jan is operations manager for a new conference center. He wants to order partitions for the exhibitor display area. The design in the figure shows exhibit locations built from 7 partitions. How many partitions will he need to build a row of 20 exhibit displays?

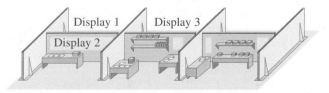

5. Mirielle has opened a restaurant. She has square tables that may be rearranged as needed. The floor plan is shown in the figure at the top of the next column. Show how she may arrange these tables to seat 12 people together.

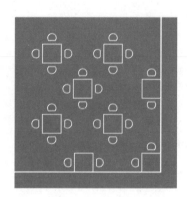

6. Mirielle hires Franck to redecorate her restaurant. He suggests buying trapezoid-shaped tables (see the figure). How many people may be seated at 1, 2, 3, and 4 of these tables if they are arranged in a long line?

7. What other shape could be made with the pet show partitions described in the Warm-up? How many partitions would be needed to set up 20 pens with this shape?

8. If the pens are all built separately, what is the total number of partitions needed?

**What assumptions do you make in the situations described in Exercises 9 and 10?**

9. You earn $20 per day. How much will you earn in a month?

10. You allow your daughter to go to a movie.

11. ***Numerical Prefixes.*** In English we have several prefixes to indicate number. The prefixes are not unique, because our language has roots in both Greek and Latin. The prefix *hex* or *sex* means "six" in *hexagon* and *sextuplets*. What word best describes each of these phrases? Here's a hint to get you started: The prefixes commonly associated with "one" are *mono* and *uni*.

a. One
   Legendary animal with one horn _____
   Ride it; one wheel; common in a circus _____
   One-lens eyepiece worn by the British _____
   Fancy letters printed on clothing or stationery
   _____

b. Two
   Two singers _____
   Two-base hit in baseball _____
   Ride it; two wheels _____
   Able to speak two languages _____
   1,000,000,000 (number with 9 zeros) _____
   Muscle in the upper arm _____
   Sea shell with two parts _____
   Glasses with two-part lenses _____
   World War I aircraft with two wings, one above the
      other _____
   Number system using only 0s and 1s _____

c. Three
   Three babies at one birth _____
   Three-sided geometric shape _____
   Stable camera stand _____
   Three-wheeled toy to ride _____

d. Four
   Group of four singers _____
   Four of these make a gallon in liquid measure _____
   Four babies at one birth _____

e. Five
   Five-sided geometric shape _____
   Olympic event (fence, swim, run, shoot, ride horse)
   _____

f. Seven
   Month when school typically begins _____
   Track event with seven activities _____

g. Eight
   Ocean animal with eight legs _____
   Full set of eight notes in music _____
   Halloween month _____
   Eight-sided geometric shape _____

h. Ten
   Ten years _____
   Last month of calendar year _____
   Dot in our number system; the _____ point
   Track event with ten activities _____

i. One hundred
   A hundred years _____
   One-hundredth of a dollar _____
   One-hundredth of a meter _____

j. One thousand
   Hans Solo's spacecraft, the _____ Falcon
   A thousand years _____

12. Find words that use a prefix meaning six or nine.

---

## 1.2   SETS OF NUMBERS AND INPUT–OUTPUT TABLES

**OBJECTIVES**   Review computation with fractions. ♦ Distinguish the mathematical meaning from the common meaning of *sum*, *difference*, *product*, and *quotient*. ♦ Read and interpret data in tabular form. ♦ Identify inputs and outputs as sets of numbers. ♦ Identify whole numbers, rational numbers, integers, and real numbers. ♦ Use a conditional table to predict outputs.

**WARM-UP**   We add, subtract, multiply, and divide $\frac{1}{3}$ and $\frac{1}{4}$ as follows.

$$\frac{1}{3} + \frac{1}{4} = \frac{4}{12} + \frac{3}{12} = \frac{7}{12}$$

$$\frac{1}{3} - \frac{1}{4} = \frac{4}{12} - \frac{3}{12} = \frac{1}{12}$$

$$\frac{1}{3} \cdot \frac{1}{4} = \frac{1}{12}$$

$$\frac{1}{3} \div \frac{1}{4} = \frac{1}{3} \cdot \frac{4}{1} = \frac{4}{3} = 1\frac{1}{3}$$

Now, add, subtract, multiply, and divide $\frac{1}{2}$ and $\frac{1}{5}$. Which operations would be meaningful if the fractions represented parts of pizzas? ♦

After an initial overview of algebra as a language, we will summarize several types of numbers and introduce input–output tables as a way of displaying number relationships or patterns.

## OVERVIEW: ALGEBRA AS A LANGUAGE

Algebra is often compared with a foreign language. It is the language used to describe numerical relationships. To learn a new language, you have to hear it in context, as well as read, speak, and write it. Speaking, hearing, reading, and writing algebra are important parts of our learning process. This textbook provides a reference and even definitions of important vocabulary, but the text does not replace your active participation in the learning process. In order to think in terms of algebra, you need to use it.

You need to be prepared for lots of detail. Algebra has its own notation, symbols, and vocabulary. The notation $2x$ is entirely different in meaning from both $x^2$ and $x_2$. Placing parentheses in an expression completely changes the meaning: $(-x)^2$ is not the same as $-x^2$. Four examples of symbols are $\sqrt{x}$, $|x|$, $x!$, and $x^2$.

To complicate matters, algebra has unique ways of using common English words.

**EXAMPLE 1**

In the first sentence of each pair, the meaning of the word in italics is not mathematical. Identify the mathematical meaning of the word in italics in the second sentence of each pair.

**a.** Luisa has a *sum* of money.
The *sum* of three and four is seven.

**b.** Bill and George had a *difference* of opinion.
The *difference* between twelve and nine is three.

**c.** The manufacturer makes a quality *product*.
The *product* of six and five is thirty.

**SOLUTION**

In mathematics, **sum**, **difference**, and **product** *refer to the answers in addition, subtraction, and multiplication.*

♦ ♦ ♦ ♦ ♦ ♦ ♦ ♦ ♦ ♦ ♦ ♦ ♦ ♦ ♦ ♦ ♦ ♦ ♦ ♦ ♦ ♦ ♦ ♦ ♦ ♦ ♦ ♦ ♦ ♦ ♦ ♦ ♦

Some words, such as *quotient*, are found only in mathematics. The quotient of ten and five is two. A **quotient** *is the answer to a division problem.*

**EXAMPLE 2**

Write each phrase using the correct arithmetic symbol and then do the calculation. Suppose the fractions in parts a and b represent small pizzas (see Figure 5). Describe the addition and subtraction within an appropriate setting—that is, make each operation into a word problem.

**a.** the sum of $2\frac{3}{4}$ and $1\frac{2}{5}$  **b.** the difference between $2\frac{3}{4}$ and $1\frac{2}{5}$
**c.** the product of $2\frac{3}{4}$ and $1\frac{2}{5}$  **d.** the quotient of $2\frac{3}{4}$ and $1\frac{2}{5}$

**SOLUTION**

The mixed numbers were changed to improper fractions before computation to show similarities among the operations.

**a.** $2\frac{3}{4} + 1\frac{2}{5} = \frac{11}{4} + \frac{7}{5} = \frac{55}{20} + \frac{28}{20} = \frac{83}{20} = 4\frac{3}{20}$

**b.** $2\frac{3}{4} - 1\frac{2}{5} = \frac{11}{4} - \frac{7}{5} = \frac{55}{20} - \frac{28}{20} = \frac{27}{20} = 1\frac{7}{20}$

**c.** $2\frac{3}{4} \cdot 1\frac{2}{5} = \frac{11}{4} \cdot \frac{7}{5} = \frac{77}{20} = 3\frac{17}{20}$

**d.** $2\frac{3}{4} \div 1\frac{2}{5} = \frac{11}{4} \div \frac{7}{5} = \frac{11}{4} \cdot \frac{5}{7} = \frac{55}{28} = 1\frac{27}{28}$

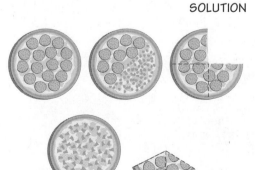

If Nguyen eats $2\frac{3}{4}$ small pizzas and Anh Lin eats $1\frac{2}{5}$ small pizzas, the sum represents what they eat together. The difference represents how much more Nguyen eats than Anh Lin.

*Figure 5*

♦ ♦ ♦ ♦ ♦ ♦ ♦ ♦ ♦ ♦ ♦ ♦ ♦ ♦ ♦ ♦ ♦ ♦ ♦ ♦ ♦ ♦ ♦ ♦ ♦ ♦ ♦ ♦ ♦ ♦ ♦ ♦ ♦ ♦ ♦ ♦

## SETS OF NUMBERS

Example 2 illustrated operations with fractions. Fractions are just one type of number in our system of real numbers. **Real numbers** *include all whole numbers, decimals, fractions, and mixed numbers, as well as other types of numbers* to be introduced later.

A **set** *is a collection of objects or numbers.* A listing of the contents of a set is usually placed in braces, { }. Brackets, [ ], or parentheses, ( ), or combinations of these are also used to describe sets.

**Rational numbers** *are the set of real numbers that may be written as a fraction in the form* $\frac{a}{b}$. The letters $a$ and $b$ may represent positive or negative numbers. Only $a$ may be zero. If $b$ is zero, then *the fraction has no meaning* and we say the fraction is **undefined**.

---

Rational numbers play an important role in algebra because almost all division problems are written as fractions.

---

A rounded or terminating decimal may be written as a fraction by placing the decimal part over a denominator equivalent to the place value of the last decimal position. For example,

$$0.75 = \frac{75}{100} \qquad \text{or} \qquad 0.0035 = \frac{35}{10,000}$$

The **numerator** *is the top number in a fraction;* the **denominator** *is the bottom number.* "Numerator" is like the word *number*, saying *how much.* "Denominator" is like the word *denomination*, indicating *what kind.* In the fraction $\frac{2}{3}$, the 3 indicates that the size, or kind, of fraction is thirds. The 2 indicates that there are two parts, each of size one-third.

Descriptions of other number sets will be included as we consider input–output tables.

## INPUT–OUTPUT TABLES

We will use input–output tables throughout this course to describe or summarize numerical relationships. Think of the left, or input, side of Table 2 as a source of numbers and the right, or output, side as the number matched with that input by some rule, pattern, or operation. In Section 1.1 we used such a table to summarize the number of pens as inputs and the partitions needed as outputs. We sought to find the **input–output relationship** (the relationship between the input and the output) and use it to predict the number of partitions for 20 pens.

College tuition provides an example of a known relationship between numbers. Table 2 shows the tuition cost for 1 to 4 credit hours. Read across the table: 1 credit costs $24, 2 credits cost $48, and so on.

| Input: Credit Hours | Output: Tuition Paid |
|:---:|:---:|
| 1 | $24 |
| 2 | $48 |
| 3 | $72 |
| 4 | $96 |

*Table 2* Tuition at Low-Cost College

E X A M P L E  3    Tuition    Refer to Table 2:

**a.** What is the cost of a 3-hour writing course?

**b.** What is the cost of a 4-hour mathematics course?

**c.** What is the cost of a 5-hour chemistry course?

SOLUTION    **a.** From Table 2, a 3-hour writing course costs $72.

**b.** From Table 2, a 4-hour mathematics course costs $96.

**c.** We have at least two ways to answer the third question.

*Method 1:* The first and each additional credit costs $24. If the fifth credit costs another $24, 5 credits would be $96 for the first 4 credits and $24 for the fifth credit, for a total of $120.

*Method 2:* If we divide each tuition amount by the number of credit hours, we find each credit to be worth $24. The product of 5 credits at $24 per credit is $120.

   The second method illustrates that the multiplication of the input (credit hours) by a cost per credit hour gives the output (tuition). Both methods assume that tuition rates continue in the same pattern for any possible number of credit hours.

♦ ♦ ♦ ♦ ♦ ♦ ♦ ♦ ♦ ♦ ♦ ♦ ♦ ♦ ♦ ♦ ♦ ♦ ♦ ♦ ♦ ♦ ♦ ♦ ♦ ♦ ♦ ♦ ♦ ♦ ♦ ♦ ♦ ♦ ♦

   The input and output numbers in Examples 3 and 4 come from the set of **whole numbers**, $\{0, 1, 2, 3, 4, \ldots\}$. The dots after the 4 indicate that the number list continues without end. In the tuition example, however, it is likely that the number of credits is less than 30.

E X A M P L E  4    Tuition and fees    Suppose that, in addition to the tuition, there is a $10 fee charged to each student, regardless of the number of hours taken.

**a.** Predict how Table 2 will change.

**b.** What will 9 credits cost?

**c.** If the tuition and fees were $250, how many credit hours were taken?

SOLUTION    **a.** Each output should increase by $10, as shown in Table 3.

**b.** If we assume all credits cost the same amount, the total cost is $24 times the number of credit hours plus the $10 fee, or

$$\$24 \cdot 9 \text{ credits} + \$10 = \$226$$

The dot commonly *replaces a multiplication symbol.* The symbol for *is equal to* is =.

**c.** One way to think about this question is that the $250 cost is $24 more than the cost of 9 credits, so it represents 10 credits.

| Input: Credit Hours | Output: Tuition Paid |
|---|---|
| 1 | $34 |
| 2 | $58 |
| 3 | $82 |
| 4 | $106 |

*Table 3  Tuition and Fees at Low-Cost College*

♦ ♦ ♦ ♦ ♦ ♦ ♦ ♦ ♦ ♦ ♦ ♦ ♦ ♦ ♦ ♦ ♦ ♦ ♦ ♦ ♦ ♦ ♦ ♦ ♦ ♦ ♦ ♦ ♦ ♦ ♦ ♦ ♦ ♦

You may know what causes the peaks on ice cubes, why ice floats in water, and why the cup of water in Figure 6 split when placed in the freezer. But do you know that water cooled below 0°C under increased pressure can be kept as a liquid? The volume of this super-cooled water changes, but not as much as when water turns to ice.

Figure 6

E X A M P L E    5

Temperature and water volume    In Table 4, the input is the temperature of water, and the output is the volume, or amount of space, filled by the water at that temperature. Increased pressure keeps the water liquid.

| Input: Temperature (°C) | Output: Volume (cc) |
|:---:|:---:|
| −5 | 1.00070 |
| −4 | 1.00055 |
| −3 | 1.00042 |
| −2 | 1.00031 |
| −1 | 1.00021 |
| 0 | 1.00013 |
| 1 | 1.00007 |
| 2 | 1.00003 |
| 3 | 1.00001 |
| 4 | 1.00000 |
| 5 | 1.00001 |
| 6 | 1.00003 |
| 7 | 1.00007 |
| 8 | 1.00012 |
| 9 | 1.00019 |
| 10 | 1.00027 |

Table 4 *Volume of Water Near Freezing*

*Note:* The mass of 1 cubic centimeter of water at 4 degrees Celsius (°C) is 1 gram. Each of the volumes listed in the table represents 1 gram of water heated or cooled to the given temperature.

*Source:* Data reprinted with permission of CRC Press, Boca Raton, Florida, from the *Handbook of Chemistry and Physics*, 36th Edition, 1954–55, Chemical Rubber Publishing Company, 2310 Superior Ave. N.E., Cleveland, Ohio, page 1963.

The following questions may help you understand the relationship between temperature and water volume.

**a.** Where is the smallest volume in the table?

**b.** Does the volume get larger or smaller as the temperature moves away from 4°C?

**c.** Does the volume of water change the same amount for each degree change?

**d.** For the data given, where does the greatest change take place?

**SOLUTION**

**a.** The smallest volume is at 4°C.

**b.** The volume gets larger as we move away from 4°C in either direction.

**c.** No, the volume changes by a different amount for each degree change.

**d.** For the data shown, the greatest volume change is by 0.00015 of a cubic centimeter (cc) between −4°C and −5°C.

♦ ♦ ♦ ♦ ♦ ♦ ♦ ♦ ♦ ♦ ♦ ♦ ♦ ♦ ♦ ♦ ♦ ♦ ♦ ♦ ♦ ♦ ♦ ♦ ♦ ♦ ♦ ♦ ♦ ♦ ♦ ♦ ♦ ♦ ♦

The outputs in the table are rational numbers in decimal form. The decimals are rounded to five places, or to the nearest hundred-thousandth. The inputs in Table 4 introduce another set of numbers—the integers.

**Integers** *are the set of positive and negative whole numbers and zero.* The set of integers is summarized in listed form as, $\{\ldots, -3, -2, -1, 0, 1, 2, 3, \ldots\}$. Integers may describe changes in value (+5 dollars, −10 dollars), changes in position (+6 yards, −3 yards), or changes in elevation relative to sea level (+100 feet, −85 feet). They are the numbers on thermometers. Integers are used in chemistry to describe ions (atoms or molecules with a positive or negative charge). Integers are also the numbers written on the number line in Figure 7, although there are positions on the number line for all real numbers.

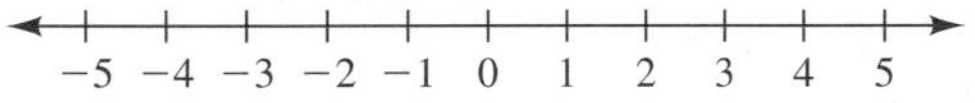

In the tuition example one rule described the entire table. We could use it to predict other input and output pairs. In the water temperature and volume example the rule is not evident. In Example 6, based on Table 5, the output rule depends on the value of the input. Throughout this text we will use the phrase **conditional rule** *to describe a rule that, although clearly relating input and output values in a table, changes for different input values.*

The inputs in Table 5 are grouped in intervals. An **interval** *is a set containing all the numbers between its endpoints and including one endpoint, both endpoints, or neither endpoint.*

*Figure 7*

**EXAMPLE  6**

*Payment schedules*    Table 5 gives the payment due on a charge balance at the end of the month. The input values are dollar amounts in intervals. The output rules change as the charge balance gets larger. What is the payment due for each of these amounts?

**a.** $15      **b.** $25      **c.** $200      **d.** $400      **e.** $550

| Input: Charge Balance | Output: Payment Due |
|---|---|
| $0 to $20 | Full amount |
| $20.01 to $500 | 10% or $20, whichever is greater |
| $500.01 or more | $50 plus the amount in excess of $500 |

*Table 5 1988 Chevron Credit Card Payment Schedule*

**SOLUTION**

**a.** $15, which is the full amount

**b.** $20, which is more than 10% of $25

**c.** $20, which is exactly 10% of $200

**d.** $40, which is 10% of $400 and is more than $20

**e.** $100, which is the $50 plus the $50 in excess of $500

♦ ♦ ♦ ♦ ♦ ♦ ♦ ♦ ♦ ♦ ♦ ♦ ♦ ♦ ♦ ♦ ♦ ♦ ♦ ♦ ♦ ♦ ♦ ♦ ♦ ♦ ♦ ♦ ♦ ♦ ♦ ♦ ♦ ♦ ♦

Example 6 had three different rules, and its complexity illustrates why the rules are printed only on the information brochure and not on the credit card monthly statement.

## EXERCISES 1.2

Exercises 1 to 32 are based on examples, tables, and definitions in the text. Refer back as necessary. Write the phrases in Exercises 1 to 12 using the correct operation symbol $(+, -, \cdot, \div)$, and then do the operation.

1. The product of 4 and $\frac{1}{2}$

2. The quotient of 5 and $\frac{1}{2}$

3. The sum of $\frac{1}{3}$ and $\frac{1}{2}$

4. The difference between $\frac{3}{4}$ and $\frac{1}{3}$

5. The quotient of 5 and 15

6. The sum of 0.25 and $\frac{3}{4}$

7. The product of $\frac{3}{5}$ and $\frac{4}{9}$

8. The difference between $\frac{7}{5}$ and $\frac{3}{4}$

9. The difference between $2\frac{2}{3}$ and $1\frac{1}{4}$

10. The product of $2\frac{2}{3}$ and $1\frac{1}{4}$

11. The quotient of $2\frac{2}{3}$ and $1\frac{1}{4}$

12. The sum of $2\frac{2}{3}$ and $1\frac{1}{4}$

13. In Example 3, if the cost per hour is assumed to remain the same for all credit totals, what is the cost of taking 12 credit hours?

14. In Example 4, what is the cost of taking 12 credit hours?

15. In Example 4, we were to find how many credits were taken if the total of tuition and fees was $250. What other ways might we calculate the number of credits? Use complete sentences to describe your method. Try your method on a total cost of $394.

16. Make a tuition and fees table for your school or a nearby college or university.

For Exercises 17 to 22 use Table 4. What is the change in volume as the temperatures change? Is the volume getting larger or smaller in each case?

17. 4°C to 10°C
18. 4°C to −2°C
19. −2°C to 10°C
20. 10°C to −5°C
21. −5°C to 2°C
22. −4°C to 4°C

23. How does Table 4 indicate the need for a fill line on freezer cartons?

24. How does Table 4 suggest the need to open a can of soup before heating?

25. If the charge balance at the end of the month is $450, what is the payment due? Use Table 5.

26. If the payment was $50, what was the charge balance before the payment was made? Use Table 5.

In Exercises 27 to 32 use each word in two sentences. In the first sentence of each pair, the meaning of the word should not be mathematical. In the second sentence, the meaning of the word should be mathematical.

27. set

28. value

29. input

30. round

31. interval

32. rational

In Exercises 33 to 36 use the table given.

| Input: Shirt Size | Output: Fabric Needed |
|---|---|
| Small: 34–36 | $1\frac{1}{2}$ yards |
| Medium: 38–40 | $1\frac{3}{4}$ yards |
| Large: 42–44 | $1\frac{3}{4}$ yards |
| Extra Large: 46–48 | $2\frac{1}{8}$ yards |

33. Find the total yards of fabric needed to make 1 size 38 shirt and 1 size 48 shirt.

34. How many more yards of fabric are needed to make a size 46 shirt than a size 40 shirt?

35. If you buy fabric for a size 46 shirt and then make only a size 34 shirt, how much fabric should you have remaining?

36. If you buy fabric for a size 44 shirt and then make only a size 34, how much fabric should you have remaining?

In the primary grades we always subtracted the smaller number from the larger number because there was "no answer" if we tried to subtract the other way—negative numbers had not yet been introduced. Throughout history, mathematicians have created new numbers whenever the existing numbers were inadequate. In Exercises 37 to 40 identify the number or set of numbers invented to answer the questions. There may be more than one correct answer.

37. What number remains when we subtract 5 from 5? (Neither the Greeks nor the Romans had one of these.)

38. What is the result of division of a number by a larger number?

39. How do we write "three of four equal parts"?

40. How do we write elevations below sea level?

**For Exercises 41 and 42 make an input–output table for the conditional situations. The *even numbers are the set of integers that are divisible by two*. The set of even numbers is described by {. . . , −4, −2, 0, 2, 4, . . .}. The *odd numbers are those integers that are not divisible by two*, {. . . , −3, −1, 1, 3, 5, . . .}.**

41. The output is 5 if the input is an even number. The output is twice the input if the input is an odd number. Use integer inputs {0, 1, 2, . . . , 8}.

42. The output is the sum of the input and 2 if the input is even. The output is the product of the input and 2 if the input is odd. Use integer inputs 0 through 8.

*PROBLEM SOLVING*

**Use the following hint to do Exercises 43 to 48: Guess a pair of numbers that fits the first condition; see how closely your guess fits the second condition.**

43. What two numbers have a sum of 6 and a product of 8?

44. What two numbers have a sum of 11 and a product of 30?

45. What two numbers have a sum of 12 and a difference of 6?

46. What two numbers have a difference of 5 and a sum of 7?

47. What two numbers have a product of 12 and a sum of 13?

48. What two numbers have a quotient of 5 and a sum of 30?

*COMPUTATION REVIEW*

**Exercises 49 and 50 contain modified input–output tables. Use the two input values to calculate the outputs, according to the rule given at the top of each output column. (The dot between $a$ and $b$ means to multiply.)**

49.

| Input $a$ | Input $b$ | Output $a + b$ | Output $a - b$ | Output $a \cdot b$ | Output $a \div b$ |
|---|---|---|---|---|---|
| 15 | 5 | | | 75 | |
| $\frac{3}{4}$ | $\frac{2}{5}$ | | | | |
| 0.36 | 0.06 | | | | |
| 5.6 | 0.7 | | | | |
| 2.25 | 1.5 | | | | |

50.

| Input $a$ | Input $b$ | Output $a + b$ | Output $a - b$ | Output $a \cdot b$ | Output $a \div b$ |
|---|---|---|---|---|---|
| 9 | 6 | | | 54 | |
| $\frac{2}{3}$ | $\frac{1}{6}$ | | | | |
| 0.5 | 0.25 | | | | |
| 49 | 0.7 | | | | |
| 6.25 | 2.5 | | | | |

## MID-CHAPTER 1 TEST

1. Jake's assistant, Kelly, suggests making the pot-bellied pigs' pens in a triangular shape. Use the top view of four completed pens in the figure to start the table. Look for a pattern, and predict how many partitions are needed to build 20 pens.

| Pens | Partitions |
|---|---|
| 1 | |
| 2 | |
| 3 | |
| 4 | |
| 5 | |
| 20 | |

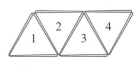

2. If the statement is true, give an example; if it is false, explain why.

   a. A number may be both an integer and a rational number.

   b. A number may be rational and a fraction.

   c. A number may be a fraction and a decimal.

   d. A number may be a whole number and a fraction.

   e. A number may be negative and rational.

3. a. What is the product of 1 and 12?

   b. What is the product of 2 and 11?

   c. What is the largest possible product of two whole numbers whose sum is 13?

   d. What is the largest possible product of two rational numbers whose sum is 13?

4. Use the table to find the cost of the following long-distance calls. Discounts are subtracted from the daytime cost.

   a. 3 minutes on Saturday

   b. $5\frac{1}{2}$ minutes Wednesday at noon

   c. $4\frac{1}{2}$ minutes Thursday evening

| Time of Day | Charge |
|---|---|
| Daytime: Monday through Friday, 8 A.M. to 5 P.M. | $1.50 first minute or portion<br>$1.00 each additional minute or portion |
| Evening: Sunday through Thursday, 5 P.M. to 8 A.M. | 10% discount |
| Weekends: Friday, 5 P.M., to Sunday, 5 P.M. | 30% discount |

**1.3**    TABLES, PATTERNS, AND ALGEBRAIC EXPRESSIONS

**OBJECTIVES**    Determine a pattern or rule from an input–output table. ♦ Express the rules for an input–output table in words. ♦ Express rules involving addition and multiplication in symbols. ♦ Identify variables, constants, and expressions. ♦ Translate word phrases into symbolic expressions and vice versa. ♦ Make input–output tables for conditional descriptions.

**WARM-UP**    Complete the tables.

1.

| Input | Output: Six Times Input |
|-------|--------------------------|
| 0 | |
| 1 | |
| 2 | |
| 3 | |
| 4 | |

2.

| Input | Output: Six More Than Input |
|-------|------------------------------|
| 0 | |
| 1 | |
| 2 | |
| 3 | |
| 4 | |

♦

> **STUDY HINT**
>
> Make 3″-by-5″ flash cards for notation, symbols, vocabulary, and se-
> lected exercises to help with learning and remembering new material.

In this section we introduce the symbolic notation needed to describe rules for visual patterns and their input–output tables. We use both words and symbolic expressions to describe the relationships. We close the section by translating word statements into symbolic expressions.

DESCRIBE PATTERNS USING WORDS

As mentioned in Example 5 of the previous section (the water temperature and volume example), one of the many uses of algebra is finding and describing a pattern.

**QUESTION  1**    What is the next number?

Pattern 1    Paper Clips: 2, 4, 6, 8, __ , in Figure 8.

Pattern 2    Polygon Sides: 3, 4, 5, 6, __ , in Figure 9.

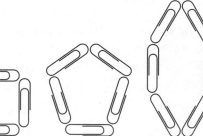

*Figure 8*                                                    *Figure 9*

If you said 10 for Pattern 1 and 7 for Pattern 2, you found the most likely answers. Other answers may also be possible.

♦ ♦ ♦ ♦ ♦ ♦ ♦ ♦ ♦ ♦ ♦ ♦ ♦ ♦ ♦ ♦ ♦ ♦ ♦ ♦ ♦ ♦ ♦ ♦ ♦ ♦ ♦ ♦ ♦ ♦ ♦ ♦ ♦

So long as we had a list of numbers, we could continue giving the next number. But what if we wanted to jump directly to the 50th or 100th number in each pattern? This is where algebra fits in.

As a first step in describing patterns using algebra, we match each design in the pattern (output) with its position (input). An effective way to show this match is with an input–output table. Table 6 lists Pattern 1, including the output 10 and rows for the new inputs, 50 and 100.

As a second step, we look for a pattern between the input and output numbers. **A key to input–output patterns is to compare numbers across the table**, not down the table. This focuses our attention on the way the output relates to the input.

| Input | Output |
|-------|--------|
| 1     | 2      |
| 2     | 4      |
| 3     | 6      |
| 4     | 8      |
| 5     | 10     |
| 50    |        |
| 100   |        |

*Table 6  Pattern 1*

**QUESTION  2**   What rule describes the input–output table for Pattern 1? Cover up the hints and answer the question before continuing.

HINT   In Pattern 1 the input 1 matches with the output 2, the input 2 matches with the output 4, 3 → 6, and 4 → 8, so 5 → 10 seems reasonable. (We use the arrow, →, in place of the phrase *matches with*.) Now predict the outputs for 50 and 100.

If you said 50 → 100 and 100 → 200, you gave reasonable answers! Next, describe in words how to get the input–output pairs 50 → 100 and 100 → 200.

To check your thinking, consider this question: All of the following describe the rule for Pattern 1 *except* which statement?

**a.** Add the input to itself to get the output.

**b.** Multiply the input by itself to get the output.

**c.** The output is twice the input.

**d.** Double the input to get the output.

**e.** The output is two times the input.

(The answer is b.)

♦ ♦ ♦ ♦ ♦ ♦ ♦ ♦ ♦ ♦ ♦ ♦ ♦ ♦ ♦ ♦ ♦ ♦ ♦ ♦ ♦ ♦ ♦ ♦ ♦ ♦ ♦ ♦ ♦ ♦ ♦ ♦ ♦

QUESTION 3    What rule describes the input–output table for Pattern 2?

| Input | Output |
|-------|--------|
| 1 | 3 |
| 2 | 4 |
| 3 | 5 |
| 4 | 6 |
| 5 | 7 |
| 50 | |
| 100 | |

*Table 7 Pattern 2*

HINT   In Pattern 2 the first polygon is a triangle and has 3 sides, so the input 1 matches with the output 3. The second polygon, a square, has 4 sides, so the input 2 matches with the output 4. The third polygon is a pentagon and has 5 sides, so $3 \rightarrow 5$. Similarly, $4 \rightarrow 6$ and $5 \rightarrow 7$. Table 7 lists Pattern 2 as well as rows for the new inputs, 50 and 100. Predict the outputs for 50 and 100.

We next describe in words how to get the output for inputs 50 and 100. We might look back at Figure 9 as a strategy. For the polygons in Figure 9 the number of sides is 2 more than the position number. We might also look at a number line. On the number line in Figure 10 we draw an arrow from each input to its output. This pattern makes jumps of 2.

What do Figure 9 and Figure 10 have in common? To check your thinking, consider this question: All of the following describe Pattern 2, the rule for both figures, *except* which statement?

**a.** Add 2 to the input to get the output.

**b.** The output is 2 greater than the input.

**c.** The output is 2 more than the input.

**d.** The input less 2 gives the output.

**e.** The input plus 2 gives the output.

(The answer is d.)

*Figure 10*

♦ ♦ ♦ ♦ ♦ ♦ ♦ ♦ ♦ ♦ ♦ ♦ ♦ ♦ ♦ ♦ ♦ ♦ ♦ ♦ ♦ ♦ ♦ ♦ ♦ ♦ ♦ ♦ ♦ ♦ ♦ ♦ ♦

DESCRIBE PATTERNS USING ALGEBRA

First we matched each number in the pattern (output) with its position (input). Then we looked for a pattern between the input and output numbers and described it in words. Now we will let the letter *n* represent *any input number* and translate our word phrase.

In Pattern 1 the phrase *adding the input to itself* is written as $n + n$. The phrases *two times the input*, *double the input*, and *twice the input* are written 2 times *n*, or $2n$.

Multiplication may be written several ways: $2n$, $2 \cdot n$, $2(n)$, or $(2)(n)$. For a number and letter together, however, $2n$ is the most common form. $2 \times n$ is not used because there may be confusion as to whether the $\times$ represents the multiplication operation or a letter to be multiplied.

We add another row to the bottom of our input–output table and write both expressions, $n + n$ and $2n$, as the output for input $n$. To save space, only the last three rows for Pattern 1 are shown in Table 8.

| Input | Output |
|-------|--------|
| 50 | 100 |
| 100 | 200 |
| $n$ | $n + n$, or $2n$ |

Table 8 Pattern 1, Continued

In Pattern 2 the output is *add 2 to the input*, *2 greater than the input*, *2 more than the input*, and *the input plus 2*. Both *greater than* and *more than* imply addition, so all the words may be written as $2 + n$ or $n + 2$. We record both in a new row as the output for input $n$ in Table 9.

| Input | Output |
|-------|--------|
| 50 | 52 |
| 100 | 102 |
| $n$ | $n + 2$, or $2 + n$ |

Table 9 Pattern 2, Continued

In both Pattern 1 and Pattern 2 we have two ways to write the words symbolically. Because both $n + n$ and $2n$ describe Pattern 1, it is reasonable for them to be the same value. Mathematicians assume there to be a 1 in front of the $n$ and write $n + n = 1n + 1n = 2n$. In Pattern 2 the expression $n + 2$ is equal to $2 + n$ for all numbers $n$. There will be more on the equality of these symbols in Section 2.1.

We have used several important ideas, which we will now formally define.

Definition

A **variable** is a letter or symbol that represents any number in the input (or output) set.

The letter $n$ was our input variable. Any number appropriate to the input set may replace the variable. In $6 + x$, $5 - \square$, and $\alpha + 90$, the $x$, $\square$, and $\alpha$ are variables, like $n$.

Definition

A **constant** is a number, letter, or symbol whose value is fixed.

The 2 in $2n$ is a constant for multiplication. The 2 in $n + 2$ is a constant for addition.

The outputs $2n$, $n + n$, $n + 2$, and $2 + n$ are all symbolic expressions.

Definition

An **expression** is any combination of numbers, constants, and variables with operations such as addition, subtraction, or multiplication.

E X A M P L E   **1**

Complete Table 10, and describe its input–output rule, Pattern 3, in words. Write an expression to describe the pattern.

| Input | Output |
|-------|--------|
| 1 | 1 |
| 2 | 3 |
| 3 | 5 |
| 4 | 7 |
| 5 | |
| 50 | |
| 100 | |
| n | |

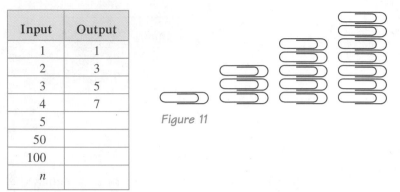

*Figure 11*

*Table 10  Pattern 3*

HINT    It may be easy to see that the output column in Pattern 3 is forming the set of odd numbers, $1, 3, 5, 7, 9, \ldots$. It is not as easy to predict the outputs for 50 and 100. Keep in mind that we want to match input with output and describe how to get from the input to the output. *A common strategy is to compare with a known idea.* Compare the paper clip pattern in Figure 11 with the even-number pattern, $2n$, in Figure 8. Also compare the outputs in Table 10 with the outputs in Table 6. How are the outputs related?

SOLUTION

Pattern 1 is twice the input, or $2n$. Each output in Table 11 is 1 less than the corresponding output in Table 6, so we describe Pattern 3 as 1 less than twice the input, or $2n - 1$.

| Input | Output |
|-------|--------|
| 1 | 1 |
| 2 | 3 |
| 3 | 5 |
| 4 | 7 |
| 5 | 9 |
| 50 | 99 |
| 100 | 199 |
| n | $2n - 1$ |

*Table 11  Pattern 3, Continued*

◆ ◆ ◆ ◆ ◆ ◆ ◆ ◆ ◆ ◆ ◆ ◆ ◆ ◆ ◆ ◆ ◆ ◆ ◆ ◆ ◆ ◆ ◆ ◆ ◆ ◆ ◆ ◆ ◆ ◆ ◆ ◆ ◆ ◆ ◆

Figure 12 compares the paper clip arrangement in Figure 11 to that in Figure 8 of Question 1. It too shows that the output rule is $2n - 1$.

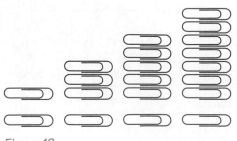

*Figure 12*

PHRASES AND EXPRESSIONS

In Section 1.2 we defined *sum, difference, product,* and *quotient* as the answers in addition, subtraction, multiplication, and division, respectively. There are also several words and phrases that indicate each operation. You may want to keep a list and add your own words to those listed here. Some words and phrases that indicate addition are *plus, added to, greater than, more than,* and *increased by.* Subtraction words include *less, less than, fewer, minus,* and *decreased by.* Generally, if the word *than* appears, the order of the numbers is the reverse of the word order.

EXAMPLE  2  Translate these subtractions into symbols:

  **a.** The difference in age between 6 and 2

  **b.** Fifteen dollars less ten dollars

  **c.** Six coins less than the number *n* coins

  **d.** Four fewer students than *x*

  **e.** *x* decreased by 5 tickets

SOLUTION  **a.** $6 - 2$   **b.** $15 - 10$   **c.** $n - 6$   **d.** $x - 4$   **e.** $x - 5$

♦ ♦ ♦ ♦ ♦ ♦ ♦ ♦ ♦ ♦ ♦ ♦ ♦ ♦ ♦ ♦ ♦ ♦ ♦ ♦ ♦ ♦ ♦ ♦ ♦ ♦ ♦ ♦ ♦ ♦ ♦ ♦

Multiplication and division words may be harder to recognize. The word *at* describes multiplication in the phrase "5 hamburgers at $3.95." *Double, twice,* and *triple* imply multiplication by a specific number. The word *of* implies multiplication in fraction work. The word *half* may be troublesome. *Half of ten* means one-half times ten, or the value five. *Divide in half* means division by two, not division by one-half.

EXAMPLE  3  Let *n* be the input number. Translate these phrases to expressions:

  **a.** One less than twice the input

  **b.** The difference between half the input and five

  **c.** The quotient of three and the input

  **d.** The product of four and the input is decreased by one

SOLUTION  **a.** $2n - 1$

  **b.** $\dfrac{1}{2}n - 5$ or $\dfrac{n}{2} - 5$

  **c.** $3 \div n$ is correct, but fraction form is preferred: $\dfrac{3}{n}$

  **d.** $4n - 1$

♦ ♦ ♦ ♦ ♦ ♦ ♦ ♦ ♦ ♦ ♦ ♦ ♦ ♦ ♦ ♦ ♦ ♦ ♦ ♦ ♦ ♦ ♦ ♦ ♦ ♦ ♦ ♦ ♦ ♦ ♦ ♦

EXAMPLE  4  Write a description for the output:

  **a.** The input, *n*, is the number of payments. Each payment is worth $35. The output is the total cost of *n* payments.

  **b.** The input, *n*, is the price of a book. The sales tax rate is 8% of the price. The output is the sum of the price of the book and the sales tax. Write the percent as a decimal.

SOLUTION  **a.** The output is $35n.    **b.** The output is $n + 0.08n$.

♦ ♦ ♦ ♦ ♦ ♦ ♦ ♦ ♦ ♦ ♦ ♦ ♦ ♦ ♦ ♦ ♦ ♦ ♦ ♦ ♦ ♦ ♦ ♦ ♦ ♦ ♦ ♦ ♦ ♦ ♦ ♦

Because fractions typically show division in algebra, the *per* in rates or units of measurement also means division. The rate "15 miles per hour" is frequently written as a fraction with 15 miles in the numerator and 1 hour in the denominator. *Each* is sometimes used in place of *per* to indicate a division or fraction: "2 round trips each day" is the fraction 2 trips divided by 1 day.

EXAMPLE 5

Write these rates or units of measurement in fraction form:

**a.** 15 miles per hour      **b.** $80 per day      **c.** 36 miles per gallon

**d.** 2 trips each day      **e.** $250 each term

SOLUTION   **a.** $\dfrac{15 \text{ miles}}{1 \text{ hour}}$   **b.** $\dfrac{\$80}{1 \text{ day}}$   **c.** $\dfrac{36 \text{ miles}}{1 \text{ gallon}}$   **d.** $\dfrac{2 \text{ trips}}{1 \text{ day}}$   **e.** $\dfrac{\$250}{1 \text{ term}}$

♦ ♦ ♦ ♦ ♦ ♦ ♦ ♦ ♦ ♦ ♦ ♦ ♦ ♦ ♦ ♦ ♦ ♦ ♦ ♦ ♦ ♦ ♦ ♦ ♦ ♦ ♦ ♦ ♦ ♦ ♦ ♦ ♦

*Percent* indicates division by 100. Change a percent to a decimal before writing it with a variable or in a formula.

EXAMPLE 6

Translate each percent phrase into an expression:

**a.** 20% of a number $n$      **b.** 75% of a number $n$

**c.** $62\frac{1}{2}\%$ of a number $n$      **d.** $\frac{1}{2}\%$ of a number $n$

SOLUTION   In parts c and d the numerator and denominator of the fractions are multiplied by 10 to clear the decimal point in 62.5 and 0.5.

**a.** 20% of $n = \frac{20}{100}n$ or $0.20n$

**b.** 75% of $n = \frac{75}{100}n$ or $0.75n$

**c.** $62\frac{1}{2}\%$ of $n = 62.5\%$ of $n = \frac{62.5}{100}n = \frac{625}{1000}n$ or $0.625n$

**d.** $\frac{1}{2}\%$ of $n = 0.5\%$ of $n = \frac{0.5}{100}n = \frac{5}{1000}n$ or $0.005n$

♦ ♦ ♦ ♦ ♦ ♦ ♦ ♦ ♦ ♦ ♦ ♦ ♦ ♦ ♦ ♦ ♦ ♦ ♦ ♦ ♦ ♦ ♦ ♦ ♦ ♦ ♦ ♦ ♦ ♦ ♦ ♦ ♦

## EXERCISES 1.3

In Exercises 1 to 4, match the table to one of the figures shown in a to d. Then identify the rule used to generate the table, and state the rule in words and with an expression. Finally, complete the four incomplete rows of the table.

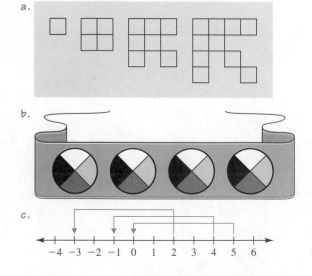

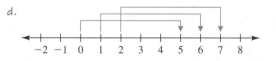

| 1. Input | Output |
|---|---|
| 0 | 5 |
| 1 | 6 |
| 2 | 7 |
| 3 | 8 |
| 4 | 9 |
| 5 |  |
| 50 |  |
| 100 |  |
| $n$ |  |

| 2. Input | Output |
|---|---|
| 0 | 1 |
| 1 | 4 |
| 2 |  |
| 3 | 10 |
| 4 | 13 |
| 5 | 16 |
| 50 |  |
| 100 |  |
| $n$ |  |

**3.**

| Input | Output |
|-------|--------|
| 1     | 4      |
| 2     | 8      |
| 3     |        |
| 4     | 16     |
| 5     | 20     |
| 6     | 24     |
| 50    |        |
| 100   |        |
| x     |        |

**4.**

| Input | Output |
|-------|--------|
| 2     | −3     |
| 3     |        |
| 4     | −1     |
| 5     | 0      |
| 6     | 1      |
| 7     | 2      |
| 50    |        |
| 100   |        |
| x     |        |

**Translate the word phrases in Exercises 5 and 6 into expressions. Let *n* be the input.**

5. a. Multiply the input by itself

   b. Three times the input, less two

   c. Add two to twice the input

   d. The product of seven and half the input

   e. The quotient of the input and eight

   f. The quotient of two and the input, decreased by three

   g. The difference between the input and four

6. a. Three plus twice the input

   b. The difference between four and triple the input

   c. The product of three and the input

   d. The quotient of eight and the input

   e. The product of the input and seven, increased by four

   f. The quotient of the input and three, decreased by two

   g. The difference between four and the input

**Write each percent statement in Exercises 7 and 8 as an expression.**

7. a. 10% of a number      b. $12\frac{1}{2}\%$ of a number

   c. $6\frac{1}{2}\%$ of a number      d. 150% of a number

8. a. 25% of a number      b. $37\frac{1}{2}\%$ of a number

   c. $87\frac{1}{2}\%$ of a number      d. 108% of a number

**In Exercises 9 to 14 write each word in two sentences. In the first sentence of each pair, the meaning of the word should not be mathematical. In the second sentence, the meaning of the word should be mathematical.**

9. constant      10. odd      11. expression

12. pattern      13. variable      14. exceed

**Write each of the units of measure given in Exercises 15 and 16 in fraction form.**

15. a. 6 stitches per inch      b. 10 miles per day

    c. 12 eggs per dozen      d. $120 a cord for firewood

16. a. 8 feet per second      b. 12 watts per day

    c. $10 each set      d. 1-a-day multivitamins

**Build an input–output table for each of the situations in Exercises 17 to 20.**

17. Input: number of cans, *n*
    Output: total cost, *C*, in dollars
    Rule: Cost is $0.86 per can.
    Use $n = \{1, 2, 3\}$ and *n*.

18. Input: number of hours, *t*
    Output: distance traveled, *D*, in miles
    Rule: $D = 25t$
    Use $t = \{1, 2, 3\}$ and *t*.

19. Input: number of miles, *m*. Let $m = \{0, 100, 200, \ldots,$ 500 miles$\}$.
    Output: cost, *C*, of rental car in dollars
    Rule: Cost is $35 plus $0.05 per mile.

20. Input: number of miles, *m*. Let $m = \{0, 1, 2, \ldots, 5 \text{ miles}\}$.
    Output: cost of taxi, *C*, in dollars
    Rule: $2.50 flagdrop (when passenger gets into taxi) plus $0.15 each tenth mile

**Write an expression for each output in Exercises 21 to 24.**

21. The input, *n*, is the speed of the car. The speed limit is 35 miles per hour. The output is the amount the car exceeds the speed limit.

22. The input, *n*, is the number of credit hours taken. Each credit hour costs $30. The output is the total cost of the credits taken.

23. The input, *n*, is the price of the car. The sales tax rate is $6\frac{1}{2}\%$. The output is the sum of the price of the car and the sales tax.

24. The input, *n*, is the credit card balance. The output is $50 plus the amount of the balance in excess of $500.

### CONDITIONAL RULES

**Complete the pattern in the tables in Exercises 25 and 26.**

**25.**

| Input  | Output |
|--------|--------|
| 1      | 1      |
| 2      | 0      |
| 3      | 3      |
| 4      | 0      |
| 5      | 5      |
| 6      |        |
| 7      |        |
| 8      |        |
| 50     |        |
| 101    |        |
| Even *n* |      |
| Odd *n*  |      |

**26.**

| Input  | Output |
|--------|--------|
| 1      | 2      |
| 2      | 3      |
| 3      | 6      |
| 4      | 5      |
| 5      | 10     |
| 6      | 7      |
| 7      | 14     |
| 8      |        |
| 50     |        |
| 101    |        |
| Even *n* |      |
| Odd *n*  |      |

**Make an input–output table for the conditional situations in Exercises 27 to 29.**

27. The output is the input number if the input is positive or zero. The output is the opposite value if the input is negative (5 is the opposite of −5). Use integer inputs $n = \{-3, -2, -1, 0, 1, 2, 3\}$.

28. The output is 0 if the input is a multiple of 3. The output is 2 otherwise. Use integer inputs $n = \{-6, -5, -4, \ldots, 6\}$.

29. The output equals the input for numbers larger than 0. The output is 0 for inputs smaller than or equal to 0. Use integers −3 to 3 as inputs. (This is a rule used in mechanical engineering.)

30. Write each of the output rules for the Chevron credit card payment schedule, repeated in the table. Use $n$ as the input amount. The output rule for $20.01 to $500 needs two descriptions separated by *or* and followed by *whichever is greater.*

| Input, $n$: Charge Balance | Output: Payment Due | Output Rule (Use $n$ as input.) |
|---|---|---|
| $0 to $20 | Full amount | |
| $20.01 to $500 | 10% or $20, whichever is greater | |
| $500.01 or more | $50 plus the amount in excess of $500 | |

*PROBLEM SOLVING*

31. ***Building Pens.*** Jake's pot-bellied pigs' pens require the partitions pattern shown in the following table. Three pens are shown in the figure. Use paper clips or toothpicks to investigate the pattern for the number of partitions needed for $n$ pens. Describe your pattern in words and with an expression. Use it to predict the number of partitions needed for 20 and 50 pens.

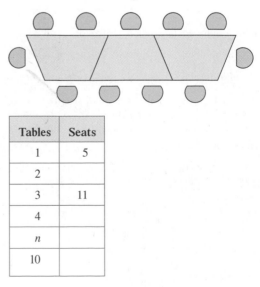

| Pens | Partitions |
|---|---|
| 1 | 4 |
| 2 | 7 |
| 3 | 10 |
| 4 | |
| $n$ | |
| 20 | |
| 50 | |

32. ***Building Pens in Pairs.*** Cassandra suggests that Jake build pens in pairs, as shown in the figure. The pairs and partitions pattern is shown in the following table. Use paper clips or toothpicks to investigate the pattern for the number

of partitions needed for $x$ pairs of pens. Describe your pattern in words and with an expression. Use it to predict the number of partitions needed for 10 and 25 pairs of pens.

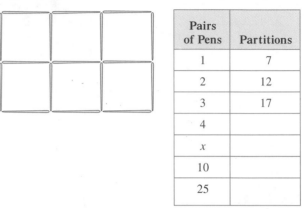

| Pairs of Pens | Partitions |
|---|---|
| 1 | 7 |
| 2 | 12 |
| 3 | 17 |
| 4 | |
| $x$ | |
| 10 | |
| 25 | |

33. ***Seating at Square Tables.*** The seating at 3 square tables placed next to each other at Mirielle's restaurant is shown in the figure. Use small tiles or cut squares from paper to find the numbers to complete the following table. Include an expression for $n$ square tables placed in a long row.

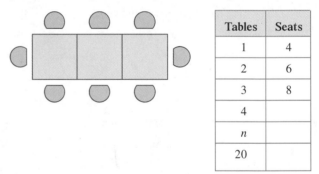

| Tables | Seats |
|---|---|
| 1 | 4 |
| 2 | 6 |
| 3 | 8 |
| 4 | |
| $n$ | |
| 20 | |

34. ***Seating at Trapezoidal Tables.*** Three of Franck's trapezoid-shaped restaurant tables are shown in the figure. Complete the following table. Include an expression for the number of seats at $n$ tables in a long row.

| Tables | Seats |
|---|---|
| 1 | 5 |
| 2 | |
| 3 | 11 |
| 4 | |
| $n$ | |
| 10 | |

Use guessing in Exercises 35 and 36, but to be efficient, keep a list of your guesses.

35. List ten pairs of numbers with a sum of 12. What is the maximum product of two whole numbers whose sum is 12?

36. What is the maximum (largest) product of two whole numbers whose sum is 10? Start by listing several pairs of numbers that add to 10.

### COMPUTATION REVIEW

Exercises 37 and 38 contain modified input–output tables. Use the two input values to calculate the outputs according to the rule given at the top of each output column. You may need to find one or both inputs. After completing the table, write a sentence to describe what $xy$ and $x + y$ mean. Use two of the appropriate words from this list in your sentence: *sum, difference, product, quotient.*

37.

| Input $x$ | Input $y$ | Output $xy$ | Output $x + y$ |
|-----------|-----------|-------------|----------------|
| 2 | 6 | 12 | |
| 10 | 2 | | |
| | | 15 | 8 |
| 1 | 15 | | |
| | 2 | 18 | |
| 4 | | 4 | |

38.

| Input $x$ | Input $y$ | Output $xy$ | Output $x + y$ |
|-----------|-----------|-------------|----------------|
| 4 | 3 | 12 | |
| | | 12 | 13 |
| 5 | 4 | | |
| | 1 | 20 | |
| | | 4 | 4 |
| | | 18 | 9 |

---

## 1.4   TABLES AND RECTANGULAR COORDINATE GRAPHS

**OBJECTIVES**   Identify $x$- and $y$-axes, quadrants, origin, and coordinates.  ◆ Graph coordinates (ordered pairs).  ◆ Read and interpret data from graphs.  ◆ Use graphs to compare values.  ◆ Plot step graphs from tables.  ◆ Plot curved graphs from tables.  ◆ Use a calculator to find square roots.  ◆ Identify irrational numbers.  ◆ Graph from a conditional input–output table or set of rules.

**WARM-UP**   Complete these input–output tables. Note that $x^2$ means the input times itself.

1.

| Input $x$ | Output $2x + 3$ |
|-----------|-----------------|
| 0 | |
| 0.5 | |
| 1.0 | |
| 1.5 | |
| 2.0 | |
| 3.0 | |
| 4.0 | |

2.

| Input $x$ | Output $x^2 + 3$ |
|-----------|------------------|
| 0.0 | |
| 0.5 | |
| 1.0 | |
| 1.5 | |
| 2.0 | |
| 3.0 | |
| 4.0 | |

◆

The pattern in a set of data or the rule for an input–output table is sometimes easier to determine if the information is displayed on a coordinate graph. This section introduces coordinate graphs. First we will examine the application of coordinates in street design and city maps, and then we will look at coordinates as a display of data from input–output tables. We will then practice reading coordinate graphs and examine questions that help us build graphs from tables.

### COORDINATE GRAPHS

A **coordinate graph** *identifies a position by two (or more) numbers or letters.* Frenchman Pierre Charles L'Enfant designed the national capital as a coordinate graph in 1791. Washington, D.C., is a square divided into four quadrants. The dome of the U.S. Capitol building is the center. Andrew Ellicott surveyed the city, in which north and south streets are numbered from the Capitol and east and west streets are lettered from the Capitol. A portion of the city map is shown in Figure 13. To permit us to focus on the basic map design, the diagonal streets (which bear the names of states) as well as the correct names of streets A and B have been omitted.

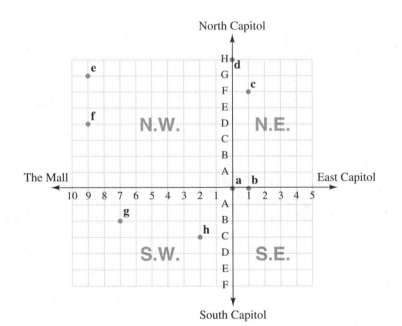

*Figure 13*

To practice reading a map based on coordinates, find the following buildings. Verify that each is located at the intersection of the indicated streets.

**a.** U.S. Capitol building

**b.** Supreme Court, 1st and East Capitol, N.E.

**c.** Union Station, 1st and F, N.E.

**d.** Veterans Administration, North Capitol and H

**e.** Martin Luther King Memorial Library, 9th and G, N.W.

**f.** Federal Bureau of Investigation, 9th and D, N.W.

**g.** National Aeronautics and Space Administration, 7th and B, S.W.

**h.** Food and Drug Administration, 2nd and C, S.W.

A major disadvantage of the Washington, D.C., map is that if someone forgets to identify the quadrant (N.E., N.W., S.W., S.E.) for an address, there are four possible locations for it. Find the four locations described by 2nd and C. The Food and Drug Administration is located at one of the four. While reading the next few paragraphs, look for how mathematicians avoid duplication of addresses.

When two number lines are placed at right angles so that they cross at zero, they form four **quadrants**, just like the map. The quadrants (see Figure 14) are numbered counterclockwise because of applications in engineering and physics. The number lines are called **axes**, and we observe that they are **perpendicular** in that they cross at a 90° angle, or square corner. The horizontal number line is the **x-axis**. The vertical number line is the **y-axis.**

The axes lie on a flat surface called the **coordinate plane**. Unlike points on the map, *each point in the coordinate plane is uniquely identified by a pair of numbers, $(x, y)$.* These numbers, called **coordinates**, may be either positive or negative. The points are also called Cartesian coordinates after René Descartes (1596–1650).

The $(x, y)$ coordinates replace the number and letter street addresses on the map. The first number, $x$, in the coordinate $(x, y)$ determines movement in the x-axis direction (east and west). The second number, $y$, determines movement in the y-axis direction (north and south). The positive and negative signs eliminate the need for the N.E., N.W., S.E., and S.W. quadrant designation. Another name for coordinates, which emphasizes that the order of the pair of numbers is important, is **ordered pairs**.

The **origin** is the name given to *the intersection of the two axes.* The word *origin* may refer to starting, or original, point. On the map of Washington, D.C., the origin corresponds to the U.S. Capitol building.

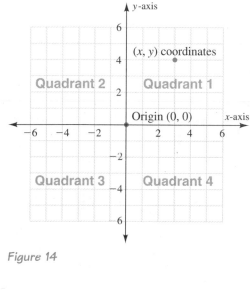

Figure 14

E X A M P L E    *1*

Graph these points on the same set of axes: $(4, 2)$, $(-2, 3)$, $(3, -2)$, $(0, 5)$, $(2, 0)$, and $(-5, -2)$. Indicate in which quadrant or on which axis each is located.

**SOLUTION**  Begin by drawing and/or labeling a set of axes.

To graph $(4, 2)$, move to 4 on the x-axis and go up 2 units in the $y$, or vertical, direction, as shown in Figure 15. (The word *unit* is used to describe one space in either the $x$ or the $y$ direction.) The point $(4, 2)$ is in the first quadrant.

Observe that $(-2, 3)$ and $(3, -2)$ are not the same point. Go to $-2$ on the x-axis and then move up 3 units in the $y$ direction to graph $(-2, 3)$ in the second quadrant. Go to 3 on the x-axis and go down 2 units in the $y$ direction to graph $(3, -2)$ in the fourth quadrant.

A zero in either the $x$ or the $y$ position means that the point is on one of the axes. The point $(0, 5)$ is on the y-axis, 5 units up from the origin. The point $(2, 0)$ is on the x-axis, 2 units to the right of the origin.

The point $(-5, -2)$ is in the third quadrant. Go to $-5$ on the x-axis and then go down 2 units to graph $(-5, -2)$.

◆ ◆ ◆ ◆ ◆ ◆ ◆ ◆ ◆ ◆ ◆ ◆ ◆ ◆ ◆ ◆ ◆ ◆ ◆ ◆ ◆ ◆ ◆ ◆ ◆ ◆ ◆ ◆ ◆ ◆ ◆ ◆ ◆ ◆ ◆ ◆ ◆

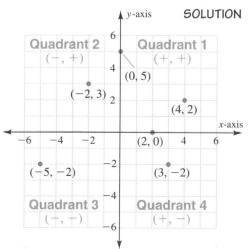

Figure 15

Each of the quadrants in Figure 15 is labeled with a pair of signs. Compare the labeled signs with the coordinates plotted in that quadrant. All coordinates in a quadrant have the same set of positive and negative signs.

The word *graph* is used as both a noun and a verb. Used as a noun, **graph** *refers to the points that have been plotted or the line or curve drawn through those points.* As a verb, **graph** *means to place coordinates, or ordered pairs, on the coordinate plane.*

It is customary to identify inputs with the x-axis and outputs with the y-axis. If the data have units (inches, dollars, liters, years, etc.), the axes are labeled clearly with the units. The graph in the next example has input in pounds and output in dollars.

### READING GRAPHS

Tables and graphs help answer these important questions: What is the output value when the input is given? What is the input value when the output is given? Graphs are particularly useful in estimating numbers between the entries on a table and in observing patterns in tabular data.

EXAMPLE  2        Bulk food purchases    Brown rice costs $0.89 per pound in the bulk food section of a grocery store. Table 14 and its graph (Figure 16) show the cost of the brown rice.

**a.** Which point (A, B, C, D, etc.) shows that 2 pounds of rice costs $1.78?

**b.** What fact from the table does point B represent?

**c.** Which point (A, B, C, D, etc.) would be used to determine the cost of 2.5 pounds of rice? Estimate the cost from the graph, calculate the exact cost, and compare your results.

**d.** Which point shows how much rice could be purchased for $4.00? Estimate the weight from the graph, calculate the exact weight, and compare your results.

| Input: Pounds | Output: Cost | Coordinate Point (x, y) |
|---|---|---|
| 1 | $0.89 | (1, 0.89) |
| 2 | 1.78 | (2, 1.78) |
| 3 | 2.67 | (3, 2.67) |
| 4 | 3.56 | (4, 3.56) |

Table 14  Cost of Brown Rice

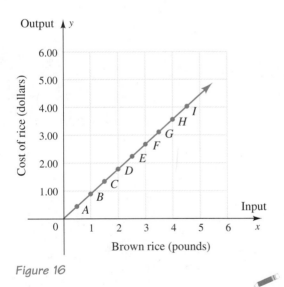

Figure 16

SOLUTION    **a.** Point D shows that 2 pounds of rice costs $1.78.

**b.** Point B represents the $0.89 cost of 1 pound of rice.

**c.** Point E would be used. Reading to the left of E, we see that the cost is between $2.00 and $2.50. To calculate exactly, we multiply 2.5 pounds times $0.89 per pound and round to the nearest cent: $2.23.

**d.** Point I shows that about 4.5 pounds could be purchased for $4.00. To calculate exactly, we divide $4.00 by $0.89 per pound to get 4.49 pounds. Remember that *per* also means division.

$$\$4.00 \div \frac{\$0.89}{\text{pound}} = \$4.00 \cdot \frac{\text{pound}}{\$0.89} = 4.49 \text{ pound}$$

◆ ◆ ◆ ◆ ◆ ◆ ◆ ◆ ◆ ◆ ◆ ◆ ◆ ◆ ◆ ◆ ◆ ◆ ◆ ◆ ◆ ◆ ◆ ◆ ◆ ◆ ◆ ◆ ◆ ◆ ◆ ◆ ◆ ◆ ◆ ◆ ◆ ◆ ◆ ◆

To make the next graph easier to read, we have labeled the *y*-axis with the exact postage costs rather than the evenly spaced numbers found in prior graphs. The graph in Figure 17 has flat parts where the postage costs stay the same. It is therefore called a **step graph**.

EXAMPLE 3

**First-class postage, 1995** The solid dot at the right end of each step in the graph in Figure 17 indicates the postage cost for each whole number of ounces.

**a.** Use the graph to complete Table 15, listing the entries as pairs of coordinates.

**b.** What may be concluded about the meaning of the circle in the graph above 0 ounces?

**c.** What is the cost for up to 1 ounce?

**d.** What is the change in cost between 1 ounce and 2 ounces?

**e.** What is the change in cost between 2 ounces and 3 ounces?

**f.** Predict the cost of an 8-ounce letter.

**g.** What rule determined the first-class postage rate in 1995?

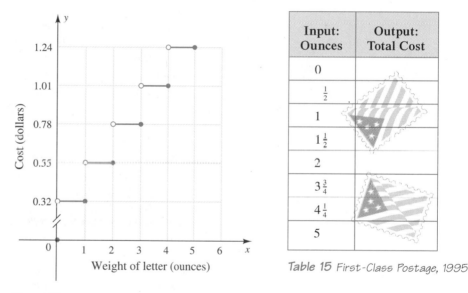

| Input: Ounces | Output: Total Cost |
|---|---|
| 0 | |
| $\frac{1}{2}$ | |
| 1 | |
| $1\frac{1}{2}$ | |
| 2 | |
| $3\frac{3}{4}$ | |
| $4\frac{1}{4}$ | |
| 5 | |

Table 15 *First-Class Postage, 1995*

Figure 17

SOLUTION **a.** The table entries are $(0,0)$, $\left(\frac{1}{2}, 0.32\right)$, $(1, 0.32)$, $\left(1\frac{1}{2}, 0.55\right)$, $(2, 0.55)$, $\left(3\frac{3}{4}, 1.01\right)$, $\left(4\frac{1}{4}, 1.24\right)$, and $(5, 1.24)$.

**b.** The circle means that 0 ounces does *not* cost $0.32. The cost of zero ounces is 0, because no ounces implies no letter. The dot at the origin indicates this fact.

**c.** $0.32

**d.** $0.23

**e.** $0.23

**f.** After the first ounce the cost rises $0.23 each ounce. To form a prediction we might add $0.23 each for the 6th, 7th, and 8th ounces to $1.24 (the cost for 5 ounces), to get $1.93.

**g.** Using the traditional way of describing postal rates, we can state the rule as $0.32 for the first ounce and $0.23 for each ounce thereafter. In 1995 the upper weight limit for a first-class letter was 11 ounces. This description is a conditional rule.

◆ ◆ ◆ ◆ ◆ ◆ ◆ ◆ ◆ ◆ ◆ ◆ ◆ ◆ ◆ ◆ ◆ ◆ ◆ ◆ ◆ ◆ ◆ ◆ ◆ ◆ ◆ ◆ ◆ ◆ ◆ ◆ ◆ ◆ ◆ ◆

We introduced two new graphing symbols in Example 3:

---

The *double slash, //,* on the *y*-axis between 0 and 0.32 *indicates that the spacing is different* between 0 and 0.32 than between other numbers on the *y*-axis. The answers to questions c, d, and e indicate this difference. Always use the double slash to indicate a change in the numbering on an axis.

The *circle* on a line segment or elsewhere in a graph *means the circled coordinate point is not part of the graph.* The input–output pair for that point does not fit the rule for the graph.

---

In Example 3 the circle above the 1-ounce mark means 1 ounce does not cost $0.55. Instead, we use the dot at $0.32 for 1 ounce. The flat line leaving the circle at $0.55 means the cost for a letter weighing just over 1 ounce is $0.55. We say a circle *excludes* the input number, whereas a dot *includes* the input number.

In the prior examples the numbers on the axes either were given in the problem or were integers. *When we decide what numbers should go on the axes, we say we are finding the* **scales** *for the axes.* To estimate the input scale, we subtract the high and low values for the input and divide by 10. To label the axis, round the scale estimate to the nearest easy counting value, such as 1, 5, 10, 20, 25, 50, 100, and so on. Repeat the process, as needed, to estimate an output scale. A **formula** *is a rule or principle written in mathematical language.* The scale formula described here is for approximation only and may be summarized as follows:

$$\text{Scale} = \frac{\text{Highest value} - \text{Lowest value}}{10}$$

Then round to the nearest reasonable number.

## CREATING GRAPHS: THE FOUR-STEP PROCESS

Let's return to the four problem-solving strategies and apply them to graphing. The strategies are listed below as steps, along with some questions to guide your thinking process. The questions focus your attention on important information and features. Finding the answers becomes easier with experience. Being able to answer the questions should be a long-term goal in your mathematics career.

1. **Understand the problem.**
   Which information is the input?
   Which information is the output?
   Keep in mind that the output depends on the input—that is, *y* depends on *x*.

2. **Plan the graph.**
   Which quadrants are appropriate to the problem situation?
   What scales should be placed on the axes? Use the formula as needed.
   Do fractional or decimal inputs make sense?

3. **Do the graph.**
   Make a table if needed.
   Graph the points.
   What shape do the points make? Look for a pattern, from left to right.

4. **Check the graph, and extend the results as needed.**
   Do all the data points lie on the graph?
   Do other points on the graph make sense in the problem situation?
   How might new data or information change the graph?

Example 4 introduces the principal square root graph. The **principal square root** of a number is the positive number that, multiplied by itself, produces the given number. Thus $\sqrt{49} = 7$ because $7 \cdot 7 = 49$, and $\sqrt{25} = 5$ because $5 \cdot 5 = 25$. The square root symbol, $\sqrt{\phantom{x}}$, is also called a **radical**.

EXAMPLE 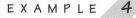 **4**

Square roots    Graph the data in Table 16, using the four steps and questions to guide your thinking.

| Input $x$ | Output $\sqrt{x}$ | Coordinate Point $(x, y)$ |
|-----------|-------------------|----------------------------|
| 0 | 0 | $(0, 0)$ |
| 0.5 | 0.707 | $(0.5, 0.707)$ |
| 1 | 1 | $(1, 1)$ |
| 1.5 | 1.225 | $(1.5, 1.225)$ |
| 2 | 1.414 | $(2, 1.414)$ |
| 3 | 1.732 | $(3, 1.732)$ |
| 4 | 2 | $(4, 2)$ |
| 5 | 2.236 | $(5, 2.236)$ |

Table 16  *Square Roots*

SOLUTION    *Understand*: The inputs and outputs are given.

*Plan*: The numbers in Table 16 are in the first quadrant. The given inputs are on the interval 0 to 5, so the scale formula gives $\frac{1}{2}$.

$$\frac{\text{High} - \text{Low}}{10} = \frac{5 - 0}{10} = \frac{5}{10} = \frac{1}{2}$$

For simplicity, we label only the whole numbers. Because the numbers involved are small, it is reasonable to use integers on the $y$-axis. Fractions and decimals are possible inputs.

*Do the graph*: The shape of the resulting square root graph (Figure 18) is a curve that starts at the origin and gradually rises to the right.

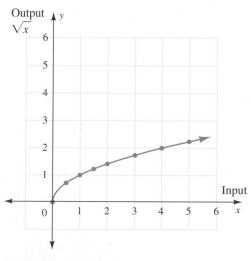

Figure 18

*Check and extend*: This graph can be checked on a graphing calculator, if available. Make sure the graph passes through $(0, 0)$, $(1, 1)$, and $(4, 2)$. We have no information at this point about the graph for negative inputs. ✔

Verify the outputs in Table 16. The outputs are rounded to three decimal places (to the nearest thousandth). To find square roots, some calculators require that $\boxed{\sqrt{\phantom{x}}}$ be entered first, whereas others need $\boxed{\sqrt{\phantom{x}}}$ entered after the number. Still others require a shift, $\boxed{\text{INV}}$ or $\boxed{\text{2nd}}$, to get the square root. Experiment with your calculator until you get a correct answer.

The outputs in Table 16 are two types of numbers: rational and irrational. The whole number outputs are rational. The expressions $\sqrt{0.5}$, $\sqrt{1.5}$, $\sqrt{2}$, $\sqrt{3}$, and $\sqrt{5}$ are irrational numbers. **Irrational numbers** *are real numbers that cannot be written as fractions.* We rounded the irrational numbers to write them in Table 16 and in doing so approximated each irrational number with a rational number. If we want an exact answer, we leave these irrational numbers in their radical form.

Attempt to take the square root of a negative number on a scientific calculator. The calculator should signal an error, usually with a small E. The keystrokes for entering a negative number are 3 $\boxed{\pm}$ or $\boxed{\pm}$ 3. The error message is telling us that the square root of a negative number is *undefined* in the set of real numbers. A few scientific graphing calculators give an answer for the square root of a negative number, but the answer is not a real number and should be disregarded at this time.

Because we graph only real numbers, the input may not be negative and so there are no points on the graph to the left of the origin in Figure 18. Thus the graph of the square root of $x$, $\sqrt{x}$, starts at the origin.

In Example 5 we will graph water temperature and volume data. This graph shows another reason to graph. A pattern emerges that may not be apparent from the data in tabular form.

E X A M P L E   **5**    *Temperature and water volume*    Graph the data in Table 17, which repeats the information from Section 1.2. Use the four steps as a guide.

| Input: Temperature (°C) | Output: Volume (cc) |
|:---:|:---:|
| −5 | 1.00070 |
| −4 | 1.00055 |
| −3 | 1.00042 |
| −2 | 1.00031 |
| −1 | 1.00021 |
| 0 | 1.00013 |
| 1 | 1.00007 |
| 2 | 1.00003 |
| 3 | 1.00001 |
| 4 | 1.00000 |
| 5 | 1.00001 |
| 6 | 1.00003 |
| 7 | 1.00007 |
| 8 | 1.00012 |
| 9 | 1.00019 |
| 10 | 1.00027 |

*Table 17 Volume of Water Near Freezing*

SOLUTION    *Understand*: The inputs and outputs are given, implying that the volume depends on the temperature.

*Plan*: The table contains both negative and positive inputs. The outputs are all positive. Thus the graph lies in both the first and the second quadrant. The integer input makes a natural $x$-axis scale, although we will count by twos. Fractional and decimal values between $-5°C$ and $10°C$ make sensible inputs. The outputs require a scale calculation:

$$y\text{-scale} = \frac{1.00070 - 1.00000}{10} = \frac{0.00070}{10} = 0.00007$$

and 0.00007 rounds to 0.00010. Because the lowest output number is 1.00000, we will start the scale on the $y$-axis at that number and use a double slash between 0 and 1.00000.

*Graph*: The volume of water graph (Figure 19) is roughly a $\cup$ shape. The shape suggests a smooth curve, which may be approximated by a graph called a **parabola**. Many parabolas have roughly a $\cap$ shape, like the path of a stream of water from a drinking fountain.

*Check*: Several pairs of inputs (1 and 7, 2 and 6, and 3 and 5) have the same output. This repetition of output numbers gives the graph its distinctive shape and confirms that we have graphed correctly. If all the outputs paired in this way, the graph would be *line symmetric*, or *have the property of folding onto itself* over the point $(4, 1)$. ✔

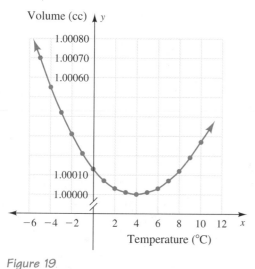

Figure 19

✦ ✦ ✦ ✦ ✦ ✦ ✦ ✦ ✦ ✦ ✦ ✦ ✦ ✦ ✦ ✦ ✦ ✦ ✦ ✦ ✦ ✦ ✦ ✦ ✦ ✦ ✦ ✦ ✦ ✦ ✦ ✦ ✦ ✦ ✦

The postage graph in Figure 17 is a conditional graph—the output rule depends on the input values. We now return to another conditional situation, the Chevron credit card payment.

EXAMPLE **6**    *Payment schedule*    Use Table 18 to verify the inputs and outputs in Table 19. Apply the four problem-solving steps in graphing the data in Table 19.

| Input: Charge Balance | Output: Payment Due |
|---|---|
| $0 to $20 | Full amount |
| $20.01 to $500 | 10% or $20, whichever is greater |
| $500.01 or more | $50 plus the amount in excess of $500 |

Table 18 *Credit Card Payment Schedule*

| Input: Balance | Output: Payment |
|---|---|
| $ 0 | $ 0 |
| 10 | 10 |
| 20 | 20 |
| 30 | 20 |
| 50 | 20 |
| 100 | 20 |
| 200 | 20 |
| 300 | 30 |
| 400 | 40 |
| 500 | 50 |
| 600 | 150 |

Table 19 *Credit Card Payments*

SOLUTION  ***Understand***: Because the inputs and outputs are given, we assume that the payment due depends on the charge balance. Both axes in Figure 20 are labeled with dollars.

***Plan***: Because no negative money values are reasonable, this is a first quadrant graph. Decimals and fractions of a dollar make sensible inputs. The inputs lie on the interval 0 to 600. This interval implies a scale of 60, but we will round to a scale of 100. The output interval is 0 to 150, for a scale of 15. We round this scale to 20, which is a convenient scale because a number of outputs are multiples of 20.

***Graph***: The graph is in four line segments.

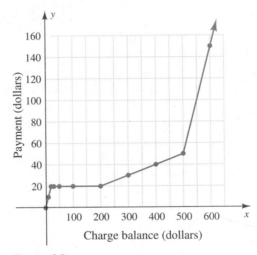

**Figure 20**

***Check***: The data in the tables all fit on the graph. The four separate line segments may be surprising because there were only three parts to the original table. ✔

♦ ♦ ♦ ♦ ♦ ♦ ♦ ♦ ♦ ♦ ♦ ♦ ♦ ♦ ♦ ♦ ♦ ♦ ♦ ♦ ♦ ♦ ♦ ♦ ♦ ♦ ♦ ♦ ♦ ♦ ♦ ♦ ♦ ♦ ♦

## EXERCISES 1.4

**Write the coordinates of points *A* to *I* in Exercises 1 and 2.**

1.

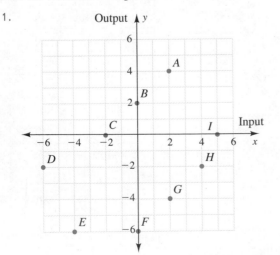

2.

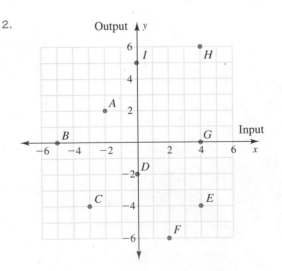

3. Explain in complete sentences how you determine the coordinates of a point.

**For Exercises 4 and 5 determine the quadrant in which each point is located.**

4. a. $(4, -3)$   b. $(-3, 2)$   c. $(-2, -3)$   d. $(2, -3)$

5. a. $(-4, 3)$   b. $(3, -4)$   c. $(-2, -4)$   d. $(-3, -2)$

6. How do you determine in which quadrant a pair of coordinates is located?

**For Exercises 7 and 8 determine on which axis each point is located.**

7. a. $(0, 4)$   b. $(-2, 0)$   c. $(-3, 0)$

8. a. $(0, -4)$   b. $(0, -2)$   c. $(3, 0)$

9. Explain in complete sentences how you determine on which axis a pair of coordinates is located.

10. Trace or make a sketch of the Washington, D.C., map (Figure 13). Expand the map to the west and north as needed. Locate the following places or buildings. (The letters a to h are not used because they are already on the map.)

   i. Lincoln Memorial, 23rd and The Mall

   j. Federal Aviation Building, 7th and C, S.W.

   k. The White House, 16th and F, N.W.

   l. Vietnam Memorial, 21st and The Mall

   m. National Geographic Society, 16th and M, N.W.

**For Exercises 11 and 12 use the graph in the figure to complete the table.**

11.

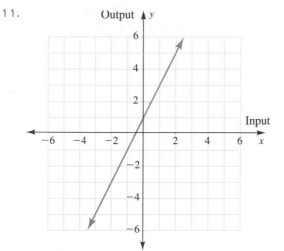

| Input x | Output y |
|---------|----------|
| 0       |          |
|         | 0        |
| 2       |          |
|         | 7        |
|         | 3        |
| -2      |          |
|         | 2        |

12.

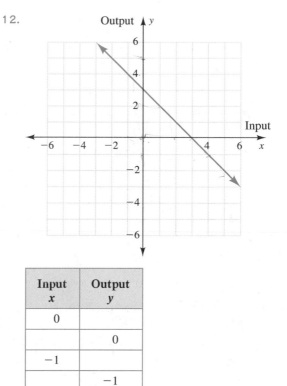

| Input x | Output y |
|---------|----------|
| 0       |          |
|         | 0        |
| -1      |          |
|         | -1       |
|         | 2        |
|         | 5        |
| 2       |          |

**In Exercises 13 to 20 graph the pair of points, and connect them with a straight line. Write the coordinates of two other points on the same line.**

13. $(3, 3), (5, 5)$   14. $(1, 2), (3, 6)$

15. $(3, 1), (9, 3)$   16. $(2, -1), (8, -4)$

17. $(4, -4), (2, -2)$   18. $(8, 4), (2, 1)$

19. $(3, 5), (5, 7)$   20. $(1, 3), (2, 6)$

21. From Exercises 13 to 20 choose the exercise whose coordinates match each rule.
   a. $y = \frac{1}{2}x$   b. $y = x$   c. $y = 3x$

22. From Exercises 13 to 20 choose the exercise whose coordinates match each rule.
   a. $y = \frac{1}{3}x$   b. $y = 2x$   c. $y = x + 2$

23. In Example 2, how would a higher price per pound for rice change the graph? Write your answer as a complete sentence.

24. Refer to Example 3.
   a. What is the cost of a $4\frac{1}{2}$-ounce letter?
   b. What is the cost of a $5\frac{1}{2}$-ounce letter?

25. Suppose the bulk food department in Example 2 offers reusable containers for $1.00. If the total cost rises by $1.00, how will the graph change?

26. In Example 5, what does the table indicate to be the reason a full container might burst when cooled below 4°C?

27. In Example 5, as the volume increases, we say the water is expanding. In the direction for temperature change, does the expansion happen faster from 1° to −2° or from 7° to 10°?

28. Refer to Example 6.

   a. Why is the graph level between the inputs of $20 and $200?

   b. Why is there a change in the graph at $200?

**Write each word in Exercises 29 and 30 in two sentences. In the first sentence of each pair, the meaning of the word should not be mathematical. In the second sentence, the meaning of the word should be mathematical.**

29. a. radical      b. irrational

30. a. segment      b. scale

31. **Bulk Food Purchases.** In the bulk foods department candy costs $1.29 per pound.

   a. Use the four-step approach to build a table and graph for 0 to 4 pounds. Label your steps.

   b. Compare your graph with that of Example 2. Do they have the same scale? From left to right, does your graph rise faster than the graph of the cost of rice? Why or why not? Use complete sentences.

   c. Use your graph to estimate the cost of these candy purchases: $2\frac{1}{2}$ pounds, $1\frac{3}{4}$ pounds, $3\frac{1}{4}$ pounds.

   d. Two and a half pounds of the same candy is available in packages for $3.98. Plot this information as a data point. Which is a better buy, bulk candy or packaged candy?

32. **Bulk Food Purchases, Continued.** Mixed nuts in the bulk foods department cost $6.50 per pound.

   a. Use the four-step approach to build a table and graph for 0 to 4 pounds. Label your steps.

   b. Compare your graph with that of Example 2. From left to right, does your graph rise faster than the graph of the cost of rice? Why or why not? Use complete sentences.

   c. Use your graph to estimate the cost of these purchases: $2\frac{1}{2}$ pounds, $1\frac{3}{4}$ pounds, $3\frac{1}{4}$ pounds.

   d. One and a half pounds of the same nut mixture is available in packages for $9.49. Plot this information as a data point. Which is a better buy, bulk or packaged nuts?

**In Exercises 33 and 34 the tables are displayed in a horizontal format. This saves space on the page. Horizontal tables have the inputs on the top row and the outputs on the bottom row.**

33. **Stock Prices**

   a. Graph the stock prices in the table. Use the four steps.

   b. Describe the general trend of the stock prices. Use a complete sentence.

| Day | 1 | 2 | 3 | 4 | 5 | 6 | 7 | |
|-----|---|---|---|---|---|---|---|---|
| Price | $8\frac{1}{2}$ | 9 | $8\frac{1}{2}$ | $8\frac{1}{4}$ | $8\frac{1}{4}$ | $8\frac{1}{2}$ | 8 | |
| Day | 8 | 9 | 10 | 11 | 12 | 13 | 14 | 15 |
| Price | $7\frac{3}{4}$ | $8\frac{1}{2}$ | $8\frac{1}{4}$ | 8 | $7\frac{1}{2}$ | $7\frac{3}{4}$ | 8 | $7\frac{1}{2}$ |

34. **Stock Prices, Continued**

   a. Graph the stock prices in the table. Use the four steps.

   b. Describe the general trend of the stock prices. Use a complete sentence.

| Day | 1 | 2 | 3 | 4 | 5 | 6 | 7 | |
|-----|---|---|---|---|---|---|---|---|
| Price | $8\frac{1}{2}$ | $8\frac{3}{4}$ | $8\frac{1}{4}$ | $8\frac{1}{4}$ | 8 | $7\frac{1}{4}$ | 8 | |
| Day | 8 | 9 | 10 | 11 | 12 | 13 | 14 | 15 |
| Price | $7\frac{1}{2}$ | $7\frac{3}{4}$ | 8 | $7\frac{3}{4}$ | 8 | $8\frac{1}{2}$ | $8\frac{3}{4}$ | $8\frac{1}{2}$ |

35. **Graphing Your Travel.** Do the following for an "average" school day.

   a. Make a table with 24 inputs, showing the time of day (by hours, midnight to midnight) as the input and the distance you are from home as the output. Graph the table.

   b. What would the graph look like if you stayed home all day?

   c. What would the graph look like if you spent an hour in one place 10 miles from home?

   d. How does the graph for traveling toward home compare with the graph for traveling away from home?

   e. If you travel at a faster rate of speed, how will the graph change?

36. **Shirt Fabric.** The data in the table for Exercises 33–36 of Section 1.2 (page 12) show the amount of fabric needed for a shirt when the fabric is 58 to 60 inches wide. The following table shows the amount of fabric needed when the fabric is 44 to 45 inches wide.

| Input: Shirt Size | Output: Fabric Needed |
|-------------------|-----------------------|
| Small: 34–36 | $2\frac{1}{8}$ yards |
| Medium: 38–40 | $2\frac{1}{8}$ yards |
| Large: 42–44 | $2\frac{1}{2}$ yards |
| Extra Large: 46–48 | $2\frac{1}{2}$ yards |

a. Graph the data from both tables on one set of axes. Use a different color ink or pencil for each set of data, and label the graphs clearly.

b. Compare the two graphs by position and by shape.

c. If 36-inch-wide fabric were available, predict where the output values for the graph would be located.

d. If 72-inch-wide fabric were available, predict where the graph would be located.

## COMPUTATION REVIEW

The whole number generated by placing an integer into $x^2$ is called a perfect square—such as $2^2 = 2 \cdot 2 = 4$, $4^2 = 4 \cdot 4 = 16$, and $5^2 = 5 \cdot 5 = 25$. The perfect square numbers 4, 16, and 25 end with digits 4, 6, and 5. Name three whole number perfect squares (if they exist) that satisfy the conditions in Exercises 37 to 40.

37. End with the digit 0.          38. End with the digit 1.

39. End with the digit 2.          40. End with the digit 3.

41. What last digits are found on whole number perfect squares?

42. What last digits are not found on whole number perfect squares?

In Exercises 43 and 44 round each irrational number to a rational number with three decimal places (the nearest thousandth). Identify each number in the first column as rational or irrational.

43.

| Square Root | Value | Rational or Irrational? |
|---|---|---|
| $\sqrt{36}$ | 6 | rational |
| $\sqrt{15}$ | | |
| $\sqrt{25}$ | | |
| $\sqrt{35}$ | | |
| $\sqrt{12.25}$ | | |
| $\sqrt{2.25}$ | | |
| $\sqrt{6}$ | | |
| $\sqrt{16}$ | | |
| $\sqrt{26}$ | | |

44.

| Square Root | Value | Rational or Irrational? |
|---|---|---|
| $\sqrt{4}$ | 2 | rational |
| $\sqrt{14}$ | | |
| $\sqrt{24}$ | | |
| $\sqrt{34}$ | | |
| $\sqrt{64}$ | | |
| $\sqrt{144}$ | | |
| $\sqrt{.01}$ | | |
| $\sqrt{6.25}$ | | |
| $\sqrt{1.44}$ | | |

## PROBLEM SOLVING

In Exercises 45 to 48 find the coordinates for points $A$ and $B$ from these portions of the coordinate plane. Each space equals one. The axes are hidden, so you must reason relative positions from the given point(s).

45.

(0, 4)

$B$          (2, 0)

$A$

46.

(0, −2)

$B$          $A$

47.

$A$

(4, 0)

$B$

48.

$A$

(2, 3)      $B$

## PROJECTS

49. **Stock Prices.** Record three weeks' closing prices of one share of common stock from a newspaper financial page. Make a graph of the data. Comment, in complete sentences, on any trend you observe. Most libraries carry back issues of newspapers, so it is possible to gather all three weeks' worth of data for this project at once.

50. **Water Usage.** Water charges at Everybody's Utility include a basic $5.15 fee and $0.65 for each thousand gallons used.

a. Make a table and graph the total cost of 0 to 8000 gallons.

b. How many gallons does your household use each month? Assume 30 days, and estimate usage. Use the following table, which gives domestic water use in the United States in gallons. If your water is metered, how close is your estimate to the metered use?

| | |
|---|---|
| Washing car with hose running | 180 |
| Watering lawn for 10 minutes | 75 |
| Washing machine at top level, 1 load | 60 |
| Ten-minute shower | 25–50 |
| Average bath | 36 |
| Hand-washing dishes with water running | 30 |
| Dripping faucet, per day | 25–30 |
| Shaving with water running | 20 |
| Automatic dishwasher | 10 |
| Toilet flush | 5–7 |
| Brushing teeth with water running | 2 |

*Source:* Data reprinted with permission from G. Tyler Miller, Jr., *Living in the Environment,* Wadsworth Publishing Co., © 1992, p. 356.

c. A gallon is 231 cubic inches. How big a container would be needed to hold 8000 gallons? Describe such a container's shape and dimensions (if rectangular give length, width, and height; if cylindrical give radius of base and height) in suitable units of measurement (inches, feet, yards, miles, etc.).

51. *Water Volume.* Research the change in volume water undergoes as it changes to ice under normal pressure. Summarize with numbers and a graph. Compare the results with those in Table 17, Example 5. Why do peaks form on ice cubes? Why does ice float in water?

## CHAPTER 1 SUMMARY

### Vocabulary

*For definitions and page references, see the Glossary/Index.*

| | |
|---|---|
| axes | parabola |
| conditional rule | perfect square |
| constant | perpendicular |
| coordinate graph | principal square root |
| coordinate plane | product |
| coordinates (ordered pair) | quadrant |
| denominator | quotient |
| difference | radical |
| even numbers | rational numbers |
| expression | real numbers |
| formula | set |
| graph | scale |
| input–output relationship | step graph |
| integers | sum |
| interval | undefined |
| irrational numbers | variable |
| numerator | whole numbers |
| odd numbers | *x*-axis (horizontal axis) |
| ordered pairs | *y*-axis (vertical axis) |
| origin | |

### Concepts

**Division** may be written as a fraction.

**Multiplication** may be written in several ways: $2n$, $2 \cdot n$, $2(n)$, $(2)(n)$.

The **four problem-solving strategies** are understand the problem, make a plan, carry out the plan, and check the solution.

In graphing, a **small circle** *excludes* the input number whereas a **dot** *includes* the input number.

The **scale** markings on the axes should be equally spaced and clearly numbered. When the scale does not start at zero, we make a **double slash** between the origin and the first mark on the axis to indicate a break in the numbering.

### Numerical Prefixes (Optional)

| | | | |
|---|---|---|---|
| one: | mono, uni | seven: | sept, hept |
| two: | do, duo, bi | eight: | oct |
| three: | tri | nine: | nov, non |
| four: | quad, quat, quart, tetra | ten: | decem, deci |
| five: | quin, penta | hundred: | cent |
| six: | sex, hex | thousand: | milli |

## CHAPTER 1 REVIEW EXERCISES

In Exercises 1 and 2 describe how the data from the table match the design in the figure. Write a rule for the table in words and symbols.

1.

| Input | Output |
|-------|--------|
| 0 | 1 |
| 1 | 4 |
| 2 | 7 |
| 3 | 10 |
| 4 | 13 |

2.

| Input | Output |
|-------|--------|
| 0 | 5 |
| 1 | 6 |
| 2 | 7 |
| 3 | 8 |
| 4 | 9 |

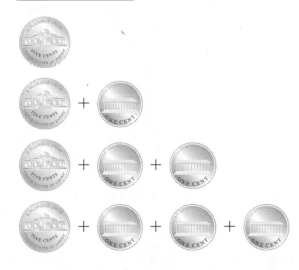

3. Use the following table to determine what discount is received on each of these purchases.

| Total Purchase | Discount |
|----------------|----------|
| $0 to $149.99 | 5% of the purchase |
| $150 to $499.99 | 6% of the purchase |
| $500 and over | $50 or 8% of the purchase, whichever is greater |

a. $65          b. $145          c. $250

d. $500          e. $550          f. $700

4. Use the following words as column headings:

Addition     Subtraction     Multiplication     Division

List each of these words, phrases, or symbols under the appropriate heading: decreased by, increased by, product, sum, more than, per, half (write it in two different ways), twice, of, fewer than, longer than, less than, difference, quotient, farther, slower than, for each, times, loses, increases, altogether, plus, combined, one third (two ways), faster than, multiplied by, $a$ less $b$, $a$ greater than $b$, $a$ diminished by $b$, $b$ subtracted from $a$, $a$ decreased by $b$, $b$ bigger than $a$, $a$ exceeds $b$ by 3, the fraction bar, $a \cdot b$, $a - b$, $(a)(b)$, $a/b$, $ab$, $a \div b$, $b/a$, $a + b$, $a(b)$, $a \times b$

5. Write each of the following words or phrases next to as many different numbers as it correctly describes: real number, rational number, irrational number, integer, whole number.

a. 1.5          b. $\sqrt{3}$          c. $\sqrt{9}$

d. $\frac{1}{2}$          e. $\frac{1}{3}$          f. $-5$

6. Complete the following table. Round irrational numbers to rational numbers with three decimal places.

| Square Root | Value | Rational or Irrational? |
|-------------|-------|-------------------------|
| $\sqrt{9}$ | 3 | rational |
| $\sqrt{19}$ | 4.3 | ir |
| $\sqrt{29}$ | 5.3 | ir |
| $\sqrt{39}$ | 6.2 | ir |
| $\sqrt{49}$ | 7 | rat |
| $\sqrt{1.21}$ | 1.1 | ir |
| $\sqrt{.09}$ | .3 | rat |
| $\sqrt{20.25}$ | 4.5 | ir |
| $\sqrt{169}$ | 13 | rat |

7. Translate into words.

a. $3x - 4$          b. $x^2 + 3$          c. $x \div 3$

8. Translate into expressions. Let $x$ be the input.

a. Four more than the product of two and the input.

b. The difference between five and the square of the input.

c. Fifteen percent of the input.

9. In the bulk foods department, pinto beans cost $0.37 per pound. Several packaged pinto bean choices are also available: 1 pound at $0.49, 2 pounds at $0.74, 4 pounds at $1.66, and 8 pounds at $2.88. Make a table and graph by answering the questions in the four-step process (Section 1.4). In Step 4 (check and extend), plot the cost of the packaged choices as individual data points. Discuss which might be the best buy and under what circumstances. What is the rule describing the bulk purchase graph? Let $x$ be the input variable.

10. In the early 1980s the federal government gave away surplus cheese, butter, and powdered milk. Eligibility guidelines for family monthly income were as follows: one person, $507; two, $682; three, $857; four, $1,032; five, $1,207; six, $1,382; seven, $1,557; eight, $1,732. Make a table and graph by following the four-step process. In Step 4 (check and extend), how much does each additional person add to the income level? If the points were connected and continued to the left, where would the graph cross the *y*-axis? Write a rule in words and in symbols that describes the guidelines.

11. The following graphs represent four different dieting experiences. Describe each experience in words. Use complete sentences.

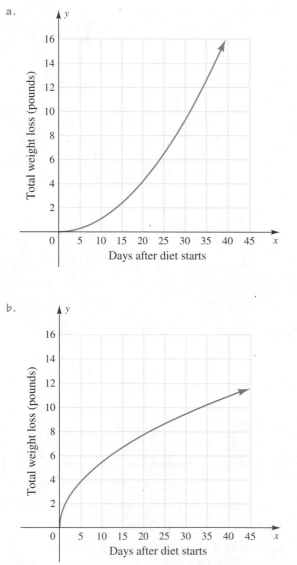

a.

b.

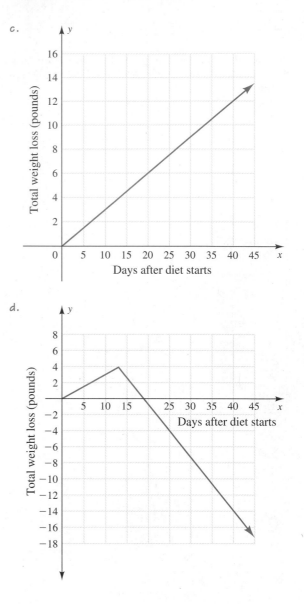

c.

d.

12. Describe how to obtain *y* from *x* in (4, −4) and (2, −2), with a rule in words or in symbols.

13. Describe how to obtain *y* from *x* in (2, −1) and (8, −4), with a rule in words or in symbols.

14. Some number operations require two numbers (addition, subtraction). Other number operations need only one number. Give an example of an operation that needs only one number.

**Make an input–output table for Exercises 15 and 16.**

15. The input, *h*, is the number of kilowatthours of electricity used. The output is $5.00 plus $0.04 per kilowatthour used.

16. The input, *n*, is the number of pounds of rice purchased. The rice costs $0.89 per pound. The output is the total cost, including a $0.10 "use your own container" refund.

**For Exercises 17 and 18 use guessing, but to be efficient, keep a list of your guesses.**

17. a. What is the maximum product of two whole numbers whose sum is 13?

   b. What fraction or decimal value gives a larger product than the whole numbers?

   c. Describe how to use the *sum* number to obtain the maximum product number.

18. a. What is the maximum product of two whole numbers whose sum is 15?

   b. What fraction or decimal value gives a larger product than the whole numbers?

   c. Describe how to use the *sum* number to obtain the maximum product number.

19. Coordinates may be applied to navigation if we describe position by directions relative to the origin. Which point in the figure matches each of the following descriptions?

   a. E3 S1      b. W1 N1      c. W2 S1      d. E2 N2

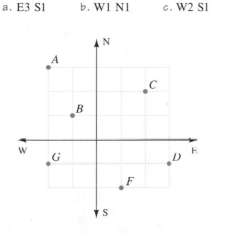

20. Find the words in the following list that match the given definitions: bifocal, quadrille, quintet, binomial, bimetal, triad, biathlon, tricolor, quadrilateral.

   a. Two thin metal strips fastened together and used to detect small changes in temperature

   b. Set of three musical notes

   c. Olympic Games event involving skiing and shooting

   d. Flag with three colors in stripes, such as the French flag

   e. Eyeglasses with two parts (top and bottom)

   f. Square dance for four couples

**For Exercises 21 and 22 assume that the coordinate axes are temporarily invisible and the scale on each graph is one space equals one unit. Give the coordinates of *A* and *B* in each.**

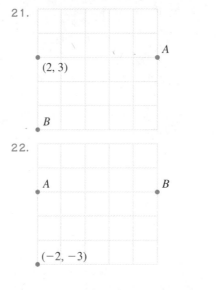

21.

22.

## CHAPTER 1 TEST

1. The answer to a subtraction problem is the _____ .

2. A listing of numbers in braces, { }, or a collection of objects is a _____ .

3. Use the figure as an aid in identifying the names of the indicated locations.

   a. The starting point

   b. The vertical line

   c. The horizontal line

   d. $(-4, 4)$ is a _____ point.

   e. The region where both members of the ordered pairs are negative is the third _____ .

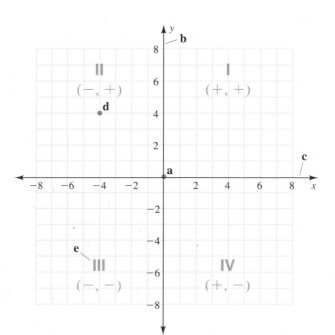

4. Make an input–output table with three rows for the following rule: Output is twice the input plus 3. Graph the data.

5. A car rental costs $150 for one week plus $0.10 each mile. Use the four-step process to make a table and graph for one week's rental.

   a. **Understand**: Which information is the input, $x$? the output? What units go on the axes?

   b. **Plan**: Explain how you determine quadrants, scales, and appropriate inputs.

   c. **Do the graph**: Make a table, and draw a graph from the table.

   d. **Check and extend**: Another company offers the same car for $200 a week with no per-mile charge. For what number of miles per week is the second car a better deal? What is an expression for the original rental?

6. List five pairs of integers that add to 19.

   a. What is the product of each pair?

   b. What is the maximum product from pairs of integers that add to 19?

   c. Is there any other pair of numbers adding to 19 that gives a larger product? Explain.

7. Use your calculator to find the following square roots. Indicate which answers are rational numbers. Describe patterns that you observe.

   a. $\sqrt{1}$    b. $\sqrt{10}$

   c. $\sqrt{100}$    d. $\sqrt{1000}$

   e. $\sqrt{10,000}$    f. $\sqrt{1,000,000}$

8. The vertical and horizontal axes have been omitted from the figure below. Use the given point to identify the other two points:

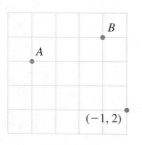

9. Make an input–output table with eight rows for this rule: If the input is an even number, the output is half the input; if the input is an odd number, the output is double the input.

10. As your twenty-first birthday approaches, your grandmother offers you a choice of gifts. Explain which option you should take.

    **Option 1:** On your twenty-first birthday, she will give you $1. On your twenty-second birthday, she will double the prior gift and give you $2. On your twenty-third birthday, she will again double the prior gift and give you $4. She repeats the pattern until you are 30.

    **Option 2:** On your twenty-first birthday, she will give you $21. On your twenty-second birthday, she will give you $22. She repeats the pattern until you are 30.

11. Explain how the rabbit-tracks design in the figure fits the input–output relationship in the following table. Complete the table and write a rule in words and symbols.

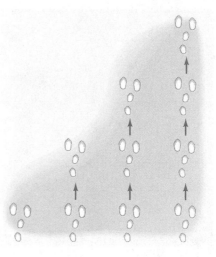

| Input | Output |
|-------|--------|
| 1 | 4 |
| 2 | 8 |
| 3 | 12 |
| 4 | 16 |
| 5 | |

# Operations with Real Numbers

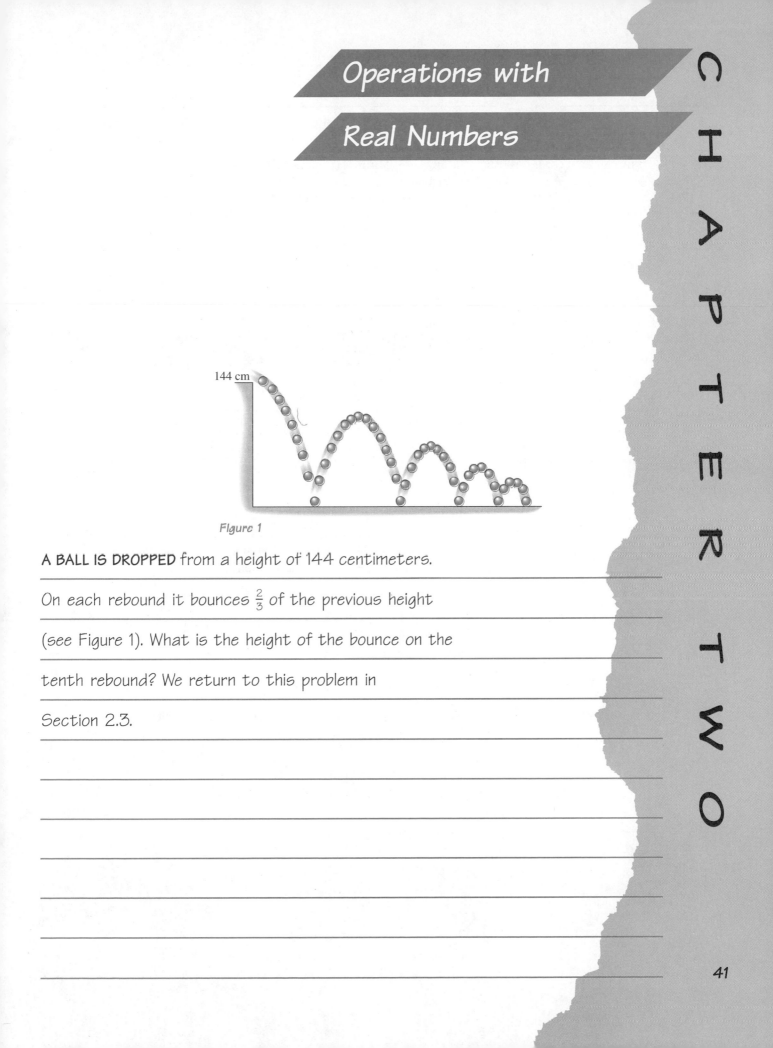

144 cm

Figure 1

**A BALL IS DROPPED** from a height of 144 centimeters.

On each rebound it bounces $\frac{2}{3}$ of the previous height

(see Figure 1). What is the height of the bounce on the

tenth rebound? We return to this problem in

Section 2.3.

## 2.1  ADDITION AND MULTIPLICATION WITH REAL NUMBERS

OBJECTIVES    Use the ion model to add positive and negative integers. ♦ Use the checkbook model to add positive and negative decimals. ♦ Find the absolute value of a number. ♦ Apply absolute value in applications such as distance and cash-over and cash-under. ♦ Make tables of absolute value expressions and graph the data. ♦ Interpret multiplication as repeated addition. ♦ Use tables and graphs to find patterns in multiplication of positive and negative numbers. ♦ Identify the commutative and associative properties. ♦ Use the commutative and associative properties for addition and multiplication of real numbers.

WARM-UP    Complete the following tables, and look for patterns in the completed tables.

**1.**

| Input $x$ | Output $x + 2$ |
|---|---|
| 2 | |
| 1 | |
| 0 | |
| −1 | |
| −2 | |

**2.**

| Input $x$ | Output $x + 1$ |
|---|---|
| 2 | |
| 1 | |
| 0 | |
| −1 | |
| −2 | |

♦

In this section we investigate ways to think about addition and multiplication with integers, as well as ways to do these operations. We summarize our addition rules with the concept of absolute value. Finally, we examine properties that make addition and multiplication easier.

### ADDITION OF REAL NUMBERS

**The Ion Model.**    The circles in Example 1 hold charged particles. Each particle is worth 1 unit. The + means a charge of positive one. The − means a charge of negative one. *One positive charge neutralizes one negative charge*; that is, $+1 + (−1) = 0$. The net charge in the circle is the charge that remains after all positive and negative pairs have been neutralized by adding to zero.

EXAMPLE    1    *Charges on ions*    What is the net charge on each circle?

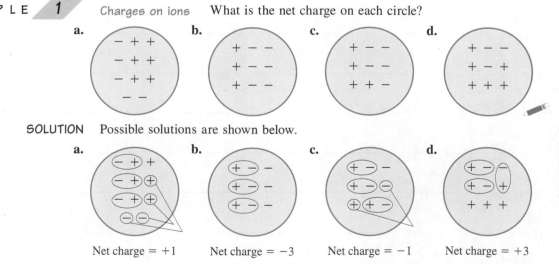

SOLUTION    Possible solutions are shown below.

Net charge = +1    Net charge = −3    Net charge = −1    Net charge = +3

♦ ♦ ♦ ♦ ♦ ♦ ♦ ♦ ♦ ♦ ♦ ♦ ♦ ♦ ♦ ♦ ♦ ♦ ♦ ♦ ♦ ♦ ♦ ♦ ♦ ♦ ♦ ♦ ♦ ♦ ♦

If there are more negative particles, the answer is negative. If there are more positive particles, the answer is positive. This way of thinking about positive and negative numbers is called the *ion model* because it is closely related to work with charged atoms and molecules in chemistry. The ion model works well for adding positive and negative integers.

EXAMPLE 2

*Charges on ions, continued*   Write each ion problem in Example 1 as an addition fact.

SOLUTION

**a.** $-5 + (+6) = +1$   **b.** $+3 + (-6) = -3$

**c.** $+4 + (-5) = -1$   **d.** $+6 + (-3) = +3$

♦ ♦ ♦ ♦ ♦ ♦ ♦ ♦ ♦ ♦ ♦ ♦ ♦ ♦ ♦ ♦ ♦ ♦ ♦ ♦ ♦ ♦ ♦ ♦ ♦ ♦ ♦ ♦ ♦ ♦ ♦ ♦ ♦ ♦ ♦ ♦

Check the solutions in Example 2 with a scientific calculator. Enter a negative number into the calculator with the $\boxed{\pm}$ or $\boxed{(-)}$ key. Do not use the subtraction operation for a negative sign. Some calculators require that $\boxed{\pm}$ be entered before the number; others require that it be entered immediately after the number. Practice until your answers agree with the solutions.

**The Checking Account Model.**   A checking account provides a way of thinking about the addition of positive and negative decimals. Suppose we consider making a deposit into the checking account as a positive number and writing a check as a negative number. Unless otherwise stated, the account starts with a zero balance.

EXAMPLE 3

*Account balances*   Use the checking account model to calculate the account balances for these situations. Start each problem with a zero balance.

**a.** Deposit $50.00, and write a check for $20.98.

**b.** Write a check for $20.50 and another for $15.00.

**c.** Deposit $50.00, and write a check for $55.50.

SOLUTION

**a.** $\$50.00 + (-\$20.98) = \$29.02$. You probably observed that writing a check is the same as a subtraction; thus $\$50.00 + (-\$20.98) = \$50.00 - \$20.98$. You have recognized an important concept.

> For all real numbers, $a + (-b)$ is the same as $a - b$.

Thus subtracting a positive and adding a negative have the same result. Check with a calculator that

$$\$50.00 + (-\$20.98) = \$50.00 - \$20.98$$

**b.** If the account started with zero, then the two checks can be added and the total overdraft is

$$(-\$20.50) + (-\$15.00) = -\$35.50$$

Of course, this does not count the charges made by the bank for "bouncing" the checks.

**c.** $\$50.00 + (-\$55.50) = -\$5.50$, implying an overdrawn account. Mathematically, this result is not as obvious as those in parts a and b.

♦ ♦ ♦ ♦ ♦ ♦ ♦ ♦ ♦ ♦ ♦ ♦ ♦ ♦ ♦ ♦ ♦ ♦ ♦ ♦ ♦ ♦ ♦ ♦ ♦ ♦ ♦ ♦ ♦ ♦ ♦ ♦ ♦ ♦ ♦ ♦

In part c of Example 3 we subtracted the two numbers and put a negative sign on the result because the check was larger than the deposit. It is tempting to write a rule for addition of positive and negative numbers that says "If the signs are different, subtract the numbers and put the sign of the larger number on the answer." This works for checking accounts and numbers of ions but not for mathematicians. The reason is the word *larger*. In mathematics *the positive numbers are all larger than the negative numbers because larger is defined as being to the right on the number line.* Thus we have a conflict between the common usage of *larger* and mathematicians' usage of *larger*. We avoid the word *larger* by using the distance from zero in the rules for addition of positive and negative numbers.

**Addition of Positive and Negative Real Numbers**

> ♦ If signs are alike, add the numbers and place the common sign on the answer.
>
> ♦ If the signs are different, subtract the number portion and place the sign of the number farthest from zero on the answer.

A check and a deposit have opposite signs, so we use the second rule. Because −$55.50 is considered farther from zero than $50.00, we select its sign and then subtract the two numbers: 55.50 − 50.00. Mathematicians have a special name for distance from zero.

## ABSOLUTE VALUE

*The distance a number is from zero is its* **absolute value**. The symbol for absolute value is two vertical lines placed around a number; the absolute value of three is written $|3|$.

**E X A M P L E   4**

Explain how the number line graph in Figure 2 shows the absolute value of

**a.** −3     **b.** +4

**SOLUTION**

**a.** The absolute value of −3 is $|-3| = 3$, because −3 is a distance of three units from zero.

**b.** The absolute value of +4 is $|+4| = 4$, because +4 is a distance of four units from zero.

♦ ♦ ♦ ♦ ♦ ♦ ♦ ♦ ♦ ♦ ♦ ♦ ♦ ♦ ♦ ♦ ♦ ♦ ♦ ♦ ♦ ♦ ♦ ♦ ♦ ♦ ♦ ♦ ♦ ♦

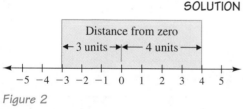

*Figure 2*

We will now use absolute value to write a formal description of the addition of positive and negative numbers. Compare these rules with those given earlier, and use the rules for addition that make the most sense to you.

**Addition of Positive and Negative Numbers Using Absolute Value**

> ♦ To add numbers with the same sign, add their absolute values and place the common sign on the answer.
>
> ♦ To add numbers with opposite signs, subtract their absolute values and place the sign from the number with the larger absolute value on the answer.

Other applications of absolute value include cash-over and cash-under, illustrated in Example 5, and distance, demonstrated in Example 6.

Checkers at grocery stores and tellers at banks and credit unions must balance the cash drawer at the close of their shift. An error in either direction, cash-over (too much money) or cash-under (too little), may indicate careless work habits. Managers often keep track of the total cash-over and cash-under by adding the absolute values of the errors.

EXAMPLE **5**   *Cash-over and cash-under*   What is the total week's cash-over and cash-under for this teller?

| Monday | Tuesday | Wednesday | Thursday | Friday |
|--------|---------|-----------|----------|--------|
| 10 over | balance | 15 under | 5 under | 10 over |

SOLUTION   $$|+10| + |-15| + |-5| + |+10| = 40, \text{ or } \$40.00$$

Compare this solution to the sum without absolute values:

$$+10 + (-15) + (-5) + (+10) = 0$$

The absolute value is needed to indicate that the teller regularly made errors, a fact that might be important to customer satisfaction and to the bank examiner.

◆ ◆ ◆ ◆ ◆ ◆ ◆ ◆ ◆ ◆ ◆ ◆ ◆ ◆ ◆ ◆ ◆ ◆ ◆ ◆ ◆ ◆ ◆ ◆ ◆ ◆ ◆ ◆ ◆ ◆ ◆ ◆ ◆ ◆

Certain quantities, such as the distance between two cities, are always positive.

EXAMPLE **6**   *Distance*   If one freeway exit is numbered 195 (miles from the southern border of the state) and a later exit is numbered 303, what is the distance between them?

SOLUTION   $$|195 - 303| = |-108| = 108 \text{ miles}$$

◆ ◆ ◆ ◆ ◆ ◆ ◆ ◆ ◆ ◆ ◆ ◆ ◆ ◆ ◆ ◆ ◆ ◆ ◆ ◆ ◆ ◆ ◆ ◆ ◆ ◆ ◆ ◆ ◆ ◆ ◆ ◆ ◆ ◆

The absolute distance between two points should not depend on the order in which the subtraction is done. Absolute value gives us this flexibility.

An absolute value key $\boxed{\text{abs}}$ is available on some calculators. The key is needed in calculator programs, in entering and evaluating expressions, and in graphing.

The absolute value operation creates some unusually shaped graphs. In Example 7 we use input–output tables to list coordinates for graphing.

EXAMPLE **7**   Make an input–output table with integers $-3$ to $+3$ as inputs. Graph output $y = |x|$.

SOLUTION   Table 1 is graphed in Figure 3.

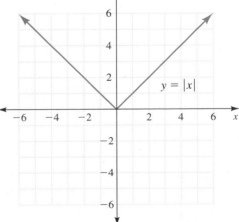

| Input $x$ | Output $y = |x|$ |
|-----------|------------------|
| $-3$ | 3 |
| $-2$ | 2 |
| $-1$ | 1 |
| 0 | 0 |
| 1 | 1 |
| 2 | 2 |
| 3 | 3 |

*Table 1*                    *Figure 3*

◆ ◆ ◆ ◆ ◆ ◆ ◆ ◆ ◆ ◆ ◆ ◆ ◆ ◆ ◆ ◆ ◆ ◆ ◆ ◆ ◆ ◆ ◆ ◆ ◆ ◆ ◆ ◆ ◆ ◆ ◆ ◆ ◆ ◆

The absolute value graph is a V-shape composed of two straight line segments. Because it contains two distinct parts, we write the absolute value rule as a conditional statement. We will return to this idea in later sections.

## MULTIPLICATION OF REAL NUMBERS

The common interpretation of multiplication is repeated addition; for example,

$$8 \cdot 5 = 5 + 5 + 5 + 5 + 5 + 5 + 5 + 5 = 40$$

$$5 \cdot 8 = 8 + 8 + 8 + 8 + 8 = 40$$

Most multiplication of positive and negative numbers may be interpreted as repeated addition: $2 \cdot (-3) = (-3) + (-3) = -6$.

The product $-3 \cdot 2$ is more difficult to interpret as repeated addition because we cannot write something negative three times. However, we can exchange the numbers, thereby changing the product $(-3) \cdot 2$ to $2 \cdot (-3)$, and obtain $-6$ as our answer. This exchange uses the commutative property of multiplication, explained later in this section.

Table 2 shows selected products of positive and negative numbers. The output is the product of $+2$ and each input. Observe that *a positive times a negative gives a negative.*

| Input <br> *x* | Output <br> $+2x$ |
|:---:|:---:|
| $-3$ | $(+2)(-3) = -6$ |
| $-2$ | $(+2)(-2) = -4$ |
| $-1$ | $(+2)(-1) = -2$ |
| $0$ | $(+2)(0) = 0$ |
| $+1$ | $(+2)(1) = +2$ |
| $+2$ | $(+2)(2) = +4$ |

*Table 2*

Refer to Table 2 as you practice multiplying positive and negative numbers on a calculator. Determine whether the keystroke sequence for your calculator is 2 · 3 $\boxed{\pm}$ $\boxed{=}$ or 2 · $\boxed{\pm}$ 3 $\boxed{=}$ . The sign change key ( $\boxed{\pm}$ , $\boxed{+/-}$ , or $\boxed{(-)}$ ) is entered *after* the number on many scientific calculators.

In Table 2, when the input increases by 1, the output increases by 2. This pattern holds for the entire table.

In Example 8 we multiply $-3$ by several inputs. We start with 2, 1, and 0 as inputs because we already know how to calculate their product with $-3$. We then look for a pattern, as in Table 2, to predict the product of two negative numbers.

EXAMPLE **8**

Use the first three rows of Table 3 to find a pattern in the outputs. Use the pattern to calculate the remaining outputs.

| Input $x$ | Output $-3x$ |
|-----------|--------------|
| +2 | $(-3)(+2) = -6$ |
| +1 | $(-3)(+1) = -3$ |
| 0 | $(-3)(0) = 0$ |
| −1 | $(-3)(-1) = \underline{\quad}$ |
| −2 | $(-3)(-2) = \underline{\quad}$ |
| −3 | $(-3)(-3) = \underline{\quad}$ |

Table 3

SOLUTION    Each time the input goes down by 1, the output goes up by 3. Thus each output is 3 larger (that is, it is located to the right on the number line) than the prior output.

$$(-3) \cdot (2) = -6$$
$$(-3) \cdot (1) = -3$$
$$(-3) \cdot (0) = 0$$

In order for this pattern to continue, the next three outputs would be 3, 6, and 9. Thus the pattern indicates that

$$(-3) \cdot (-1) = 3$$
$$(-3) \cdot (-2) = 6$$
$$(-3) \cdot (-3) = 9$$

♦ ♦ ♦ ♦ ♦ ♦ ♦ ♦ ♦ ♦ ♦ ♦ ♦ ♦ ♦ ♦ ♦ ♦ ♦ ♦ ♦ ♦ ♦ ♦ ♦ ♦ ♦ ♦ ♦ ♦ ♦ ♦ ♦

The pattern in Example 8 suggests that *a negative multiplied by a negative gives a positive.* Confirm this fact on a scientific calculator. We also had a positive result when multiplying two positives. We summarize the multiplication rules in three statements:

**Multiplication of Positive and Negative Numbers**

> ♦ If two numbers have like signs, their product is positive.
> ♦ If two numbers have unlike signs, their product is negative.
> ♦ The product of zero and any number is zero.

EXAMPLE   9    Apply the addition models within a multiplication problem.

**a.** Find the value of ♥ + ♥ + ♥ + ♥ + ♦ + ♦ if ♥ = −3 and ♦ = +2.

**b.** Use the checking account model to find the account balance with an initial deposit of $150 and three checks written for $35 each.

SOLUTION   **a.** 4 ♥ + 2 ♦ = 4(−3) + 2(+2) = −12 + 4 = −8

**b.** $150 + 3(−$35) = $150 + (−$105) = $150 − $105 = $45

In 4(−3), 2(+2), and 3(−35), the parentheses indicate multiplication.

♦ ♦ ♦ ♦ ♦ ♦ ♦ ♦ ♦ ♦ ♦ ♦ ♦ ♦ ♦ ♦ ♦ ♦ ♦ ♦ ♦ ♦ ♦ ♦ ♦ ♦ ♦ ♦ ♦ ♦ ♦ ♦ ♦

### THE COMMUTATIVE AND ASSOCIATIVE PROPERTIES

We close this section with properties of real numbers: the commutative properties and the associative properties. In Section 1.3 we mentioned that $n + 2$ is equal to $2 + n$. Earlier in this section we stated that $(-3) \cdot 2$ is equal to $2(-3)$. These are examples of the commutative properties. The **commutative properties** *refer to the order in which an action or operation is done.* The word *commutative* comes from the word *commute*, which means "to exchange" or "to change position or order." On the coordinate map of Washington, D.C., moving a pencil north 3 blocks and then east 5 blocks gives the same destination as moving it east 5 blocks and then north 3 blocks.

The commutative property applies to both addition and multiplication, so we have these two statements:

**The Commutative Property of Addition**

$a + b = b + a$ for all real numbers.

**The Commutative Property of Multiplication**

$a \cdot b = b \cdot a$ for all real numbers.

In arithmetic the commutative property allows us to rearrange numbers for easier computation. We often rearrange numbers so that they add to 10 or multiples of 10. We also change the order of products of numbers to take advantage of multiplication by 10 or 20.

**EXAMPLE 10**

Change the order of the following numbers to place together numbers that add to 10 or a multiple of 10; add the numbers to take advantage of the sums to 10:

$$8 + 4 + 9 + 6 + 1 + 2$$

**SOLUTION**   Here is one way. You may think of others.

$$8 + 4 + 9 + 6 + 1 + 2 = 8 + 2 + 4 + 6 + 9 + 1$$
$$= (8 + 2) + (4 + 6) + (9 + 1)$$
$$= 10 + 10 + 10$$
$$= 30$$

The **associative property** allows us to add the 8 and 2 separately from the 4 and 6 or 9 and 1. The term *associative property* is based on the word *associate*, meaning "to select groups" or "to choose connections." The associative property applies to both addition and multiplication.

**The Associative Property of Addition**

$a + (b + c) = (a + b) + c$ for all real numbers.

**The Associative Property of Multiplication**

$a \cdot (b \cdot c) = (a \cdot b) \cdot c$ for all real numbers.

By using both the commutative and the associative properties, we may rearrange the numbers in an expression to find convenient pairs to add or multiply.

E X A M P L E  **11**    Use the associative and commutative properties to change the order and select groups, making it easy to do these multiplications mentally.

**a.** $2 \cdot 7 \cdot 5$    **b.** $5 \cdot 13 \cdot 4$

SOLUTION    **a.** $2 \cdot 7 \cdot 5 = \boxed{2 \cdot 5} \cdot 7 = 10 \cdot 7 = 70$

**b.** $5 \cdot 13 \cdot 4 = 13 \cdot \boxed{5 \cdot 4} = 13 \cdot 20 = 260$

◆ ◆ ◆ ◆ ◆ ◆ ◆ ◆ ◆ ◆ ◆ ◆ ◆ ◆ ◆ ◆ ◆ ◆ ◆ ◆ ◆ ◆ ◆ ◆ ◆ ◆ ◆ ◆ ◆ ◆ ◆ ◆ ◆ ◆ ◆

Not all real-world and mathematical operations are commutative and associative. If you put on your socks and then your shoes, your outcome is entirely different than if you put on your shoes followed by your socks. You will be investigating whether or not these properties apply to subtraction and division in the exercises.

### EXERCISES 2.1

What addition fact does each ion circle in Exercises 1 to 6 represent? What is the net charge?

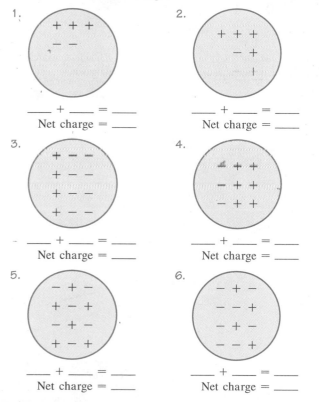

1.
____ + ____ = ____
Net charge = ____

2.
____ + ____ = ____
Net charge = ____

3.
____ + ____ = ____
Net charge = ____

4.
____ + ____ = ____
Net charge = ____

5.
____ + ____ = ____
Net charge = ____

6.
____ + ____ = ____
Net charge = ____

For Exercises 7 to 10 use the ion model to represent the problem, and then solve the problem.

7. $-3 + (+2) = $ ____    8. $+6 + (-4) = $ ____

9. $+3 + (-5) = $ ____    10. $-4 + (-3) = $ ____

For Exercises 11 to 14 use the checking account model to write the problem as addition of positive numbers (deposits) and negative numbers (checks). Use a calculator to solve.

11. Start with zero. Deposit $15.10. Write a check for $9.25. Deposit $8.00. Write a check for $12.58. Find the balance.

12. Start with $25.42. Deposit $53.20. Write checks for $39.70, $14.20, and $18.63. Find the balance.

13. Initial balance is $15. Write a check for $19.98. Deposit $50. Write a check for $45.76. Find the balance.

14. Initial balance is $29.89. Deposit $15.72. Write checks for $18.86 and $14.92. Find the balance.

**In Exercises 15 to 18 use either the ion model or the checking account model to add.**

15. $-12$
$+ 6$
$-15$
$+ 7$

16. $+15$
$-16$
$+17$
$- 9$

17. $-6 + (-7) + (-8) + 20$

18. $-18 + 7 + (-9) + (-12)$

**Calculate the expressions in Exercises 19 and 20.**

19. a. $|-4|$    b. $|5|$    c. $|4 - 9|$    d. $-|2 + 5|$

20. a. $|7|$    b. $|-6|$    c. $|3 - 9|$    d. $-|3 + 4|$

21. The following table lists possible exit numbers for cities along Route 35. Exit numbers are determined by the number of miles from the exit to the southern border of Texas. Determine the driving distance.

| Laredo | 0 |
|---|---|
| San Antonio | 153 |
| Austin | 230 |
| Dallas | 423 |
| Gainesville | 482 |

a. From San Antonio to Dallas

b. From Austin to Gainesville

c. From Gainesville to San Antonio

Describe how you calculated your answers to assure a positive distance.

**22.** The following table lists possible exit numbers for cities along Route 1 and Route 95, as determined by the number of miles from the exit to the southernmost tip of Florida. Determine the driving distance.

| | |
|---|---|
| Key West | 0 |
| Key Largo | 108 |
| Miami | 161 |
| Vero Beach | 298 |
| Cocoa | 353 |
| Daytona Beach | 419 |
| Jacksonville | 510 |

**a.** From Key Largo to Jacksonville

**b.** From Daytona Beach to Miami

**c.** From Vero Beach to Jacksonville

Describe how you calculated your answers to assure a positive distance.

**What is the total cash-over/cash-under for the tellers in Exercises 23 and 24?**

**23.** Monday, $15 under; Tuesday, $10 under; Wednesday, balance; Thursday, $5 under; Friday, $10 over

**24.** Monday, $5 under; Tuesday, $5 over; Wednesday, $15 over; Thursday, $5 under; Friday, balance

**25.** A friend calls on the telephone and wants you to fax instructions on how to find the absolute value of a number *n*. Write a set of instructions using words and symbols. Your friend recognizes positive and negative numbers and does addition, subtraction, multiplication, and division with both positive and negative numbers. (*Hint*: You might start with the phrase "If the number is positive . . . .")

**26.** A friend calls on the telephone and wants instructions on how to find the absolute value of a number *n* from a number line. Explain in words because you are talking on the phone and a picture or fax is not available.

**27.** Fill in the missing numbers.

**a.** $-3 \cdot \underline{\hspace{1cm}} = -12$    **b.** $-4 \cdot \underline{\hspace{1cm}} = -12$

**c.** $-4 \cdot \underline{\hspace{1cm}} = 12$    **d.** $3 \cdot \underline{\hspace{1cm}} = -12$

**e.** $4 \cdot \underline{\hspace{1cm}} = -12$    **f.** $-3 \cdot \underline{\hspace{1cm}} = 12$

**28.** Fill in the missing numbers.

**a.** $-2 \cdot \underline{\hspace{1cm}} = -12$    **b.** $-6 \cdot \underline{\hspace{1cm}} = 12$

**c.** $2 \cdot \underline{\hspace{1cm}} = -12$    **d.** $-6 \cdot \underline{\hspace{1cm}} = -12$

**e.** $-2 \cdot \underline{\hspace{1cm}} = 12$    **f.** $6 \cdot \underline{\hspace{1cm}} = -12$

**In Exercises 29 to 34 the symbols represent different charges, as designated below. Write the exercises using numbers and multiplication and addition signs. Then find the net charge in each set.**

$$☺ = 4 \qquad ♦ = -3 \qquad ♣ = 2 \qquad ▢ = -5$$

**29.** ☺ + ☺ + ☺ + ♦ + ♦ + ♦ + ♦

**30.** ♣ + ▢ + ▢ + ▢ + ▢ + ▢ + ▢

**31.** ▢ + ▢ + ▢ + ☺ + ☺ + ♦ + ♦ + ♦ + ♦

**32.** ♣ + ♣ + ♣ + ♦ + ♦ + ♦ + ♦ + ♦

**33.** ♦ + ♦ + ♦ + ▢ + ▢ + ♣

**34.** ♣ + ♣ + ♣ + ▢ + ▢ + ♦

**35.** Multiply these expressions.

**a.** $-3\left(\frac{1}{3}\right)$    **b.** $-3\left(-\frac{1}{3}\right)$    **c.** $2\left(-\frac{1}{2}\right)$    **d.** $-\frac{1}{4}(-4)$

**36.** Multiply these expressions.

**a.** $-2\left(-\frac{1}{2}\right)$    **b.** $3\left(-\frac{1}{3}\right)$    **c.** $-4\left(\frac{1}{4}\right)$    **d.** $\frac{1}{2}(-2)$

**Complete the tables in Exercises 37 to 40. Then match the table with line a, b, c, or d graphed in the figure.**

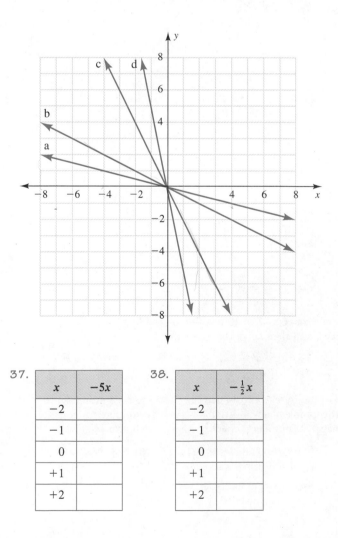

**37.**

| $x$ | $-5x$ |
|---|---|
| $-2$ | |
| $-1$ | |
| $0$ | |
| $+1$ | |
| $+2$ | |

**38.**

| $x$ | $-\frac{1}{2}x$ |
|---|---|
| $-2$ | |
| $-1$ | |
| $0$ | |
| $+1$ | |
| $+2$ | |

39.

| $x$ | $-\frac{1}{4}x$ |
|-----|-----------------|
| $-8$ | |
| $-4$ | |
| $0$ | |
| $+4$ | |
| $+8$ | |

40.

| $x$ | $-2x$ |
|-----|-------|
| $-2$ | |
| $-1$ | |
| $0$ | |
| $+1$ | |
| $+2$ | |

41. True or false:

a. $2 - 3 = 3 - 2$     b. $4 \div 5 = 5 \div 4$

c. $8 \div 2 = 2 \div 8$     d. $6 - 5 = 5 - 6$

What may be concluded about subtraction and the commutative property? What may be concluded about division and the commutative property?

42. Which of these activities are commutative? Give situations that might change your answers.

a. get dressed, eat breakfast

b. start car, fasten seatbelt

c. put key in ignition, start car

d. turn right, walk five steps forward

43. What is each of the following worth if you do the operation in the parentheses, ( ), first? Are any of the expressions equal?

a. $8 - (5 - 3)$     b. $(8 - 5) - 3$

c. $16 \div (4 \div 2)$     d. $(16 \div 4) \div 2$

What may be concluded about subtraction and the associative property? What may be concluded about division and the associative property?

44. Part of the multiplication table is shown below.

|   | 1 | 2 | 3 | 4 | 5 | 6 | 7 | 8 |
|---|---|---|---|---|---|---|---|---|
| 1 | 1 | 2 | 3 | 4 | 5 | 6 | 7 | 8 |
| 2 | 2 | 4 | 6 | 8 | 10 | 12 | 14 | 16 |
| 3 | 3 | 6 | 9 | 12 | 15 | 18 | 21 | 24 |
| 4 | 4 | 8 | 12 | 16 | 20 | 24 | 28 | 32 |
| 5 | 5 | 10 | 15 | 20 | 25 | 30 | 35 | 40 |
| 6 | 6 | 12 | 18 | 24 | 30 | 36 | 42 | 48 |
| 7 | 7 | 14 | 21 | 28 | 35 | 42 | 49 | 56 |
| 8 | 8 | 16 | 24 | 32 | 40 | 48 | 56 | 64 |

a. The fact that $4 \cdot 3 = 3 \cdot 4$ illustrates the commutative property of multiplication. What effect does the commutative property have on the multiplication table?

b. Where are the numbers $n \cdot n = n^2$ located on the multiplication table?

c. Copy the table, and extend it to $12 \times 12$. In the answer section of the table (all but the top row and the left-hand column) shade all 4s with one color, all 12s with a second color, and all 24s with a third color. Comment on any patterns.

**In Exercises 45 and 46 make an input–output table for the conditional rules, with the integers −3 to 3 as inputs.**

45. If the input is zero or positive, the output equals the input. If the input is negative, the output is negative one times the input (a rule used in mathematics).

46. The output is negative one times the input if the input is even. The output equals the input if the input is odd.

**In Exercises 47 to 50 make an input–output table for each absolute value rule, using the integers −4 to 4 as inputs. Then graph the input–output ordered pairs.**

47. Input: $x$
Output: $|x + 3|$

48. Input: $x$
Output: $|x - 2|$

49. Input: $x$
Output: $|x| - 2$

50. Input: $x$
Output: $|x| + 3$

COMPUTATION REVIEW
**For Exercises 51 and 52 complete the table.**

51.

| Input $x$ | Input $y$ | Output $xy$ | Output $x + y$ |
|-----------|-----------|-------------|----------------|
| 3 | 4 | 12 | |
| −2 | 3 | | |
| −7 | | 21 | |
| −3 | 5 | | |
| 2 | | −6 | |
| 4 | | | 1 |
| | −2 | | −3 |
| 3 | | | 2 |

52.

| Input $x$ | Input $y$ | Output $xy$ | Output $x + y$ |
|-----------|-----------|-------------|----------------|
| −2 | 4 | | |
| −3 | 1 | | |
| 2 | | −6 | |
| −3 | | 6 | |
| −1 | | | −7 |
| | −2 | | −7 |
| | −2 | | 1 |
| 2 | | | 7 |

PROJECT

53. **Wave Action.** Waves are associated with sound, water, vibration, and even earthquakes. Two waves from different sources passing through the same body add together. The waves illustrated here represent water waves, and the numbers are in feet. In each pair of graphs, the inputs (on the x-axis) are the same. The outputs represent the height of the wave at each input position.

For each input, add the outputs for the pair of waves and form a new graph, as shown in the example figure. It is possible to build the new graph directly from the wave graphs. You can also, however, copy data from the graphs onto a table such as the one shown, add the data in the table, and then graph the sum as output.

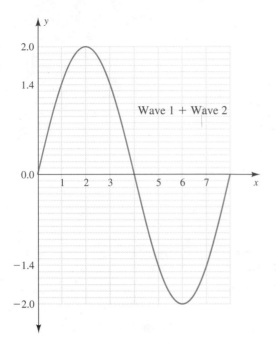

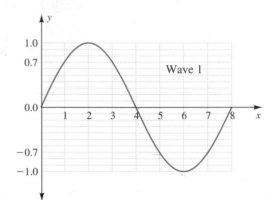

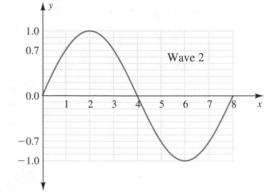

| Input (feet) | 0 | 1 | 2 | 3 | 4 | 5 | 6 | 7 | 8 |
|---|---|---|---|---|---|---|---|---|---|
| Wave 1, height | 0 | 0.7 | 1.0 | 0.7 | 0 | −0.7 | −1 | −0.7 | 0 |
| Wave 2, height | 0 | 0.7 | 1.0 | 0.7 | 0 | −0.7 | −1 | −0.7 | 0 |
| Wave 1 + Wave 2 | 0 | 1.4 | 2 | 1.4 | 0 | −1.4 | −2 | −1.4 | 0 |

Add these wave forms, either directly or using tables.

a.

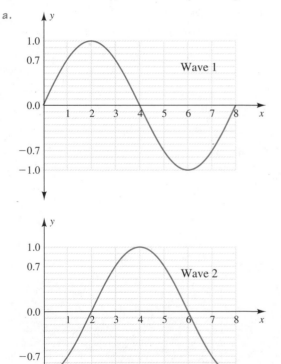

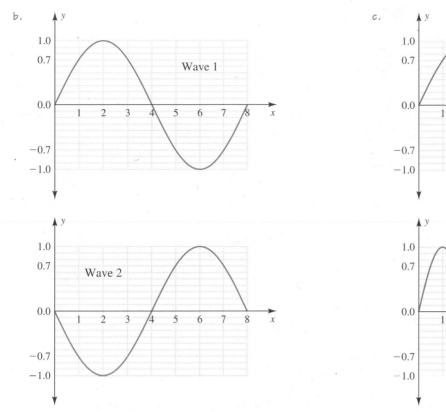

*Note:* The waves in part b add to zero. The zero sum is the principle behind expensive hearing protectors: An electronic circuit takes the incoming sound wave and produces an opposite wave. The sum of the two waves is zero and results in sound suppression.

## 2.2 SUBTRACTION AND DIVISION WITH REAL NUMBERS

**OBJECTIVES** Identify opposites (additive inverses) of real numbers. ♦ Identify reciprocals (multiplicative inverses) of real numbers. ♦ Apply the simplification property of fractions to expressions. ♦ Identify prime and composite numbers. ♦ Use reciprocals to divide fractions. ♦ Interpret subtraction as the opposite, or inverse operation, of addition. ♦ Use a calculator to subtract real numbers. ♦ Use subtraction to model removing objects from a set (ion model). ♦ Use subtraction to find differences in elevation.

**WARM-UP** What are the missing numbers?

**1.** $-3 +$ ____ $= 0$  **2.** ____ $+ 5 = 0$  **3.** ____ $+ (-2) = 0$

**4.** $4 +$ ____ $= 0$  **5.** $5 \cdot$ ____ $= 1$  **6.** $\frac{4}{7} \cdot$ ____ $= 1$

**7.** $\frac{-1}{2} \cdot$ ____ $= 1$  **8.** $\frac{-5}{3} \cdot$ ____ $= 1$  ♦

I n this section we examine inverses: opposites, reciprocals, and inverse operations. **Inverse** *means reversed in position or direction.* We look at the additive inverse first. We then relate inverses to division and subtraction, and we close by applying the inverse concept to a model for subtraction of integers.

ADDITIVE INVERSE, OR OPPOSITE

Inverses for addition, or *additive inverses*, are commonly known as **opposites**. The first four problems in the Warm-up illustrate opposites. We define opposite numbers in terms of the number line. The numbers 3 and −3 are on opposite sides of zero on the number line and are each three units from zero.

Definition

> **Additive inverses**, or **opposites**, are numbers that are the same distance from zero on the number line but are on different sides of zero.

E X A M P L E   **1**

Find the opposite of each number. Then find the sum of the number and its opposite.

**a.**  −3      **b.** 5      **c.** $-\frac{1}{2}$

SOLUTION   **a.**  The opposite of −3 is 3. The sum of −3 and 3 is 0.
   **b.**  The opposite of 5 is −5. The sum of 5 and −5 is 0.
   **c.**  The opposite of $-\frac{1}{2}$ is $\frac{1}{2}$. The sum of $-\frac{1}{2}$ and $\frac{1}{2}$ is 0.

♦ ♦ ♦ ♦ ♦ ♦ ♦ ♦ ♦ ♦ ♦ ♦ ♦ ♦ ♦ ♦ ♦ ♦ ♦ ♦ ♦ ♦ ♦ ♦ ♦ ♦ ♦ ♦ ♦ ♦ ♦

Example 1 leads us to an important property of opposites:

> A number and its opposite add to zero.

It is the addition of opposites that makes the ion model work. As indicated in Section 2.1, the positive and negative ions refer to a +1 charge and a −1 charge and are opposites. They add to zero and neutralize each other.

E X A M P L E   **2**

Charges on ions      What is the net charge in the ion circle?

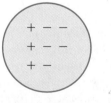

SOLUTION   To calculate the net charge, we combine positive and negative ions to get zero and then count what remains, as shown below.

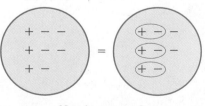

Net charge = −2

Using opposites, we have

$$3 + (-5) = 3 + (-3) + (-2) = 0 + (-2) = -2$$

♦ ♦ ♦ ♦ ♦ ♦ ♦ ♦ ♦ ♦ ♦ ♦ ♦ ♦ ♦ ♦ ♦ ♦ ♦ ♦ ♦ ♦ ♦ ♦ ♦ ♦ ♦ ♦ ♦ ♦ ♦

Expressions also have opposites. The opposite of $a$ is $-a$; the opposite of $-x$ is $x$; the opposite of $2x^2$ is $-2x^2$. The same addition-to-zero property holds for opposite expressions.

> An expression and its opposite add to zero.
>
> $$a + -a = 0 \qquad -x + x = 0 \qquad -2x^2 + 2x^2 = 0$$

 The opposite key on a calculator is labeled $\boxed{+/-}$, $\boxed{(-)}$, or $\boxed{\pm}$. This key also obtains a negative number. Determine whether your calculator requires that this key be pressed before or after a number or expression is entered.

### MULTIPLICATIVE INVERSE, OR RECIPROCAL

The last four problems in the Warm-up illustrate **reciprocals**. The reciprocal is the common name for the *multiplicative inverse*.

**Definition**

> The **reciprocal**, or **multiplicative inverse**, of a number $n$ is the number that, when multiplied by $n$, gives 1. Thus for all real numbers except zero,
>
> $$n \cdot \frac{1}{n} = 1$$

Because the product of a number and its reciprocal is positive one, the number and its reciprocal have the same sign; either both are positive or both are negative.

**EXAMPLE  3**

Find the reciprocal of each number. Check by multiplying the number and its reciprocal.

**a.** $-\frac{2}{3}$    **b.** $1\frac{1}{4}$    **c.** $0.75$

**SOLUTION**

**a.** The reciprocal of $-\frac{2}{3}$ is $-\frac{3}{2}$ or $-1\frac{1}{2}$.

*Check*               $-\frac{2}{3} \cdot -\frac{3}{2} = +\frac{6}{6} = 1$ ✔

**b.** Because $1\frac{1}{4} = \frac{5}{4}$, its reciprocal is $\frac{4}{5}$.

*Check*               $\frac{5}{4} \cdot \frac{4}{5} = \frac{20}{20} = 1$ ✔

**c.** Because $0.75 = \frac{3}{4}$, its reciprocal is $\frac{4}{3}$ or $1\frac{1}{3}$.

*Check*               $\frac{3}{4} \cdot \frac{4}{3} = \frac{12}{12} = 1$ ✔

◆ ◆ ◆ ◆ ◆ ◆ ◆ ◆ ◆ ◆ ◆ ◆ ◆ ◆ ◆ ◆ ◆ ◆ ◆ ◆ ◆ ◆ ◆ ◆ ◆ ◆ ◆ ◆ ◆ ◆ ◆ ◆ ◆

The calculator reciprocal key is $\boxed{1/x}$ or $\boxed{x^{-1}}$. Practice with this key by entering each number in Example 3 and finding its reciprocal.

*Simplifying Fractions.* The expression $\frac{a}{a} = a \cdot \frac{1}{a} = 1$ leads us to the **simplification property of fractions**:

**Simplification Property of Fractions**

> For all real numbers, $a$ not zero and $c$ not zero,
>
> $$\frac{ab}{ac} = \frac{a}{a} \cdot \frac{b}{c} = 1 \cdot \frac{b}{c} = \frac{b}{c}$$

In words, this property says that if the numerator and denominator of any fraction are multiplied (or divided) by the same nonzero quantity, the value of the fraction does not change.

The simplification property depends on the fact that $\frac{a}{a} = 1$. Figure 4 reminds us why, if $a \neq 0$,

$$\frac{1}{1} = \frac{2}{2} = \frac{3}{3} = \frac{4}{4} = \frac{5}{5} = \frac{a}{a} = 1$$

When we write $ab$, we need to recall that the two letters are being multiplied. We give the name **factors** to the letters being multiplied. *If a and b multiply together, then a and b are factors.* When we simplify a fraction, we change the numerator and denominator to factors and look for common factors, such as $\frac{a}{a}$. If there are no common factors, the fraction may not be simplified.

The word *simplify* is a shorter way to write instructions. In this section **simplify** *means to use the simplification property of fractions.* We will learn other meanings when we consider exponents and the order of operations in Section 2.3, ratios and rates in Chapter 4, and rational expressions in Chapter 8.

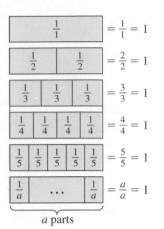

Figure 4

EXAMPLE **4**

Identify the factors in the numerator and denominator, and simplify.

**a.** $\dfrac{6x}{4x}$    **b.** $\dfrac{12ac}{15bc}$

SOLUTION

**a.** The factors of $6x$ are 2, 3, and $x$. The factors of $4x$ are 2, 2, and $x$.

$$\frac{6x}{4x} = \frac{2 \cdot 3 \cdot x}{2 \cdot 2 \cdot x} = \frac{2}{2} \cdot \frac{3}{2} \cdot \frac{x}{x} = 1 \cdot \frac{3}{2} \cdot 1 = \frac{3}{2}$$

**b.** The factors of $12ac$ are 2, 2, 3, $a$, and $c$. The factors of $15bc$ are 3, 5, $b$, and $c$.

$$\frac{12ac}{15bc} = \frac{2 \cdot 2 \cdot 3 \cdot a \cdot c}{3 \cdot 5 \cdot b \cdot c} = \frac{2 \cdot 2}{5} \cdot \frac{3}{3} \cdot \frac{a}{b} \cdot \frac{c}{c} = \frac{2 \cdot 2}{5} \cdot 1 \cdot \frac{a}{b} \cdot 1 = \frac{4a}{5b}$$

♦ ♦ ♦ ♦ ♦ ♦ ♦ ♦ ♦ ♦ ♦ ♦ ♦ ♦ ♦ ♦ ♦ ♦ ♦ ♦ ♦ ♦ ♦ ♦ ♦ ♦ ♦ ♦ ♦ ♦

In Example 4a, when we found the factors for 6 and 4, we were *factoring* them into prime numbers. **Prime numbers** *have no integral factors except 1 and the number itself.* The first prime number is 2. **Composite numbers** *have factors other than themselves and 1.* The number 1 is neither prime nor composite. Many students do not need to change fractions into primes in order to simplify; use primes as needed.

### DIVISION OF REAL NUMBERS

A third use of inverses is in the operations of addition, subtraction, multiplication, and division. What is $100 \div 4 \cdot 4$ worth? What is $8 - 3 + 3$ worth? In each case the operations nullify each other, and we return to the starting number. Thus we have the following:

♦ Addition and subtraction are inverse operations.
♦ Multiplication and division are inverse operations.

Recall that division by $n$ is accomplished by multiplication by the reciprocal, $\frac{1}{n}$. The same rule works for division of positive and negative real numbers.

EXAMPLE **5**  Divide:

**a.** $-3 \div 4$   **b.** $-3 \div (-2)$   **c.** $\frac{1}{3} \div \left(-\frac{1}{4}\right)$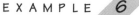

SOLUTION  **a.** $-3 \div 4 = -3 \cdot \frac{1}{4} = \frac{-3 \cdot 1}{4} = \frac{-3}{4}$

**b.** $-3 \div (-2) = -3 \cdot \left(-\frac{1}{2}\right) = \frac{+3}{2}$

**c.** $\frac{1}{3} \div \left(-\frac{1}{4}\right) = \frac{1}{3} \cdot \left(-\frac{4}{1}\right) = \frac{-4}{3} = -1\frac{1}{3}$

◆ ◆ ◆ ◆ ◆ ◆ ◆ ◆ ◆ ◆ ◆ ◆ ◆ ◆ ◆ ◆ ◆ ◆ ◆ ◆ ◆ ◆ ◆ ◆ ◆ ◆ ◆ ◆ ◆ ◆ ◆ ◆ ◆ ◆ ◆ ◆ ◆ ◆ ◆

Because the negative sign does not change when we find the reciprocal of a negative number, the rules for multiplication of positive and negative numbers apply to division.

> In multiplication and division of two real numbers, if the signs are alike, the answer is positive; if the signs are different, the answer is negative.

The placement of signs on fractions is a closely related idea and is noted here to prevent confusion. One negative sign may be placed either in front of a fraction, in the numerator, or in the denominator, wherever it is most convenient.

> For all real numbers, $b$ not zero,
> $$-\frac{a}{b} = \frac{-a}{b} = \frac{a}{-b}$$

### SUBTRACTION OF REAL NUMBERS

Consider these statements about division and subtraction:

To divide, we multiply by the reciprocal.

To subtract, we add the opposite.

These statements are based on the inverse nature of the operations.

EXAMPLE **6**  Change each subtraction to an addition by adding the opposite, and then do the addition.

**a.** $6 - 5$   **b.** $14 - 21$   **c.** $-1 - (+2)$   **d.** $2 - (+3)$

**e.** $2 - (-2)$   **f.** $-5 - (-3)$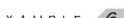

SOLUTION  **a.** $6 - 5 = 6 + (-5) = 1$    **b.** $14 - 21 = 14 + (-21) = -7$

**c.** $-1 - (+2) = -1 + (-2) = -3$   **d.** $2 - (+3) = 2 + (-3) = -1$

**e.** $2 - (-2) = 2 + (+2) = 4$    **f.** $-5 - (-3) = -5 + (+3) = -2$

◆ ◆ ◆ ◆ ◆ ◆ ◆ ◆ ◆ ◆ ◆ ◆ ◆ ◆ ◆ ◆ ◆ ◆ ◆ ◆ ◆ ◆ ◆ ◆ ◆ ◆ ◆ ◆ ◆ ◆ ◆ ◆ ◆ ◆ ◆ ◆ ◆ ◆ ◆

     Rework each part of Example 6 on a scientific calculator. Use the $\boxed{\pm}$ or $\boxed{(-)}$ key to obtain a negative. Doing the subtractions in Example 6 may also be described as *simplifying expressions* and may be summarized as follows:

---

For all real numbers,

$$a - b = a + (-b)$$

For all real numbers,

$$a - (-b) = a + b$$

---

**Subtraction Applications.**   The remaining examples give models or applications of subtraction. Subtraction may mean removing objects from a set, finding a difference in values, or changing a position. As a result, no single model works well for all subtraction with positive and negative numbers. Look for ways in which "subtraction by addition of the opposite" appears in the examples.

E X A M P L E  **7**

**Charges on ions, continued**   The ion model illustrates subtraction interpreted as removing objects from a set.

Three ion circles are shown. Determine the initial charge in the circle, and then remove (subtract) the circled particles. What is the ending charge in each circle?

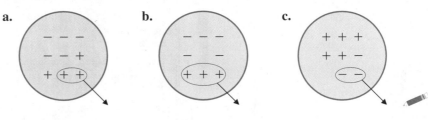

SOLUTION   **a.** $(-1)$ initial charge $- (+2)$ removed $= -3$ ending charge.
Using addition of opposites, we have $(-1) - (+2) = (-1) + (-2) = -3$.

**b.** $(-2)$ initial charge $- (+3)$ removed $= -5$ ending charge.
Using addition of opposites, we have $(-2) - (+3) = (-2) + (-3) = -5$.

**c.** $(+2)$ initial charge $- (-2)$ removed $= +4$ ending charge.
Using addition of opposites, we have $(+2) - (-2) = (+2) + (+2) = 4$.

◆ ◆ ◆ ◆ ◆ ◆ ◆ ◆ ◆ ◆ ◆ ◆ ◆ ◆ ◆ ◆ ◆ ◆ ◆ ◆ ◆ ◆ ◆ ◆ ◆ ◆ ◆ ◆ ◆ ◆ ◆ ◆ ◆ ◆ ◆

We frequently associate subtraction with a change. However, in elevation descriptions, subtraction finds the difference in vertical position.

E X A M P L E  **8**

**Elevations**   Use the data in Table 4 to find the difference in elevation between the two places given.

**a.** Mt. McKinley in Alaska and Pike's Peak in Colorado

**b.** Mt. Everest and the Mariana Trench in the Pacific Ocean (south of Japan and east of the Philippines)

**c.** Death Valley in California and the Dead Sea between Israel and Jordan

SOLUTION   **a.** The elevation of Mt. McKinley minus the elevation of Pike's Peak is

$$20{,}320 - 14{,}110 = 6{,}210 \text{ feet}$$

**b.** The elevation of Mt. Everest minus the elevation of Mariana Trench is

$$29,028 - (-35,840) = 29,028 + (+35,840) = 64,868 \text{ feet}$$

**c.** The elevation of Death Valley minus the elevation of the Dead Sea is

$$-282 - (-1312) = -282 + (+1312) = 1030 \text{ feet}$$

| | |
|---|---|
| Mt. Everest | +29,028 |
| Mt. McKinley | +20,320 |
| Pike's Peak | +14,110 |
| Mauna Kea | +13,710 |
| Sea Level | 0 |
| Death Valley | −282 |
| Dead Sea | −1,312 |
| Ocean Floor, near Hawaii | −16,400 |
| Mariana Trench | −35,840 |

*Table 4 Elevation in Feet Relative to Sea Level*

◆ ◆ ◆ ◆ ◆ ◆ ◆ ◆ ◆ ◆ ◆ ◆ ◆ ◆ ◆ ◆ ◆ ◆ ◆ ◆ ◆ ◆ ◆ ◆ ◆ ◆ ◆ ◆ ◆ ◆ ◆ ◆ ◆ ◆ ◆ ◆ ◆ ◆ ◆ ◆

## EXERCISES 2.2

**Give the opposite of each number or expression in Exercises 1 and 2.**

1. a. 5    b. $-\frac{1}{2}$    c. 0.4    d. $x$    e. $-2x$

2. a. −5    b. $\frac{2}{3}$    c. 2.5    d. $a^2b$    e. $-ab$

**Write the reciprocal of each number or expression in Exercises 3 to 6.**

3. a. 4    b. −2    c. $\frac{1}{2}$    d. $-\frac{3}{4}$    e. 0.5

4. a. −3    b. 6    c. $-\frac{1}{3}$    d. $\frac{2}{3}$    e. 0.25

5. a. $3\frac{1}{3}$    b. 6.5    c. $x$    d. $\frac{a}{b}$    e. $-x$

6. a. $2\frac{3}{4}$    b. 8.2    c. $n$    d. $\frac{x}{y}$    e. $-y$

**Simplify the expressions in Exercises 7 and 8. Assume that no variable is zero.**

7. a. $\dfrac{ab}{bc}$    b. $\dfrac{ab}{ac}$    c. $\dfrac{ay}{by}$

8. a. $\dfrac{ac}{bc}$    b. $\dfrac{xy}{xyz}$    c. $\dfrac{bx}{xy}$

**For Exercises 9 and 10 give three inputs that result in whole number answers.**

9. a. $\frac{1}{2}x$    b. $\frac{2}{3}x$

10. a. $\frac{1}{5}x$    b. $\frac{3}{4}x$

**Simplify the expressions in Exercises 11 to 14.**

11. a. $8 - 11$    b. $-8 - (-11)$    c. $-16 - 3$
    d. $0 - 5$    e. $-17 - 4$

12. a. $12 - 7$    b. $-12 - 7$    c. $-17 - (-4)$
    d. $17 - 4$    e. $16 - (-3)$

13. a. $-4 - (-4)$    b. $14 - (10)$    c. $0 - (-5)$
    d. $8 - 9$    e. $-4 - 5$

14. a. $10 - 14$    b. $-8 - (-9)$    c. $-7 - 6$
    d. $-3 - (-3)$    e. $4 - (-5)$

**State the subtraction problem illustrated by each of the ion circles in Exercises 15 to 18, and then work the problem.**

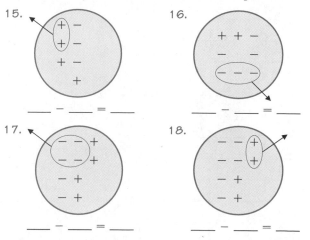

15. ____ − ____ = ____

16. ____ − ____ = ____

17. ____ − ____ = ____

18. ____ − ____ = ____

Make an ion circle that shows each problem in Exercises 19 to 22, and solve the problem. Is it possible for various students to have different circle contents but the same answer?

19. $-4 - (-3)$    20. $-1 - (-3)$

21. $-3 - (+2)$    22. $8 - 4$

Change subtraction to addition of the opposite and use the commutative and associative properties to simplify the expressions in Exercises 23 and 24.

23. $-68 - 74 - 26 - 32 - 14$

24. $-16 - 18 - 35 - 12 - 15 - 24$

Simplify the expressions in Exercises 25 and 26.

25. a. $\dfrac{-2x}{-6x}$    b. $\dfrac{-14a}{21a}$    c. $\dfrac{6x}{15xyz}$

26. a. $\dfrac{2xy}{-10xyz}$    b. $\dfrac{-15ab}{9ac}$    c. $\dfrac{-6ac}{-27cx}$

In Exercises 27 to 30 complete the tables.

27.

| $x$ | $y$ | $x \cdot y$ | $x + y$ |
|-----|-----|-------------|---------|
| 2 | $-2$ | | |
| 3 | | $-6$ | |
| $-4$ | | $-12$ | |
| | 3 | | 0 |
| 2 | | | $-1$ |

28.

| $x$ | $y$ | $x \cdot y$ | $x + y$ |
|-----|-----|-------------|---------|
| $-4$ | | 12 | |
| | | $-16$ | 0 |
| | | $-15$ | 2 |
| $-5$ | | | $-8$ |
| $-1$ | | 6 | |

29.

| $x$ | $\dfrac{3-x}{x-3}$ | $\dfrac{x-2}{2-x}$ |
|-----|--------------------|--------------------|
| 6 | | |
| 4 | | |
| 0 | | |
| $-3$ | | |
| $-5$ | | |

30.

| $x$ | $\dfrac{1-x}{x-1}$ | $\dfrac{x-5}{5-x}$ |
|-----|--------------------|--------------------|
| 6 | | |
| 4 | | |
| 0 | | |
| $-3$ | | |
| $-5$ | | |

In Exercises 31 and 32 calculate the divisions, and simplify answers.

31. a. $15 \div \frac{1}{3}$    b. $\frac{2}{3} \div \frac{4}{5}$    c. $\dfrac{-5}{8} \div \dfrac{2}{3}$

d. $\dfrac{3}{4} \div \dfrac{-1}{4}$    e. $\dfrac{-5}{8} \div \dfrac{3}{2}$    f. $\dfrac{x}{y} \div \dfrac{-x}{y}$

32. a. $-16 \div \frac{8}{5}$    b. $24 \div \dfrac{-3}{4}$    c. $\frac{3}{4} \div \frac{6}{7}$

d. $\dfrac{a}{b} \div \dfrac{a}{b}$    e. $\dfrac{2}{3} \div \dfrac{-5}{6}$    f. $\dfrac{x}{y} \div \dfrac{-y}{x}$

33. When we divide by fractions, we often use the phrase *invert the fraction and multiply*. Use complete sentences to discuss this phrase in terms of the meaning of the verb *invert* and reciprocals, or multiplicative inverses.

34. Complete the table. Write a sentence about patterns you observe.

| $a$ | $b$ | $-\left(\dfrac{b}{a}\right)$ | $\dfrac{-b}{a}$ | $\dfrac{b}{-a}$ |
|-----|-----|------------------------------|-----------------|-----------------|
| 5 | 35 | | | |
| $-27$ | 3 | | | |
| $-2$ | 8 | | | |
| $-20$ | $-4$ | | | |

35. Match the symbols $-\dfrac{x}{y}$, $\dfrac{-x}{y}$, and $\dfrac{x}{-y}$ with the word descriptions.

a. The opposite of the quotient of $x$ and $y$.

b. The quotient of $x$ and the opposite of $y$.

c. The quotient of the opposite of $x$ and $y$.

36. What two expressions are correctly described by "the opposite of $b$ divided by $a$"?

37. The Mariana Trench (Pacific Ocean) is 10,930 meters below sea level. The Puerto Rico Trench (Atlantic Ocean) is 8600 meters below sea level. Write an appropriate subtraction, and determine how much deeper the Mariana Trench is than the Puerto Rico Trench.

In Exercises 38 to 40 what is the difference in elevation between the highest and lowest point on each continent?

38. South America: Mt. Aconcagua, 6960 meters, and Valdes Peninsula, $-40$ meters

39. Africa: Kilimanjaro, 5896 meters, and Lake Assal, $-156$ meters

40. Australia: Mt. Kosciusko, 2228 meters, and Lake Eyre, $-16$ meters

# EXPONENTS AND ORDER OF OPERATIONS

**OBJECTIVES**   Identify the exponent and base of an expression in exponential form. ♦ Write expressions with and without exponents. ♦ Identify squared and cubed as geometric interpretations of the second and third powers of a variable or number. ♦ Use the positive integer definition of exponents to simplify expressions involving multiplication, division, and powers of exponential expressions. ♦ Simplify exponential expressions with a calculator. ♦ Find the square and the square root of a number. ♦ Identify grouping symbols ( ), [ ], and { }. ♦ Simplify expressions with the order of operations. ♦ Simplify expressions containing special grouping symbols: fraction bar, absolute value, and square root.

**WARM-UP**   Use the properties of positive and negative numbers to simplify the following:

**1.** $(-2)(-2)(-2)(-2)$  **2.** $(-2)(2)(2)(2)$  **3.** $(-6)(-6)(-6)\left(\frac{1}{2}\right)\left(\frac{1}{2}\right)$

**4.** Look again at the picture in Figure 1 at the beginning of the chapter. A superball bounces $\frac{2}{3}$ of its previous height on each bounce. If the ball is dropped from 144 centimeters, how high will it bounce on the tenth bounce?   ♦

The skills in this section are the basis for working with formulas and solving equations. We review properties of positive integer exponents and observe how the order in which we do operations changes the output from an expression.

## EXPONENTS

The superball in the Warm-up will bounce $\frac{2}{3}$ of the 144 centimeters on the first bounce. The height of the second bounce is $\frac{2}{3}$ of $\frac{2}{3}$ of 144. On the third bounce it rises $\frac{2}{3}$ of $\frac{2}{3}$ of $\frac{2}{3}$ of 144. We have a shorter way to write this expression:

$$\left(\frac{2}{3}\right)^3 \cdot 144$$

If this expression represents the third bounce, what expression might represent the tenth bounce? Compare your answer with the solution to Example 2b.

The expression $\left(\frac{2}{3}\right)^3 \cdot 144$ is in exponential form because it contains an expression with an exponent. In $\left(\frac{2}{3}\right)^3 \cdot 144$, $\left(\frac{2}{3}\right)$ is the **base** and 3 is the **exponent**. Having an exponent 3 means the base is used as a factor three times. Thus $\left(\frac{2}{3}\right)^3$ means $\left(\frac{2}{3}\right) \cdot \left(\frac{2}{3}\right) \cdot \left(\frac{2}{3}\right)$.

> In the expression $x^n$, the *positive integer exponent n* indicates the number of factors of the base $x$. That is,
>
> $$x^n = x \cdot x \cdot x \cdot x \cdots \cdot x$$
>
> with $x$ written $n$ times.

**EXAMPLE 1**   Identify the bases and exponents in these formulas:

**a.** $M = \dfrac{mgl^3}{4sa^3b}$   **b.** $d = -\frac{1}{2}gt^2 + vt + s$   **c.** $Q = catT^4$

**SOLUTION**   **a.** The base $l$ has exponent 3, and the base $a$ has exponent 3. All other bases have exponent 1.

**b.** The base $t$ has exponent 2. All other bases have exponent 1.

**c.** The base $T$ has exponent 4. All other bases have exponent 1.

♦ ♦ ♦ ♦ ♦ ♦ ♦ ♦ ♦ ♦ ♦ ♦ ♦ ♦ ♦ ♦ ♦ ♦ ♦ ♦ ♦ ♦ ♦ ♦ ♦ ♦ ♦ ♦ ♦ ♦ ♦

Take special note that the definition of exponents given above is limited to positive integers. Other numbers—such as fractions, decimals, and negative and irrational numbers—may be used as exponents but require other definitions. A calculator project in Exercise 45 suggests problems to explore with these other exponents. For now, we will work only with positive integer exponents.

E X A M P L E   **2**

Identify each base; then write the expression as factors and multiply.

**a.** $5^3$     **b.** $144\left(\frac{2}{3}\right)^{10}$     **c.** $(-2)^4$     **d.** $-2^4$     **e.** $(-6)^3\left(\frac{1}{2}\right)^2$

SOLUTION

**a.** The base is 5; $5^3 = 5 \cdot 5 \cdot 5 = 125$.

**b.** The base is $\frac{2}{3}$; $144\left(\frac{2}{3}\right)^{10} = 144\left(\frac{2}{3}\right)\left(\frac{2}{3}\right)\left(\frac{2}{3}\right)\left(\frac{2}{3}\right)\left(\frac{2}{3}\right)\left(\frac{2}{3}\right)\left(\frac{2}{3}\right)\left(\frac{2}{3}\right)\left(\frac{2}{3}\right)\left(\frac{2}{3}\right) \approx 2.5$.

**c.** The base is $(-2)$; $(-2)^4 = (-2)(-2)(-2)(-2) = 16$.

**d.** The base is 2 not $-2$; $-(2^4) = -(2 \cdot 2 \cdot 2 \cdot 2) = -16$.

**e.** The bases are $(-6)$ and $\frac{1}{2}$; $(-6)(-6)(-6)\left(\frac{1}{2}\right)\left(\frac{1}{2}\right) = -\frac{216}{4} = -54$.

◆ ◆ ◆ ◆ ◆ ◆ ◆ ◆ ◆ ◆ ◆ ◆ ◆ ◆ ◆ ◆ ◆ ◆ ◆ ◆ ◆ ◆ ◆ ◆ ◆ ◆ ◆ ◆ ◆ ◆ ◆ ◆ ◆ ◆ ◆

Mathematicians have agreed to place bases with negative signs within parentheses, as in parts c and e of Example 2. Thus in part d the negative sign on $-2^4$ is not part of the base. The expression $-2^4$ means *the opposite of* $2^4$. The base 2 is raised to the exponent 4 first, and then the negative sign is applied: $-2^4 = -(2^4)$, *not* $(-2)^4$.

Multiplication and division operations with exponential numbers follow the same rules for all exponents; however, the operations are most easily understood with positive integer exponents. Look for patterns in the multiplications and divisions in the next two examples to see whether you can predict the rules.

E X A M P L E   **3**

Write these expressions with a single base and exponent:

**a.** $x^4 \cdot x^3$     **b.** $x^1 \cdot x^6$

SOLUTION

**a.** $x^4 \cdot x^3 = xxxx \cdot xxx = x^7$     **b.** $x^1 \cdot x^6 = x \cdot xxxxxx = x^7$

◆ ◆ ◆ ◆ ◆ ◆ ◆ ◆ ◆ ◆ ◆ ◆ ◆ ◆ ◆ ◆ ◆ ◆ ◆ ◆ ◆ ◆ ◆ ◆ ◆ ◆ ◆ ◆ ◆ ◆ ◆ ◆ ◆ ◆ ◆ ◆ ◆

What other exponent expressions would make $x^7$? What happens to the exponents when we multiply?

E X A M P L E   **4**

Write these expressions with a single base and exponent. (*Hint:* Use the simplification property of fractions.)

**a.** $\dfrac{x^5}{x^3}$     **b.** $\dfrac{x^7}{x^5}$     **c.** $\dfrac{x^4}{x^2}$

SOLUTION

**a.** $\dfrac{x^5}{x^3} = \dfrac{xxxxx}{xxx} = \dfrac{x}{x} \cdot \dfrac{x}{x} \cdot \dfrac{x}{x} \cdot x \cdot x = 1 \cdot 1 \cdot 1 \cdot x \cdot x = x^2$

**b.** $\dfrac{x^7}{x^5} = \dfrac{xxxxxxx}{xxxxx} = \dfrac{x}{x} \cdot \dfrac{x}{x} \cdot \dfrac{x}{x} \cdot \dfrac{x}{x} \cdot \dfrac{x}{x} \cdot x \cdot x = 1 \cdot 1 \cdot 1 \cdot 1 \cdot 1 \cdot x \cdot x = x^2$

**c.** $\dfrac{x^4}{x^2} = \dfrac{xxxx}{xx} = \dfrac{x}{x} \cdot \dfrac{x}{x} \cdot x \cdot x = 1 \cdot 1 \cdot x \cdot x = x^2$

◆ ◆ ◆ ◆ ◆ ◆ ◆ ◆ ◆ ◆ ◆ ◆ ◆ ◆ ◆ ◆ ◆ ◆ ◆ ◆ ◆ ◆ ◆ ◆ ◆ ◆ ◆ ◆ ◆ ◆ ◆ ◆ ◆ ◆ ◆ ◆

What other exponent expressions would make $x^2$? What happens to the exponents when we divide?

In Example 5 we note that if the base itself is in exponential form, such as $(a^m)^n$, then the entire base is used as a factor $n$ times.

EXAMPLE **5**

Use the definition of positive integer exponents to write each of these expressions without parentheses.

a. $(2x^2)^3$   b. $(0.5y^4)^2$   c. $\left(\frac{2}{3}x^5\right)^2$   d. $3(2x)^2(3y)^1$   e. $3(2x)^1(3y)^2$

SOLUTION   a. $(2x^2)^3 = (2x^2)(2x^2)(2x^2) = 2 \cdot 2 \cdot 2 \cdot x^2 \cdot x^2 \cdot x^2 = 8x^6$

b. $(0.5y^4)^2 = (0.5y^4)(0.5y^4) = 0.25y^8$

c. $\left(\frac{2}{3}x^5\right)^2 = \left(\frac{2}{3}x^5\right)\left(\frac{2}{3}x^5\right) = \frac{4}{9}x^{10}$

d. $3(2x)^2(3y)^1 = 3(2x)(2x)(3y) = 3 \cdot 2 \cdot 2 \cdot 3 \cdot x \cdot x \cdot y = 36x^2y$

e. $3(2x)^1(3y)^2 = 3(2x)(3y)(3y) = 3 \cdot 2 \cdot 3 \cdot 3 \cdot x \cdot y \cdot y = 54xy^2$

◆ ◆ ◆ ◆ ◆ ◆ ◆ ◆ ◆ ◆ ◆ ◆ ◆ ◆ ◆ ◆ ◆ ◆ ◆ ◆ ◆ ◆ ◆ ◆ ◆ ◆ ◆ ◆ ◆ ◆ ◆ ◆ ◆ ◆ ◆ ◆ ◆ ◆ ◆ ◆ ◆ ◆

EXAMPLE **6**

Rewrite these exponent expressions by combining expressions with like bases or by multiplying out the expressions in parentheses.

a. $(2a)^2(2b)^3$   b. $10(2x)^2(3y)^3$   c. $\left(\frac{3b}{2a}\right)^3$   d. $\left(\frac{1}{2}x^2\right)^3$

SOLUTION   a. $(2a)^2(2b)^3 = 2a \cdot 2a \cdot 2b \cdot 2b \cdot 2b - 32a^2b^3$

b. $10(2x)^2(3y)^3 = 10 \cdot 2x \cdot 2x \cdot 3y \cdot 3y \cdot 3y = 10 \cdot 4 \cdot 27x^2y^3 = 1080x^2y^3$

c. $\left(\frac{3b}{2a}\right)^3 = \frac{3b}{2a} \cdot \frac{3b}{2a} \cdot \frac{3b}{2a} = \frac{27b^3}{8a^3}$

d. $\left(\frac{1}{2}x^2\right)^3 = \frac{1}{2}x^2 \cdot \frac{1}{2}x^2 \cdot \frac{1}{2}x^2 = \frac{1}{8}x^6$

◆ ◆ ◆ ◆ ◆ ◆ ◆ ◆ ◆ ◆ ◆ ◆ ◆ ◆ ◆ ◆ ◆ ◆ ◆ ◆ ◆ ◆ ◆ ◆ ◆ ◆ ◆ ◆ ◆ ◆ ◆ ◆ ◆ ◆ ◆ ◆ ◆ ◆ ◆ ◆ ◆ ◆

The formula for the volume of a sphere—such as a baseball or a basketball—is $\frac{4}{3}\pi r^3$. (For more information on volume, see Section 2.4.)

EXAMPLE **7**

The radius of a basketball is about 3.5 times the radius of a baseball. If we compared their volumes by division, we would have the expression

$$\frac{\text{Volume of basketball}}{\text{Volume of baseball}} = \frac{\frac{4}{3}\pi(3.5r)^3}{\frac{4}{3}\pi r^3}$$

How many times bigger is the volume of the basketball?

SOLUTION   We may simplify expressions of this form, as in Example 4, by canceling $\frac{a}{a}$ to obtain 1.

$$\frac{\text{Volume of basketball}}{\text{Volume of baseball}} = \frac{\frac{4}{3}\pi(3.5r)^3}{\frac{4}{3}\pi r^3}$$

We line up factors that are alike and then cancel.

$$= \frac{\frac{4}{3}}{\frac{4}{3}} \cdot \frac{\pi}{\pi}(3.5)^3 \cdot \frac{r^3}{r^3}$$

$$= 1 \cdot 1 \cdot (3.5)^3 \cdot 1$$

$$\approx 43$$

The volume of the basketball is approximately 43 times that of the baseball.

◆ ◆ ◆ ◆ ◆ ◆ ◆ ◆ ◆ ◆ ◆ ◆ ◆ ◆ ◆ ◆ ◆ ◆ ◆ ◆ ◆ ◆ ◆ ◆ ◆ ◆ ◆ ◆ ◆ ◆ ◆ ◆ ◆ ◆ ◆ ◆ ◆ ◆ ◆ ◆ ◆ ◆

We now summarize the operations with exponents.

**Operations with Exponents**

> To multiply numbers with like bases, add the exponents:
>
> $$x^a \cdot x^b = x^{a+b}$$
>
> To divide numbers with like bases, subtract the exponents:
>
> $$\frac{x^a}{x^b} = x^{a-b}$$
>
> To apply an exponent to an expression in exponential form, multiply exponents:
>
> $$(x^a)^b = x^{a \cdot b}$$
>
> An exponent outside the parentheses applies to all parts of a product or quotient inside the parentheses:
>
> $$(x \cdot y)^a = x^a \cdot y^a \qquad \left(\frac{x}{y}\right)^a = \frac{x^a}{y^a}$$

Square with side $x$

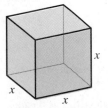

Cube with side $x$

Figure 4

**Squares and Cubes.**    When 2 is used as an exponent, we say the base is *squared*. When 3 is an exponent, the base is *cubed*. These words come from the measurement of area and volume. Because of the roles of area and volume in mathematics applications, it is important to be familiar with their meanings and their relationships to exponents.

The **area** of a square is the space covered by the square. *To find the area of a square, multiply the lengths of the two perpendicular sides.* Thus in exponents, "taking the square of a base" means multiplying the base by itself. In Figure 4 the area of the square is $x \cdot x = x^2$, or $x$ squared.

The **volume** of a cube is the space filled by the cube. *To find the volume of a cube, multiply the lengths of the three perpendicular sides at one corner (length, width, and height).* Thus in exponents, "taking the cube of a base" means multiplying three factors, each containing the base. In Figure 4 the volume of the cube is $x \cdot x \cdot x = x^3$, or $x$ cubed.

We now turn to the order in which we do multiple operations in a problem.

## ORDER OF OPERATIONS

The next several examples are based on this puzzle: *What different numbers can we come up with using any operations on the digits 1, 9, 9, and 8?*

For instance, if we calculate the expression $1 - 9 + 9 \cdot 8$ from left to right, our answer will be 8. If we use a four-function calculator (a calculator that does only addition, subtraction, multiplication, and division), we will also get 8.

We get a different answer, however, with a scientific calculator. Enter the numbers and operations exactly as they appear: $1 - 9 + 9 \cdot 8$. This time the answer is 64.

Obtaining different answers is not satisfactory. We need an agreed-upon set of rules in algebra so that everyone comes out with the same answer. This set of rules is called the **order of operations**, and it produces answers in agreement with those from scientific calculators.

**The Order of Operations**

> 1. Calculate expressions within parentheses and other grouping symbols.
> 2. Calculate exponents and square roots.
> 3. Do the remaining multiplication and division, left to right.
> 4. Do the remaining addition and subtraction, left to right.

**E X A M P L E   8**   Evaluate $1 - 9 + 9 \cdot 8$, applying the order of operations.

**SOLUTION**   The multiplication of 9 and 8 is done before any addition or subtraction:

$$1 - 9 + 9 \cdot 8 \quad \text{Multiply first.}$$
$$1 - 9 + \quad 72 \quad \text{Subtract next; subtraction is left of addition.}$$
$$-8 \quad + \quad 72 \quad \text{Add last.}$$
$$64$$

Thus

$$1 - 9 + 9 \cdot 8 = 64$$

◆ ◆ ◆ ◆ ◆ ◆ ◆ ◆ ◆ ◆ ◆ ◆ ◆ ◆ ◆ ◆ ◆ ◆ ◆ ◆ ◆ ◆ ◆ ◆ ◆ ◆ ◆ ◆ ◆ ◆ ◆ ◆ ◆ ◆ ◆ ◆ ◆ ◆ ◆

The order of operations is used in at least three ways: finding values of expressions (as seen here), evaluating formulas (in the next section), and solving equations (in Chapter 3). The order of operations is often memorized by means of an acronym such as PEMDAS (parentheses, exponents, multiplication, division, addition, subtraction) or a saying such as "Please Excuse My Dear Aunt Sally." Ask your instructor for other favorite aids to memorizing the order of operations.

We now expand the meaning of the word *simplify*. **Simplify** *directs us to use the order of operations, to calculate an expression, to apply the properties of exponents, or to use the associative and commutative properties in working with expressions.*

**E X A M P L E   9**   Use the order of operations to simplify these expressions containing parentheses. Describe the order of working each problem.

**a.** $1 - 9 + (9 \cdot 8)$   **b.** $(1 - 9 + 9) \cdot 8$

**c.** $1 - (9 + 9) \cdot 8$   **d.** $1 - (9 + 9 \cdot 8)$

**SOLUTION**   **a.** The order of operations requires that the operation in parentheses be done first. The subtraction and addition are then done, left to right, to give the answer: 64. The parentheses give the same outcome as Example 8 because they duplicate the priority given to the multiplication of 9 and 8.

**b.** $(1 - 9 + 9) \cdot 8 = 1 \cdot 8 = 8$. We first combine the numbers inside the parentheses, in order from left to right, and then multiply by 8. The parentheses give us the same result as the four-function calculator did.

**c.** $1 - (9 + 9) \cdot 8 = 1 - 18 \cdot 8 = 1 - 144 = -143$. The parentheses are done first, then the multiplication, and finally the subtraction.

**d.** $1 - (9 + 9 \cdot 8) = 1 - (9 + 72) = 1 - 81 = -80$. Within the parentheses, the multiplication is done before the addition. The subtraction is done last.

◆ ◆ ◆ ◆ ◆ ◆ ◆ ◆ ◆ ◆ ◆ ◆ ◆ ◆ ◆ ◆ ◆ ◆ ◆ ◆ ◆ ◆ ◆ ◆ ◆ ◆ ◆ ◆ ◆ ◆ ◆ ◆ ◆ ◆ ◆ ◆ ◆

**Grouping Symbols.**   When parentheses within parentheses are needed, we sometimes use different symbols, such as brackets or braces. *The preferred way to write multiple grouping symbols is* {[()]}, *with parentheses on the inside, the square-shaped brackets used next, and braces (like little wires) used outermost.* If the context is not confusing, double parentheses, (( )), are acceptable.

If an expression contains exponents or square roots, they are calculated immediately after the parentheses but before multiplication and division.

EXAMPLE **10**    Simplify the following expressions. Work the innermost parentheses, ( ), first; then the brackets, [ ]; and finally the braces, { }.

$$\textbf{a. } \{-1 + [(9 - 9) + 8]\}^2 \qquad \textbf{b. } (1 - 9)^2 + 9 \cdot 8$$

$$\textbf{c. } -1 + [9 - (9 + 8)] \qquad \textbf{d. } 1 - \sqrt{9} + 9 \cdot 8$$

$$\textbf{e. } \{-1 \cdot [9 \div (9 \cdot 8)]\}^2$$

SOLUTION    **a.** $\{-1 + [(9 - 9) + 8]\}^2 = \{-1 + [0 + 8]\}^2 = \{7\}^2 = 49$

**b.** $(1 - 9)^2 + 9 \cdot 8 = (-8)^2 + 9 \cdot 8 = 64 + 72 = 136$

**c.** $-1 + [9 - (9 + 8)] = -1 + [9 - 17] = -1 + [-8] = -9$

**d.** $1 - \sqrt{9} + 9 \cdot 8 = 1 - 3 + 9 \cdot 8 = 1 - 3 + 72 = -2 + 72 = 70$

**e.** $\{-1 \cdot [9 \div (9 \cdot 8)]\}^2 = \{-1 \cdot [9 \div 72]\}^2 = \left\{-1 \cdot \left[\frac{1}{8}\right]\right\}^2 = \frac{1}{64}$

♦ ♦ ♦ ♦ ♦ ♦ ♦ ♦ ♦ ♦ ♦ ♦ ♦ ♦ ♦ ♦ ♦ ♦ ♦ ♦ ♦ ♦ ♦ ♦ ♦ ♦ ♦ ♦ ♦ ♦ ♦ ♦ ♦ ♦ ♦ ♦

Parentheses, brackets, and braces are just one type of **grouping symbol**; other grouping symbols include the absolute value symbol, the square root symbol, and the horizontal fraction bar.

The *absolute value* symbol may act as a grouping symbol if it contains an expression, such as $|1 + 9|$. First calculate the expression inside the absolute value, and then find the absolute value.

The *square root* symbol, or *radical* sign, is a grouping symbol if it contains an expression rather than a single number, as in $\sqrt{8 + 1}$. The expression inside should be calculated before taking the square root. In writing an expression under a radical sign it is important to draw the overbar over the whole expression.

On a calculator it is not possible to draw the overbar of a radical sign over the whole expression, so we must use parentheses around the expression inside the radical sign. The expression $\sqrt{(5^2 + 12^2)}$ is written in calculator form.

The *horizontal fraction bar* acts as a grouping symbol. The numerator and denominator of the fraction are calculated separately. Because division in algebra is almost always written as a fraction, we frequently encounter fractions in expressions (see Example 6). When using a calculator to simplify fractions that contain additions or subtractions, we use parentheses around the numerator or denominator expression.

EXAMPLE **11**    Simplify these expressions:

$$\textbf{a. } \frac{3 + 4}{6 - 2} \qquad\qquad \textbf{b. } \frac{-1 + \sqrt{49}}{2} \qquad\qquad \textbf{c. } |3 - 8| - |4 - 6|$$

$$\textbf{d. } \sqrt{(0 - 3)^2 + (4 - 8)^2} \qquad \textbf{e. } \frac{4 - \sqrt{4^2 + 48}}{2}$$

SOLUTION    **a.** $\dfrac{3 + 4}{6 - 2} = \dfrac{7}{4} = 1.75$

**b.** $\dfrac{-1 + \sqrt{49}}{2} = \dfrac{-1 + 7}{2} = \dfrac{6}{2} = 3$

**c.** $|3 - 8| - |4 - 6| = |-5| - |-2| = 5 - 2 = 3$

**d.** $\sqrt{(0 - 3)^2 + (4 - 8)^2} = \sqrt{(-3)^2 + (-4)^2} = \sqrt{9 + 16} = \sqrt{25} = 5$

**e.** $\dfrac{4 - \sqrt{4^2 + 48}}{2} = \dfrac{4 - \sqrt{16 + 48}}{2} = \dfrac{4 - \sqrt{64}}{2} = \dfrac{4 - 8}{2} = \dfrac{-4}{2} = -2$

♦ ♦ ♦ ♦ ♦ ♦ ♦ ♦ ♦ ♦ ♦ ♦ ♦ ♦ ♦ ♦ ♦ ♦ ♦ ♦ ♦ ♦ ♦ ♦ ♦ ♦ ♦ ♦ ♦ ♦ ♦ ♦ ♦ ♦ ♦ ♦

CAUTION  Many calculators have a squaring key, $\boxed{x^2}$, but a number may also be squared with the exponent key: $\boxed{y^x}$, $\boxed{x^y}$, or $\boxed{\land}$. Caution is needed in using these keys: on TI 30 calculators, and possibly others, identical keystrokes give different answers, depending on how rapidly the keystrokes are made.

　　To calculate $2^5 \cdot 2^3$ on the TI 30 we use the keystrokes 2 $\boxed{y^x}$ 5 $\boxed{\times}$ 2 $\boxed{y^x}$ 3. The calculator answer may be 256 or 32,768 or 262144. Obviously, not all can be correct. The correct answer is found if we pause after entering the multiplication sign or use the $\boxed{=}$ key after the exponent 5.  ♦

## EXERCISES 2.3

In Exercises 1 and 2 identify the base for the exponent 2, and write the expression using factors.

1. a. $3x^2$　　　　　　　　　　b. $-3x^2$

   c. $(-3x)^2$　　　　　　　　d. $ax^2$

   e. $-x^2$　　　　　　　　　f. $(-x)^2$

2. a. $-b^2$　　　　　　　　　b. $(-b)^2$

   c. $ab^2$　　　　　　　　　d. $mn^2$

   e. $(ab)^2$　　　　　　　　f. $-2x^2$

Simplify the exponential expressions in Exercises 3 and 4.

3. a. $3^5$　　b. $2^6$　　c. $(-2)^2$　　d. $(-3)^3$

   e. $\left(\frac{2}{3}\right)^2$　　f. $\left(-\frac{1}{2}\right)^2$　　g. $\left(\frac{1}{3}\right)^3$

4. a. $4^4$　　b. $5^3$　　c. $(-2)^3$　　d. $(-3)^2$

   e. $\left(\frac{1}{3}\right)^3$　　f. $\left(\frac{4}{5}\right)^3$　　g. $\left(-\frac{2}{3}\right)^3$

5. What exponent $n$ makes $\left(\frac{2}{5}\right)^n = \frac{8}{125}$?

6. What exponents $n$ make $(-1)^n$ a positive number?

Simplify the expressions in Exercises 7 and 8.

7. a. $m^3n^5$　　b. $n^4n^4$　　c. $a^6a^2$　　d. $a^7a^1$

8. a. $m^3m^3$　　b. $m^1m^5$　　c. $b^2b^4$　　d. $m^2n^3$

9. What pairs of exponents make $x^\square x^\bigcirc = x^{12}$?

10. What pairs of exponents make $x^\square x^\bigcirc = x^{10}$?

Simplify the expressions in Exercises 11 to 16.

11. a. $(2a)^3(2a)^6$　　b. $(2a)^2(3b)^3$　　c. $4(3a)^3(-1)^4$

    d. $4(3a)^1(-1)^3$　　e. $5(2x)^4(3y)^1$

12. a. $(2b)^2(2b)^4$　　b. $(3b)^3(2b)^2$　　c. $6(3a)^2(-1)^2$

    d. $3(2x)^2(3y)^2$　　e. $10(2x)^3(3y)^2$

13. a. $\dfrac{x^5}{x^2}$　　b. $\dfrac{a^8}{a^5}$　　c. $\left(\dfrac{x}{y}\right)^2$　　d. $\left(\dfrac{x}{y^2}\right)^2$

14. a. $\dfrac{x^6}{x^3}$　　b. $\dfrac{b^8}{b^4}$　　c. $\left(\dfrac{x^2}{y}\right)^2$　　d. $\left(\dfrac{2a}{3b}\right)^3$

15. a. $(x^2)^2$　　　　b. $(2x^2)^3$　　　　c. $\left(\frac{1}{4}a^2\right)^2$

    d. $5(x^2)^4(2y^2)^1$　　e. $10(x^2)^2(2y^2)^3$

16. a. $(y^2)^3$　　　　b. $(3y^3)^2$　　　　c. $\left(\frac{1}{3}y^3\right)^2$

    d. $10(x^2)^3(2y^2)^2$　　e. $5(x^2)(2y^2)^4$

For Exercises 17 and 18 find the area of the squares and the volume of the cubes.

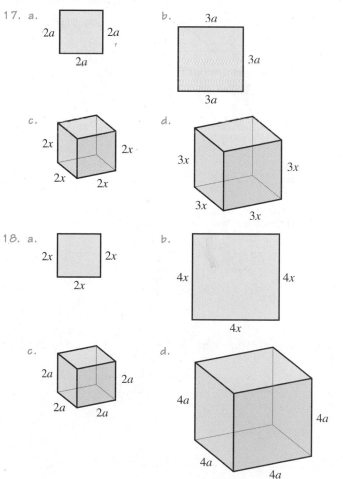

17. a. $2a$ × $2a$ × $2a$　　b. $3a$ × $3a$

    c. $2x$ cube　　d. $3x$ cube

18. a. $2x$ × $2x$ × $2x$　　b. $4x$ × $4x$

    c. $2a$ cube　　d. $4a$ cube

19. Using a fraction, compare the areas in Exercise 17a and b. Using a fraction, compare the volumes in Exercise 17c and d.

20. Using a fraction, compare the areas in Exercise 18a and b. Using a fraction, compare the volumes in Exercise 18c and d.

21. A bouncing ball bounces half the previous height with each bounce. What is the height of the fifth bounce if the ball starts from 96 centimeters?

22. A basketball tournament starts with 64 teams. The teams play on Wednesdays, Saturdays, and Sundays. How many game days are needed to get down to the final game if half the teams are eliminated each day? On which game day will the final game be played?

23. A bacteria population doubles every 2 hours. If there are 1,000,000 bacteria to start with, how many will there be in 24 hours?

24. Suppose a gambling machine returns an average of 75% of the coins placed in it. If someone starts with $20 in nickels and sets aside the money returned from each round of playing, how many rounds will the person play before all the money will be gone?

**Simplify the expressions in Exercises 25 and 26. What steps in the order of operations make the answers different?**

25. a. $7 - 2(4 - 1)$     b. $(7 - 2)(4 - 1)$

26. a. $5 - 2(6 + 2)$     b. $(5 - 2)(6 + 2)$

**Simplify the expressions in Exercises 27 and 28.**

27. a. $1 + 9 \cdot 9 - 8$     b. $1 + \sqrt{9} \cdot 9 - 8$

    c. $(1 + \sqrt{9}) \cdot 9 - 8$

28. a. $1 - \sqrt{9} \cdot (9 - 8)$     b. $1 + 9 \cdot (\sqrt{9} - 8)$

    c. $(1 + 9) \cdot \sqrt{9} - 8$

**Simplify the expressions in Exercises 29 and 30 using a calculator. Estimate the answer first. Round to 3 decimal places.**

29. a. $\dfrac{2 - \sqrt{2}}{2}$     b. $\dfrac{3 + \sqrt{6}}{3}$     c. $\dfrac{3\sqrt{6}}{3}$

30. a. $\dfrac{2 + \sqrt{2}}{2}$     b. $\dfrac{3 - \sqrt{6}}{3}$     c. $\dfrac{4\sqrt{8}}{4}$

**Simplify the expressions in Exercises 31 to 34 without a calculator.**

31. a. $\dfrac{-(-4) - \sqrt{(-4)^2 - 4 \cdot (1) \cdot (-12)}}{2 \cdot 1}$

    b. $\dfrac{-(-4) + \sqrt{(-4)^2 - 4 \cdot (1) \cdot (-12)}}{2 \cdot 1}$

32. a. $\dfrac{-(-5) + \sqrt{(-5)^2 - 4(6)(1)}}{2(6)}$

    b. $\dfrac{-(-5) - \sqrt{(-5)^2 - 4(6)(1)}}{2(6)}$

33. a. $|4 - 6| + |6 - 4|$     b. $|3 - 7| - |7 - 3|$

34. a. $|5 - 2| + |2 - 5|$     b. $|3 - 8| - |8 - 5|$

35. In Example 5a, what property allowed us to multiply first $2 \cdot 2 \cdot 2$ and then $x^2 \cdot x^2 \cdot x^2$?

36. In Example 5a, what property allowed us to write $2x^2 \cdot 2x^2 \cdot 2x^2 = 2 \cdot 2 \cdot 2 \cdot x^2 \cdot x^2 \cdot x^2$?

37. If a fraction is simplified, will it be possible to simplify the square of the fraction?

38. The statement "$x^2$ is larger than $x$" may be both true and false.

    a. Give an example that makes it true.

    b. Give an example that makes it false.

*ERROR ANALYSIS*

39. Which student has the correct answer? Write a sentence or two to explain to the other students what they did wrong.

    Student #1 has $(3x^3)^2 = 9x^9$.
    Student #2 has $(3x^3)^2 = 6x^6$.
    Student #3 has $(3x^3)^2 = 9x^6$.

40. Which student has the correct answer? Write a sentence or two to explain to the other students what they did wrong.

    Student #1 has $(-0.2x^3)^2 = 0.4x^9$.
    Student #2 has $(-0.2x^3)^2 = -0.4x^6$.
    Student #3 has $(-0.2x^3)^2 = 0.04x^9$.
    Student #4 has $(-0.2x^3)^2 = 0.04x^6$.

41. a. Complete the table.

| $a$ | $b$ | $a^2 + b^2$ | $\sqrt{(a^2 + b^2)}$ | $\sqrt{a^2} + \sqrt{b^2}$ |
|---|---|---|---|---|
| 3 | 4 | $9 + 16 = \_\_$ | | |
| 9 | 12 | | 15 | |
| 5 | 12 | | | |
| 6 | 8 | | | |
| 9 | 40 | | | |

    b. Compare the values of $\sqrt{a^2 + b^2}$ and $\sqrt{a^2} + \sqrt{b^2}$. Describe an error someone might make in working with $\sqrt{a^2 + b^2}$.

*PROJECTS*

42. **Acronyms and Other Learning Tools.** Ask three students from other countries to share how they learned the order of operations in their language. Did they use an acronym such as PEMDAS or a phrase such as "Please Excuse My Dear Aunt Sally"? Record and explain the method.

43. **Operations on Digits**

    a. Check these problems. One is not correct. Change one operation sign to make it true.

      $1 \cdot 9 + 9 - 8 = 10$
      $1 - 9 + 9 + 8 = 9$
      $1 \cdot [9 - (9 - 8)] = 8$
      $-1 + 9 - 9 - 8 = 7$
      $-1 - 9 \div 9 + 8 = 6$

b. Check these problems. One is not correct. Change one operation sign to make it true.

$-1 + (9 + 9) \div 9 = 1$
$1 \cdot (9 + 9) \div 9 = 2$
$1 + (9 + 9) \div 9 = 3$
$1 \cdot (9 \div 9) + \sqrt{9} = 4$
$1 + (9 \div 9) - \sqrt{9} = 5$

c. Using the digits in the year of your birth, obtain ten outcomes from 1 to 20.

44. **Bouncing Balls.** Use a yardstick or meterstick and a super-ball to estimate the height of each bounce for 5 bounces.

a. Describe your experiment.

b. Record your results clearly in a table.

c. Summarize your findings. Include in your findings an estimated fraction of the original height you think the ball bounces.

d. Use your fraction to predict the height for 10 bounces.

45. **Exponents That Are Not Positive Integers**

a. Find the pattern, and complete the blanks in both directions:

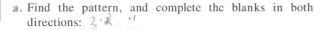
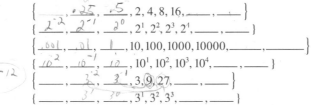

$\{\_\_\_, .25, .5, 2, 4, 8, 16, \_\_\_, \_\_\_\}$
$\{2^{-2}, 2^{-1}, 2^0, 2^1, 2^2, 2^3, 2^4, \_\_\_, \_\_\_\}$
$\{.001, .01, .1, 10, 100, 1000, 10000, \_\_\_, \_\_\_\}$
$\{10^{-2}, 10^{-1}, 10, 10^1, 10^2, 10^3, 10^4, \_\_\_, \_\_\_\}$
$\{\_\_\_, 3^{-2}, 3^{-1}, 3, 9, 27, \_\_\_, \_\_\_\}$
$\{\_\_\_, 3^{-1}, 3^0, 3^1, 3^2, 3^3, \_\_\_, \_\_\_\}$

b. Use your calculator to explore the effect of 0 as an exponent. Look for a pattern. Look for an exception.

$115^0 \qquad 4^0 \qquad 1^0$
$0^0 \qquad (-1)^0 \qquad (-2)^0$
$2.5^0 \qquad \pi^0 \qquad 3^0$

Put negative numbers, such as $-2$, into the calculator in parentheses. Otherwise you will get an incorrect answer. The expression $-2^0$ is the opposite of $2^0$, whereas $(-2)^0$ is $(-2)$ raised to the exponent 0.

Which number above produced an error message when entered into the calculator?

Complete the statement: $x^0 = \_\_\_$ for all numbers $x$ except $\_\_\_$.

c. Use your calculator to explore the effect of $-1$ as an exponent. The calculator will give decimals. Write the decimals as fractions to see the effect of a $-1$ exponent more clearly.

$5^{-1} \qquad (-4)^{-1} \qquad 2^{-1}$
$(-25)^{-1} \qquad 20^{-1} \qquad (-50)^{-1}$
$\left(\tfrac{1}{2}\right)^{-1} \qquad \left(\tfrac{1}{4}\right)^{-1} \qquad \left(-\tfrac{1}{5}\right)^{-1}$

Describe the pattern in a sentence or two.

d. Use your calculator to explore the meaning of $-2$ as an exponent.

$\left(\tfrac{1}{2}\right)^{-2} \qquad 2^{-2} \qquad 3^{-2}$
$\left(\tfrac{1}{3}\right)^{-2} \qquad \left(-\tfrac{1}{4}\right)^{-2} \qquad \left(\tfrac{1}{5}\right)^{-2}$

Write a sentence or two to describe the pattern.

e. Investigate $x^{-3}$.

f. In general, what does a negative exponent do?

g. We are not limited to integers as exponents. Explore the answers to these problems:

$1^{0.5} \qquad 2^{0.5} \qquad 3^{0.5}$
$4^{0.5} \qquad 9^{0.5} \qquad 16^{0.5}$
$49^{0.5} \qquad 6.25^{0.5} \qquad 0.01^{0.5}$

Describe the pattern that emerges. What is the meaning of 0.5, or 1/2, as an exponent?

## MID-CHAPTER 2 TEST

1. Add or subtract in these problems, as indicated.

a. $2 - 5$       b. $-3 + 5$       c. $-3 - (-5)$

d. $3 - (-4)$       e. $-2 + (-5)$       f. $6 + (-2.5)$

State the subtraction problem illustrated by the ion circle figure in Exercises 2 and 3, and then work the problem.

2.
3.

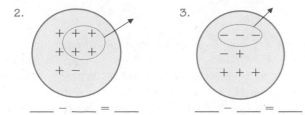

$\_\_\_ - \_\_\_ = \_\_\_$        $\_\_\_ - \_\_\_ = \_\_\_$

Make ion circles that illustrate the problems in Exercises 4 and 5, and work the problems.

4. $2 - (-2)$                5. $0 - 3$

Describe the problems in Exercises 6 and 7 using a checking account model. Assume you start with a zero balance.

6. $-5.50 + (18.98) - 12.76$       7. $-3.89 - 42.39 + 50.00$

For Exercises 8 to 11 simplify the expressions in your head. Describe any shortcuts or properties used.

8. $3 + 6 + 9 + 12 + 15 + 18 + 21 + 24 + 27$

9. $-6 + 12 + 18 - 24 + 3$

10. $13 \cdot 4 \cdot 5 \cdot 3$          11. $7 \cdot 5 \cdot 20 \cdot \tfrac{1}{2}$

**For Exercises 12 to 15 make input–output tables with $x$ as integers from $-2$ to 3. Graph the $(x, y)$ pairs on coordinate axes.**

12. $y = x^2$             13. $y = -x^2$

14. $y = |x + 3|$          15. $y = 3 + |x|$

**In Exercises 16 and 17 what is the difference in elevation between the highest and lowest point on each continent?**

16. Asia: Mt. Everest, 8850 meters, and the Dead Sea, $-400$ meters

17. North America: Mt. McKinley, 6194 meters, and Death Valley, $-86$ meters

18. Mt. Everest is recognized as the highest mountain in the world. Mauna Kea is the inactive volcano on the island of Hawaii. Mauna Kea's volcanic base actually rises from the ocean floor. Using the data in the following table on elevation in feet relative to sea level, find which mountain is actually "taller" and by how much.

| Mt. Everest | +29,028 |
|---|---|
| Mauna Kea | +13,710 |
| Sea Level | 0 |
| Ocean Floor, near Hawaii | −16,400 |

19. Simplify these exponential expressions.

   a. $3^4$
   b. $\left(-\frac{1}{2}\right)^4$
   c. $(3m)^2(3n)^3$

   d. $5(2x)^1(3y)^4$
   e. $(3x^2)^4$
   f. $\left(-\frac{3}{4}b^2\right)^2$

   g. $\left(\dfrac{2x}{3x^2}\right)^2$
   h. $\left(-\dfrac{0.2x}{y^3}\right)^3$

**Simplify the expressions in Exercises 20 and 21. What steps in the order of operations make the answers different?**

20. $4 - 3(3 - 5)$        21. $(4 - 3)(3 - 5)$

**Simplify the expressions in Exercises 22 to 25.**

22. $\dfrac{-5 + \sqrt{5^2 - 4(2)(-12)}}{2(2)}$

23. Use a calculator to find $\dfrac{4 - \sqrt{8}}{4}$.

24. $|6 - 1| - |3 - 9|$

25. $|7 - 2| + |4 - 1|$

---

## 2.4   UNIT ANALYSIS AND FORMULAS

**OBJECTIVES**   Change from one unit of measure to another.   ♦   Calculate perimeter, area, surface area, and volume.   ♦   Simplify units in formula work using unit analysis.   ♦   Evaluate geometric formulas involving the order of operations.   ♦   Store numbers for formula evaluation in a calculator's memory.

**WARM-UP**   Simplify these fractions before multiplying.

1. $\dfrac{\$6}{h} \cdot \dfrac{8h}{d} \cdot \dfrac{5d}{w} \cdot \dfrac{52w}{y}$             2. $\dfrac{6s}{i} \cdot \dfrac{12i}{f} \cdot \dfrac{3f}{y} \cdot \dfrac{2y}{b}$

3. $\dfrac{4000f}{t} \cdot \dfrac{m}{5280f} \cdot \dfrac{2t}{h} \cdot \dfrac{8h}{s}$                                                                    ♦

We now introduce unit analysis, a skill that will help you work with units of measurement, organize your thinking, and present your solutions clearly. We close the section with practice in using a variety of geometric formulas. Both units of measurement and order of operations are especially important in formula work.

### UNIT ANALYSIS

**Unit analysis** is a method of changing from one unit of measurement to another. We change units by placing our measurement facts in fraction form, building a product of fractions, and canceling unwanted units. Where possible, we want to use our problem-solving steps: understand, plan, do, and check. In the first example we use unit analysis to justify a calculation shortcut.

EXAMPLE 1  Yearly income   In business and labor negotiations, multiplying the hourly wage by 2000 estimates full-time annual (yearly) income. Use unit analysis to show that $6 per hour is approximately $12,000 per year.

*Useful facts:*

1 day = 8 hours
1 week = 5 working days
1 year = 52 weeks

SOLUTION   *Understand*: We need to change hours into years, using the given facts.

*Plan*: We are verifying an estimate, so no further estimate of the result is needed. We start with $6 per hour and have a goal of dollars per year. The word *per* means division, so hours go in the denominator. Arrange the facts into fractions so that each unit of time appears once in a numerator and once in a denominator.

*Do*: Start with $6 per hour. The next fraction should be 8 hours over 1 day. The third fraction contains 5 days over 1 week, and the last needs weeks on top, so 52 weeks goes over 1 year. The hours, days, and weeks cancel, leaving dollars per year.

$$\frac{\$6}{1 \text{ hour}} \cdot \frac{8 \text{ hours}}{1 \text{ day}} \cdot \frac{5 \text{ days}}{1 \text{ week}} \cdot \frac{52 \text{ weeks}}{1 \text{ year}} = \frac{\$6(8)(5)(52)}{1 \text{ year}} = \frac{\$12,480}{\text{year}}$$

*Check*: The $12,480 is quite close to $6 · 2000 = $12,000. If we assume only 50 weeks a year, instead of 52, then (8)(5)(50) = 2000, which indicates the source of the estimate.

♦ ♦ ♦ ♦ ♦ ♦ ♦ ♦ ♦ ♦ ♦ ♦ ♦ ♦ ♦ ♦ ♦ ♦ ♦ ♦ ♦ ♦ ♦ ♦ ♦ ♦ ♦ ♦ ♦ ♦ ♦ ♦

Each fact creates a fraction worth 1, because the numerator and denominator represent different ways of saying the same unit of measure. Remember to set up the product of fractions so that each unit of measurement appears once in a numerator and once in a denominator.

---

A key idea in unit analysis is that the units of measurement cancel.

---

EXAMPLE 2  Sewing stitches   Arrange the facts into fractions to find how many sewing machine stitches are required to attach a decorative trim along the bottom edge of a gathered skirt (see Figure 5). (If, because of a mistake, you have to rip out the stitches, you care how many stitches there are!)

*Useful facts*:

Bottom edge = 2 yards
6 stitches per inch
1 foot = 12 inches
1 yard = 3 feet

*Figure 5*

SOLUTION   *Understand*: Find the number of stitches along the bottom edge.

*Plan*: We estimate that there are well over 100 stitches because there are a large number of inches in each yard. Start with stitches per inch, and have a goal of stitches per bottom edge. Write each fact as a fraction.

*Do*:   $$\frac{6 \text{ stitches}}{1 \text{ inch}} \cdot \frac{12 \text{ inches}}{1 \text{ foot}} \cdot \frac{3 \text{ feet}}{1 \text{ yard}} \cdot \frac{2 \text{ yards}}{\text{Bottom edge}} = \frac{432 \text{ stitches}}{\text{Bottom edge}}$$

*Check*: We find over 400 stitches, which confirms our estimate that there would be quite a large number.

♦ ♦ ♦ ♦ ♦ ♦ ♦ ♦ ♦ ♦ ♦ ♦ ♦ ♦ ♦ ♦ ♦ ♦ ♦ ♦ ♦ ♦ ♦ ♦ ♦ ♦ ♦ ♦ ♦ ♦ ♦ ♦

In Example 3 we change facts in both the numerator and the denominator.

EXAMPLE  3

Guarding a hazardous waste site    Suppose a guard at a hazardous waste site is required to make 2 trips around the site each hour. The site is 4000 feet around. What is the total distance, in miles, walked during an 8-hour shift?

*Useful fact*: 5280 feet = 1 mile

SOLUTION

*Understand*: We need to convert both feet to miles and trips around to shift.

*Plan*: First determine the facts. The distance around the site is 4000 feet, which means 4000 feet per trip. Two other facts are stated. The 8-hour shift means 8 hours per shift, and 2 trips each hour means 2 trips per hour. As an estimate, 1 trip is less than a mile, so the guard should walk between 8 and 16 miles in an 8-hour shift.

   We should start with the distance around, 4000 feet per trip, and end with a number of miles per shift.

*Do*: If we start with 4000 feet per trip, placing 5280 feet in the next denominator (and 1 mile in the numerator) cancels the feet. The other fractions are set up to cancel trips and, finally, hours.

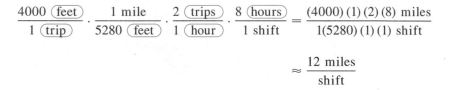

$$\frac{4000 \; \text{feet}}{1 \; \text{trip}} \cdot \frac{1 \; \text{mile}}{5280 \; \text{feet}} \cdot \frac{2 \; \text{trips}}{1 \; \text{hour}} \cdot \frac{8 \; \text{hours}}{1 \; \text{shift}} = \frac{(4000)(1)(2)(8) \; \text{miles}}{1(5280)(1)(1) \; \text{shift}}$$

$$\approx \frac{12 \; \text{miles}}{\text{shift}}$$

*Check*: Our answer is within our estimate of 8 to 16 miles. The final answer is rounded to the nearest mile and thus follows the symbol ≈, which means *approximately equal*.  ✔

♦ ♦ ♦ ♦ ♦ ♦ ♦ ♦ ♦ ♦ ♦ ♦ ♦ ♦ ♦ ♦ ♦ ♦ ♦ ♦ ♦ ♦ ♦ ♦ ♦ ♦ ♦ ♦ ♦ ♦ ♦ ♦ ♦ ♦ ♦ ♦ ♦

Unit Analysis Summary

> **1.** Identify the units of measurement that you want to change. Identify the units needed in the answer.
>
> **2.** List facts that contain the starting units of measurement and the ending units.
>
> **3.** Write the starting fact as a fraction. Set up a product of fractions using your list of facts so that each unit of measurement appears once in a numerator and once in a denominator.
>
> **4.** Cancel the units of measurement where possible, and calculate numbers.

We now apply work with units of measurement to geometric formulas.

FORMULAS

Perimeter and Area.    The formulas in Table 5 and Table 6 summarize many common geometric formulas. These formulas, along with order of operations, the commutative property, and the calculator, will be used in the following examples.

| Triangle | Area $= \frac{1}{2}$ base $\cdot$ height $= \frac{1}{2}bh$ | |
| Square | Area $=$ side $\cdot$ side $= s^2$ <br> Perimeter $= 4 \cdot$ side $= 4s$ | |
| Rectangle | Area $=$ length $\cdot$ width $= lw$ <br> Perimeter $= 2l + 2w$ | |
| Parallelogram | Area $=$ base $\cdot$ height $= bh$ | |
| Trapezoid | Area $= \frac{1}{2}$ height $\cdot$ (sum of parallel sides) <br> $= \frac{1}{2}h(a + b)$ | |
| Circle | Area $= \pi r^2$, $r =$ radius <br> Circumference $= 2\pi r = \pi d$ <br> Diameter $= d = 2r$ | |

*Table 5* *Selected Geometric Formulas for Two-Dimensional Figures*

*Note:* In all formulas the base and height (or length and width) refer to dimensions that are perpendicular.

In Section 2.3 we shortened instructions by using the word *simplify* to direct us to use the order of operations, to calculate, to apply the properties of exponents, or to use the associative and commutative properties in working with expressions. In this section we will use another instruction word: **evaluate**. *Evaluate formulas and expressions by substituting numbers in place of the variables.*

EXAMPLE 4

Guarding a hazardous waste site, continued    Suppose the guard in Example 3 walks around the rectangular waste site shown in Figure 6. Which formula in Table 5 determines the distance once around? To confirm your choice, evaluate the formula and describe the order of operations needed to obtain the distance.

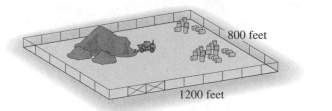

*Figure 6*

SOLUTION    The distance around is 2 lengths and 2 widths, which is the same as the perimeter formula:

Perimeter $= 2l + 2w = 2(1200 \text{ ft}) + 2(800 \text{ ft}) = 2400 \text{ ft} + 1600 \text{ ft} = 4000 \text{ ft}$

The length and width are each multiplied by 2, and then the two products are added together. The distance around is 4000 feet, which confirms the data in Example 3.

◆ ◆ ◆ ◆ ◆ ◆ ◆ ◆ ◆ ◆ ◆ ◆ ◆ ◆ ◆ ◆ ◆ ◆ ◆ ◆ ◆ ◆ ◆ ◆ ◆ ◆ ◆ ◆ ◆ ◆ ◆ ◆ ◆ ◆ ◆ ◆ ◆ ◆

Example 4 found the perimeter of a rectangle. The **perimeter** *is the distance around the outside of a flat object*. The perimeter of a circle is called the **circumference**. Several formulas, including the one for the circumference of a circle, contain the constant pi. **Pi** *is the number found by dividing the circumference of any circle by its diameter*. The symbol for pi is $\pi$. Use either the $\boxed{\pi}$ key on your calculator or 3.14 to approximate pi. The calculator key will give slightly more accurate results.

Although there are formulas for the perimeter of a rectangle and the circumference of a circle, the formulas should not replace understanding that perimeter is the sum of the lengths of the sides. Example 5 is typical of many problems in that it requires a plan: calculating parts of the perimeter separately and then adding the results.

EXAMPLE 5    Track around a soccer field    What is the perimeter of a track placed around a rectangular soccer pitch (playing field), as shown in Figure 7? Assume that the curved parts are half-circles.

SOLUTION    This perimeter is the sum of the circumference of the curved parts (totaling a complete circle) and the two straight sides. From Table 5, the circumference of a circle is $\pi d$, or pi times the diameter. The perimeter is

$$P = \pi d + 2l$$
$$= \pi(73 \text{ m}) + 2(100 \text{ m})$$
$$\approx 3.14(73 \text{ m}) + (200 \text{ m})$$
$$\approx (229.22 \text{ m}) + (200 \text{ m})$$
$$\approx 429.22 \text{ m}$$

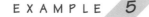

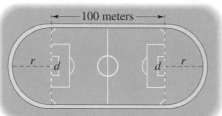

$d = 73$ meters

*Figure 7*

◆ ◆ ◆ ◆ ◆ ◆ ◆ ◆ ◆ ◆ ◆ ◆ ◆ ◆ ◆ ◆ ◆ ◆ ◆ ◆ ◆ ◆ ◆ ◆ ◆ ◆ ◆ ◆ ◆ ◆ ◆ ◆ ◆ ◆ ◆ ◆ ◆ ◆

Abbreviations of metric units—such as meters—are written without periods. As a result, 73 m may be confused with $73m$, the product of 73 and $m$. If you are in doubt, look for the space between the number and the letter. Also, in typeset material, the variable $m$ will usually be in italics.

The **area** of a figure *is how much surface it covers*. Area is described in square units such as square inches, square feet, square centimeters, and square meters. In choosing measurements for calculating area, make sure the base and height (or length and width) are perpendicular. Look for a small square in the corner indicating perpendicular lines.

EXAMPLE 6    Trapezoidal wall hanging    Ray has a trapezoidal area, shown in Figure 8, of a hooked wall hanging to complete with yellow yarn.

**a.** What is the area in square inches?

**b.** If each square inch requires 49 pieces of yarn, how many pieces will be needed to complete the trapezoid?

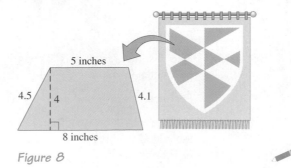

*Figure 8*

SOLUTION   **a.**              Area $= \frac{1}{2}h(a + b)$, where $a$ and $b$ are the parallel sides

$$= \left(\tfrac{1}{2} \cdot 4 \text{ in.}\right)(5 \text{ in.} + 8 \text{ in.})$$

$$= (2 \text{ in.})(13 \text{ in.})$$

$$= 26 \text{ sq in.}$$

$$= 26 \text{ in}^2$$

Add the numbers in the parentheses first, and then multiply from left to right. We abbreviate square inches as in$^2$.

**b.**   $\dfrac{49 \text{ pieces of yarn}}{1 \text{ in}^2} \cdot \dfrac{26 \text{ in}^2}{\text{trapezoidal area}} = \dfrac{1274 \text{ pieces of yarn}}{\text{trapezoidal area}}$

◆ ◆ ◆ ◆ ◆ ◆ ◆ ◆ ◆ ◆ ◆ ◆ ◆ ◆ ◆ ◆ ◆ ◆ ◆ ◆ ◆ ◆ ◆ ◆ ◆ ◆ ◆ ◆ ◆ ◆ ◆ ◆ ◆ ◆ ◆ ◆

EXAMPLE   **7**   *Pizza sizes*   One pizza has a radius of 5 inches. Another has a radius of 7 inches. How many times larger is the 7-inch pizza than the 5-inch pizza?

SOLUTION   We divide the larger area by the smaller area to find how many times larger the pizza is. We assume the pizzas are round and of the same thickness.

$$\frac{\text{Area 7-in. pizza}}{\text{Area 5-in. pizza}} = \frac{\pi \cdot 7^2}{\pi \cdot 5^2} = \frac{\pi \cdot 49}{\pi \cdot 25} = 1.96$$

The area of the 7-inch pizza is almost double that of the 5-inch pizza.

◆ ◆ ◆ ◆ ◆ ◆ ◆ ◆ ◆ ◆ ◆ ◆ ◆ ◆ ◆ ◆ ◆ ◆ ◆ ◆ ◆ ◆ ◆ ◆ ◆ ◆ ◆ ◆ ◆ ◆ ◆ ◆ ◆ ◆ ◆ ◆

*Surface Area and Volume.*   **Surface area** *is the amount of area needed to cover the outside of a three-dimensional object.* Because the thickness is negligible, the metal forming a home fuel tank or a can of food is the surface area of a cylinder. The soap film forming a bubble floating in the air and the fabric forming a basketball are examples of the surface area for a sphere. Three surface area formulas are shown in Table 6.

| | |
|---|---|
| Rectangular prism (box) | Surface area $= 2lw + 2hl + 2hw$ <br> Volume $= lwh$ <br><br> 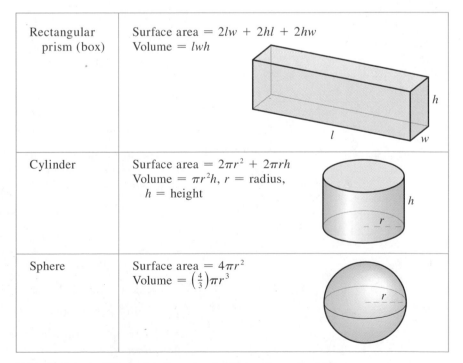 |
| Cylinder | Surface area $= 2\pi r^2 + 2\pi rh$ <br> Volume $= \pi r^2 h$, $r =$ radius, <br> $\quad h =$ height |
| Sphere | Surface area $= 4\pi r^2$ <br> Volume $= \left(\tfrac{4}{3}\right)\pi r^3$ |

*Table 6 Selected Geometric Formulas for Three-Dimensional Figures*

EXAMPLE 8

**Surface area of a storage tank**    If we wanted to paint a cylindrical storage tank, we would need to know the surface area to determine the amount of paint needed. Find the surface area of the tank shown in Figure 9. Describe the order of operations needed to evaluate the formula.

SOLUTION    Note that the height need not be shown in the vertical position. The "height" of this cylinder, 10 feet, is the distance between the circular ends. The radius is 2 feet.

10 feet

2 feet

$$\text{Surface area} = 2\pi r^2 + 2\pi rh$$

$$\approx 2(3.14)(2 \text{ ft})^2 + 2(3.14)(2 \text{ ft})(10 \text{ ft})$$

$$\approx 2(3.14)(4 \text{ ft}^2) + 2(3.14)2(10 \text{ ft}^2)$$

$$\approx (25.12 \text{ ft}^2) + (125.6 \text{ ft}^2)$$

$$\approx 150.72 \text{ ft}^2$$

Figure 9

Calculate the exponent expression first, then do the multiplications, and finally do the addition.

♦ ♦ ♦ ♦ ♦ ♦ ♦ ♦ ♦ ♦ ♦ ♦ ♦ ♦ ♦ ♦ ♦ ♦ ♦ ♦ ♦ ♦ ♦ ♦ ♦ ♦ ♦ ♦ ♦ ♦ ♦ ♦ ♦ ♦

   **Volume** *describes how much space an object takes up.* Volume is measured in cubic units such as cubic inches, cubic millimeters, and cubic meters. The air inside a balloon is a volume. A doctor prescribes an injection with cc's (cubic centimeters). We order cubic yards of gravel and concrete. Bales of peat moss are labeled with cubic feet. Three common volume formulas are shown in Table 6.

EXAMPLE 9

**Inflating balloons**    Suppose a balloon approximates a sphere (see Figure 10). Could someone inflate

**a.** an empty balloon to a radius of 3 inches in one breath?

**b.** an empty balloon to a radius of 4 inches in one breath?

**c.** a balloon from a radius of 4 inches to a radius of 5 inches?

*Useful facts*: Full capacity for human lungs is about 350 cubic inches. A normal breath at rest is about 30 cubic inches. A forced exhalation with no deep breath is only about 90 cubic inches. A nonsmoker in excellent physical condition, using deep inhalation and forced exhalation, might reach 250 to 275 cubic inches. Some air must remain or the lungs will collapse.

Figure 10

SOLUTION    We need to compare volumes by evaluating the formula with several radii (plural of *radius*). We will *round* our answers to the nearest whole number, and we will use $\pi \approx 3.14$. The volume of a sphere is given by $\frac{4}{3}\pi r^3$.

$$\text{For radius } r = 3 \text{ in., } V = \tfrac{4}{3}\pi(3 \text{ in.})^3 \approx 113 \text{ in}^3$$

$$\text{For radius } r = 4 \text{ in., } V = \tfrac{4}{3}\pi(4 \text{ in.})^3 \approx 268 \text{ in}^3$$

$$\text{For radius } r = 5 \text{ in., } V = \tfrac{4}{3}\pi(5 \text{ in.})^3 \approx 523 \text{ in}^3$$

**a.** Reaching a radius of 3 inches is likely.

**b.** Only a fit person could reach a radius of 4 inches.

**c.** A fit person could inflate from a radius of 4 inches to a radius of 5 inches. The *change in volume* is $\Delta V = 523 - 268 = 255$ cubic inches.

♦ ♦ ♦ ♦ ♦ ♦ ♦ ♦ ♦ ♦ ♦ ♦ ♦ ♦ ♦ ♦ ♦ ♦ ♦ ♦ ♦ ♦ ♦ ♦ ♦ ♦ ♦ ♦ ♦ ♦ ♦ ♦ ♦ ♦

The small triangle used in part c above is the capital Greek letter *delta*. Delta, $\Delta$, indicates *a change in the value of the variable that follows it*. The symbol delta emphasizes change.

CAUTION   The numbers in Example 9 do not take into account the resistance of the balloon against air going into it. You may want to seek medical advice before trying Example 9 as an experiment.   ♦

When a portion of a formula is used repeatedly, using the calculator memory saves time and improves accuracy. A number is stored in memory with the key $\boxed{\text{STO}}$, $\boxed{\text{M in}}$, or $\boxed{x \rightarrow \text{M}}$. With multiple storage locations either $\boxed{x \rightarrow \text{M}}$ or $\boxed{\text{STO} \blacktriangleright}$ is followed by a letter or number. To recall the stored number, use $\boxed{\text{RCL}}$, $\boxed{\text{M out}}$, or $\boxed{\text{M} \rightarrow x}$. With multiple memory locations a letter or number is entered after the recall command.

E X A M P L E   **10**   *Using memory for repeated calculation*   In Example 9, $\frac{4}{3}\pi$ appeared in each volume expression. List the keystrokes needed to store $\frac{4}{3}\pi$ and to recall it for each volume calculation.

SOLUTION   Following is a sample solution for TI scientific calculators:
To store $\frac{4}{3}\pi$, enter

$$4 \boxed{\div} 3 \boxed{=} \boxed{\times} 3.14 \boxed{=} \boxed{\text{STO}}$$

To calculate volume, enter

$$3 \boxed{y^x} 3 \boxed{\times} \boxed{\text{RCL}} \boxed{=}$$
$$4 \boxed{y^x} 3 \boxed{\times} \boxed{\text{RCL}} \boxed{=}$$
$$5 \boxed{y^x} 3 \boxed{\times} \boxed{\text{RCL}} \boxed{=}$$

♦ ♦ ♦ ♦ ♦ ♦ ♦ ♦ ♦ ♦ ♦ ♦ ♦ ♦ ♦ ♦ ♦ ♦ ♦ ♦ ♦ ♦ ♦ ♦ ♦ ♦ ♦ ♦ ♦ ♦ ♦ ♦

When adding to or subtracting from memory, use the $\boxed{\text{SUM}}$, $\boxed{\text{M+}}$, $\boxed{\text{+M}}$, or $\boxed{\text{M}-}$ key. Using both $\boxed{\pm}$ and $\boxed{\text{SUM}}$ keys will subtract from memory if the calculator does not have an $\boxed{\text{M}-}$ key.

**Graphing Calculator**

> On most graphing calculators you enter the entire expression before evaluating. If you make an error in keying in the expression, pressing $\boxed{\text{2nd}}$ $\boxed{\text{ENTER}}$ or $\boxed{\text{replay}}$ will restore the expression to the display for editing. This feature also permits you to recall and change the formula for a new radius, thus replacing the need for memory, as was used in Example 10.

## EXERCISES 2.4

**In Exercises 1 to 4 use unit analysis to convert each salary per year to dollars per hour. Use a 40-hour week and a 52-week year.**

1. $12,000

2. $30,000

3. $80,000

4. $150,000

5. Set up a unit analysis to determine the annual (yearly) cost of these two rental options. Use 52 weeks per year and 12 months per year. Explain, in complete sentences, which plan might be better. Why might some people think the rents are the same?
   Rental #1:   Pay $125 per week
   Rental #2:   Pay $500 per month

6. Set up a unit analysis to determine the annual (yearly) cost of each payment plan. Use 52 weeks per year and 12 months per year. Explain, in complete sentences, which plan might be better. Why might some people think the plans cost exactly the same on an annual basis?
   Payment Plan A:   Pay $15 per week
   Payment Plan B:   Pay $60 per month

**In Exercises 7 to 10 convert miles per hour to feet per second. Start with a fraction showing miles per hour, and then set up the unit analysis product of fractions.**

*Useful facts:*  5280 feet = 1 mile
                  60 seconds = 1 minute
                  60 minutes = 1 hour

7. 30 miles per hour      8. 55 miles per hour

9. 80 miles per hour     10. 500 miles per hour

**In Exercises 11 to 14 convert feet per second to miles per hour. Start with a fraction showing feet per second, and then set up the unit analysis product of fractions.**

11. 88 feet per second      12. 66 feet per second

13. 220 feet per second     14. 4.4 feet per second

15. Use unit analysis to find the number of feet you have driven in a 4-second reaction time at 60 miles per hour. Guess before you start: Will it be more than, equal to, or less than a residential city block of length 334 feet?

16. The difference in elevation between Mt. Everest and the Mariana Trench is 64,868 feet. Use unit analysis to determine how many miles this elevation difference represents.

17. The Cheez-It® snack crackers box weighs 16 ounces. The serving size is 12 crackers. There are 140 calories in each ounce. A box contains 32 servings. Use unit analysis to calculate the number of calories per cracker.

18. A 15-ounce bag of Diane's Tortilla Chips® contains 15 servings. There are 80 milligrams (mg) of sodium in each serving. How many milligrams of sodium are consumed in eating a third of a bag of chips? Use unit analysis.

19. We need to replace the fence around the hazardous waste site in Example 4. Suppose the fencing manufacturer uses metric dimensions. Our only reference book indicates that 1 meter is 39.37 inches. Use unit analysis to convert 4000 feet to meters.

20. We want to decorate the edges of a square tablecloth with beads. There are $1\frac{1}{2}$ yards of fabric on each side of the cloth. If we get 4 beads per inch, use unit analysis to find the total number of beads needed for the cloth.

21. Find the perimeter and area of each shape.

a.
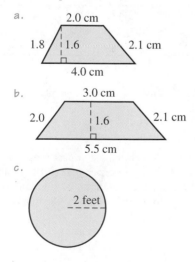
2.0 cm
1.8 | 1.6 | 2.1 cm
4.0 cm

b.
3.0 cm
2.0 | 1.6 | 2.1 cm
5.5 cm

c.
2 feet

d.
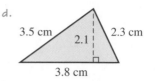
3.5 cm   2.1   2.3 cm
3.8 cm

22. Find the perimeter and area of each shape.

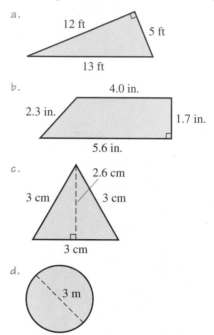

a.
12 ft    5 ft
13 ft

b.
4.0 in.
2.3 in.    1.7 in.
5.6 in.

c.
2.6 cm
3 cm   3 cm
3 cm

d.
3 m

**Measure the shapes in Exercises 23 and 24 using millimeters. Calculate the perimeter and area of each. Your answers should be within 5 mm of the perimeter and 10 mm² of the area.**

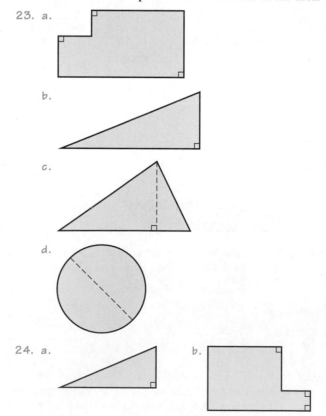

23. a.

b.

c.

d.

24. a.                    b.

c. 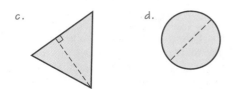 d.

**What are the surface area and volume of each shape in Exercises 35 and 36? Use $\pi \approx 3.14$.**

25. What other dimensions would have given a 4000-foot perimeter for the hazardous waste site in Example 4? What are the advantages and disadvantages of each shape?

26. What is the area of the hazardous waste site in Example 4?

27. What is the perimeter of the trapezoid in Example 6?

28. What is the area inside the track in Example 5?

29. Find the area of the shaded part in each figure, using $\pi \approx 3.14$.

a.

3 cm

3 cm

b.

2 cm

2 cm

c.

4 cm

2 cm

30. All of the parallelograms below have the same length sides. Find the perimeter and the area of each parallelogram. Then comment on any patterns you observe.

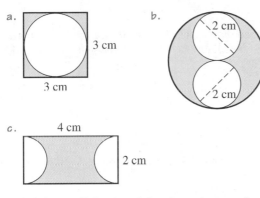

a.

5

7

b.

4

5

7

c.

5

3

7

d.

5

2

7

**For Exercises 31 and 32 assume the pizzas are round and of the same thickness.**

31. How many times larger is a pizza with a 9-inch radius than a pizza with a 6-inch radius?

32. How many times larger is a pizza with a radius of $2x$ inches than a pizza with a radius of $x$ inches?

33. What is the volume of an athletic shoe box $4\frac{1}{2}$ inches in height, 7 inches wide, and $13\frac{1}{2}$ inches long? How many cubic feet would it take to store 1000 such boxes?

34. What is the volume of a cereal box with depth 7 centimeters, width 21 centimeters, and height 30.5 centimeters? How many cubic meters are needed to store 1000 such boxes?

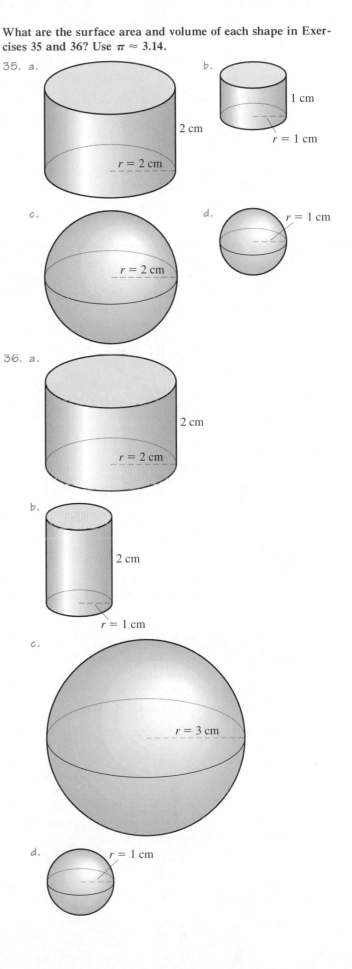

35. a.

2 cm

$r = 2$ cm

b.

1 cm

$r = 1$ cm

c.

$r = 2$ cm

d.

$r = 1$ cm

36. a.

2 cm

$r = 2$ cm

b.

2 cm

$r = 1$ cm

c.

$r = 3$ cm

d.

$r = 1$ cm

37. ***Moving Boxes.*** Adam Moving Company decides to double the dimensions (length, width, and height) of its book boxes so that movers can put more books in each box. The original book boxes were 1.5 feet on each side.

    a. Determine the new length, width, and height.

    b. What is the volume of the old box? ($V = l \cdot w \cdot h$)

    c. What is the volume of the new "double" box?

    d. If the original box held 30 pounds of books, what would the new box, full of books, weigh?

    e. What happened to the employee who thought of this idea?

38. ***Soup Cans.*** The marketing manager for the Bell Soup Company wanted more visibility on grocery shelves for the company's soup. He reasoned that doubling the radius and the height of the soup can would double the shelf space and make the product easier to see. The original can was a cylinder with diameter 7.5 centimeters and height 10.5 centimeters.

    a. What is the volume of the old can?

    b. Determine the diameter and height of the new can.

    c. What is the volume of the new can?

    d. How many times larger is the new can than the old can?

    e. The marketing manager recommended that the price also be doubled. Was he correct?

    f. If the old can sold for $0.69, what should be the price of the new can? Justify your answer.

    g. Do you think that the new can, priced as in part f, will sell as well as the old can?

39. ***Balloons Revisited***

    a. Complete the table by calculating volumes for the first two radii and copying the remaining volumes from Example 9.

| Radius of Sphere | Volume of Sphere |
|---|---|
| 1 | |
| 2 | |
| 3 | |
| 4 | |
| 5 | |

    b. Graph the data from the table.

    c. Draw a horizontal line at the volume that approximates full human lung capacity.

    d. Use the graph to predict the volume of a sphere with radius 4.5 inches.

    e. Use the graph to predict the volume of a sphere with radius 6 inches.

PROBLEM SOLVING

40. A rectangle has base $b$ and height $h$. When we draw the diagonals in the rectangle, four triangles are formed (see the figure). What is the area of each triangle? State any assumptions you make in finding the areas.

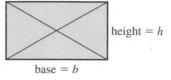

base = b, height = h

41. Suppose we wish to make a 400-meter Olympic track, using the shape in Example 5. What diameter is needed for the half-circles that are added on to the rectangular field of length 100 meters so that the perimeter of the track is 400 meters?

42. Evaluate the formula for the surface area of a sphere with the given radius, and then finish the sentence that follows in part d.

    a. $r = 1$

    b. $r = 2$

    c. $r = 4$

    d. If the radius of a sphere doubles, the surface area of the sphere _____ .

PROJECTS

43. ***Measuring Items***

    a. Match each of the seven units of measurement listed below with the appropriate one of these items: fabric, concrete, paint (when purchased), paint (when applied), rope, peat moss, carpeting.

      gallons or liters

      square yards or square meters

      cubic feet

      feet or meters

      square feet

      yards or meters

      cubic yards or cubic meters

    b. List two other items that are measured with each unit of measurement.

44. ***Body Surface Area.*** Body surface area estimates are particularly important in dosages for children's medication and in burn treatment. Body surface area charts, referencing both height and weight, are found in medical dosage calculation textbooks in prenursing programs. Books such as the American Medical Association's *Encyclopedia of Medicine* (© 1989) list percent of body area by part under "Burns." Using one of these sources or something similar, pose a body surface area problem and solve it.

45. ***Fixed Perimeter and Area.*** Cut a piece of string 30 inches long. Form it into a rectangle with width 1 inch.

a. Measure the length of the rectangle, and record it in the table below. Calculate the area of the rectangle.

| Width | Length | Area |
|-------|--------|------|
| 1 in. | | |
| 2 in. | | |
| 3 in. | | |
| 4 in. | | |
| 5 in. | | |
| 6 in. | | |
| | | |
| | | |

b. Form another rectangle with width 2 inches. Measure the length of the new rectangle, and record it. Calculate the area.

c. Repeat this procedure to complete the table. Add more rows until you reach a largest possible area.

d. Graph the length and width pairs, using length as $x$ and width as $y$.

e. Graph the length and area pairs, using length as $x$ and area as $y$.

f. Are there fractional widths and lengths that give an area larger than 56 square inches? How is this number related to the length of the string (30 inches)?

g. Where on the length and area graph is the point representing the largest area?

46. **Volume of Box.** Cut, if necessary, a piece of paper to 8 by $10\frac{1}{2}$ inches. Cut a $\frac{1}{2}$-inch square from each corner, as shown in the figure. Fold the paper on the dotted lines to make a box without a top.

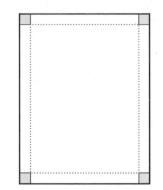

a. Measure the length, width, and height of the box. Place the data in the table, and calculate the volume of the box.

| Corner Square | Length (in.) | Width (in.) | Height (in.) | Volume (in.³) |
|---------------|--------------|-------------|--------------|---------------|
| $\frac{1}{2}$ | | | | |
| 1 | | | | |
| $1\frac{1}{2}$ | | | | |
| 2 | | | | |
| $2\frac{1}{2}$ | | | | |
| 3 | | | | |
| $3\frac{1}{2}$ | | | | |

b. Repeat the calculations for a 1-inch square cut from each corner, then a $1\frac{1}{2}$-inch square, and so forth, until the table is complete.

c. What is the largest volume of the box?

d. Draw a graph with the corner square's side length on the $x$-axis and the volume on the $y$-axis.

e. Is it possible to choose a different corner square that would yield a larger volume?

---

# 2.5  INEQUALITIES AND INTERVALS

**OBJECTIVES**   Compare two numbers using inequality signs. ◆ Write sets of numbers using inequalities. ◆ Write sets of numbers using intervals. ◆ Graph intervals and inequalities on the number line. ◆ Use inequalities and intervals to describe inputs in conditional situations.

**WARM-UP**   Complete the table by finding $n$ percent of each number or expression in the top row.

| $n$ percent | $1.00 | $5.00 | $10.00 | $x$ |
|-------------|-------|-------|--------|-----|
| 6% | | | | |
| 10% | | | | |
| 25% | | | | |
| 100% | | | | |
| 150% | | | | |

◆

I inequalities and intervals are special ways of writing sets of numbers. They are useful in describing input and output sets and, consequently, take on considerable importance in work with calculators and computers. We will return to the credit card payment schedule because its inputs are sets of numbers. We start with inequalities, go on to intervals, and then relate the two with line graphs.

### INEQUALITIES

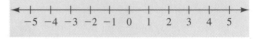

Figure 11

An **inequality** *is a statement that one quantity is greater than or less than another quantity.* Inequality symbols are used to compare the relative position of two numbers on a number line (see Figure 11) or the relative size of two or more expressions.

| Inequality Symbol | Meaning |
|---|---|
| < | is less than (left of another number on the number line) |
| > | is greater than (right of another number on the number line) |
| ≤ | is less than or equal to |
| ≥ | is greater than or equal to |

E X A M P L E    1

Compare the relative position of these numbers on a number line by identifying which sign(s) make true statements when placed between them.

**a.** $3 \_\_ -5$      **b.** $-5 \_\_ 2$      **c.** $-2 \_\_ -2$

SOLUTION

**a.** $3 > -5$ or $3 \geq -5$; 3 is to the right of $-5$ on the number line. Thus, 3 is greater than $-5$.

**b.** $-5 < 2$ or $-5 \leq 2$; $-5$ is to the left of 2 on the number line. Thus, $-5$ is less than 2.

**c.** $-2 = -2$, $-2 \leq -2$, or $-2 \geq -2$; $-2$ is equal to itself. Thus, any symbol that includes equality is correct. The *or* in the meanings of $\leq$ and $\geq$ allows the symbols to be placed between equal statements.

♦ ♦ ♦ ♦ ♦ ♦ ♦ ♦ ♦ ♦ ♦ ♦ ♦ ♦ ♦ ♦ ♦ ♦ ♦ ♦ ♦ ♦ ♦ ♦ ♦ ♦ ♦ ♦ ♦ ♦ ♦ ♦ ♦ ♦ ♦ ♦ ♦

We use a variable with an inequality to obtain sets of numbers. Suppose we want to describe the set of input numbers from $0 to $20 in the credit card example. If $x$ represents possible inputs, then $0 \leq x$ says 0 is smaller than or equal to $x$. Writing $x \leq 20$ indicates $x$ is smaller than or equal to 20. We write these two inequalities together in one statement as a *continued inequality*, $0 \leq x \leq 20$. This statement is read as all the numbers between 0 and 20, including 0 and 20.

E X A M P L E    2

*Credit card payments*    Use inequalities to describe the inputs of the credit card payment schedule in Table 7.

| Input: Charge Balance | Output: Payment Due |
|---|---|
| $0 to $20 | Full amount |
| $20.01 to $500 | 10% or $20, whichever is greater |
| $500.01 or more | $50 plus the amount in excess of $500 |

*Table 7* *Credit Card Payment Schedule*

SOLUTION    Table 8 shows inequalities replacing the inputs in Table 7. If we want to include both 0 and 20, we write the inequality $0 \le x \le 20$. (The inequalities with a line under them, $\le$ and $\ge$, include the endpoint.) We read this inequality as "the set of numbers between zero and twenty, inclusive."

The inequality $20.01 \le x \le 500$ is the set of numbers between 20.01 and 500, inclusive.

The inequality $x \ge 500.01$ is all the numbers greater than or equal to 500.01.

| Input, $x$: Charge Balance (dollars) | Output: Payment Due |
|---|---|
| $0 \le x \le 20$ | Full amount |
| $20.01 \le x \le 500$ | 10% or $20, whichever is greater |
| $x \ge 500.01$ | $50 plus the amount in excess of $500 |

*Table 8 Credit Card Payment Schedule*

♦ ♦ ♦ ♦ ♦ ♦ ♦ ♦ ♦ ♦ ♦ ♦ ♦ ♦ ♦ ♦ ♦ ♦ ♦ ♦ ♦ ♦ ♦ ♦ ♦ ♦ ♦ ♦ ♦ ♦ ♦ ♦ ♦ ♦ ♦ ♦

The inequalities in Table 8 are satisfactory in business and finance, but not in mathematics. There are number gaps between the given input sets. Between $20 and $20.01 are all the fraction and decimal portions of one cent: $20.00$\frac{1}{2}$, $20.005, or even $20.009. There is another gap between $500 and $500.01. In mathematics, we prefer to include all numbers in our intervals, leaving no gaps.

EXAMPLE    3    *Credit card payments, continued*    Use inequalities to describe the inputs of the credit card payment schedule in Table 8 without any gaps between the input inequalities.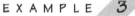

SOLUTION    The first inequality, $0 \le x \le 20$, remains the same.

The second inequality changes, because we start at 20 but do not wish to include 20. We do, however, want to include 500. The continued inequality becomes $20 < x \le 500$, which excludes 20 but includes 500.

The third inequality also changes, because we eliminate the gap between 500 and 500.01. The inequality becomes $x > 500$, which excludes the endpoint 500. The results are summarized in Table 9, Example 5.

♦ ♦ ♦ ♦ ♦ ♦ ♦ ♦ ♦ ♦ ♦ ♦ ♦ ♦ ♦ ♦ ♦ ♦ ♦ ♦ ♦ ♦ ♦ ♦ ♦ ♦ ♦ ♦ ♦ ♦ ♦ ♦ ♦ ♦ ♦ ♦

EXAMPLE    4    What set of numbers is described by each inequality? Write each in words.

**a.** $-2 < x \le 4$    **b.** $-3 \le x < 4$    **c.** $x < 4$    **d.** $x \ge -2$

SOLUTION    **a.** The inequality $-2 < x \le 4$ describes the set of numbers between $-2$ and 4. It excludes the left endpoint, $-2$, and includes the right endpoint, 4.

**b.** The inequality $-3 \le x < 4$ describes the set of numbers between $-3$ and 4. It includes the left endpoint, $-3$, and excludes the right endpoint, 4.

**c.** The inequality $x < 4$ is the set of all numbers smaller than 4. The set of numbers decreases without bound to the left of 4 on the number line.

**d.** The inequality $x \ge -2$ is the set of all numbers larger than or equal to $-2$. The set of numbers increases without bound to the right of $-2$ on the number line.

♦ ♦ ♦ ♦ ♦ ♦ ♦ ♦ ♦ ♦ ♦ ♦ ♦ ♦ ♦ ♦ ♦ ♦ ♦ ♦ ♦ ♦ ♦ ♦ ♦ ♦ ♦ ♦ ♦ ♦ ♦ ♦ ♦ ♦ ♦ ♦

## INTERVALS

We may also state our inequalities as intervals. We used intervals when we first described sets of inputs in Section 1.2. Recall that an **interval** *is a set containing all the numbers between its endpoints as well as one endpoint, both endpoints, or neither endpoint.*

Intervals have several advantages over inequalities: They are shorter to write, they show just the endpoints, and they indicate inclusion or exclusion of endpoints by the use of brackets or parentheses. The brackets, [ ], around an interval tell where to start and end. They also indicate that the endpoints are included in the set. Parentheses, ( ), are used when endpoints are excluded from the set. We may mix brackets and parentheses in one interval.

EXAMPLE **5**

**Credit card payments revisited**    Use intervals to describe the inputs of the credit card payment schedule in Table 9, without any gaps between the input intervals.

| Input, $x$: Charge Balance (dollars) | Output: Payment Due |
|---|---|
| $0 \le x \le 20$ | Full amount |
| $20 < x \le 500$ | 10% or $20, whichever is greater |
| $x > 500$ | $50 plus the amount in excess of $500 |

*Table 9*   Credit Card Payment Schedule

SOLUTION

To *include* both 0 and 20, we write $x$ in [0, 20]. The brackets, [ ], include the endpoints.

To describe the set of numbers from 20 to 500 but not including 20, we write $x$ in (20, 500]. We *exclude* the endpoint 20 by using a parenthesis.

The last interval, $x$ in (500, $+\infty$), describes all numbers larger than 500. There is no greatest number, so we need a symbol to say that the numbers get large without bound. We use an **infinity sign**, $\infty$. Realistically, the credit card company objects to this much spending and puts a limit, say $1000, on the card. The intervals are summarized in Table 10.

| Input, $x$: Charge Balance (dollars) | Output: Payment Due |
|---|---|
| $x$ in [0, 20] | Full amount |
| $x$ in (20, 500] | 10% or $20, whichever is greater |
| $x$ in (500, $+\infty$) | $50 plus the amount in excess of $500 |

*Table 10*   Credit Card Payment Schedule

◆ ◆ ◆ ◆ ◆ ◆ ◆ ◆ ◆ ◆ ◆ ◆ ◆ ◆ ◆ ◆ ◆ ◆ ◆ ◆ ◆ ◆ ◆ ◆ ◆ ◆ ◆ ◆ ◆ ◆ ◆ ◆ ◆

**Infinite** *means without bound.* Infinity describes a concept, not a number. The axes on our coordinate graph and the number lines all have arrows on their ends because the lines go on without bound. The numbers on the number line go to the right or left forever. Placing a positive sign before the infinity sign means infinite to the right on the number line; a negative sign indicates infinite to the left (see Figure 12).

We generally write inequalities with the smallest number on the left. Use of the greater than symbol places the larger number on the left and is sometimes confusing or hard to read. We *always* write intervals with the smallest number on the left.

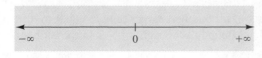

*Figure 12*

Figure 13

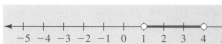

Figure 14

LINE GRAPHS

It is sometimes useful to draw a picture of an interval or inequality. We use a line graph to do so. The graph of the inequality $1 \le x \le 4$ or interval [1, 4] is a number line with dots at 1 and 4 and a line segment connecting them (Figure 13). (Using brackets on the number line instead of dots is also acceptable.)

The graph of $1 < x < 4$ or its interval (1, 4) has *small* (*open*) *circles* as endpoints at 1 and 4 and a line segment connecting them (Figure 14). (Using parentheses instead of the small circles is also acceptable.)

E X A M P L E   **6**

Write each inequality as an interval, and make a line graph.

**a.** $-2 < x \le 4$      **b.** $-3 \le x < 4$

SOLUTION   **a.** The inequality $-2 < x \le 4$ is written $(-2, 4]$. It describes the numbers between $-2$ and 4, with only the endpoint 4 included. Both endpoint options for the line graph are shown in Figure 15.

**b.** The inequality $-3 \le x < 4$ is written $[-3, 4)$. It describes the numbers between $-3$ and 4, with only the endpoint $-3$ included. Both endpoint forms of the line graph are shown in Figure 16.

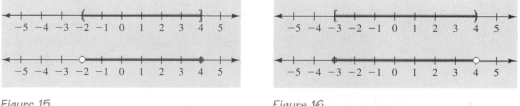

Figure 15                                    Figure 16

♦ ♦ ♦ ♦ ♦ ♦ ♦ ♦ ♦ ♦ ♦ ♦ ♦ ♦ ♦ ♦ ♦ ♦ ♦ ♦ ♦ ♦ ♦ ♦ ♦ ♦ ♦ ♦ ♦ ♦ ♦ ♦ ♦

Expressions such as $x \ge -2$ may be awkward to read. It is not always obvious whether the graph is to the left or right of the number. To be sure that we have the line graph drawn correctly, we select a *test point* on the number line. If the test point makes the inequality true, then the graph should pass through the point. If the test point makes the inequality false, then the graph goes in the opposite direction.

---

Zero is a convenient test point.

---

E X A M P L E   **7**

Write each inequality as an interval, and make a line graph. Use a test point to check the graph.

**a.** $x < 4$      **b.** $x \ge -2$

SOLUTION   **a.** The inequality $x < 4$ is the set of all numbers smaller than 4 and describes the interval $(-\infty, 4)$. The test point 0 gives $0 < 4$, which is true. Thus the graph goes through 0, as shown in Figure 17. Either the parenthesis or the small circle on the 4 excludes 4 from the graph.

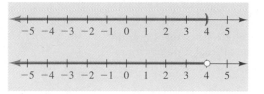

Figure 17

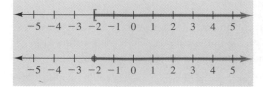

**Figure 18**

**b.** The inequality $x \geq -2$ is the set of all numbers larger than or equal to $-2$, and the interval is $[-2, +\infty)$. The test point 0 gives $0 \geq -2$, which is true. The graph goes through 0, as shown in Figure 18. The dot or bracket at $-2$ shows the inclusion of $-2$ in the set.

♦ ♦ ♦ ♦ ♦ ♦ ♦ ♦ ♦ ♦ ♦ ♦ ♦ ♦ ♦ ♦ ♦ ♦ ♦ ♦ ♦ ♦ ♦ ♦ ♦ ♦ ♦ ♦ ♦ ♦ ♦ ♦ ♦ ♦ ♦ ♦ ♦ ♦

In reading the summary chart in Table 11, observe the use of parentheses with $<$, $>$, or the infinity sign ($\infty$) and the use of brackets with $\leq$ or $\geq$.

| Symbol | Meaning | Graph Notation | Interval Notation |
|--------|---------|----------------|-------------------|
| $<$ | is less than | small circle or ( ) | ( ) |
| $>$ | is greater than | small circle or ( ) | ( ) |
| $=$ | is equal to | dot | |
| $\leq$ | is less than or equal to | dot or [ ] | [´ ] |
| $\geq$ | is greater than or equal to | dot or [ ] | [ ] |
| $+\infty$ | positive infinity | $\rightarrow$ | , $+\infty$) |
| $-\infty$ | negative infinity | $\leftarrow$ | $(-\infty,$ |

**Table 11** *Symbols Used in Inequalities and Intervals*

CAUTION   Interval notation may look like the coordinates of a point. Read carefully when you see (1, 4) to determine whether the reference is to a coordinate point $(x, y) = (1, 4)$ or an interval (1, 4) describing the set $1 < x < 4$.   ♦

## EXERCISES 2.5

**In Exercises 1 and 2 what is the correct sign between each pair of numbers? Use $<$, $=$, or $>$.**

1. a. $-8$ ___ $-3$     b. $+4$ ___ $-9$

   c. $(-3)^2$ ___ $3^2$     d. $0.5$ ___ $0.5^2$

   e. $6$ ___ $-5$     f. $(-2)(6)$ ___ $(-2)(-5)$

   g. $-6$ ___ $-5$     h. $(-2)(-6)$ ___ $(-2)(-5)$

2. a. $-7$ ___ $5$     b. $2(-7)$ ___ $2(5)$

   c. $-2(-7)$ ___ $-2(5)$     d. $0.2^2$ ___ $0.2$

   e. $1.5$ ___ $1.5^2$     f. $\frac{3}{4}$ ___ $\left(\frac{3}{4}\right)^2$

   g. $3(-7)$ ___ $3(-6)$     h. $-3(-7)$ ___ $-3(-6)$

**For Exercises 3 to 6 find a number that makes each statement true and another that makes each statement false.**

3. $-x > x$   4. $x^2 > x$   5. $x > x^2$   6. $\dfrac{1}{x} > x$

7. Why is $2 \leq 2$ a true statement?

8. Why is $3 \geq 3$ a true statement?

**Translate the conditional inputs in Exercises 9 to 14 into inequalities and into intervals. Assume all real numbers (numbers from $-\infty$ to $+\infty$) may be used. Do not leave any gaps between your intervals.**

9. a. Not over 2000     b. Over 2000 but not over 5000

   c. Over 5000

10. a. 10 or less     b. Greater than 10 and less than 50

    c. 50 or more

11. a. $-5$ or less     b. Greater than $-5$ and less than 5

    c. 5 or larger

12. a. Less than $-1$     b. $-1$ to 1

    c. Larger than $+1$

13. a. Less than five     b. Five to fifty

    c. Larger than fifty

14. a. Less than eighteen

    b. Eighteen and above but less than twenty-one

    c. Twenty-one and above

**Choose one listed inequality and one listed interval that describe each line graph in Exercises 15 to 20.**

| Inequality | Interval |
|------------|----------|
| a. $-3 < x < 3$ | p. $(-\infty, -3)$ |
| b. $-3 \leq x \leq 3$ | q. $(-\infty, 3]$ |
| c. $x \geq -3$ | r. $[-3, 3]$ |
| d. $x > -3$ | s. $(-\infty, 3)$ |
| e. $x \geq 3$ | t. $[-3, +\infty)$ |
| f. $x < 3$ | u. $(-3, 3)$ |
| g. $x > 3$ | v. $(-3, +\infty)$ |
| h. $x \leq 3$ | w. $[3, +\infty)$ |

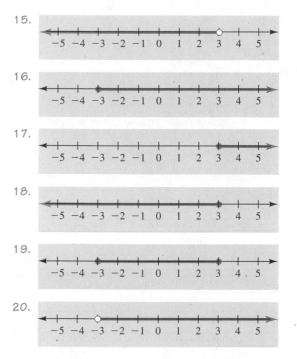

15.

16.

17.

18.

19.

20.

**In Exercises 21 to 24, make a table with integer inputs on the interval $[-5, 5]$; and graph the results on coordinate axes.**

21. If $x \leq 3$, output is $3 - x$. If $x > 3$, output is $x - 3$.

22. If $x < -3$, output is $-x - 3$. If $x \geq -3$, output is $x + 3$.

23. If $x < -2$, output is $x + 2$. If $x \geq -2$, output is $-x - 2$.

24. If $x < \frac{1}{2}$, output is $2x - 1$. If $x > \frac{1}{2}$, output is $1 - 2x$.

### GRAPHING CALCULATOR

**In Exercises 25 to 28 use a graphing calculator to compare the choices with the indicated exercise.**

25. Which of these absolute value equations is another name for the graph in Exercise 21?

   a. $y = |x + 3|$    b. $y = -|x - 3|$    c. $y = |x - 3|$

26. Which of these absolute value equations is another name for the graph in Exercise 22?

   a. $y = |x + 3|$    b. $y = -|x - 3|$    c. $y = |3 - x|$

27. Which of these absolute value equations is another name for the graph in Exercise 23?

   a. $y = -|x + 2|$    b. $y = |x - 2|$    c. $y = |x + 2|$

28. Which of these absolute value equations is another name for the graph in Exercise 24?

   a. $y = -|2x - 1|$    b. $y = |2x - 1|$    c. $y = |1 + 2x|$

### CONDITIONAL RULES

**In Exercises 29 and 30**

   a. Make an input–output table with six entries (at least two for each condition).

   b. Use the input–output table to help draw a conditional graph.

   c. Describe the inputs for each condition using both an interval and an inequality.

29. Oregon income tax, 1991, for single taxpayer:

| If your taxable income is | Your tax is |
|---|---|
| Not over $2000 | 5% of taxable income |
| Over $2000 but not over $5000 | $100 plus 7% of excess over $2000 |
| Over $5000 | $310 plus 9% of excess over $5000 |

30. Federal income tax, 1991, for single taxpayer:

| If your taxable income is | Your tax is |
|---|---|
| $0 to $20,350 | 15% of taxable income |
| Over $20,350 but not over $49,300 | $3052.50 + 28% of amount over $20,350 |
| Over $49,300 | $11,158.50 + 31% of amount over $49,300 |

### PROJECTS

31. *Credit Card Payments.* Find a current credit card payment schedule. Determine whether the schedule is conditional. If it is conditional, write the input conditions in both interval and inequality form.

32. *Utility Payments.* Call a local utility (electricity or water) and obtain a current cost schedule. Determine whether the cost is conditional. If it is conditional, write input conditions in both interval and inequality form.

33. *Income Tax.* Find a current income tax schedule. Determine whether the tax rates are conditional. If they are conditional, write the input conditions in both interval and inequality form.

## CHAPTER 2 SUMMARY

### Vocabulary

*For definitions and page references, see the Glossary/Index.*

| | | | |
|---|---|---|---|
| absolute value | composite numbers | infinite | prime numbers |
| area | evaluate | infinity sign | reciprocal |
| associative properties | exponent | interval | simplification property of fractions |
| base | factors | inverse | simplify |
| circumference | grouping symbols | opposite | surface area |
| commutative properties | inequality | order of operations | unit analysis |
| | | perimeter | volume |
| | | pi | |

## Properties of Operations

(The properties marked * are new and are added at this time to provide a more complete listing.)

**Associative property of addition:** $(a + b) + c = a + (b + c)$

**Associative property of multiplication:** $(a \cdot b) \cdot c = a \cdot (b \cdot c)$

**Commutative property of addition:** $a + b = b + a$

**Commutative property of multiplication:** $a \cdot b = b \cdot a$

*The sum of a number $n$ and 0 is $n$.

The sum of a number $n$ and its opposite, $-n$, is 0.

Addition and subtraction are **inverse** operations.

Subtraction is equivalent to adding the **opposite** (additive inverse):
$$a - b = a + (-b)$$
$$a - (-b) = a + b$$

*The product of a number $n$ and 1 is $n$.

*The product of a number $n$ and 0 is 0.

Multiplication and division are **inverse** operations.

Division by $n$ is equivalent to multiplication by the **reciprocal**, $\dfrac{1}{n}$.

The order of operations for algebra and scientific calculators:

1. Calculate expressions in parentheses and other **grouping symbols** first.
2. Calculate exponents, roots, and absolute value expressions next.
3. Do remaining multiplication and division, left to right.
4. Do remaining addition and subtraction, left to right.

## Operations with Signed Numbers

If signs are alike, add the numbers and place the common sign on the answer.

If the signs are different, subtract the number portion and place the sign of the number farthest from zero on the answer.

Change subtraction to addition by adding the opposite.

In multiplication and division of two real numbers, if the signs are alike, the answer is positive. If the signs are different, the answer is negative.

## Selected Geometric Formulas

See Table 5 and Table 6.

---

## CHAPTER 2 REVIEW EXERCISES

1. a. Complete the table.

   b. Graph the data points.

   c. Explain how you determined the numbers to place on the axes of the graph.

   | Input $x$ | Output $-x - 2$ |
   |-----------|-----------------|
   | $-2$ | 0 |
   | $-1$ | $-1$ |
   | $0$ | $-2$ |
   | $1$ | $-3$ |
   | $2$ | $-4$ |

2. Add or subtract, as indicated.

   a. $-4 + 9$   b. $-5 + 11$   c. $5 + (+6)$

   d. $-3 - (-7)$   e. $-1.0 + 0.6$   f. $-1.3 + 0.8$

   g. $2.6 - (-1.3)$   h. $-1.9 - 4.7$   i. $-0.3 + (-0.4)$

   j. $-0.7 + (-1.3)$   k. $-\frac{1}{4} + \left(-\frac{3}{4}\right)$   l. $\frac{5}{3} + \left(-\frac{2}{3}\right)$

   m. $-\frac{2}{3} - \left(-\frac{3}{4}\right)$   n. $-\frac{4}{5} + \frac{5}{3}$

3. Multiply or divide, as indicated.

   a. $(-9)(6)$   b. $(-9)(-6)$   c. $(-18)(-3)$

   d. $(-18)(3)$   e. $(-8)(-7)$   f. $8(-7)$

   g. $4 \cdot (-14)$   h. $(-4)(-14)$   i. $(-48) \div (-24)$

   j. $(-48) \div (12)$   k. $48 \div (-6)$   l. $(-48) \div (-6)$

   m. $-48 \div (-3)$   n. $-48 \div 3$

4. Multiply or divide, as indicated.

   a. $-1.0(0.6)$   b. $(-1.3)(-0.8)$

   c. $(2.6) \div (-1.3)$   d. $(-1.7) \div (5.1)$

   e. $-0.3(-0.4)$   f. $(0.7) \div (-1.4)$

   g. $\left(-\frac{1}{4}\right)\left(-\frac{3}{4}\right)$   h. $\left(\frac{5}{3}\right)\left(-\frac{2}{3}\right)$

   i. $\left(-\frac{2}{3}\right) \div \left(-\frac{4}{3}\right)$   j. $\left(-\frac{4}{5}\right) \div \left(\frac{5}{3}\right)$

5. Simplify.

   a. $-(-2)^2$   b. $4 - (-2) + (-2)^2$

   c. $5 - (-3) + (-2)^2$   d. $-(-3)^2$

   e. $\sqrt{(3^2 + 4^2)}$   f. $\sqrt{(8^2 + 6^2)}$

   g. $\sqrt{(25^2 - 20^2)}$   h. $(ab)^2$

   i. $\sqrt{(1.5^2 + 2^2)}$   j. $(2ab^2)^2$

   k. $(-ab)^2$   l. $\sqrt{(15^2 - 12^2)}$

   m. $(-2ab^2)^2$   n. $\sqrt{(10^2 - 6^2)}$

   o. $\left(\dfrac{4x}{y}\right)^2$   p. $\left(\dfrac{x}{3y}\right)^2$

   q. $m^4 m^5$   r. $m^2 m^7$

   s. $m^5 \div m^2$   t. $m^7 \div m^4$

6. Make an input–output table for each of the following equations, with $x$ as integers in the interval $[-2, 3]$. Graph the $(x, y)$ pairs on coordinate axes.

   a. $y = (-x)^2$   b. $y = -(x^2)$

   c. $y = |x - 2|$   d. $y = |x| - 2$

7. Simplify these expressions in your head. Describe any shortcuts or properties used.

   a. $1 + 2 + 3 + 4 + 5$

   b. $2 + 4 + 6 + 8 + 10 + 12 + 14 + 16 + 18 + 20$

   c. $(-10) + (-8) + (-6) + (-4) + (-2)$

   d. $4 + 8 + 12 + 16 + 20$

8. Which property allows the rearranging of numbers in an addition or multiplication problem?

9. Which property allows the choice of convenient groupings of numbers in an addition or multiplication problem?

10. Simplify. Round decimals to nearest thousandth.

    a. $6 - (-2) + 4 \cdot 3 + 7$

    b. $24 \div 3 \cdot 2 - 2 + 5 \cdot 6$

    c. $\dfrac{-5 - \sqrt{5^2 - 4(2)(-12)}}{2(2)}$

    d. $\dfrac{4 + \sqrt{8}}{4}$

11. Discuss the difference between pages per minute and minutes per page in complete sentences. Give at least two situations where pages per minute is appropriate and two situations where minutes per page is appropriate.

12. An avalanche travels up to 200 miles per hour.

    a. How many yards will the avalanche travel in 10.4 seconds? (This was once a world record for the 100-yard dash.)

    b. How many seconds would it take the avalanche to travel the length of a city block (500 feet)?

    You will need to list your own facts and convert them into a fractional form in order to do the unit analysis. Round answers to nearest tenth.

13. Evaluate these formulas for $r = 1, 2, 4,$ and $8$. Summarize your work in a table, with $r$ as input and $V$ as output.

    a. Volume of a sphere: $V = \frac{4}{3}\pi r^3$

    b. Volume of a cylinder: $V = \pi r^2 h$; let $h = 3$.

14. Evaluate these formulas for $h = 1, 2, 4,$ and $8$. Summarize your work in a table, with $h$ as input and $A$ as output.

    a. Area of a trapezoid: $A = \frac{1}{2}h(a + b)$; let $a = 5$ and $b = 8$.

    b. Area of a triangle: $A = \frac{1}{2}bh$; let $b = 8$.

15. Explain the role of inverse operations in subtraction of negatives and division of fractions.

16. Write three sentences that illustrate the difference in meaning among *base of triangle*, *base of exponent*, and any other use of *base*.

17. Write five sentences that illustrate the difference in meaning among *set of points*, *set of numbers*, *set of rules*, *set of line segments*, and *set of axes*.

18. Each row in the table below contains equivalent statements. Fill in the blanks for each row.

| Number Line | Inequality | Interval | Words |
|---|---|---|---|
| | $x \leq 5$ | | $x$ less than or equal to 5 |
| | $-3 \leq x < 4$ | | |
| | | $[-3, 5]$ | |
| | | $(2, 4)$ | $x$ between 2 and 4 |
| | | $(-5, 0]$ | |
| −5  0 | | | $x$ greater than −5 |
| −2  0      6 | | | |
| 0 1 | | | |
| | | *x>0* | $x$ is positive |
| | | *x<0* *−∞,0* | $x$ is negative |

19. Write the inputs for the table using inequalities, intervals, and number lines.

| Input: Cost of Item | Output: Discount |
|---|---|
| Less than $50 | 2% off price tag |
| $50 to $500 | 5% off price tag |
| Over $500 | 10% off price tag |

$x < 50$

$(\infty, 50)$

20. Describe how the commutative and associative properties of addition can be used to simplify the addition of the numbers 1 to 100, allowing you to solve it in your head.

21. If we let the top parallel line, *a*, of a trapezoid be zero, then what happens to the formula for area, $A = \frac{1}{2}h(a + b)$? Why might this result make sense?

---

## CHAPTER 2 TEST

1. Complete the input–output table. Graph the data points, and connect them.

| $x$ | $|x - 2|$ |
|---|---|
| $-2$ | |
| $0$ | |
| $2$ | |
| $4$ | |

2. Make an input–output table for $y = -2x + 3$. Let $y =$ output. Graph the data.

3. What property allows us to change $1 + 4 + 9 + 16 + 25$ to $1 + 9 + 4 + 16 + 25$?

4. What property allows us to add the problem in Exercise 3 as $(1 + 9) + (4 + 16) + 25$ instead of following the usual left-to-right order in addition?

5. Simplify.

   a. $-5 + 9$     b. $-1.4 + 2.5 - 3.6$

   c. $-4 - (-3)$     d. $(-3)(4)(-5)$

   e. $8 - (-3)^2$     f. $\sqrt{(26^2 - 24^2)}$

   g. $m^2 m^9$     h. $m^7 \div m^3$

   i. $36 \div 2 \cdot 2 - 3 + (3^2 - 5)$

6. Are $x^2$ and $-x^2$ the same? Explain why or why not in complete sentences.

7. The area formula for a circle is $A = \pi r^2$. If the radius of the circle is multiplied by 10, how many times larger is the resulting area than $\pi r^2$?

8. Concrete is sold by the cubic yard. We wish to make a sidewalk 40 inches wide, 8 inches thick, and 20 feet long. How many cubic yards of concrete will be needed?

9. A party punch is to be made from cranberry juice and ginger ale. So as to not dilute the punch, the ice cubes will be made of orange juice. Determine how many ice cube trays are needed for each can of orange juice.

   1 cup = 8 fluid ounces
   2 cups = 1 ice cube tray
   1 can orange juice = 12 fluid ounces
   Use 3 cans of water with 1 can of orange juice.

10. The oval shape shown is called an *ellipse*. The formula for the area of an ellipse is $A = \pi r R$.

   a. Find the area of an ellipse with $R = 4$ and $r = 3$.

   b. If $r = R$, what shape is created? Is the formula for the area of an ellipse consistent with the shape where $r = R$? Explain in complete sentences.

**In Exercises 11 and 12, complete the table.**

11.

|  | Inequality | Interval |
|---|---|---|
| 0 to 20 | | |
| More than 20 and less than 50 | $20 < x < 50$ | $(20, 50)$ |
| 50 or greater | | |

12.

| $x$ | $y$ | $x + y$ | $x - y$ | $x \cdot y$ | $x \div y$ |
|---|---|---|---|---|---|
| $-8$ | $4$ | | | | |
| $-6$ | | $-8$ | | | |
| | | | | $9$ | $-18$ |

C
H
A
P
T
E
R

T
H
R
E
E

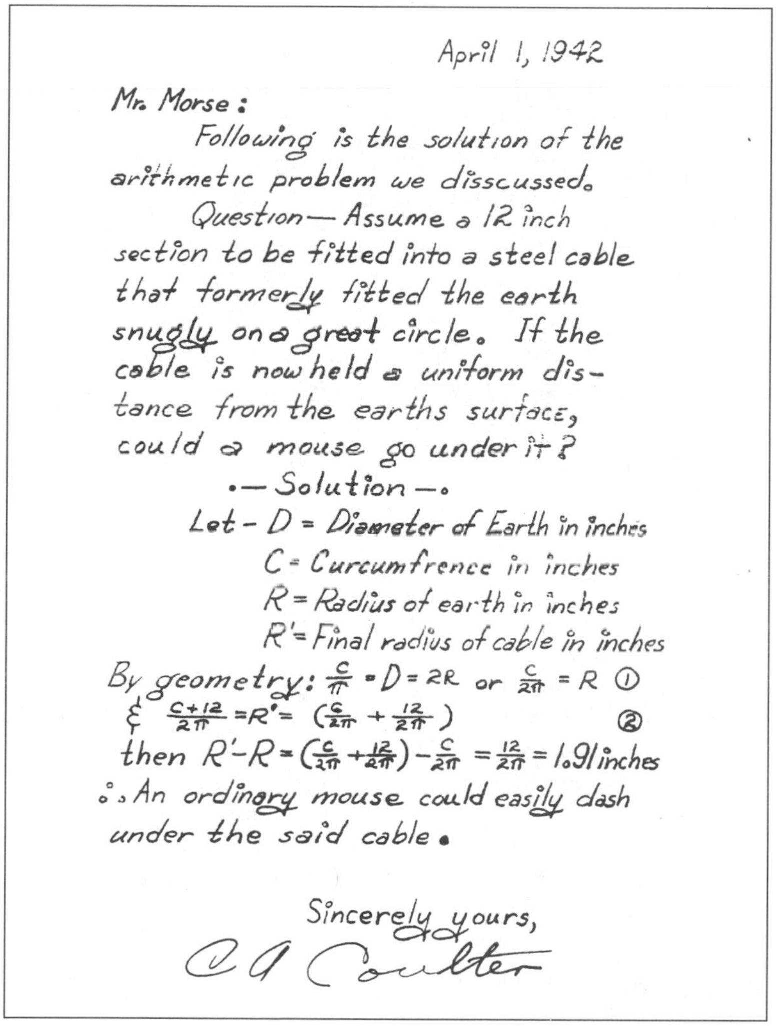

April 1, 1942

Mr. Morse:

Following is the solution of the arithmetic problem we disscussed.

Question— Assume a 12 inch section to be fitted into a steel cable that formerly fitted the earth snugly on a great circle. If the cable is now held a uniform distance from the earths surface, could a mouse go under it?

•— Solution —•

Let - D = Diameter of Earth in inches

C = Curcumfrence in inches

R = Radius of earth in inches

R'= Final radius of cable in inches

By geometry: $\frac{C}{\pi} = D = 2R$ or $\frac{C}{2\pi} = R$ ①

& $\frac{C+12}{2\pi} = R' = \left(\frac{C}{2\pi} + \frac{12}{2\pi}\right)$ ②

then $R'-R = \left(\frac{C}{2\pi} + \frac{12}{2\pi}\right) - \frac{C}{2\pi} = \frac{12}{2\pi} = 1.91$ inches

∴ An ordinary mouse could easily dash under the said cable.

Sincerely yours,

C. A. Coulter

**Figure 1** *Letter from C. A. Coulter*

**DOES THE WRITER** of the letter in Figure 1 lose credibility because of the spelling errors? We will work with formulas in Section 3.5 to determine whether this letter accurately describes the solution to the problem.

SOLVING EQUATIONS WITH TABLES AND GRAPHS

OBJECTIVES   Translate word sentences into equations.  ♦  Find the solution to an equation from a table.  ♦  Find the solution to an equation from a graph.  ♦  Identify graphs forming a straight line as linear.  ♦  Use tables and graphs of linear quantities to predict inputs and outputs.  ♦  Interpret the meaning of the intersection of the graph with the axes. ♦  Identify independent and dependent variables.

WARM-UP   **1.** Complete the table for the given inputs.

| $x$ | $2 - 6x$ | $4x + 1$ | $3x - 2$ |
|-----|----------|----------|----------|
| $-1$ |          |          |          |
| $0$ |          |          |          |
| $1$ |          |          |          |
| $2$ |          |          |          |

**2.** What is the rule describing the number of squares in the pattern below? Let $n$ be the number of the figure in the pattern sequence.

♦

I n this section we return to the rules that tell us how to get from the input to the output. We state our rules with equations. We use tables and graphs to solve equations. Some of these equations will be formed by matching an expression with a particular output, $y$. Other equations will come from application settings. Because much of introductory algebra involves linear equations, we will close with two linear equation applications and related vocabulary. In the next section we will solve linear equations by using formal symbolic operations.

EQUATIONS

An **equation** *is a statement of equality between two expressions.* The rule relating inputs, $x$, and outputs, $y$, may be a single equation that holds for all possible inputs, as in each column of the table in the Warm-up: $y = 2 - 6x$, $y = 4x + 1$, or $y = 3x - 2$. If $y$ describes an output and $x$ describes an input, then the rule *the output is 2 less than 3 times the input* has the equation $y = 3x - 2$. This is the rule for the pattern in the figure in the Warm-up. How is it the same and yet different from the third column of the table?

The rule relating inputs and outputs may be several equations, conditional on the inputs, as in the Chevron credit card payment schedule repeated in Example 1.

EXAMPLE 1

Credit card payments    Use the output descriptions in Table 1 to write equations. Let $x$ represent the inputs in dollars and $y$ represent the outputs in dollars.

| Input, x: Charge Balance | Output, y: Payment Due |
|---|---|
| [0, 20] | Full amount |
| (20, 500] | 10% or $20, whichever is greater |
| (500, +∞) | $50 plus the amount in excess of $500 |

Table 1 *Credit Card Payment Schedule*

SOLUTION

**a.** The equations are shown in Table 2. The phrase *full amount* means the payment due exactly equals the charge balance. Because output equals input, the equation is $y = x$.

**b.** Ten percent means 0.10 times the input, or $y = 0.10x$. Twenty dollars as output means $y = \$20$, regardless of the input in this interval. The two outputs need to be compared and the highest output paid.

**c.** The rule for the last set of inputs requires subtracting $500 from the input and adding $50. The output equation is $y = (x - \$500) + \$50$.

| Input, x: Charge Balance | Output, y: Payment Due |
|---|---|
| [0, 20] | $y = x$ |
| (20, 500] | $y = 0.10x$ or $y = \$20$, whichever is greater |
| (500, +∞) | $y = (x - \$500) + \$50$ |

Table 2 *Credit Card Payment Schedule*

♦ ♦ ♦ ♦ ♦ ♦ ♦ ♦ ♦ ♦ ♦ ♦ ♦ ♦ ♦ ♦ ♦ ♦ ♦ ♦ ♦ ♦ ♦ ♦ ♦ ♦ ♦ ♦ ♦ ♦ ♦ ♦ ♦ ♦ ♦ ♦ ♦

Two items in the Example 1 equations are noteworthy. First, we write percents in decimal form in equations as well as in expressions. Second, the parentheses are not needed in the third equation, but they serve to remind us to find the *excess* amount first.

In problem situations look for the output, $y$, depending on the input, $x$. Write equations so that $y$ *depends on x.*

EXAMPLE 2

Write an equation that describes each of these problem situations.

**a.** What is the total cost of bulk rice at $0.89 per pound with a $0.50 deposit on a reusable container that holds up to 10 pounds?

**b.** What is the total cost of tuition at $24 per credit hour plus $10 in fees?

**c.** What is the sales tax on a purchase in a city where taxes are 6% of the price?

SOLUTION

In each equation $x$ is the input and $y$ is the output.

**a.** Cost depends on pounds purchased, so pounds are the input, $x$. The total cost is $y = 0.89x + 0.50$ if we use dollars or $y = 89x + 50$ if we use cents.

**b.** Cost depends on the credits taken, so credits are the input, $x$. The total tuition, in dollars, is $y = 24x + 10$.

**c.** Tax depends on price, so price is the input, $x$. Tax, in dollars, is $y = 0.06x$. The phrase *of the price* reminds us to multiply the 6% by the input $x$.

♦ ♦ ♦ ♦ ♦ ♦ ♦ ♦ ♦ ♦ ♦ ♦ ♦ ♦ ♦ ♦ ♦ ♦ ♦ ♦ ♦ ♦ ♦ ♦ ♦ ♦ ♦ ♦ ♦ ♦ ♦ ♦ ♦ ♦ ♦ ♦ ♦

SOLUTIONS TO EQUATIONS

A **solution** *is the input that makes an equation true.* **Solving an equation** *is the process of finding that input.* We are able to solve equations from tables and graphs because all ordered pairs of the input–output table make the equation true, as do all points on the graph created by plotting the ordered pairs and connecting the points.

When we are given an input–output rule in equation form, we should write it both in the output column of the table and on the graph. If the equation has more than one number that makes it true for a given output, we may wish to *list all the solutions in a set*, the **solution set**.

SOLVING EQUATIONS FROM TABLES

In Example 3 we use the input–output table to solve $y = 3x - 2$ for an input, $x$. Although the 0 and $-1$ inputs are meaningless in the pattern (Figure 2), the pattern's equation, $y = 3x - 2$, is valid for all inputs. If the solution needed is not shown on the table, we estimate the solution.

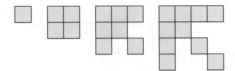

*Figure 2*

EXAMPLE **3**

Use Table 3 to answer each question.

**a.** For what input $x$ is $3x - 2 = 4$?

**b.** For what input $x$ is $3x - 2 = -5$?

| Input: $x$ | Output: $y = 3x - 2$ |
|:---:|:---:|
| $-1$ | $-5$ |
| 0 | $-2$ |
| 1 | 1 |
| 2 | 4 |

*Table 3*

**c.** Describe the input where $3x - 2 = 0$ by writing an inequality and an interval that contain $x$.

**d.** Describe the input where $3x - 2 = 9$ by writing an inequality and an interval that contain $x$.

SOLUTION    **a.** $x = 2$

***Check***: To check, we replace the $x$ in $y = 3x - 2$ with the input 2. Write the checking step as

$$3(2) - 2 \stackrel{?}{=} 4$$

The symbol $\stackrel{?}{=}$ asks the question "Does it equal?"  ✔

**b.** $x = -1$

***Check***: $3(-1) - 2 \stackrel{?}{=} -5$  ✔

**c.** $x$ is between 0 and 1, so $0 < x < 1$ and $x$ is in the interval (0, 1).

*Check*: An input of 0 gives output $-2$; an input of 1 gives output 1. The interval is sensible because the output 0 lies between $-2$ and 1.   ✔

**d.**  Look at the $\Delta x$ (change in $x$) and $\Delta y$ (change in $y$) pattern in Table 4. The output increases by 3 for every input increase of 1. Continuing in the same pattern indicates that an output of 9 will have $x$ between 3 and 4, so $3 < x < 4$ and $x$ is in the interval (3, 4).

*Check*: 3 and 4 give an output just below and above 9.   ✔

| $\Delta x$ | $x$ | $y$ | $\Delta y$ |
|:---:|:---:|:---:|:---:|
| | $-1$ | $-5$ | |
| $+1$ | | | $+3$ |
| | 0 | $-2$ | |
| $+1$ | | | $+3$ |
| | 1 | 1 | |
| $+1$ | | | $+3$ |
| | 2 | 4 | |
| $+1$ | | | $+3$ |
| | 3 | 7 | |
| $+1$ | | | $+3$ |
| | 4 | 10 | |

*Table 4*  Change in x and in y

◆ ◆ ◆ ◆ ◆ ◆ ◆ ◆ ◆ ◆ ◆ ◆ ◆ ◆ ◆ ◆ ◆ ◆ ◆ ◆ ◆ ◆ ◆ ◆ ◆ ◆ ◆ ◆ ◆ ◆ ◆ ◆ ◆ ◆

**Graphing Calculator Technique**

**Evaluating an Expression or Checking a Solution**

*Method 1, TEST key (TI 81 and 82)*: Enter $3(2) - 2 = 4$ using the equals sign from [ 2nd ] [ TEST ]. If the equation is true, the calculator will return a 1. If the equation is false, the calculator will return a 0. The equals sign under [ TEST ] behaves the same way as the $\overset{?}{=}$ we use in checking equations.

*Method 2, Evaluation (TI 81 and 82)*: Evaluate $3x - 2$ for $x = 2$ to determine whether the result is 4. Enter $Y_1 = 3X - 2$ using the [ y = ] key. [ 2nd ] [ QUIT ] will return to the computation window. Store 2 in the X variable with 2 [ STO▶ ] X [ ENTER ]. Evaluate $Y_1$ with [ 2nd ] [ Y VARS ] [ ENTER ] [ ENTER ] [ ENTER ]. (The third [ ENTER ] is needed only on the TI 82.) The result should be 4.

*Method 3, Table (TI 82 only)*: Enter $Y_1 = 3X - 2$. Go to the table set-up options with [ 2nd ] [ TblSet ]. Move the cursor to "Ask" after the independent variable, "Indpnt," and press [ ENTER ]. Display the table with [ 2nd ] [ TABLE ]. Press 2 [ ENTER ] to evaluate $3(2) - 2$. The calculator will display 2 in the input column under X and 4 in the output column under $Y_1$. Return to [ 2nd ] [ TblSet ] and change the "Indpnt" option back to "Auto." Look again at [ 2nd ] [ TABLE ].

## SOLVING EQUATIONS FROM GRAPHS

We use graphs to find the input that gives any particular output. Because graphs may include fraction and decimal inputs, they give us more complete sets of inputs and outputs than tables do. With computer or calculator graphing technology we can find accurate solutions to equations that cannot be solved by algebra. If you have the technology available and you are not otherwise directed, you may use calculators or computers to solve any of the examples or exercises.

Graphing Calculator Technique

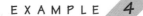

> **Solving an Equation**
> One way to solve an equation on the graphing calculator is to graph each side as a separate equation and trace to the $x$-value of the intersection of the two graphs.
> **Example**: Solve $3x - 2 = 9$.
>     Enter $Y_1 = 3X - 2$ and $Y_2 = 9$. Set the viewing window or range using the interval $[-10, 10]$ for $x$ and $[-10, 10]$ for $y$. Graph, and trace on $y = 3x - 2$ to $y = 9$. Zoom in. Trace on $3x - 2$ again to $y = 9$, zoom in, and trace to $y = 9$ again. Read the corresponding $x$ value, $x \approx 3.67$ $\left(\text{which may be recognized as } 3\frac{2}{3}\right)$. Check by storing 3.67 in $x$ and evaluating $y = 3x - 2$ (see Evaluation in the previous Graphing Calculator Technique box).
> **Example**: Solve $3x - 2 = 0$.
>     Enter $Y_1 = 3X - 2$. Set the same viewing window as above, and graph. The line $y = 0$ is the $x$-axis and is automatically graphed. Because $y = 0$, we are looking for the $x$-intercept. Trace on $y = 3x - 2$ to a point near the $x$-intercept. Zoom in. Trace to the intercept again, zoom in, and trace a third time. The $x$-value corresponding to $y = 0$ is approximately 0.67 $\left(\text{or the fraction } \frac{2}{3}\right)$. Check the solution by storing 0.67 in $x$ and evaluating $y = 3x - 2$.

The next two examples illustrate the use of graphs to solve equations. The process is the same as finding input–output pairs in Chapter 1 because we are assuming $x$ is the input and the number on the right side of the equation is $y$, or the output.

EXAMPLE  **4**     Use the graph in Figure 3 to solve these equations.

**a.** $3x - 2 = 4$     **b.** $3x - 2 = -5$     **c.** $3x - 2 = 0$     **d.** $3x - 2 = 9$

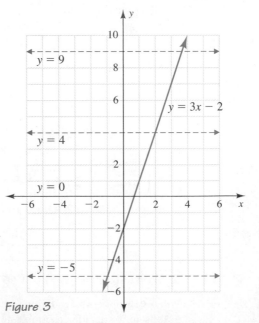

Figure 3

SOLUTION     **a.** To solve $3x - 2 = 4$, trace along the horizontal line $y = 4$ to where it intersects $y = 3x - 2$. Look below this intersection to the $x$-axis. The input $x = 2$ lies directly below the intersection. Thus $3x - 2 = 4$ at the point where $x = 2$.

***Check***: $3(2) - 2 \overset{?}{=} 4$ ✔

**b.** To solve $3x - 2 = -5$ we trace along the horizontal line $y = -5$ to where it intersects $y = 3x - 2$. The $x$-coordinate directly above that intersection is $x = -1$. Thus $3x - 2 = -5$ at the point where $x = -1$.

*Check*: $3(-1) - 2 \overset{?}{=} -5$ ✔

**c.** To solve $3x - 2 = 0$ we need a horizontal line that represents $y = 0$. The line $y = 0$ is the $x$-axis. The solution to $3x - 2 = 0$ is in the interval $(0, 1)$.

*Check*: See Example 3c. ✔

**d.** To solve $3x - 2 = 9$ we trace along the horizontal line $y = 9$. The lines $y = 3x - 2$ and $y = 9$ cross between $x = 3$ and $x = 4$, so the solution is in the interval $(3, 4)$.

*Check*: See Example 3d. ✔

♦ ♦ ♦ ♦ ♦ ♦ ♦ ♦ ♦ ♦ ♦ ♦ ♦ ♦ ♦ ♦ ♦ ♦ ♦ ♦ ♦ ♦ ♦ ♦ ♦ ♦ ♦ ♦ ♦ ♦ ♦ ♦ ♦ ♦ ♦

Example 4c solved for the $x$-intercept. The **$x$-intercept** *is the point where a graph crosses the $x$-axis. It is where the output is zero*, $y = 0$. The $x$-intercept has considerable significance in application problems. For example, it is where a thrown object will strike the ground, where business loss turns to profit, and where a vibrating object may be expected to stabilize, to name just a few.

The equations in Example 4, had either one solution or a solution on an interval. In Example 5 we have a curved graph, illustrating that some equations may have more than one solution or no solution at all.

EXAMPLE   **5**

Use the graph of $y = x^2$ in Figure 4 to determine the solutions to the equations. Use sets to describe the solutions.

**a.** $x^2 = 9$

**b.** $x^2 = 2$

**c.** $x^2 = 0$. What is special about this point on the graph of $y = x^2$?

**d.** $x^2 = -2$

SOLUTION   **a.** Look for the inputs that give $y = 9$. For $x^2 = 9$ there are two solutions: $x = 3$ and $x = -3$. Thus the solution set is $\{-3, 3\}$.

*Check*: $(-3)^2 \overset{?}{=} 9$    and    $3^2 \overset{?}{=} 9$ ✔

**b.** Look for inputs that give $y = 2$. For $x^2 = 2$ there are two solutions: $x \approx 1.4$ and $x \approx -1.4$. Thus the approximate solution set is $\{-1.4, 1.4\}$.

*Check*: $(-1.4)^2 \overset{?}{=} 2$ and $(1.4)^2 \overset{?}{=} 2$ if answers are rounded to the nearest tenth. ✔

**c.** This point is the lowest point on the curve. Look for the inputs that give $y = 0$. For $x^2 = 0$ there is only one solution: $x = 0$. The solution set is $\{0\}$. Note that this is also the $x$-intercept and the lowest point on the curve $y = x^2$.

*Check*: $0^2 \overset{?}{=} 0$ ✔

**d.** Look for the inputs that give $y = -2$. There are no solutions to $x^2 = -2$, because the curve never goes below the $x$-axis. We say that $x^2 = -2$ has no real number solution.

♦ ♦ ♦ ♦ ♦ ♦ ♦ ♦ ♦ ♦ ♦ ♦ ♦ ♦ ♦ ♦ ♦ ♦ ♦ ♦ ♦ ♦ ♦ ♦ ♦ ♦ ♦ ♦ ♦ ♦ ♦ ♦ ♦ ♦ ♦

When an equation such as that in Example 5d has no real number solution, we say the solution set is empty and write the symbol $\{\ \}$ or $\varnothing$. *The symbol $\{\ \}$ shows a set with nothing in it, the* **empty set**. The other symbol, $\varnothing$, is also common. Choose either symbol.

CAUTION   Compare 5c and 5d carefully; $\{0\}$ and $\{\ \}$ are not the same result.

♦

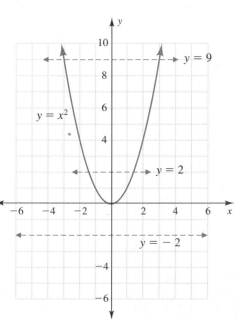

*Figure 4*

### VOCABULARY AND APPLICATIONS

A **linear equation** *describes the graph of input–output pairs (x, y) that make a straight line*. Examples 3 and 4 were the same linear equation, $y = 3x - 2$. The tuition and bulk purchase examples in Chapter 1 were also linear. The credit card, postal rates, and absolute value graphs combined linear segments. **Nonlinear equation** *is the generic name given to equations whose graphs do not form straight lines*. The graph, $y = x^2$, in Figure 4 is nonlinear. It is a parabola. The square root curve, $y = \sqrt{x}$, and the temperature-volume relationship for water are two examples of nonlinear equations found in Chapter 1.

Because computer and calculator technologies are rapidly replacing many traditional symbolic skills, learning to compare graphs and to understand the relationship between an application and its graph becomes more important. The next three examples stress how the graphs of linear equations describe certain aspects of applications.

EXAMPLE 6

Confirm that the relationship $D = r \cdot t$ is a linear equation if either the rate $r$ or the time $t$ is constant. Driving 55 miles per hour for a period of time is an example of constant rate, $D = 55t$. Driving for 3 hours gives $D = 3r$, an example of constant time.

a. Make a table and graph for $D = 55t$.

b. Make a table and graph for $D = 3r$.

c. Compare the graphs. Are both linear?

SOLUTION

a. We construct Table 5 and Figure 5.

b. We construct Table 6 and Figure 6.

c. The graphs appear to be straight, so we assume the equations are linear.

| Time: $t$ | Distance: $D = 55t$ |
|-----------|---------------------|
| 1 hr | 55 mi |
| 2 hr | 110 mi |
| 3 hr | 165 mi |
| 4 hr | 220 mi |

**Table 5** *Distance at Constant Rate*

| Rate: $r$ | Distance: $D = 3r$ |
|-----------|--------------------|
| 10 mph | 30 mi |
| 20 mph | 60 mi |
| 30 mph | 90 mi |
| 40 mph | 120 mi |

**Table 6** *Distance in Constant Time*

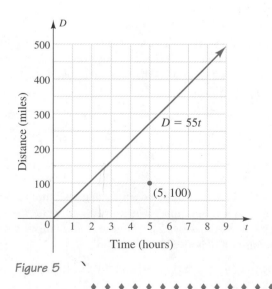

Figure 5

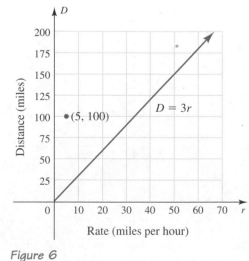

Figure 6

The graphs in Example 6 look alike; however, other than that they both pass through the origin (0, 0), the graphs are different. *The similarity in appearance is due to the choice of scales on the axes.* The point (5, 100) is shown on each set of axes. In Figure 5, (5, 100) is below the line. In Figure 6, (5, 100) is above the line. Because the scales on the axes affect the appearance of the graph, we use a number called *slope* to describe the steepness of a line. We will return to slope and steepness in Section 3.6.

The linear graphs in Example 6 passed through the origin (0, 0). In Example 4 and Example 7 the graph does not pass through the origin. Try to determine what it is about the problems that makes this difference.

EXAMPLE   7

**Electric bill**   West Coast Electric charges $5.00 per month basic fee and $0.03715 per kilowatt hour of electricity used.

**a.** Make a table for the cost each month of 0 to 3000 kilowatt hours. Graph the data from the table.

**b.** Examine the table, and explain why the graph does not pass through the origin (0, 0).

**c.** What is the cost if 1500 kilowatt hours is used?

**d.** What is the cost if 2500 kilowatt hours is used?

**e.** How much electricity is used if the total cost is $75?

**f.** Write the equation for the monthly cost. Let $x$ be the number of kilowatt hours used and $y$ be total monthly cost. Write part e as an equation.

SOLUTION   **a.** We construct Table 7 and Figure 7.

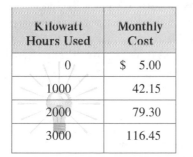

| Kilowatt Hours Used | Monthly Cost |
|---|---|
| 0 | $  5.00 |
| 1000 | 42.15 |
| 2000 | 79.30 |
| 3000 | 116.45 |

*Table 7* *Monthly Electric Cost*

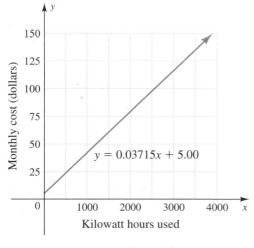

$y = 0.03715x + 5.00$

*Figure 7*

**b.** The graph does not pass through the origin because the cost is $5.00 even if no electricity is used.

**c.** 1500 kilowatt hours costs approximately $60.

**d.** 2500 kilowatt hours costs approximately $100.

**e.** $75 buys between 1750 and 2000 kilowatt hours.

**f.** The cost (output) is 0.03715 times the kilowatt hours (input) plus the $5.00 fee, or $y = 0.03715x + 5.00$. Part e sets the cost at $75, so the equation in part f is $75 = 0.03715x + 5.00$.

♦ ♦ ♦ ♦ ♦ ♦ ♦ ♦ ♦ ♦ ♦ ♦ ♦ ♦ ♦ ♦ ♦ ♦ ♦ ♦ ♦ ♦ ♦ ♦ ♦ ♦ ♦ ♦ ♦ ♦ ♦ ♦ ♦ ♦ ♦

Another way to identify linear equations besides graphing is to arrange equations into a certain form. The linear equations in this section included $y = 3x - 2$, $D = 55t$, $D = 3r$, and $y = 0.03715x + 5.00$. *Each linear equation has the input multiplied by a constant and then a constant added or subtracted.* The constant added may be zero. This describes a general form for all linear equations.

---

♦ A **linear equation in one variable** may be written in the form

$$mx + b = 0$$

where $x$ is the variable and $m$ and $b$ are constants.

♦ A **linear equation in two variables** may be written in the form

$$y = mx + b$$

where $x$ and $y$ are variables and $m$ and $b$ are constants.

---

Think of the constant $m$ as the *m*ultiplier on the input $x$, and think of the letter $b$ as *b*eing added.

In an expression, *a quantity being added or subtracted is called a* **term**. A term frequently contains a number (positive or negative) that may be multiplied by one or more variables. In the electric bill the *output y* is described by the sum of the two terms, $0.03715x$ and $5.00$. Distance traveled, $D = rt$, was described with one term, $rt$. The correct use of *term* will be especially important in Section 3.3.

The letter $x$ is usually the input variable and is called the **independent variable**. We place the independent variable on the horizontal axis. The letter $y$ is usually the output, the **dependent variable**. We place the dependent variable on the vertical axis. We say *y depends on x* and write equations for *y in terms of x*. The use of *terms* here is exactly the same as in the paragraph above.

---

## EXERCISES 3.1

**For Exercises 1 to 4 use the graph to estimate the input $x$ that gives the indicated output $y$. Write the inputs and outputs as coordinate pairs.**

1. $y = \frac{1}{2}x + 3$; estimate $x$ for $y = 1$, $y = 0$, and $y = -2$.

2. $y = \frac{1}{2}x - 2$; estimate $x$ for $y = 3$, $y = 0$, and $y = -4$.

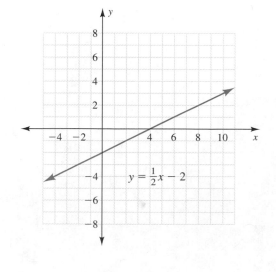

3. $y = 4.2x - 3$; estimate $x$ for $y = 4$, $y = 0$, and $y = -8$.

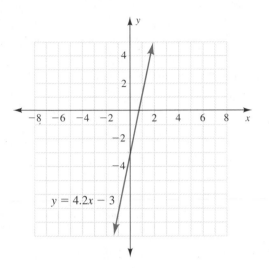

4. $y = 2.5x + 2$; estimate $x$ for $y = 4$, $y = 0$, and $y = -4$.

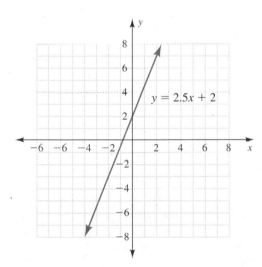

In Exercises 5 to 8 make an input–output table for the first equation in each set. Use the table to estimate the input, $x$, that solves the remaining three equations. Write the inputs and outputs as coordinate pairs.

5. $y = 10 - x$; estimate $x$ for $10 - x = 3$, $10 - x = 0$, and $10 - x = -2$.

6. $y = 4 - 2x$; estimate $x$ for $4 - 2x = 10$, $4 - 2x = 0$, and $4 - 2x = -3$.

7. $y = 3 - 3x$; estimate $x$ for $3 - 3x = 9$, $3 - 3x = 0$, and $3 - 3x = -4$.

8. $y = 2 - 4x$; estimate $x$ for $2 - 4x = 6$, $2 - 4x = 0$, and $2 - 4x = -1$.

In Exercises 9 to 16 determine whether the given input $x$ is a solution to the equation.

9. $3x - 4 = -16$; $\quad x = -4$

10. $4x - 3 = -9$; $\quad x = -3$

11. $-8x + 5 = 1$; $\quad x = \frac{1}{2}$

12. $-6 - 10x = -11$; $\quad x = \frac{1}{2}$

13. $-4x + 5 = -3$; $\quad x = -2$

14. $-5x + 4 = 9$; $\quad x = -1$

15. $2.5x - 3 = -1$; $\quad x = 0.8$

16. $8 - 7.5x = 5$; $\quad x = 0.4$

For Exercises 17 to 22 copy or trace graphs a and b.

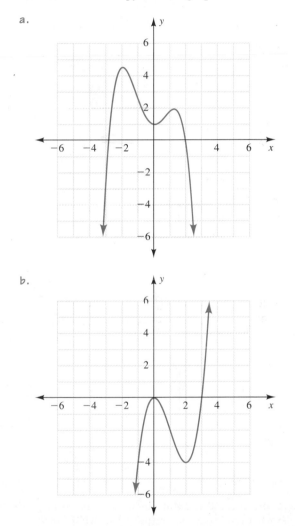

On each graph draw a horizontal line, and label it with the appropriate equation, $y = $ a number, to identify an output where the indicated number of solutions can be found. If it is not possible to achieve such an output, say so.

17. Zero solutions

18. One solution

19. Two solutions

20. Three solutions

21. Four solutions

22. Five solutions

Use the graphs to estimate the solutions to the equations in Exercises 23 and 24. The equations may be entered on a graphing calculator and the solutions found by tracing.

23.

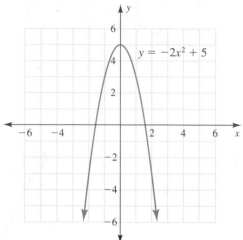

a. $-2x^2 + 5 = -3$     b. $-2x^2 + 5 = 0$

c. $-2x^2 + 5 = 5$

24.

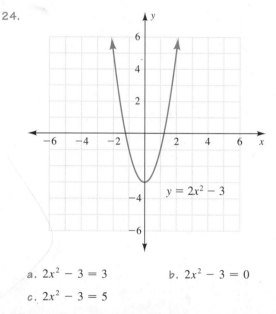

a. $2x^2 - 3 = 3$     b. $2x^2 - 3 = 0$

c. $2x^2 - 3 = 5$

In Exercises 25 to 28 translate the sentences into equations. Describe what the input and output variables mean.

25. a. The distance, $D$, traveled at 35 miles per hour for $x$ hours

b. The total distance, $D$, traveled in a day if you drive 200 miles in the morning and then 35 miles per hour during $x$ hours in the afternoon

26. a. The distance, $D$, traveled at $x$ miles per hour for 5 hours

b. The total distance, $D$, traveled in a day if you drive 200 miles in the morning and then $x$ miles per hour for 5 hours in the afternoon

27. a. The 8% sales tax due on a purchase of $x$ dollars

b. The total cost of a purchase of $x$ dollars with an 8% sales tax on that purchase (Think carefully!)

28. a. The cost of $x$ thousand gallons of water at $1.15 per thousand gallons

b. The total monthly cost of $x$ thousand gallons of water at $1.15 per thousand gallons, with a $5.35 monthly service charge

29. Refer to the situation in Example 6a.

a. How many miles would be driven in $2\frac{1}{2}$ hours?

b. How many hours would it take to drive 400 miles?

c. What is a reasonable number of hours as input if there is only one driver involved?

d. What speed is a driver traveling if she goes 100 miles in 5 hours?

e. Write a sentence that explains the meaning of the point (0, 0).

30. Refer to the situation in Example 6b.

a. What speed would a bicyclist travel to cover a distance of 81 miles?

b. What average speed would a race car maintain to cover 500 miles?

c. How long will a driver travel if he went 100 miles at 5 miles per hour?

d. Write a sentence that explains the meaning of the point (0, 0).

31. How will the graph in Example 7 change if the cost rises to $0.04 per kilowatt hour?

32. How will the graph in Example 7 change if the basic monthly charge rises to $10?

The graphs in Exercises 33–35 appear to have the same steepness, but because of the difference in scales on the axes, they represent different linear equations. Assuming the axes are in feet, determine which graph might represent each of the following:

a. a wheelchair access ramp for a home

b. a ladder or cliff

c. a steep set of stairs

33.

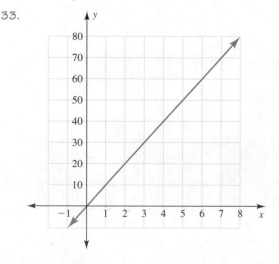

34.

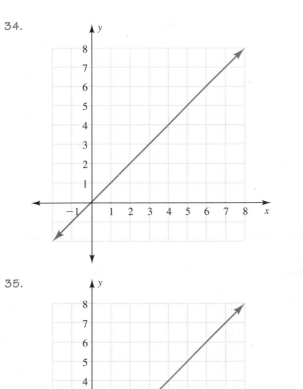

35.

b.

c.

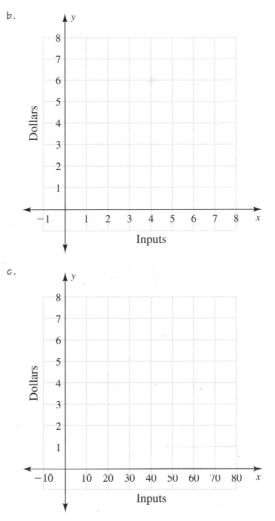

For Exercises 36 to 38 find the equation for the indicated graph from Exercises 33 to 35 by listing four or more ordered pairs (coordinates) in a table and looking for a pattern. Write the equation with *x* as input and *y* as output.

36. The graph in Exercise 33

37. The graph in Exercise 34

38. The graph in Exercise 35

For Exercises 39 to 41 select the figure below that contains the best scales for graphing the problem, and explain your choice.

a.

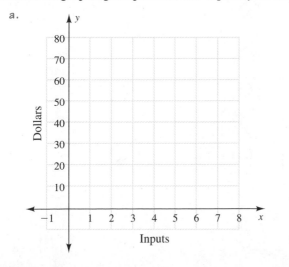

39. Inputs are gallons of gasoline at \$1.50 per gallon, and outputs are the total cost.

40. Inputs are gallons of gasoline at 36¢ per gallon (1970), and outputs are the total cost.

41. Inputs are the cost of a meal, and outputs are the tips at 15% of the meal's cost.

42. Explain the difference in meaning between the symbols { } and {0}.

43. Which equations pass through the origin? Why?

   a. $y = 3x$    b. $y = \frac{1}{2}x$    c. $y = x + 2$    d. $y = 2x$

44. Is it possible to tell, from looking at an equation, where it will cross the *y*-axis? (*Hint*: Look carefully at the equations and graphs in the examples.)

GRAPHING CALCULATOR

45. Which graphs pass through (0, 2)? How do you know from the equation?

   a. $y = 2x$        b. $y = x + 2$
   c. $y = x^2 + 2$    d. $y = \sqrt{x} + 2$

46. Which graphs pass through (0, −3)? How do you know from the equation?

   a. $y = -3x$        b. $y = -3x^2$
   c. $y = x - 3$      d. $y = x^2 - 3$

47. **Distinguishing Intervals from Coordinates.** The expression in parentheses (2, 4) could describe an interval or an ordered pair (coordinate point), because both use the same notation. Identify which meaning is intended in each of these examples. (*Hint*: Sometimes drawing a picture is helpful.)

a. The postage doubled for packages with weight (2, 4).

b. (2, 4) makes the equation $y = 2x$ true.

c. The graphs $y = x + 2$ and $y = x^2$ cross at (2, 4) and (−1, 1).

d. The graph of $y = x^2 - 6x + 8$ is below the x-axis for (2, 4).

e. Make an appointment in the afternoon, (2, 4).

f. The line $y = 4$ passes through (2, 4) and is parallel to the x-axis.

g. The solution to the equation $0 = x - 3$ is in (2, 4).

h. A solution to the equation $y = 3x - 2$ is (2, 4).

---

## 3.2   SOLVING EQUATIONS WITH SYMBOLS

OBJECTIVES
Solve $y = ax + b$ using the reverse order of operations and inverse operations. ◆ Identify equivalent equations. ◆ Identify the addition property of equations. ◆ Identify the multiplication property of equations.

WARM-UP
Evaluate each expression. Describe in a complete sentence the order of operations you use to evaluate each. (*Hint*: Start with, "Take the number replacing x and . . .")

1. $3x - 2$ for $x = 1$       2. $2 - 6x$ for $x = -\frac{1}{2}$

3. $4x + 1$ for $x = 2.25$    4. $\frac{2}{3}x + 4$ for $x = 27$

Rewrite each exercise using an inverse operation and then simplify.

**Example:** $3 - (-4) = 3 + (+4) = 7$

5. $3 - 4$                    6. $-3 - (-4)$

7. $6 \div \frac{1}{2}$       8. $-4 \div \frac{2}{3}$

9. $9 \div \frac{3}{4}$

10. Use the graph in Figure 8 to complete Table 8.

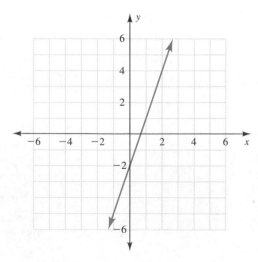

| Input x | Output y |
|---------|----------|
| 1       |          |
| 0       |          |
|         | −5       |
|         | 7        |

Table 8

Figure 8

In this section we examine where to start and what steps to follow in solving an equation. Some equations can be solved quite quickly with symbols, others take a lot of work, and still others cannot be solved symbolically. We need to do enough symbolic equation solving to understand the process. If we understand the process, then we can use the algebra in applications, determine whether computer or calculator answers are reasonable, and learn related mathematical concepts.

## WORKING BACKWARD

Working backward from the answer is a common problem-solving strategy. (How often have you not known what to do with a problem and been helped by looking at the answer in the back of the book?) In problem 10 of the Warm-up we worked backward with opposites. To evaluate $y$ for a given $x$—say, $x = 1$—we started at 1 on the $x$-axis, traced vertically to the graph, and then traced horizontally to the output, $y = 1$. When we were given $y$ and asked to find $x$, we worked backward and moved in opposite directions on the graph. We started at the output—say, $y = -5$—traced horizontally to the graph, and then traced vertically to the input, $x = -1$. This process corresponds to that used to "solve for $x$" in Section 3.1.

**EXAMPLE 1** I am thinking of a number. When I triple it and subtract 2, I get 4. What is my number?

**SOLUTION** The strategy is to work backward!

If the result was 4 and it was found by subtracting 2, then the prior number must have been $4 + 2$, or 6. If the answer after tripling the original number was 6, what was tripled to get 6? Since 2 is one-third of 6, our answer is 2.

***Check***: Triple 2 and subtract 2 gives 4: $3(2) - 2 - 4$. ✔

♦ ♦ ♦ ♦ ♦ ♦ ♦ ♦ ♦ ♦ ♦ ♦ ♦ ♦ ♦ ♦ ♦ ♦ ♦ ♦ ♦ ♦ ♦ ♦ ♦ ♦ ♦ ♦ ♦ ♦ ♦ ♦ ♦ ♦ ♦ ♦ ♦ ♦ ♦

We also work backward to solve an equation with algebra. We focus on the order of operations. Instead of moving in opposite directions (as on the graph), we do the opposite, or inverse, operations in the reverse order of operations. In Example 2 the arrows separate steps and show the direction in which we read the phrases.

**EXAMPLE 2** Solve the equation $3x - 2 = 4$.

**SOLUTION** List the order of operations as applied to $x$:

$$\textit{the number } x \to \textit{ times } 3 \to \textit{ subtract } 2 \to \textit{ gives } 4$$

Write the inverse operations, and do the sentence backward, from the right:

End                                               Start here

$$\textit{gives } x = 2 \leftarrow \textit{ divide by } 3 \leftarrow \textit{ add } 2 \leftarrow \textit{ start with } 4$$

***Check***: Substitute 2 for $x$ in the equation: $3(2) - 2 \overset{?}{=} 4$. ✔

♦ ♦ ♦ ♦ ♦ ♦ ♦ ♦ ♦ ♦ ♦ ♦ ♦ ♦ ♦ ♦ ♦ ♦ ♦ ♦ ♦ ♦ ♦ ♦ ♦ ♦ ♦ ♦ ♦ ♦ ♦ ♦ ♦ ♦ ♦ ♦ ♦ ♦ ♦

**EXAMPLE 3** Solve the equation $5 = 2 - 6x$.

**SOLUTION** List the order of operations:

$$\textit{the number } x \to \textit{ times } -6 \to \textit{ add } 2 \to \textit{ gives } 5$$

Reverse the order of operations, working from the right to the left:

End                                               Start here

$$\textit{gives } x = -\tfrac{1}{2} \leftarrow \textit{ divide by } -6 \leftarrow \textit{ subtract } 2 \leftarrow \textit{ start with } 5$$

***Check***: Substitute $-\tfrac{1}{2}$ for $x$ in the equation: $5 \overset{?}{=} 2 - 6\left(-\tfrac{1}{2}\right)$. ✔

♦ ♦ ♦ ♦ ♦ ♦ ♦ ♦ ♦ ♦ ♦ ♦ ♦ ♦ ♦ ♦ ♦ ♦ ♦ ♦ ♦ ♦ ♦ ♦ ♦ ♦ ♦ ♦ ♦ ♦ ♦ ♦ ♦ ♦ ♦ ♦ ♦ ♦ ♦

**Summary**

> ♦ The solution to the equation requires the opposite, or inverse, of each operation in the reverse order of operations.
>
> ♦ The operation of addition is the opposite of subtraction.
>
> ♦ The operation of multiplication is the opposite of division.

## USING SYMBOLS

We now streamline the reversal process and use symbols instead of words.

**EXAMPLE 4**    Solve $3x - 2 = 1$.

**SOLUTION**    Look at the order of operations, and select the last operation that is done on the left side: subtract 2. Do the inverse operation (add 2) to both sides to keep the equation in balance:

$$3x - 2 = 1$$

$$3x - 2 + 2 = 1 + 2$$

$$3x = 3$$
Look at the remaining operation on the left side: multiply by 3. Do the inverse operation (divide by 3) on both sides.

$$3x \div 3 = 3 \div 3$$

$$x = 1$$

***Check***: $3(1) - 2 \overset{?}{=} 1$  ✔

♦ ♦ ♦ ♦ ♦ ♦ ♦ ♦ ♦ ♦ ♦ ♦ ♦ ♦ ♦ ♦ ♦ ♦ ♦ ♦ ♦ ♦ ♦ ♦ ♦ ♦ ♦ ♦ ♦ ♦ ♦ ♦ ♦ ♦ ♦

**EXAMPLE 5**    Solve $4x + 1 = 10$.

**SOLUTION**    Adding 1 is the last step in the order of operations.

$$4x + 1 = 10$$

$$4x + 1 - 1 = 10 - 1$$   Subtract 1 from each side.

$$4x = 9$$

$$4x \div 4 = 9 \div 4$$   Divide both sides by 4.

$$x = \tfrac{9}{4}$$

***Check***: $4\left(\tfrac{9}{4}\right) + 1 \overset{?}{=} 10$  ✔

♦ ♦ ♦ ♦ ♦ ♦ ♦ ♦ ♦ ♦ ♦ ♦ ♦ ♦ ♦ ♦ ♦ ♦ ♦ ♦ ♦ ♦ ♦ ♦ ♦ ♦ ♦ ♦ ♦ ♦ ♦ ♦ ♦ ♦ ♦

**EXAMPLE 6**    Solve $\tfrac{2}{3}x + 4 = 22$.

**SOLUTION**    
$$\tfrac{2}{3}x + 4 = 22$$

$$\tfrac{2}{3}x + 4 - 4 = 22 - 4$$   Subtract 4.

$$\tfrac{2}{3}x = 18$$

$$\tfrac{2}{3}x \div \tfrac{2}{3} = 18 \div \tfrac{2}{3}$$   Divide both sides by $\tfrac{2}{3}$.

$$x = 18 \cdot \tfrac{3}{2}$$   Multiply by the reciprocal.

$$x = 27$$

***Check***: $\tfrac{2}{3}(27) + 4 \overset{?}{=} 22$  ✔

♦ ♦ ♦ ♦ ♦ ♦ ♦ ♦ ♦ ♦ ♦ ♦ ♦ ♦ ♦ ♦ ♦ ♦ ♦ ♦ ♦ ♦ ♦ ♦ ♦ ♦ ♦ ♦ ♦ ♦ ♦ ♦ ♦ ♦ ♦

To summarize, *we solve an equation by finding the input that makes the equation true*. The equations in each step of a solution process are said to be equivalent. **Equivalent equations** *are equations that have the same solution set*. We saw that adding the same number to both sides of the equation produced an equivalent equation. This leads to the **addition property of equations**.

**Addition Property of Equations**

> Adding the same number to both sides of an equation produces an equivalent equation.

Addition and subtraction are inverse operations. A subtraction may be made into an addition by adding the opposite number. Thus the addition property of equations applies to subtraction. We also saw that multiplying both sides of an equation by the same number produced an equivalent equation, which leads us to the **multiplication property of equations**.

**Multiplication Property of Equations**

> Multiplying both sides of an equation by the same nonzero number produces an equivalent equation.

Multiplication and division are inverse operations. A division may be made into a multiplication by multiplying by the reciprocal. Thus the multiplication property of equations applies to division.

**Alternative Solution for Fractional Equations.**    The inverse nature of multiplication and division means we can multiply by a reciprocal rather than divide by a fraction. Example 6 contains a fraction, $\frac{2}{3}$. Instead of dividing by that fraction we might multiply by the reciprocal, $\frac{3}{2}$. We show this technique in Example 7.

**E X A M P L E    7**    Solve $\frac{1}{2}x + 4 = 16$ with symbols.

**SOLUTION**    To solve with symbols we use the reverse order of operations, but we modify the process to multiply by the reciprocal, $\frac{2}{1}$, instead of dividing by $\frac{1}{2}$.

$$\frac{1}{2}x + 4 = 16$$

$$\frac{1}{2}x + 4 - 4 = 16 - 4 \quad \text{Subtract 4.}$$

$$\frac{1}{2}x = 12$$

$$\frac{2}{1}\left(\frac{1}{2}\right)x = \frac{2}{1}(12) \quad \text{Multiply by reciprocal of } \frac{1}{2}.$$

$$1x = 24$$

***Check***: $\frac{1}{2}(24) + 4 \overset{?}{=} 16$  ✔

◆ ◆ ◆ ◆ ◆ ◆ ◆ ◆ ◆ ◆ ◆ ◆ ◆ ◆ ◆ ◆ ◆ ◆ ◆ ◆ ◆ ◆ ◆ ◆ ◆ ◆ ◆ ◆ ◆ ◆ ◆ ◆ ◆ ◆ ◆ ◆ ◆ ◆

## EXERCISES 3.2

**Match each equation in Exercises 1 to 6 with the correct order of operations phrase, listed below.**

a. $x$ times $-4$ plus 2 gives 3.

b. $x$ times $\frac{1}{4}$ subtract 2 gives 3.

c. $x$ times $\frac{1}{2}$ subtract 4 gives 3.

d. $x$ times 4 plus 2 gives 3.

e. $x$ times $-2$ subtract 4 gives 3.

f. $x$ times $-2$ plus 4 gives 3.

g. $x$ times 2 plus 4 gives 3.

1. $3 = 4 + 2x$    2. $3 = 2 - 4x$    3. $3 = \frac{1}{2}x - 4$

4. $3 = 2x + 4$    5. $3 = \frac{1}{4}x - 2$    6. $3 = 4x + 2$

7. Solve Exercises 1, 3, and 5 for $x$, using the reverse operation process in words.

8. Solve Exercises 2, 4, and 6 for $x$, using the reverse operation process in words.

9. Solve Exercises 1, 3, and 5 for $x$, using the reverse operation process in symbols.

10. Solve Exercises 2, 4, and 6 for $x$, using the reverse operation process in symbols.

**Solve for $x$ in each of the equations in Exercises 11 to 28. Use symbols. In the first problem, note that $10 - x$ means $10 - 1x$.**

11. $-2 = 10 - x$      12. $10 = 4 - 2x$

13. $3 = 10 - x$      14. $-3 = 4 - 2x$

15. $0 = 3 - 3x$      16. $6 = 2 - 4x$

17. $9 = 3 - 3x$      18. $-1 = 2 - 4x$

19. $-2 = \frac{1}{2}x + 3$      20. $-4 = \frac{1}{2}x - 2$

21. $0 = \frac{1}{2}x + 3$      22. $0 = \frac{1}{2}x - 2$

23. $-5 = \frac{1}{3}x - 2$      24. $1 = \frac{1}{3}x - 2$

25. $-9.3 = 4.2x - 3$      26. $-3.5 = 2.5x + 2$

27. $7.5 = 4.2x - 3$      28. $5 = 2.5x + 2$

**Solve for $C$ in the equations in Exercises 29 and 30. Use symbols.**

29. $212 = \frac{9}{5}C + 32$      30. $32 = \frac{9}{5}C + 32$

**In Exercises 31 to 42 decide whether the pairs of equations are equivalent equations. If so, write a sentence that states which property (addition property or multiplication property) and number are used to get the second equation from the first equation.**

31. $2x = 10; x = 5$      32. $x + 4 = 6; x = 10$

33. $2x + 5 = 11; 2x = 6$      34. $3x = 6; 3x + 4 = 10$

35. $x + 5 = 12; x = 17$      36. $2x - 5 = 11; 2x = 16$

37. $3x = 12; 3x - 4 = 8$      38. $x - 7 = -3; x = 4$

39. $2x - 4 = 10; 2x = 14$      40. $\frac{1}{2}x = 10; x = 20$

41. $\frac{1}{2}x = 15; x = 7.5$      42. $\frac{1}{2}x + 4 = 8; \frac{1}{2}x = 4$

**Solve the applications in Exercises 43 to 52.**

43. $D = 55t$; solve for $t$ if $D = 200$ miles.

44. $D = 55t$; solve for $t$ if $D = 450$ miles.

45. $D = 3r$; solve for $r$ if $D = 200$ miles.

46. $D = 3r$; solve for $r$ if $D = 450$ miles.

47. The 6% tax on a purchase, $p$, is $T = 0.06p$. If $T = \$0.10$, solve for $p$.

48. The 6% tax on a purchase, $p$, is $T = 0.06p$. If $T = \$0.25$, solve for $p$.

49. Electricity cost is $C = \$5.00 + \$0.03715x$, where $x$ is the number of kilowatt hours used. If $C = \$75$, solve for $x$.

50. Electricity cost is $C = \$5.00 + \$0.03715x$, where $x$ is the number of kilowatt hours used. If $C = \$50$, solve for $x$.

51. The area of a triangle is $A = \frac{1}{2}bh$. If $A = 15$ and $b = 3$, solve for $h$.

52. The area of a triangle is $A = \frac{1}{2}bh$. If $A = 28$ and $h = 7$, solve for $b$.

**In Exercises 53 to 56,**

  a. find the intersection $(x, y)$ of the two graphs in the figure, and

  b. substitute the $x$-coordinate of the intersection into both equations to check that the same $y$-coordinate is obtained.

53. $y = 3x - 3$ and $y = x + 1$

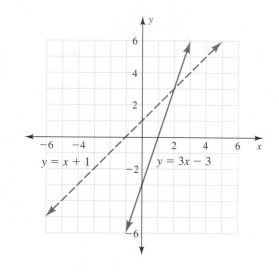

54. $y = 2x + 2$ and $y = 4x - 8$

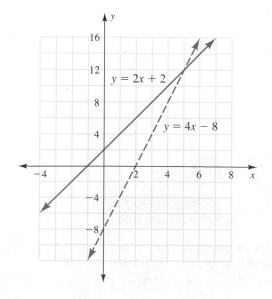

**55.** $y = 2x - 6$ and $y = -4x - 12$

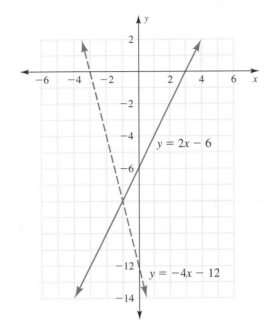

**56.** $y = 4x + 8$ and $y = 5x + 5$

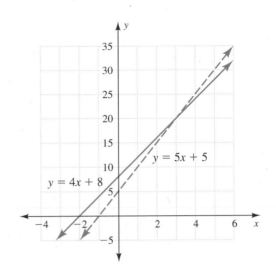

---

## 3.3    DISTRIBUTIVE PROPERTY AND FACTORING

**OBJECTIVES**
Identify the distributive property. ♦ Use the distributive property for mental arithmetic. ♦ Use multiplication (area model) to remove parentheses. ♦ Reverse the distributive property by factoring. ♦ Identify a common monomial factor. ♦ Identify the numerical coefficient.

**WARM-UP**
1. Multiply:

     **a.** $(-2)(-4)$      **b.** $(-3)(5)$      **c.** $-4 \cdot -6$      **d.** $3 \cdot -8$

     **e.** $0.5 \cdot -4$      **f.** $-\frac{1}{2}(7)$      **g.** $-\frac{1}{2} \cdot -\frac{1}{4}$      **h.** $-3(0.4)$

     **i.** $(-2)(-3)(-4)$

2. Factor into prime factors:

     **a.** 39      **b.** 28      **c.** 21      **d.** 45

     **e.** 30      **f.** 72

 ♦

I
n this section we learn the distributive property, which will give us more flexibility in simplifying expressions and in using the order of operations to solve equations.

### DISTRIBUTIVE PROPERTY

You have $6.50 and want to buy 3 pens at $0.99 and 2 notebooks at $1.89. The checkout line is long. You estimate the total cost to determine whether you have enough cash. Three times $1 plus 2 times $2 gives $7. You don't have enough cash, so you write a check.

The line still hasn't moved. You begin writing your check for the exact amount and want to mentally calculate the total cost. You figure the cost as 3 times $1.00 less $0.03 (Figure 9) plus 2 times $2.00 less $0.22 (Figure 10).

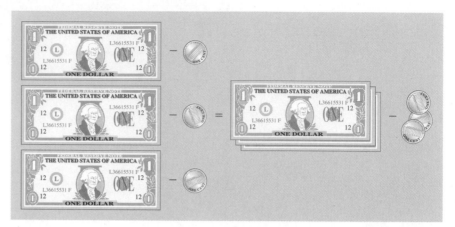

Figure 9

Figure 10

This checkout-line reasoning illustrates what mathematicians call the **distributive property**. In symbols this process is

$$3(\$0.99) = 3(\$1.00 - \$0.01) = 3(\$1.00) - 3(\$0.01) = \$3.00 - \$0.03 = \$2.97$$

$$2(\$1.89) = 2(\$2.00 - \$0.11) = 2(\$2.00) - 2(\$0.11) = \$4.00 - \$0.22 = \$3.78$$

The total is

$$\$2.97 + \$3.78 = \$6.75$$

The steps in color show the distributive property in action.

**Distributive Property**

> For all real numbers $a$, $b$, and $c$,
> $$a(b + c) = ab + ac$$

The name *distributive* may have been chosen because multiplying each term in the parentheses by $a$ is like dealing cards to each person in a game or serving cake to each guest at a wedding reception. Both dealing and serving are *distributing* actions.

CAUTION   The multiplication sign in $a(b + c)$ is omitted; it is assumed to be there. Multiplication may be shown with a dot, as in $a \cdot (b + c)$, or parentheses, as in $(a)(b + c)$.   ◆

For cost estimation the parentheses grouped the *$0.01 less than $1.00* or the *$0.11 less than $2.00*. Look carefully at how the parentheses group information in Example 1.

EXAMPLE 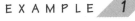 Translate each problem into an equation. Each may need parentheses and a condition.

**a.** You earn $0.27 a mile driving a truck. If you drive $x$ total miles in a day, with 175 miles driven by noon, what are the afternoon earnings, $y$?

**b.** Rental of a weed cutter is $20.00 for 3 hours plus $5.00 per additional hour. What is the cost if the cutter is used for $x$ hours?

SOLUTION    **a.** $y = 0$ if $x \le 175$, $y = 0.27(x - 175)$ if $x > 175$

**b.** $y = 20$ if $0 \le x \le 3$, $y = 20 + 5(x - 3)$ if $x > 3$

◆ ◆ ◆ ◆ ◆ ◆ ◆ ◆ ◆ ◆ ◆ ◆ ◆ ◆ ◆ ◆ ◆ ◆ ◆ ◆ ◆ ◆ ◆ ◆ ◆ ◆ ◆ ◆ ◆ ◆ ◆ ◆ ◆ ◆

**Distributive Property and Area.** Figure 11 contains rectangles marked in square units. The left and right sets of rectangles both contain 35 squares; they show that $5(4 + 3) = 5(4) + 5(3)$.

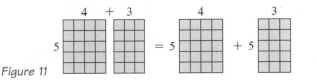

*Figure 11*

Suppose the top of the rectangle in Figure 12 is formed by two line segments of length $b$ and length $c$, respectively. The total length of the top is $b + c$. The area of the rectangle is the product of the width, $a$, and the length, $b + c$, or

$$\text{Area} = a(b + c) = ab + ac$$

The areas in Figures 11 and 12 illustrate the distributive property.

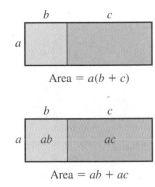

Area = $a(b + c)$

Area = $ab + ac$

*Figure 12*

EXAMPLE **2** State the total area in each figure.

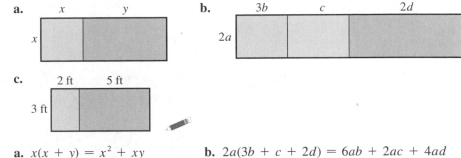

SOLUTION    **a.** $x(x + y) = x^2 + xy$          **b.** $2a(3b + c + 2d) = 6ab + 2ac + 4ad$

**c.** $3 \text{ ft}(2 \text{ ft} + 5 \text{ ft}) = 6 \text{ ft}^2 + 15 \text{ ft}^2 = 21 \text{ ft}^2$

Figure 13 shows the area of each part. In part c it is easier to add the 2 feet and 5 feet first and then multiply by 3 feet.

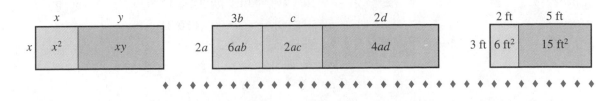

*Figure 13*

◆ ◆ ◆ ◆ ◆ ◆ ◆ ◆ ◆ ◆ ◆ ◆ ◆ ◆ ◆ ◆ ◆ ◆ ◆ ◆ ◆ ◆ ◆ ◆ ◆ ◆ ◆ ◆ ◆ ◆ ◆ ◆ ◆ ◆ ◆ ◆ ◆ ◆

In Example 2 both variables and units are multiplied, resulting in expressions containing $x^2$ and ft$^2$. Observe that variables are next to the numbers whereas units are one space away from the numbers. We will encounter these multiplications in several ways in subsequent examples and sections.

*Table Form.* Multiplication may be placed into a modified table to aid in visualizing the multiplication of each part. The term in front goes on the left side of the table, and the expression in the parentheses goes along the top row of the table. If we ignore the headings, the table looks like a rectangular area.

EXAMPLE 3

**a.** Use table form to multiply $7(1 + 3x + 4y)$.

| Multiply | 1 | +3x | +4y |
|----------|---|-----|-----|
| 7 | | | |

**b.** Use table form to multiply $3a(2a - 3b + 4c)$.

| Multiply | 2a | −3b | +4c |
|----------|----|-----|-----|
| 3a | | | |

SOLUTION   **a.** $7 + 21x + 28y$

| Multiply | 1 | +3x | +4y |
|----------|---|-----|-----|
| 7 | 7 | +21x | +28y |

**b.** $6a^2 - 9ab + 12ac$

| Multiply | 2a | −3b | +4c |
|----------|----|-----|-----|
| 3a | $6a^2$ | −9ab | +12ac |

The table form reminds us to multiply each part within the parentheses.

◆ ◆ ◆ ◆ ◆ ◆ ◆ ◆ ◆ ◆ ◆ ◆ ◆ ◆ ◆ ◆ ◆ ◆ ◆ ◆ ◆ ◆ ◆ ◆ ◆ ◆ ◆ ◆ ◆ ◆ ◆ ◆ ◆ ◆ ◆

*Other Examples.* In parts g and h of Example 4 we distribute a negative number. Look carefully at how it changes the negative inside the parentheses. For parts e, f, and h of Example 4, remember that the order of operations requires multiplication before addition and subtraction.

EXAMPLE 4

Use the distributive property to multiply these expressions:

**a.** $5(\$3.98) = 5(\$4.00 - \$0.02)$

**b.** $8(\$11.96) = 8(\$12.00 - \$0.04)$

**c.** $5(x + 3)$

**d.** $0.27(x - 175)$

**e.** $20 + 5(x - 3)$

**f.** $3 + 5(x - 2)$

**g.** $-3(x - 4)$

**h.** $13 - 4(x - 2)$ ✐

SOLUTION  **a.** $5(\$4.00 - \$0.02) = \$20.00 - \$0.10 = \$19.90$

**b.** $8(\$12.00 - \$0.04) = 8(\$12.00) - 8(\$0.04) = \$96.00 - \$0.32 = \$95.68$

**c.** $5 \cdot x + 5 \cdot 3 = 5x + 15$

**d.** $0.27(x) - (0.27)(175) = 0.27x - 47.25$

**e.** $20 + 5 \cdot x - 5 \cdot 3 = 20 + 5x - 15 = 5 + 5x$

**f.** $3 + 5 \cdot x - 5 \cdot 2 = 3 + 5x - 10 = 5x - 7$

**g.** $-3 \cdot x - (-3) \cdot 4 = -3x - (-12) = -3x + 12$

**h.** $13 - 4(x) - (-4)(2) = 13 - 4x + 8 = 21 - 4x$

◆ ◆ ◆ ◆ ◆ ◆ ◆ ◆ ◆ ◆ ◆ ◆ ◆ ◆ ◆ ◆ ◆ ◆ ◆ ◆ ◆ ◆ ◆ ◆ ◆ ◆ ◆ ◆ ◆ ◆ ◆ ◆ ◆

The distributive property also applies when the number being distributed does not show. We assume there to be a 1 in front of the parentheses in $-(-4x - 12)$. Thus

$$-(-4x - 12) = -1(-4x - 12) = (-1)(-4x) + (-1)(-12) = +4x + 12$$

EXAMPLE 5  Use the assumed 1 in these multiplications.

**a.** $-(x + 4)$    **b.** $-(x - 3)$    **c.** $-(-4)$ ✐

SOLUTION  **a.** $-1(x + 4) = (-1)(x) + (-1)(4) = -1x - 4 = -x - 4$

**b.** $-1(x - 3) = (-1)(x) + (-1)(-3) = -1x + 3 = -x + 3$

**c.** $-1(-4) = +4$

◆ ◆ ◆ ◆ ◆ ◆ ◆ ◆ ◆ ◆ ◆ ◆ ◆ ◆ ◆ ◆ ◆ ◆ ◆ ◆ ◆ ◆ ◆ ◆ ◆ ◆ ◆ ◆ ◆ ◆ ◆ ◆ ◆

---

### JUST FOR FUN

*Shopping at a supermarket for a last-minute party, I plopped a bag containing five tomatoes on the checkout counter. The cashier put the bag on the scale and started punching numbers. Then, to my surprise, she paused, took the tomatoes out of the bag, and began weighing them individually.*

*"Why don't you weigh them all at once?" I asked.*

*"Don't be silly," she replied. "They're all different sizes."\**

Suppose the five tomatoes weighed 0.83, 0.34, 0.47, 0.73, and 0.42 pounds, respectively. If tomatoes cost \$1.39 a pound, what is the total cost of the five tomatoes? How does the distributive property apply to the story?

---

### REVERSING THE DISTRIBUTIVE PROPERTY: FACTORING

The distributive property also works well in reverse. Figure 14 illustrates an example where different objects have the same price per unit.

---

*This anecdote was contributed to the *Reader's Digest* by Daniel P. White, III. Reprinted with permission from the July 1992 *Reader's Digest*. Copyright © 1992 by The Reader's Digest Assn., Inc.

EXAMPLE 6   *Grocery prices*   Find the total cost of 6 cans of frozen juice at $0.89 per can and 4 pounds of grapes at $0.89 per pound.

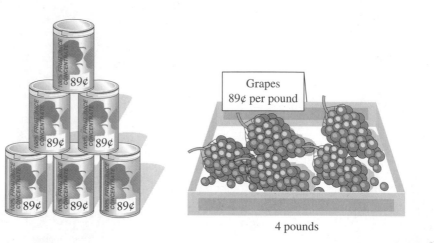

*Figure 14*

SOLUTION   The total cost is 6 cans times $0.89 per can plus 4 pounds times $0.89 per pound.

$$6 \text{ cans} \cdot \frac{\$0.89}{\text{cans}} + 4 \text{ pounds} \cdot \frac{\$0.89}{\text{pounds}} = 6(0.89) + 4(0.89)$$

$$= 0.89(6 + 4) = 0.89(10) = \$8.90$$

◆ ◆ ◆ ◆ ◆ ◆ ◆ ◆ ◆ ◆ ◆ ◆ ◆ ◆ ◆ ◆ ◆ ◆ ◆ ◆ ◆ ◆ ◆ ◆ ◆ ◆ ◆ ◆ ◆ ◆ ◆ ◆ ◆

We do not add cans and pounds. The units are multiplied, the cans and pounds cancel out, and only the dollars remain. When we write the equation

$$0.89(6) + 0.89(4) = 0.89(6 + 4)$$

we are reversing the distributive property, or **factoring**. In this example we factored out the common dollar amount, $0.89. Not surprisingly, 0.89 is called the *common factor*.

If we know the area of a rectangle and one side, we can factor to find the length and width. Example 7 shows several rectangles, with only portions of their dimensions given.

EXAMPLE 7   *Areas*   The area and one side are given for each portion of the rectangle. Use factoring to determine the other sides.

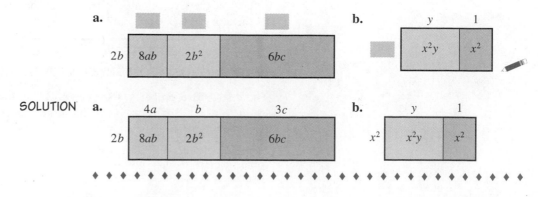

◆ ◆ ◆ ◆ ◆ ◆ ◆ ◆ ◆ ◆ ◆ ◆ ◆ ◆ ◆ ◆ ◆ ◆ ◆ ◆ ◆ ◆ ◆ ◆ ◆ ◆ ◆ ◆ ◆ ◆ ◆ ◆ ◆ ◆

Both the grocery example and the area example indicate that we may reverse the distributive process. Thus $a(b + c) = ab + ac$ is also written $ab + ac = a(b + c)$. This gives us another useful form of the distributive property:

**Factored Form of the Distributive Property**

$$ab + ac = a(b + c)$$

In factored form, $a$ is the *common factor*.

**EXAMPLE 8**

Identify the common factor or common unit, and then apply the distributive property to change each expression to factored form. Complete all additions and subtractions.

**a.** 7 cans at \$1.29 per can + 3 pounds at \$1.29 per pound

**b.** $8b + 9b$

**c.** $7x + 7y$

**d.** $12x^2 + 4x - 8$

**e.** $5a + 10b + 5c$

**f.** 5 seconds + 14 seconds + 8 seconds

**g.** 10 feet − 4.5 feet

**SOLUTION**

**a.** \$1.29 is the common factor; $\$1.29(7 + 3) = \$1.29(10) - \$12.90$. The units (cans, pounds) both cancel.

**b.** $b$ is the common factor; $b(8 + 9) = b(17) = 17b$.

**c.** 7 is the common factor; $7(x + y)$

**d.** 4 is the common factor; $4(3x^2 + x - 2)$. We might have factored out only the 2 and had $2(6x^2 + 2x - 4)$, but we should always factor out the largest common factor.

**e.** 5 is the common factor; $5(a + 2b + c)$

**f.** Seconds is the common unit; $(5 + 14 + 8)$ seconds = 27 seconds. *Seconds* is put after the parentheses because units are normally written after a number. The numbers are then added. It is unusual in mathematics to factor units; however, they may be treated as if they factor.

**g.** Feet is the common unit; $(10 - 4.5)$ feet = 5.5 feet.

♦ ♦ ♦ ♦ ♦ ♦ ♦ ♦ ♦ ♦ ♦ ♦ ♦ ♦ ♦ ♦ ♦ ♦ ♦ ♦ ♦ ♦ ♦ ♦ ♦ ♦ ♦ ♦ ♦ ♦ ♦ ♦ ♦ ♦

**EXAMPLE 9**

We also apply factoring to tables. Factor the following tables, and place the common factor on the left side.

**a.** Use table form to factor $18m - 24n + 30p$.

| Factor | | | |
|--------|------|------|------|
| | $18m$ | $-24n$ | $+30p$ |

**b.** Use table form to factor $7x + xy - xz$.

| Factor | | | |
|---|---|---|---|
| | $7x$ | $+xy$ | $-xz$ |

SOLUTION   **a.** The common factor is 6:

$$18m - 24n + 30p = 6(3m - 4n + 5p)$$

Observe the placement of $3m$, $-4n$, and $5p$ above the expression to be factored.

| Factor | $3m$ | $-4n$ | $+5p$ |
|---|---|---|---|
| 6 | $18m$ | $-24n$ | $+30p$ |

**b.** The common factor is $x$:

$$7x + xy - xz = x(7 + y - z)$$

Observe the placement of 7, $+y$, and $-z$ above the expression to be factored.

| Factor | $+7$ | $+y$ | $-z$ |
|---|---|---|---|
| $x$ | $+7x$ | $+xy$ | $-xz$ |

◆ ◆ ◆ ◆ ◆ ◆ ◆ ◆ ◆ ◆ ◆ ◆ ◆ ◆ ◆ ◆ ◆ ◆ ◆ ◆ ◆ ◆ ◆ ◆ ◆ ◆ ◆ ◆ ◆ ◆ ◆

Vocabulary.   In Examples 8 and 9 *the numbers and letters that are factored out of the expressions are called* **common monomial factors**. *A* **monomial** *is a one-term expression containing only variables with positive integer powers. The terms include the addition or subtraction signs. In Example 9a the term 18m is assumed to be positive. The +18, −24, and +30 are numerical coefficients. The* **numerical coefficient** *is the sign and number factor of a term.*

## EXERCISES 3.3

Explain, using complete sentences, how you would use the distributive property to mentally obtain the exact answer to each of the products in Exercises 1 to 4. Use your description to find the product.

1. 4 times $4.97      2. 7 times $5.99

3. 3 times $10.98      4. 6 times $7.96

Simplify the expressions in Exercises 5 and 6.

5. a. $6(x + 2)$    b. $-3(x - 3)$    c. $-6(x + 4)$

   d. $-(2 - x)$    e. $x(x - 3)$

   f. $-3(x + y - 5)$

6. a. $4(x - 4)$    b. $-4(x + 2)$    c. $-5(x - 3)$

   d. $-(x - 3)$    e. $y(4 + y)$

   f. $-5(2x - 4 + y)$

In Exercises 7 to 10 find the area of the figure.

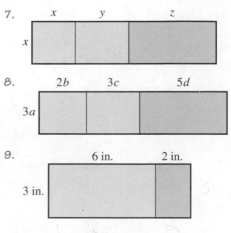

7.

8.

9.

10.

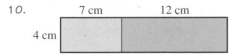

| 7 cm | 12 cm |
|------|-------|
4 cm

**In Exercises 11 to 16 complete the table.**

11.

| Multiply | 3 | −2x | +3y |
|----------|---|-----|-----|
| 8 | | | |

12.

| Multiply | 2x | −3y | +z |
|----------|----|-----|----|
| 5 | | | |

13.

| Multiply | a | −b | +c |
|----------|---|----|----|
| 2a | | | |

14.

| Multiply | 5a | −3b | +c |
|----------|----|-----|----|
| 5a | | | |

15.

| Multiply | 6a | −4b | +3c |
|----------|----|-----|-----|
| −3a | | | |

16.

| Multiply | 6x | −y | −3z |
|----------|----|----|-----|
| −2x | | | |

**Find the missing dimensions of the rectangles in Exercises 17 to 20.**

17.

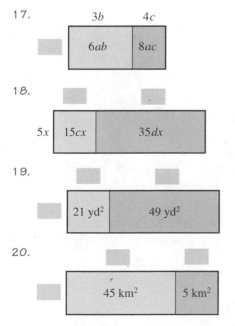

| 3b | 4c |
|-----|-----|
| 6ab | 8ac |

18.

| | | |
|-----|------|------|
| 5x | 15cx | 35dx |

19.

| | | |
|-----|---------|---------|
| | 21 yd² | 49 yd² |

20.

| | | |
|-----|----------|---------|
| | 45 km² | 5 km² |

**For Exercises 21 to 30 identify the common monomial factor, if any. Then factor, and add or subtract when possible.**

21. $5a + 5b$

22. $8x + 8y$

23.

| Factor | | |
|--------|----|-----|
| | 2x | −4y |

24.

| Factor | | |
|--------|----|-----|
| | 6x | −3y |

25. $5a + 6a$

26. $8c − 5c$

27. 5 meters + 8.5 meters

28. 6 grams + 3.4 grams

29. 5 pounds at $0.79 per pound + 4 cans at $0.79 per can

30. 4 yards at $2.59 per yard + 7 packages at $2.59 per package

**Identify the numerical coefficient in each of the terms in Exercises 31 and 32.**

31. a. $-4x$    b. $x$    c. $-x$

32. a. $-3x^2$    b. $5x$    c. $y$

**For Exercises 33 to 38 write equations containing parentheses (and conditions, as needed) for each problem situation.**

33. Markus had $150 in his checking account. He wrote a check for $x$ dollars and forgot to record it. He wrote another check for $20. What is now the checking account balance, $y$?

34. A rental car costs $65 for the first 100 miles and $0.15 per mile thereafter. If $x$ is the total miles driven, what is the total cost, $y$?

35. A pressure washer costs $30 for the first 4 hours and $10 per additional hour. If $x$ is the total hours rented, what is the total cost, $y$?

36. A transcript costs $2.00 for the first copy and $1.50 for each additional copy. What is the total cost, $y$, of $x$ copies?

37. At an outdoor concert the staff adds 3 seats to each row of $x$ seats. There are 25 rows altogether. How many total seats, $y$, are there?

38. At an outdoor wedding, the chair rental company places 3 additional seats to each row of $x$ seats. There are $x$ rows altogether. How many total seats, $y$, are there?

**In Exercises 39 and 40 explain in complete sentences the error made by the student who obtained the indicated wrong answer, and give the correct answer.**

39. $7 − 4(x − 3)$ simplifies to $3x − 9$.

40. $5 − 3(x − 1)$ simplifies to $3x − 2$.

For Exercises 41 to 46 identify the property used in each equation—associative, commutative, distributive—or label the statement false.

41. $3(4)(5) = 3(5)(4)$

42. $3(4 + 5) = 3(4) + 3(5)$

43. $3 + 4 + 5 = 4 + 3 + 5$

44. $3(4 \cdot 5) = (3 \cdot 4) \cdot 5$

45. $3 + (4 + 5) = (3 + 4) + 5$

46. $3 + 4 \cdot 5 = 3 + 4 \cdot 3 + 5$

The expression $a + b$ has two terms. Placing parentheses around the $a + b$ makes one factor, $(a + b)$. The term $x^2$ has two factors, $x \cdot x$. In Exercises 47 to 51 how many factors are there in each expression?

47. $4(a + b)(c + d)$          48. $x(x + y)$

49. $(c + d)(c + d)(c + d)(c + d)$     50. $ab(c + d)$

51. $abcd(e + f)$

In Exercises 52 to 57 how many terms are there in each expression?

52. $xxxxxxxxxxx$          53. $w + x + y + 2z$

54. $x^4 + 4x^3 + 6x^2 + 4x + 1$     55. $yyyyyy$

56. $x^2 - y^2$          57. $ax^2 + bx + c$

58. Write in words how you distinguish factors from terms.

59. Is $a(bc)$ an example of the distributive property? (*Hint*: Compare $4(5 \cdot 6)$ with $4(5) + 4(6)$ and $4(5) \cdot 4(6)$.)

60. In factoring $2x + 2y + 2$ a student writes $2(x + y)$. Explain in complete sentences what is wrong, and provide the correct answer.

PROJECT

61. ***Distinguishing Terms from Factors***

a. Make a two-column list. Head one side *Expressions with Two or More Terms* and the other *Expressions with One Term and Multiple Factors*.

b. Record each of these expressions under the appropriate heading:

| | | | |
|---|---|---|---|
| $12xy$ | $3x + 2y - 7z$ | $2x - y$ | $2 + \sqrt{3}$ |
| $abc$ | $a + b + c$ | $\dfrac{a + b}{c}$ | $\dfrac{a}{c} + \dfrac{b}{c}$ |
| $\dfrac{12xy}{4x^2}$ | $15ab$ | $\dfrac{-b}{2a}$ | $\frac{1}{2}h(a + b)$ |
| $\dfrac{a}{b}$ | $\dfrac{abc}{abc}$ | | |
| | $\frac{1}{2}ah + \frac{1}{2}bh$ | $\dfrac{-b}{2a} + \dfrac{\sqrt{b^2 - 4ac}}{2a}$ | |

c. Circle the expressions that are equivalent, and connect them with a line.

---

## 3.4  SOLVING EQUATIONS: TABLES AND GRAPHS AND THE DISTRIBUTIVE PROPERTY

**OBJECTIVES**  Solve equations containing parentheses by using tables and graphs. ♦ Solve equations with expressions on left and right sides by using tables and graphs. ♦ Identify and add like terms. ♦ Use the distributive property to solve equations containing parentheses.

**WARM-UP**  Simplify:

1. $4(x + 3)$          2. $-2(x + 1)$          3. $-4(x - 2)$

4. $-(x + 1)$          5. $13 - 4(x - 2)$

Factor:

6. $4x - 8$          7. $2x - 6$          8. $3x - 3$

9. $-4x + 8$          10. $2x + 3x - 4x$          ♦

---

**JUST FOR FUN**

There is a saying that when something is filled halfway, some will say that it is half full and others will say that it is half empty. Equating these two ideas, we have

$$\tfrac{1}{2} \text{ full} = \tfrac{1}{2} \text{ empty}$$

The multiplication property of equations states that we can multiply both sides of an equation by the same number. Multiplying both sides by two gives

$$2 \cdot \tfrac{1}{2} \text{ full} = 2 \cdot \tfrac{1}{2} \text{ empty}$$

$$\text{full} = \text{empty!}$$

In this section we use tables and graphs to solve equations with inputs $x$ on both sides and to solve equations symbolically using the distributive property and addition of like terms.

### SOLVING EQUATIONS: TABLES AND GRAPHS

We now use tables and graphs to solve equations with variables on both sides. We first solve the equation by finding a common point in the tables for the left and right sides of the equation. We then graph the left and right sides to determine the point of intersection.

**EXAMPLE 1**        Solve $3(x - 1) = x + 1$ using Tables 9 and 10 and the graph in Figure 15.

| $x$ | $3(x - 1)$ |
|-----|-----------|
| $-1$ | $-6$ |
| $0$ | $-3$ |
| $1$ | $0$ |
| $2$ | $3$ |
| $3$ | $6$ |

Table 9

| $x$ | $x + 1$ |
|-----|---------|
| $-1$ | $0$ |
| $0$ | $1$ |
| $1$ | $2$ |
| $2$ | $3$ |
| $3$ | $4$ |

Table 10

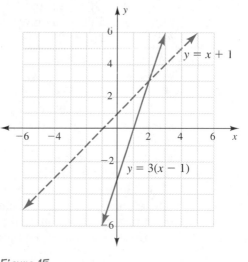

Figure 15

**SOLUTION**   The point $(2, 3)$ appears in both tables. Thus the input $x = 2$ makes the two expressions $3(x - 1)$ and $x + 1$ equal, and $x = 2$ solves the equation $3(x - 1) = x + 1$.

***Check***: $3(2 - 1) \overset{?}{=} 2 + 1$  ✔

To solve $3(x - 1) = x + 1$ from the graph, we look for the point of intersection of the two graphs, $y = 3(x - 1)$ and $y = x + 1$. The graphs intersect at $(2, 3)$. This means that the input $x = 2$ makes both $3(x - 1)$ and $x + 1$ equal to $y = 3$. Thus $x = 2$ solves the equation.

***Check***: $3(2 - 1) \overset{?}{=} 2 + 1$  ✔

EXAMPLE **2**    Solve $4(x - 2) = 2(x + 1)$ with tables and a graph.

SOLUTION    First we construct Tables 11 and 12. Extend the tables until a common coordinate is found. Both tables contain the coordinates (5, 12). At $x = 5$ both $4(x - 2)$ and $2(x + 1)$ equal 12, so $x = 5$ solves the equation $4(x - 2) = 2(x + 1)$.

***Check***: $4(5 - 2) \overset{?}{=} 2(5 + 1)$    ✔

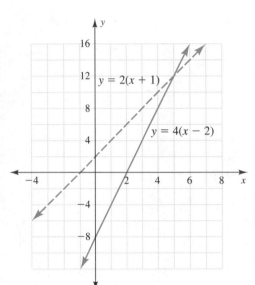

| $x$ | $4(x - 2)$ |
|-----|------------|
| $-2$ | $-16$ |
| $-1$ | $-12$ |
| $0$ | $-8$ |
| $1$ | $-4$ |
| $2$ | $0$ |
| $3$ | $4$ |
| $4$ | $8$ |
| $5$ | $12$ |

*Table 11*

| $x$ | $2(x + 1)$ |
|-----|------------|
| $-2$ | $-2$ |
| $-1$ | $0$ |
| $0$ | $2$ |
| $1$ | $4$ |
| $2$ | $6$ |
| $3$ | $8$ |
| $4$ | $10$ |
| $5$ | $12$ |

*Table 12*

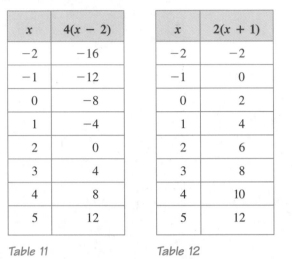

*Figure 16*

Next we plot the graph of each side of the equation, as shown in Figure 16. The coordinates (5, 12) locate the point of intersection of the two graphs, $y = 4(x - 2)$ and $y = 2(x + 1)$. Thus $x = 5$ as an input to both $4(x - 2)$ and $2(x + 1)$ gives 12 as an output, so $x = 5$ solves the equation $4(x - 2) = 2(x + 1)$.

***Check***: $4(5 - 2) \overset{?}{=} 2(5 + 1)$    ✔

♦ ♦ ♦ ♦ ♦ ♦ ♦ ♦ ♦ ♦ ♦ ♦ ♦ ♦ ♦ ♦ ♦ ♦ ♦ ♦ ♦ ♦ ♦ ♦ ♦ ♦ ♦ ♦ ♦ ♦ ♦ ♦ ♦ ♦

It is not reasonable to always solve an equation both ways. Both ways are shown here to clarify the meaning of finding a solution. In reality, people choose any one of a number of solution methods based on what seems to work best. Only the experience gained from practice will help you choose the best method for future problems.

If you are solving only by table, there is no need to write additional coordinates once the solution (a common set of coordinates) is found. If you are using the tables to graph, find at least three sets of coordinates. Using a graphing calculator or other graphing utility is also an option.

---

The intersection of the graphs of each side of an equation determines an ordered pair $(x, y)$. The input, $x$, is the number that solves the equation. The number $y$ is the output shared by both sides of the equation.

---

Example 3 illustrates how difficult it may be to identify a fractional solution.

EXAMPLE **3**    Solve $6 - 5x = 3(1 - x)$ with tables and a graph.

SOLUTION    Tables 13 and 14 have no coordinate point in common. The coordinates are closest between $x = 1$ and $x = 2$.

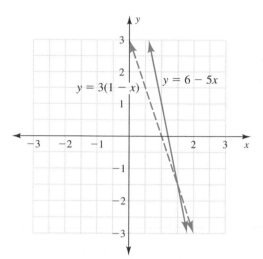

Figure 17

| $x$ | $6 - 5x$ |
|-----|----------|
| $-1$ | 11 |
| 0 | 6 |
| 1 | 1 |
| 2 | $-4$ |
| 3 | $-9$ |

Table 13

| $x$ | $3(1 - x)$ |
|-----|------------|
| $-1$ | 6 |
| 0 | 3 |
| 1 | 0 |
| 2 | $-3$ |
| 3 | $-6$ |

Table 14

The graph in Figure 17 shows the point of intersection to be a fraction for both $x$ and $y$. We guess $\left(1\frac{1}{2}, -1\frac{1}{2}\right)$ and check it.

◆ ◆ ◆ ◆ ◆ ◆ ◆ ◆ ◆ ◆ ◆ ◆ ◆ ◆ ◆ ◆ ◆ ◆ ◆ ◆ ◆ ◆ ◆ ◆ ◆ ◆ ◆ ◆ ◆ ◆ ◆ ◆ ◆ ◆

We will solve Example 3 with symbols in Example 7. For many equations like the one in Example 3, symbolic solutions are easier than tables or graphs. If the symbolic method becomes difficult or even impossible, we can turn to graphing calculators or computers for a solution. We now summarize addition of like terms, a skill needed to solve equations with symbols.

### ADDING LIKE TERMS

$3x + 4x + 2x = x(3 + 4 + 2) = 9x$

Figure 18

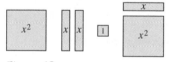

Figure 19

The distributive property allows us to add terms that are alike, or *like terms*. In Figure 18 the rectangles with area $x$ are grouped, $3x + 4x + 2x$. The shapes are also known as algebra tiles. The distributive property indicates that we factor the $x$ and add up the numerical coefficients: $3 + 4 + 2 = 9$. Thus $3x + 4x + 2x = 9x$. We verify this by counting the tiles.

**Like terms** *have identical variables.* The expression $3a + 4a - 5a$ contains three like terms because all terms contain only the variable $a$. Neither the expression $x^2 + 2x + 1$ nor the expression $2a + 2b$ contains like terms. The exponents on the variables of like terms must be identical: $3x^2 + 4x^2 - 5x^2$, for example. In Figure 19 the areas are given in each shape. If we add up the areas of the like shapes, we obtain the total area,

$$x^2 + x^2 + x + x + x + 1 = 2x^2 + 3x + 1$$

---

To add like terms, we add the numerical coefficients.

---

E X A M P L E   **4**

Identify the expressions containing like terms. Add the like terms. Factor, if possible, any expressions that do not contain like terms.

**a.** $4x + 4y$          **b.** $4x - 5x$          **c.** $3x^2 + 15x^2 - 4x^2$

**d.** $7x - 7x^2 + 7x^3$          **e.** $3xy + 4yx$

SOLUTION   **a.** Contains no like terms; it factors to $4(x + y)$.

**b.** Contains two like terms; $4x - 5x = (4 - 5)x = -1x = -x$.

**c.** Contains three like terms; $(3 + 15 - 4)x^2 = 14x^2$.

**d.** Contains no like terms; it factors to $7x(1 - x + x^2)$.

**e.** Contains like terms because $xy = yx$; $(3 + 4)xy = 7xy$.

◆ ◆ ◆ ◆ ◆ ◆ ◆ ◆ ◆ ◆ ◆ ◆ ◆ ◆ ◆ ◆ ◆ ◆ ◆ ◆ ◆ ◆ ◆ ◆ ◆ ◆ ◆ ◆ ◆ ◆ ◆ ◆ ◆ ◆

SOLVING EQUATIONS WITH SYMBOLS

Before or after using the addition and multiplication properties, we can use the distributive property and addition of like terms to simplify equations containing parentheses.

EXAMPLE  **5**

Solve for $x$:   $13 - 4(x - 2) = 1$.

SOLUTION

$13 - 4(x - 2) = 1$

$13 - 4x + 8 = 1$         Use the distributive property.

$21 - 4x = 1$         Add like terms.

$21 - 4x - 21 = 1 - 21$         Subtract 21 from both sides (addition property).

$-4x = -20$

$-4x \div -4 = -20 \div -4$         Divide both sides by $-4$ (multiplication property).

$x = 5$

*Check*: $13 - 4(5 - 2) \overset{?}{=} 1$  ✔

◆ ◆ ◆ ◆ ◆ ◆ ◆ ◆ ◆ ◆ ◆ ◆ ◆ ◆ ◆ ◆ ◆ ◆ ◆ ◆ ◆ ◆ ◆ ◆ ◆ ◆ ◆ ◆ ◆ ◆ ◆ ◆ ◆ ◆

The addition and multiplication properties of equations apply to variables as well as numbers and permit us to solve equations containing variables on both sides of the equal sign. When an equation has a variable on both sides, our first goal is to get the variable on one side only. We *make fewer errors if we try to keep the variable's coefficient positive when we add variable terms to each side.*

EXAMPLE  **6**

Solve for $x$:   $2(x - 3) = -4(x + 3)$.

SOLUTION

$2(x - 3) = -4(x + 3)$

$2x - 6 = -4x - 12$         Distributive property

$2x - 6 + 4x = -4x - 12 + 4x$         Add 4x (addition property) to keep the coefficient of the variable term positive.

$6x - 6 = -12$         Add like terms.

$6x - 6 + 6 = -12 + 6$         Add 6 (addition property).

$6x = -6$         Add like terms.

$6x \div 6 = -6 \div 6$         Divide by 6 (multiplication property).

$x = -1$

*Check*: $2(-1 - 3) \overset{?}{=} -4(-1 + 3)$  ✔

◆ ◆ ◆ ◆ ◆ ◆ ◆ ◆ ◆ ◆ ◆ ◆ ◆ ◆ ◆ ◆ ◆ ◆ ◆ ◆ ◆ ◆ ◆ ◆ ◆ ◆ ◆ ◆ ◆ ◆ ◆ ◆ ◆ ◆

Example 7 gives the symbolic solution to the equation of Example 3.

EXAMPLE  **7**

Solve for $x$:   $6 - 5x = 3(1 - x)$.

SOLUTION

$6 - 5x = 3(1 - x)$

$6 - 5x = 3 - 3x$         Distributive property

$6 - 5x + 5x = 3 - 3x + 5x$         Add 5x (addition property) to keep the coefficient positive.

$6 = 3 + 2x$         Add like terms.

$6 - 3 = 3 + 2x - 3$         Subtract 3 (addition property).

$3 = 2x$         Add like terms.

$3 \div 2 = 2x \div 2$         Divide both sides by 2 (multiplication property).

$1\frac{1}{2} = x$

*Check*: $6 - 5\left(1\frac{1}{2}\right) \overset{?}{=} 3\left(1 - 1\frac{1}{2}\right)$  ✔

◆ ◆ ◆ ◆ ◆ ◆ ◆ ◆ ◆ ◆ ◆ ◆ ◆ ◆ ◆ ◆ ◆ ◆ ◆ ◆ ◆ ◆ ◆ ◆ ◆ ◆ ◆ ◆ ◆ ◆ ◆ ◆ ◆ ◆

**EXERCISES 3.4**

**Solve the equations in Exercises 1 to 4 for *x* by completing the given tables.**

1. $5x - 8 = 2(x + 2)$

| x | 5x − 8 |
|---|---|
| −1 | |
| 0 | |
| 1 | |
| 2 | |
| 3 | |
| 4 | |

| x | 2(x + 2) |
|---|---|
| −1 | |
| 0 | |
| 1 | |
| 2 | |
| 3 | |
| 4 | |

2. $7(x - 5) = 3 - 12x$

| x | 7(x − 5) |
|---|---|
| −1 | |
| 0 | |
| 1 | |
| 2 | |
| 3 | |
| 4 | |

| x | 3 − 12x |
|---|---|
| −1 | |
| 0 | |
| 1 | |
| 2 | |
| 3 | |
| 4 | |

3. $3(x - 3) = 6(x - 2)$

| x | 3(x − 3) |
|---|---|
| −1 | |
| 0 | |
| 1 | |
| 2 | |
| 3 | |
| 4 | |

| x | 6(x − 2) |
|---|---|
| −1 | |
| 0 | |
| 1 | |
| 2 | |
| 3 | |
| 4 | |

4. $4(x + 1) = 2(3x - 1)$

| x | 4(x + 1) |
|---|---|
| −1 | |
| 0 | |
| 1 | |
| 2 | |
| 3 | |
| 4 | |

| x | 2(3x − 1) |
|---|---|
| −1 | |
| 0 | |
| 1 | |
| 2 | |
| 3 | |
| 4 | |

5. Solve $2(4 + 3x) = 3x + 2$ with the figure.

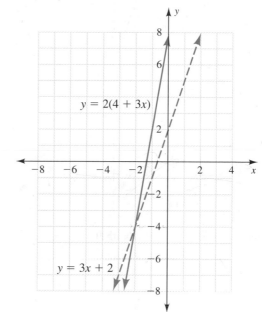

6. Solve $3(3 + x) = 2(2 - x)$ with the figure.

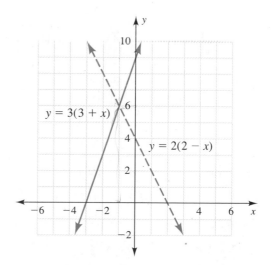

7. Solve $3(2 - x) = 4 - x$ with the figure.

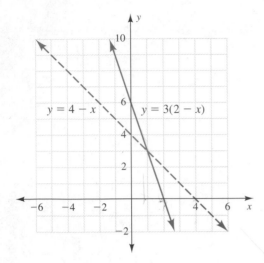

8. Solve $1 - 3x = 2(x - 2)$ with the figure.

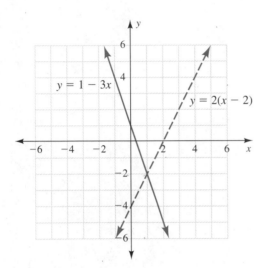

9. Use the figure in Exercise 5 to solve

   a. $2(4 + 3x) = 8$     b. $3x + 2 = 8$

10. Use the figure in Exercise 6 to solve

   a. $3(3 + x) = 0$     b. $2(2 - x) = 0$

11. Use the figure in Exercise 7 to solve

   a. $3(2 - x) = 6$     b. $4 - x = 6$

12. Use the figure in Exercise 8 to solve

   a. $2(x - 2) = 2$     b. $1 - 3x = 4$

13. Make a table for and graph $y = 8 - 3(x + 2)$. Solve $8 - 3(x + 2) = 11$ from the graph.

14. Make a table for and graph $y = 8 - 3(x + 4)$. Solve $8 - 3(x + 4) = 2$ from the graph.

15. Use your graph from Exercise 13 to solve $8 - 3(x + 2) = -3$.

16. Use your graph from Exercise 14 to solve $8 - 3(x + 4) = 0$.

The figures in Exercises 17 and 18 contain sets of algebra tiles. What expression is represented by each figure? Write the expression without parentheses and in factored form.

17.

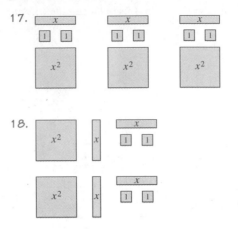

18.

**In Exercises 19 and 20 add like terms.**

19. a. $-3x + 4x - 6x + 8x$

   b. $-6y^2 + 8y^2 - 10y^2 - 3y^2$

   c. $2a + 6 + 3a + 9$

   d. $3(x + 1) - 2(x + 1)$

   e. $3(x - 4) + 4(4 - x)$

20. a. $-4y - 6y + 8y - 10y$

   b. $+7x^2 - 12x^2 - 5x^2 + 8x^2$

   c. $5x - 15 + 4x - 12$

   d. $4(x - 1) - 3(x - 1)$

   e. $7(x - 3) + 5(3 - x)$

**In Exercises 21 and 22 simplify the expressions.**

21. a. 15 seconds + 30 seconds − 45 seconds

   b. 3 inches + 11 inches − 7 inches

   c. 6.2 meters + 5.8 meters

   d. $2\frac{1}{2}$ inches + $5\frac{1}{2}$ inches

22. a. 50 seconds − 25 seconds + 75 seconds

   b. 8 inches − 6 inches + 3 inches

   c. 7.5 meters + 4.6 meters

   d. $3\frac{1}{2}$ inches + $8\frac{1}{2}$ inches

**In Exercises 23 to 26 change to like units or common denominators and add.**

23. 2 yards + 5 feet + 12 inches

24. 1 yard + 7 feet + 18 inches

25. $\frac{1}{2} + \frac{1}{4} + \frac{5}{8}$

26. $\frac{2}{3} + \frac{3}{4} + \frac{1}{2}$

**Solve the equations in Exercises 27 to 38 using any method.**

27. $2(x - 3) = 0$

28. $3(x - 1) = 2$

29. $-2(x + 1) = 5$

30. $-3(x - 2) = 0$

31. $2(3x + 1) = 5 - 3x$

32. $3(3x + 1) = 13 - 6x$

33. $4x + 6 = 2(1 + 3x) + 1$

34. $7x + 2 = 3(1 + x) + 2$

35. $7(x + 1) = 11 + x$

36. $3(x + 2) = 4 - x$

37. $5 - 2(x - 3) = 3(x - 3)$

38. $7 - 2(x - 1) = 5(x - 1)$

**For Exercises 39 to 44 explain in complete sentences what was done to get from the first equation to the second equation. What property of equations or numbers justifies the step?**

39. $2(2x + 3) = 2 + 6x + 1$ to $4x + 6 = 2 + 6x + 1$

40. $7x + 2 = 3 + 3x + 2$ to $4x + 2 = 3 + 2$

41. $7x + 7 = 11 + x$ to $7x = 4 + x$

42. $3(x + 2) = 4 - x$ to $3x + 6 = 4 - x$

43. $3x + 6 = 4 - x$ to $4x + 6 = 4$

44. $4x + 6 = 6x + 2 + 1$ to $4x + 6 = 6x + 3$

**For Exercises 45 to 50 write an equation containing parentheses for each problem situation, then solve the equation.**

45. Markus had $150 in his checking account. He wrote a check for $x$ dollars and forgot to record the amount. He wrote another for $20. What was the amount of the unknown check if the current balance is $98.75?

46. A rental car costs a basic fee of $65 for the first 100 miles and $0.15 per mile over 100 miles. How many miles, $x$, may be driven for a total of $100?

47. A pressure washer costs $30 for the first 4 hours and $10 per additional hour. What is the total hours rented, $x$, if the total cost is $70?

48. A transcript costs $2.00 for the first copy and $1.50 for each additional copy. If the total cost is $6.50, what is the number of copies made, $x$?

49. At an outdoor concert the staff adds 3 seats to each row of $x$ seats. There are 25 rows altogether, and there are 1200 seats altogether. How many seats, $x$, were in the original rows?

50. At an outdoor wedding the chair rental company places 3 additional seats at the end of each row of $x$ seats. There are 20 rows of seats and 440 seats altogether. How many seats, $x$, were in the original rows?

PROJECTS

51. *Equivalent Expressions.* When we add like terms or simplify an expression, the result must equal the original expression for *all* inputs.

a. Is the statement $2x + 3x = 5x^2$ true for all inputs, true for some inputs, or never true? (*Hint*: Try $x = 3, 7, 10$.)

b. Is the statement $5 - 2(x - 3) = 3(x - 3)$ true for all inputs, true for some inputs, or never true? (*Hint*: Try $x = 3, 7, 10$.)

c. For what two numbers is the statement in part a true? (*Hint*: Graph each side.)

d. For what number is the statement in part b true?

e. Use the distributive property or add like terms to correctly simplify the expression $2x + 3x$ and the expression $5 - 2(x - 3)$.

f. Sketch algebra tiles to show why $2x + 3x$ does not equal $5x^2$.

52. *Adding Like Terms.* Answer each question, and then describe how each question and answer relate to the concept of adding like terms. (*Hint*: Look up the source of the words *denomination* and *denominator* in the dictionary.)

a. Why do we find a common denominator when adding or subtracting fractions?

b. Why do we line up the decimal points in adding decimals?

c. Why do we change to the same units when adding 15 feet + 3 yards + 145 inches?

d. Why does a bank deposit slip (see the figure) for cash transactions provide space to write each type of coin and each denomination of currency?

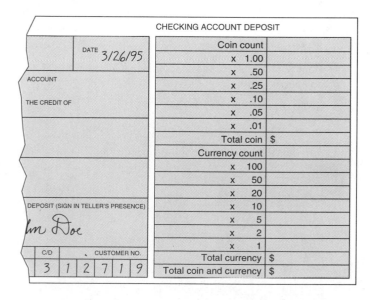

# MID-CHAPTER 3 TEST

1. Simplify with the distributive property.

   a. $4(x - 3)$

   b. $-2(x - y + 3)$

   c. $2 - (x - 3)$

   d. $4 - 2(x - 4)$

   e.
   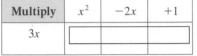

   | Multiply | $x^2$ | $-2x$ | $+1$ |
   |----------|-------|-------|------|
   | $3x$     |       |       |      |

2. Name the common factor or units in each of these, and then use the distributive property to change to factored form. Complete all additions and subtractions.

   a. $4x^2 - 8x + 12$

   b. $6x + 12y - 15$

   c. $4\frac{1}{2}$ in. $+ 2\frac{1}{2}$ in. $- 3\frac{1}{4}$ in.

   d. $6.7$ m $- 4$ m $+ 2.3$ m

   e.

   | Factor |        |          |          |
   |--------|--------|----------|----------|
   |        | $4x^2y$ | $-8x^2y^2$ | $+2xy^2$ |

3. Add like terms. Simplify expressions to remove parentheses as necessary.

   a. $-2x + 3y - 4x + 2x - 5y$

   b. $x^3 + 2x^2 - x - 3x^2 - 6x + 3$

   c. $3(x + 1) + 5(x - 2)$

   d. $8(x + 2) - 3(x + 2)$

4. Change to like units and add. (*Metric note*: 10 mm = 1 cm, 100 cm = 1 m.)

   a. 80 cm + 1.4 m + 150 cm

   b. 14 mm + 71 mm + 17 cm

5. Use the table to solve the following equations.

   | $x$ | $y = \frac{1}{2}x + 4$ |
   |-----|------------------------|
   | 2   | 5                      |
   | 4   | 6                      |
   | 6   | 7                      |
   | 8   | 8                      |

   a. $\frac{1}{2}x + 4 = 5$

   b. $\frac{1}{2}x + 4 = 8$

   c. $\frac{1}{2}x + 4 = 4$

   d. $\frac{1}{2}x + 4 = 6.5$

**Solve the equations in Exercises 6 and 7 from the figures.**

6.
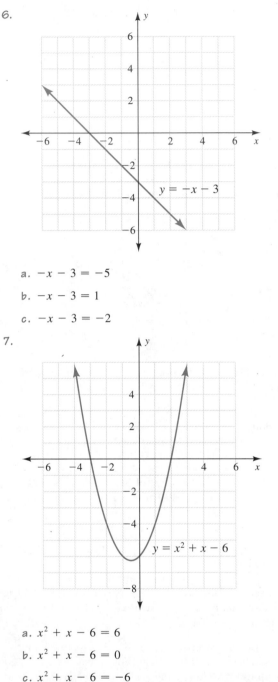

   a. $-x - 3 = -5$

   b. $-x - 3 = 1$

   c. $-x - 3 = -2$

7.

   a. $x^2 + x - 6 = 6$

   b. $x^2 + x - 6 = 0$

   c. $x^2 + x - 6 = -6$

8. a. Use the graph at the top of the next page to find the point of intersection of the two lines.

   b. Substitute the intersection point into each equation shown on the graph.

   c. Solve the equation $2x - 4 = -x - 1$.

   d. What does the graph indicate about the equation $2x - 4 = -x - 1$?

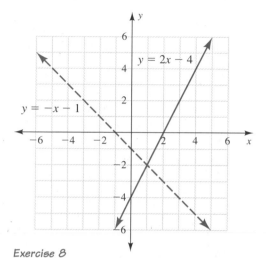

Exercise 8

Solve for $x$ in each of the equations in Exercises 9 to 12.

9. $2(x - 2) = -x - 1$  10. $4(x - 3) = 6$

11. $4 - 2(x - 4) = 2x$  12. $-2(x - 3) = -0.5(x - 6)$

**In Exercises 13 to 15 solve for the indicated letter.**

13. $-5 = \frac{1}{2}(-6) + b$ for $b$  14. $6 = 4(-1) + b$ for $b$

15. $37 = \frac{5}{9}(F - 32)$ for $F$

16. If $A = \frac{1}{2}h(a + b)$, what is side $b$ when $A = 27$, $h = 3$, and $a = 9$?

---

## 3.5 SOLVING FORMULAS

**OBJECTIVES**

Describe relationships using *in terms of*. ♦ Distinguish between the forms of answers expected in solving equations and in solving formulas. ♦ Solve formulas for a specified variable. ♦ Evaluate formulas containing variables with subscripts. ♦ Solve inequalities containing no negative variables.

**WARM-UP**

In Warm-ups 1 to 4 place $>$ or $<$ between the two expressions.

1. **a.** $4 \underline{\hspace{1cm}} 5$  **b.** $4 - 2 \underline{\hspace{1cm}} 5 - 2$  **c.** $4 + (-5) \underline{\hspace{1cm}} 5 + (-5)$

2. **a.** $-3 \underline{\hspace{1cm}} -5$  **b.** $-3 - 4 \underline{\hspace{1cm}} -5 - 4$

   **c.** $-3 + 5 \underline{\hspace{1cm}} -5 + 5$

3. **a.** $4 \underline{\hspace{1cm}} -2$  **b.** $4(5) \underline{\hspace{1cm}} -2(5)$  **c.** $4 \div 2 \underline{\hspace{1cm}} -2 \div 2$

4. **a.** $-3 \underline{\hspace{1cm}} 6$  **b.** $-3(4) \underline{\hspace{1cm}} 6(4)$  **c.** $-3 \div 3 \underline{\hspace{1cm}} 6 \div 3$

5. Does adding or subtracting a number change the inequality sign between two numbers or expressions? Does multiplying or dividing by a positive number change the inequality sign between two numbers or expressions?

In this section we practice solving formulas as we explore the steel cable question posed on the chapter-opening page. Subscripted variables are introduced, and grade calculations are explored.

### FORMULAS

We return to the question posed on the chapter-opening page:

Assume a 12-inch section is to be fitted into a steel cable that formerly fitted the earth snugly on a great circle. If the cable is now held a uniform distance from the earth's surface, could a mouse go under it?

The question was answered with formulas, not by using the actual radius and circumference of the earth. Because the letter is dated April 1, we should probably not yet trust the results as given. In Example 1 we try the problem numerically for three different radii.

EXAMPLE **1**    For each of the three radii

$$r = 6 \text{ inches}, \qquad r = 3600 \text{ inches}, \qquad r = 1{,}000{,}000 \text{ inches}$$

do the following.

**a.** Calculate the circumference. Use calculator $\pi$, not 3.14.
**b.** Add 12 inches to the circumference and find the resulting radius, $r'$.
**c.** Subtract the beginning radius from the ending radius.
**d.** Comment on the results.

SOLUTION    ***For $r = 6$ inches (the approximate radius of a globe):***

**a.** $C = 2\pi r = 2\pi(6) \approx 37.699$
**b.** New circumference $\approx 37.699 + 12 \approx 49.699$. To find the new radius we place 49.699 into $C = 2\pi r'$ and solve for $r'$:

$$C = 2\pi r'$$
$$49.699 = 2\pi r'$$
$$\frac{49.699}{2\pi} = r'$$
$$r' \approx 7.910$$

**c.** We now subtract the beginning radius from the ending radius:

$$r' - r = 7.910 - 6 = 1.91 \text{ in.}$$

**d.** Mr. Coulter's letter is true for a small globe.

***For $r = 3600$ inches (a radius the length of a football field):***

**a.** $C = 2\pi r = 2\pi(3600) \approx 22{,}619.467$
**b.** New circumference $\approx 22{,}619.467 + 12 \approx 22{,}631.467$. To find the new radius we place 22,631.437 into $C = 2\pi r'$ and solve for $r'$:

$$C = 2\pi r'$$
$$22\,631.467 = 2\pi r'$$
$$\frac{22\,631.467}{2\pi} = r'$$
$$r' \approx 3601.910$$

**c.** We now subtract the beginning radius from the ending radius:

$$r' - r = 3601.910 - 3600 = 1.91 \text{ in.}$$

**d.** Mr. Coulter's letter is still true.

***For $r = 1{,}000{,}000$ inches (still somewhat smaller than the radius of the earth):***

**a.** $C = 2\pi r = 2\pi(1{,}000{,}000) \approx 6{,}283{,}185.307$
**b.** New circumference $\approx 6{,}283{,}185.307 + 12 \approx 6{,}283{,}197.307$. To find the new radius we place 6,283,197.307 into $C = 2\pi r'$ and solve for $r'$:

$$C = 2\pi r'$$
$$6\,283\,197.307 = 2\pi r'$$
$$\frac{6\,283\,197.307}{2\pi} = r'$$
$$r' \approx 1{,}000{,}001.91$$

**c.** We subtract the beginning radius from the ending radius:

$$r' - r = 1{,}000{,}001.91 - 1{,}000{,}000 = 1.91 \text{ in.}$$

**d.** Mr. Coulter's letter is again true.

◆ ◆ ◆ ◆ ◆ ◆ ◆ ◆ ◆ ◆ ◆ ◆ ◆ ◆ ◆ ◆ ◆ ◆ ◆ ◆ ◆ ◆ ◆ ◆ ◆ ◆ ◆ ◆ ◆ ◆ ◆ ◆ ◆ ◆ ◆ ◆ ◆ ◆ ◆

We have repeated the same steps over and over in checking the 1.91-inch change in radius. In general this is not an efficient use of time nor does it really prove that Mr. Coulter is correct. A better use of our time is in solving the formulas, as Mr. Coulter did in the letter.

EXAMPLE **2**

Use equation-solving steps to do the following.

**a.** Solve the circumference formula, $C = 2\pi r$, for $r$.

**b.** Solve the new circumference, $C + 12$ in. $= 2\pi r'$, for $r'$.

**c.** Subtract $r$ from $r'$.

**d.** Use a calculator to divide 12 inches by $2\pi$. Place the $2\pi$ in parentheses to preserve the order of operations.

**e.** Comment on your results.

SOLUTION

**a.** The radius is

$$r = \frac{C}{2\pi}$$

**b.** The new radius is

$$r' = \frac{C + 12}{2\pi} = \frac{C}{2\pi} + \frac{12}{2\pi}$$

**c.** Subtract $r$ from $r'$:

$$r' - r = \frac{C}{2\pi} + \frac{12}{2\pi} - \frac{C}{2\pi} = \frac{12}{2\pi}$$

**d.** 12 in. divided by $(2\pi) \approx 1.910$ in.

**e.** Mr. Coulter's letter is true.

♦ ♦ ♦ ♦ ♦ ♦ ♦ ♦ ♦ ♦ ♦ ♦ ♦ ♦ ♦ ♦ ♦ ♦ ♦ ♦ ♦ ♦ ♦ ♦ ♦ ♦ ♦ ♦ ♦ ♦ ♦ ♦ ♦ ♦ ♦ ♦ ♦ ♦ ♦ ♦

One difference between Example 2 and our equation solving in other sections is that we have variables remaining on both sides of our answer.

---

Expect the solutions to formulas to look different from the solutions to equations. Equations generally solve to a numerical answer, whereas formulas remain in expression form.

---

**In Terms of.** When we solve a formula, we are isolating one letter. On the other side of the equal sign will be terms containing letters, symbols, and numbers. After solving an equation or formula for a particular variable, we use the phrase **in terms of** to describe the relationship.

EXAMPLE **3**

Describe each equation using the phrase *in terms of*. If you recognize the formulas, use words instead of letters.

**a.** $y = 2x + 4$     **b.** $w = A \div l$     **c.** $A = \frac{1}{2}bh$

**d.** $h = -\frac{1}{2}gt^2 + v_0 t + h_0$

SOLUTION

**a.** $y$ in terms of $x$

**b.** Width of a rectangle in terms of area and length

**c.** Area of a triangle in terms of base and height

**d.** This is the formula for height above ground level ($h$) in terms of gravity ($g$), time ($t$), initial velocity ($v_0$), and initial height ($h_0$).

♦ ♦ ♦ ♦ ♦ ♦ ♦ ♦ ♦ ♦ ♦ ♦ ♦ ♦ ♦ ♦ ♦ ♦ ♦ ♦ ♦ ♦ ♦ ♦ ♦ ♦ ♦ ♦ ♦ ♦ ♦ ♦ ♦ ♦ ♦ ♦ ♦ ♦ ♦ ♦

**Figure 23**

The small numbers to the right of the variables in Example 3d are subscripts. **Subscripts** *are a common way to label a particular item from a group of similar items.* In Example 3d the subscripts represent the starting velocity and starting height. In the nutrition field, vitamins are identified by subscripts: Vitamin $B_1$, Vitamin $B_2$, Vitamin $B_6$ (Figure 23). In music middle C is $C_4$, and a chord might be described as $G_4$-$C_5$-$E_5$. In Section 3.6 we will use subscripts in naming coordinate points.

It is important to keep track of which letter is the object of the solution process. Try marking the selected letter with a small arrow, $\downarrow$. As in other equation solving, solve for the selected letter using the reverse order of operations.

**E X A M P L E    4**    Solve $A = (bh)/2$ for $b$.

**SOLUTION**    Observe that $b$ is multiplied by $h$ and the result is divided by 2. We use the inverse operations in the reverse order: multiply by 2 and divide by $h$.

$$A = \frac{b\overset{\downarrow}{h}}{2} \qquad \text{Mark the letter with an arrow.}$$

$$2 \cdot A = 2 \cdot \frac{bh}{2} \qquad \text{Multiply both sides by 2.}$$

$$\frac{2 \cdot A}{h} = \frac{bh}{h} \qquad \text{Divide both sides by } h.$$

$$\frac{2A}{h} = b$$

◆ ◆ ◆ ◆ ◆ ◆ ◆ ◆ ◆ ◆ ◆ ◆ ◆ ◆ ◆ ◆ ◆ ◆ ◆ ◆ ◆ ◆ ◆ ◆ ◆ ◆ ◆ ◆ ◆ ◆ ◆ ◆ ◆ ◆ ◆ ◆ ◆ ◆

**E X A M P L E    5**    Solve $A = \frac{1}{2}h(a + b)$ for $a$. Describe the order of operations on $a$ and the order used to reach the solution.

**SOLUTION**    Observe that $b$ is added to $a$, then the sum is multiplied by $h$ and divided by 2. The reverse, with opposite operations, is to multiply by 2, divide by $h$, and subtract $b$.

$$A = \tfrac{1}{2}h(\overset{\downarrow}{a} + b) \qquad \text{Arrow on } a$$

$$2 \cdot A = 2 \cdot \tfrac{1}{2}h(a + b) \qquad \text{Multiply by 2.}$$

$$2A = h(a + b)$$

$$\frac{2A}{h} = \frac{h(a + b)}{h} \qquad \text{Divide by } h.$$

$$\frac{2A}{h} = a + b$$

$$\frac{2A}{h} - b = a + b - b \qquad \text{Subtract } b.$$

$$\frac{2A}{h} - b = a$$

◆ ◆ ◆ ◆ ◆ ◆ ◆ ◆ ◆ ◆ ◆ ◆ ◆ ◆ ◆ ◆ ◆ ◆ ◆ ◆ ◆ ◆ ◆ ◆ ◆ ◆ ◆ ◆ ◆ ◆ ◆ ◆ ◆ ◆ ◆ ◆ ◆ ◆

FORMULAS AND INEQUALITIES

The estimation of grades part way through a course is a familiar situation. If we include two properties of inequalities in our work with equations and formulas, we may predict what test scores are needed to achieve a desired grade.

In Examples 6 and 7 we replace the equality sign with an inequality sign. The properties of inequalities are quite similar to the addition and multiplication properties of equations, as long as we multiply or divide with positive numbers.

**Addition Property of Inequalities**

> If the same number is added (or subtracted) on both sides of an inequality, the inequality is not changed.

**Multiplication Property of Inequalities for Positive Numbers**

> If the same *positive* number is used to multiply (or divide) on both sides of an inequality, the inequality is not changed.

**EXAMPLE 6**

**Grades**  Suppose a course has three tests worth 100 points each, a project or homework assignment worth 70 points, and a final exam worth 150 points. The instructor grades on a percent basis: 90% for an A, 80% for a B, 70% for a C. Student 1 has test scores 78, 84, and 72, with full credit on homework (70 points). What score does the student need on the final exam to earn at least a B?

**SOLUTION**  The grade is based on points earned relative to total points; thus we do not think of the individual scores as fractions needing a common denominator. Instead, the points earned are added, with a variable representing the last test, and this sum is placed over the total possible points to obtain a percent. Because any percent larger than 80% will give a B, we write an inequality using $\geq 0.80$.

$$\frac{78 + 84 + 72 + 70 + x}{100 + 100 + 100 + 70 + 150} \geq 0.80 \qquad \text{Add the scores and possible points.}$$

$$\frac{304 + x}{520} \geq 0.80$$

$$(520)\frac{304 + x}{520} \geq 520(0.80) \qquad \text{Multiply by 520.}$$

$$304 + x \geq 416$$

$$304 + x - 304 \geq 416 - 304 \qquad \text{Subtract 304.}$$

$$x \geq 112$$

The student had a C+ on tests: $(78 + 84 + 72) \div 3 = 78$ average. The student needs $\frac{112}{150} = 75\%$ on the final for a B in the course. The homework helped!

♦ ♦ ♦ ♦ ♦ ♦ ♦ ♦ ♦ ♦ ♦ ♦ ♦ ♦ ♦ ♦ ♦ ♦ ♦ ♦ ♦ ♦ ♦ ♦ ♦ ♦ ♦ ♦ ♦ ♦ ♦ ♦ ♦ ♦ ♦ ♦ ♦ ♦ ♦ ♦ ♦ ♦

EXAMPLE  **7**    *More grades*    Student 2 has not done much homework (5 out of 70 points) but has the same test scores as Student 1. What score does Student 2 need on the final exam to earn a B?

SOLUTION

$$\frac{78 + 84 + 72 + 5 + x}{100 + 100 + 100 + 70 + 150} \geq 0.80$$  *Add the scores and possible points.*

$$\frac{239 + x}{520} \geq 0.80$$

$$(520)\frac{239 + x}{520} \geq 520(0.80)$$  *Multiply by 520.*

$$239 + x \geq 416$$

$$239 + x - 239 \geq 416 - 239$$  *Subtract 239.*

$$x \geq 177$$

A score of 177 is greater than the 150 points possible on the final. This student's low homework score has lowered the overall grade sufficiently to make it impossible to earn a B. Is it possible for the student to earn 70% for a C?

◆ ◆ ◆ ◆ ◆ ◆ ◆ ◆ ◆ ◆ ◆ ◆ ◆ ◆ ◆ ◆ ◆ ◆ ◆ ◆ ◆ ◆ ◆ ◆ ◆ ◆ ◆ ◆ ◆ ◆ ◆ ◆ ◆ ◆ ◆ ◆ ◆

In Examples 6 and 7 we worked with addition and subtraction or with multiplication by a positive. Multiplication or division by a negative has special impact on an inequality sign. We will return to this idea in Section 5.5.

## EXERCISES 3.5

In Exercises 1 to 6 use a full sentence and *in terms of* to describe the output of each formula or equation in terms of the input variables. If you recognize the formula, use words; otherwise use the variables themselves.

1.  $D = rt$

2.  $P = 2l + 2w$

3.  $r = \dfrac{D}{t}$

4.  $l = \dfrac{P - 2w}{2}$

5.  $G = \dfrac{T_1 + T_2 + T_3 + H + E}{P}$, where $G$ = percent earned, $T$ = test, $H$ = homework, $E$ = final exam, $P$ = total points possible

6.  The area of a trapezoid: $A = \frac{1}{2}h(a + b)$

Exercises 7 to 18 provide practice in solving for one of two variables, a skill needed to make a table or graph.

In Exercises 7 to 12 solve for $y$ in terms of $x$.

7.  $xy = -4$        8.  $xy = -6$        9.  $3x - y = 10$

10.  $2x - y = 3$       11.  $x - 2y = -5$

12.  $2x - 3y - 4 = 0$

In Exercises 13 to 18 solve for $x$ in terms of $y$.

13.  $x - 3y = -6$        14.  $4x + 3y - 12 = 0$

15.  $xy = 7$        16.  $x + y = 4$

17.  $y - x = 5$        18.  $2y - x = 6$

Write the formula described by each of the statements in Exercises 19 to 22.

19.  The length of a rectangle in terms of area and width

20.  The distance traveled in terms of rate and time

21.  The circumference of a circle in terms of radius

22.  The circumference of a circle in terms of diameter

Solve each formula in Exercises 23 to 38 for the indicated letter.

23.  $A = bh$ for $h$        24.  $D = rt$ for $r$

25.  $I = prt$ for $t$        26.  $I = prt$ for $r$

27.  $C = \pi d$ for $d$        28.  $A = \pi r^2$ for $r^2$

29.  $P = R - C$ for $R$        30.  $P = R - C$ for $C$

31.  $PV = nRT$ for $n$        32.  $PV = nRT$ for $R$

33.  $C_1 V_1 = C_2 V_2$ for $V_1$        34.  $C_1 V_1 = C_2 V_2$ for $C_2$

35. $P = a + b + c$ for $c$

36. $P = 2l + 2w$ for $w$

37. $A = \frac{1}{2}h(a + b)$ for $h$

38. $A = \frac{1}{2}h(a + b)$ for $b$

**In Exercises 39 and 40, $A$ = amount, $P$ = principal, $t$ = time in years, and $r$ = percent interest.**

39. a. Solve $A = P + Prt$ for $r$.

    b. An amount of $11,050 is received on a two-year time certificate for a $10,000 principal. What is the rate of interest, $r$?

40. a. Solve $A = P + Prt$ for $t$.

    b. An amount of $60,125 is received on a time certificate at 6.75% interest on a $50,000 principal. What is the number of years, $t$, on the certificate?

41. a. Solve $C = \frac{5}{9}(F - 32)$ for $F$.

    b. What is the Fahrenheit temperature, $F$, corresponding to 37° Celsius, $C$?

42. a. Solve $K = C + 273$ for $C$.

    b. What is the Celsius temperature corresponding to absolute zero, 0 K?

    c. Use the answer to part b and the answer to Exercise 41a to obtain the Fahrenheit temperature for absolute zero.

**In Exercises 43 and 44 use the data provided to solve for $b$ in the equation $y = mx + b$.**

43. a. $m = 2, (x, y) = (3, 4)$ (*Hint*: Use $x = 3$ and $y = 4$.)

    b. $m = -2, (x, y) = (3, 4)$

44. a. $m = \frac{1}{2}, (x, y) = (3, 4)$

    b. $m = -\frac{1}{2}, (x, y) = (3, 4)$

**Exercises 45 and 46 refer to Examples 6 and 7.**

45. Test scores are 88 out of 100, 84 out of 100, and 89 out of 100. Homework is 70 out of 70. What final exam score (200 points possible) is needed to get 90% or better?

46. Test scores are 92 out of 100, 88 out of 100, and 91 out of 100. Homework is 25 out of 70. What final exam score (200 points possible) is needed to get 90% or better?

*PROBLEM SOLVING*

**In Exercises 47 and 48, $v_0$ is initial velocity. The zero subscript is used to indicate an initial or starting number. The formula indicates the velocity at which something is falling after $t$ seconds; $g$ is the gravitational constant, 32.2 feet per second$^2$.**

47. a. Solve $v = gt + v_0$ for $t$ in terms of $v, g, v_0$. Use the formula to answer the following questions.

    b. If we drop a rock ($v_0 = 0$) from the top of a high cliff, how long will it be until the rock reaches a velocity of 66 feet per second (45 miles per hour)?

    c. If we throw the rock downward with an initial velocity $v_0 = 22$ feet per second, how long will it be until the rock reaches a velocity of 66 feet per second?

48. a. Solve $v = gt + v_0$ for $v_0$ in terms of $v, g, t$. Use the formula to answer the following questions.

    b. What is the initial velocity if the velocity after 7 seconds is 300 feet per second (204 miles per hour)?

    c. What is the initial velocity if the velocity after 10 seconds is 322 feet per second?

*PROJECTS*

49. *Equations and Graphs.* List six different equations containing the symbols $x, y, 2, =$, and an operation sign $(+, -, \times, \div)$. Make a table for each equation. Plot the points and sketch the graph for each. Which of your rules are linear?

50. *Subscript Research.* Subscripts are used in genetics to describe generations after a parent generation. The letter P represents the parent generation; $F_1$ is the next, or first filial, generation; $F_2$ is the second filial generation; $F_3$ is the third; and so forth. Research an application of subscripts in a subject of interest to you. Give several examples and some detail about the application. Explain why subscripts are necessary in the application. If you use a subscript mentioned in the text, include research from another source.

51. *The Mouse and the Earth's Circumference.* Research the radius of the earth, and do Example 1 again. *Warning*: You may need to be creative in order to have the calculator give you sufficient digits to write the radius in inches. Make a model of the problem, using a cardboard circle and pieces of cable (broken brake cables are available from most bicycle repair shops), and measure the change in radius after 12 inches is added to the circumference.

---

### 3.6   SLOPE AND RATE OF CHANGE

**OBJECTIVES**   Identify the steepness of a graph as a rate of change. ◆ Identify slope as a name given to the rate of change of a line. ◆ Interpret the meaning of slope in terms of units on the axes. ◆ Calculate the slope of a line from its graph. ◆ Confirm that the slope of a straight line is constant. ◆ Identify positive and negative slopes. ◆ Identify the slopes of a vertical line and a horizontal line. ◆ Calculate slope from a table. ◆ Calculate slope from coordinates. ◆ Match a calculated slope with the corresponding number in an equation. ◆ Recognize the slope concept in a variety of situations.

Perform the following subtractions.

**1.** $4 - (-2)$     **2.** $3 - (-4)$     **3.** $0 - (-2)$     **4.** $-1 - (-3)$

**5.** $-3 - (-5)$     **6.** $-4 - (-2)$                                                ♦

I n this section and the next we focus on parts of a linear equation: the slope, or rate of change; the *y*-intercept; and the *x*-intercept.

## INTRODUCTION

In 1990 world population was increasing at a rate of 1 million people every four days. This fact describes the rate of change, or steepness of the graph, of world population (output) relative to time (input). It does not tell us what the population was at any given time, only how rapidly the population changed.

In looking at world population in the past few centuries, we find that the rate of change, or growth rate, gives us a comparison over time. The *rate of change* in the population is the change in population during a time period divided by the number of years in that time period.

In graphing, the rate of change is a slope. *Slope* describes the steepness of a line. We define **slope** *as the vertical change divided by the horizontal change between two points on a graph.* As we saw in earlier sections, how steep or flat a line appears depends on the scales placed on the axes. Describing steepness with slope permits comparison of graphs without regard to the scale on the axes.

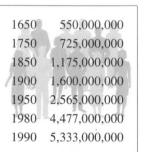

| | |
|---|---|
| 1650 | 550,000,000 |
| 1750 | 725,000,000 |
| 1850 | 1,175,000,000 |
| 1900 | 1,600,000,000 |
| 1950 | 2,565,000,000 |
| 1980 | 4,477,000,000 |
| 1990 | 5,333,000,000 |

*Table 15 Estimated World Population*

Data from *The World Almanac and Book of Facts 1992* (New York: Pharos Books, 1991), p. 822.

**EXAMPLE 1**

**World population: 1650 to 1990**    Use the data in Table 15 to answer the following questions.

**a.** What is the rate of change of population between 1650 and 1750?

**b.** What is the slope of the line connecting the population between 1980 and 1990?

**c.** Compare the results in parts a and b with the graph in Figure 21.

**SOLUTION**

**a.** rate of change $= \dfrac{\text{change in population}}{\text{change in time}}$

$= \dfrac{(725 - 550) \text{ million}}{(1750 - 1650) \text{ yr}} = \dfrac{175 \text{ million}}{100 \text{ yr}}$

$= 1.75$ million per yr

**b.** slope $= \dfrac{\text{vertical change}}{\text{horizontal change}}$

$= \dfrac{(5333 - 4477) \text{ million}}{(1990 - 1980) \text{ yr}} = \dfrac{856 \text{ million}}{10 \text{ yr}}$

$= 85.6$ million per yr

**c.** The rate of change from 1650 to 1750 is 1.75 million per year. The change from 1980 to 1990 is 85.6 million per year. In Figure 21 the graph segment for 1650 to 1750 is almost flat compared with the segment for 1980 to 1990.

If we divide 85.6 million by 1.75 million to compare parts a and b, the quotient is about 50. Thus the rate of change of population in recent years was about 50 times that of the period 1650 to 1750.

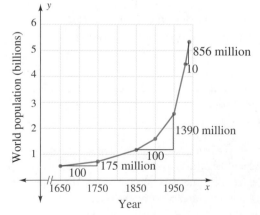

*Figure 21*

♦ ♦ ♦ ♦ ♦ ♦ ♦ ♦ ♦ ♦ ♦ ♦ ♦ ♦ ♦ ♦ ♦ ♦ ♦ ♦ ♦ ♦ ♦ ♦ ♦ ♦ ♦ ♦ ♦ ♦ ♦ ♦ ♦ ♦ ♦ ♦ ♦ ♦ ♦

In Example 1 we observe that when the rate of change, or slope, is different for various parts of the graph, the graph is not linear.

> Curved graphs have slopes that change as we move along the graph.

In Examples 2 and 3 the rate of change, or slope, is constant. The graphs in Examples 2 and 3 are straight lines and will have the same slope between any pair of points selected.

### MEANING OF SLOPE

Slope is both a numerical quantity and a comparison between the vertical and horizontal units on the graph. In Example 1 slope compared population to time. Look for the units being compared in Examples 2 and 3.

EXAMPLE  2

Taxi fare    Suppose a taxi driver charges a $2.00 fee plus $3.00 for each mile traveled.

**a.** Use Table 16 and the graph in Figure 22 to find the slope of line segment $AD$.

**b.** Find the slope of line segment $BC$.

**c.** What is the equation if $x =$ miles?

**d.** What is the meaning of the slope?

**e.** Where does the graph cross the $y$-axis, and what is the meaning of the point?

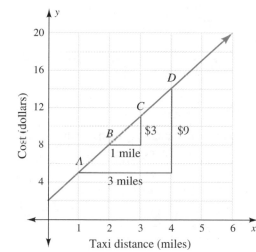

Figure 22

| Miles: $x$ | Cost in Dollars: $C$ |
|:---:|:---:|
| 1 | $ 5.00 |
| 2 | 8.00 |
| 3 | 11.00 |
| 4 | 14.00 |

Table 16  Taxi Fare

SOLUTION    Recall that slope is defined as vertical change (sometimes called *rise*) divided by horizontal change (sometimes called *run*). The triangles drawn under the line show the vertical and horizontal change for each calculation.

**a.** slope $AD = \dfrac{\text{rise}}{\text{run}} = \dfrac{\$14.00 - \$5.00}{4 \text{ mi} - 1 \text{ mi}} = \dfrac{\$9.00}{3 \text{ mi}} = \$3.00$ per mi

**b.** slope $BC = \dfrac{\text{rise}}{\text{run}} = \dfrac{\$11.00 - \$8.00}{3 \text{ mi} - 2 \text{ mi}} = \dfrac{\$3.00}{1 \text{ mi}} = \$3.00$ per mi

**c.** The taxi costs a $2.00 fee plus $3.00 for each mile, so if $x$ is the number of miles, the equation is $C = \$2.00 + x(\$3.00 \text{ per mi})$, or $C = 2 + 3x$.

**d.** The vertical change (rise) is in dollars, and the horizontal change (run) is in miles. The slope of the line $C = \$2.00 + \$3.00x$ is the $3.00 per mile.

**e.** The graph cross the $y$-axis at $x = 0$. At $x = 0$ zero miles have been traveled, and the cost is $2.00. The basic fee is sometimes listed on the taxi as "flagdrop."

EXAMPLE  3

**Photocopy costs**    A prepaid photocopy machine card costs $5.00. Each photocopy is $0.05. Table 17 lists the money remaining on the card after $x$ copies are made. Figure 23 shows the graph for the card's remaining value.

**a.** Find the slope between points $A$ and $B$.

**b.** Find the slope between points $C$ and $D$.

**c.** What is the meaning of the slope? Why is it negative?

**d.** Where does the graph cross the $y$-axis? What is the meaning of this point?

**e.** Write an equation for the value of the card after $x$ photocopies.

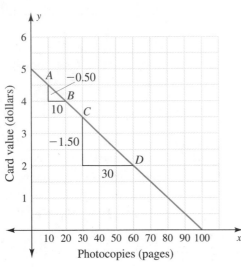

Figure 23

| Copies: $x$ | Value of Card: $y$ |
|:---:|:---:|
| 0 | $5.00 |
| 10 | 4.50 |
| 20 | 4.00 |
| 30 | 3.50 |
| 40 | 3.00 |

Table 17 *Photocopy Card Value*

SOLUTION

**a.** slope $AB = \dfrac{\text{rise}}{\text{run}} = \dfrac{\$4.00 - \$4.50}{(20 - 10) \text{ copies}} = \dfrac{-\$0.50}{10 \text{ copies}} = -\$0.05$ per copy

**b.** slope $CD = \dfrac{\text{rise}}{\text{run}} = \dfrac{\$2.00 - \$3.50}{(60 - 30) \text{ copies}} = \dfrac{-\$1.50}{30 \text{ copies}} = -\$0.05$ per copy

**c.** The slope is the cost of each photocopy deducted from the value of the card. The slope is negative because the value of the card decreases with each photocopy made.

**d.** The line crosses the $y$-axis at $5.00, the initial cost of the card.

**e.** The equation is $y = \$5.00 - x(0.05 \text{ per copy})$, or $y = 5.00 - 0.05x$.

♦ ♦ ♦ ♦ ♦ ♦ ♦ ♦ ♦ ♦ ♦ ♦ ♦ ♦ ♦ ♦ ♦ ♦ ♦ ♦ ♦ ♦ ♦ ♦ ♦ ♦ ♦ ♦ ♦ ♦ ♦ ♦ ♦ ♦

Examples 2 and 3 are linear graphs. In these examples the slope remained the same, no matter what pair of points on the line was chosen. The line in Example 2 has a positive slope—the cost of the taxi ride increased with the distance traveled. *A positive slope occurs when inputs and outputs both get larger.* We say that a positively sloped graph is *increasing* as we move from left to right.

The line in Example 3 has a negative slope—the value of the copy card decreased each time it was used. *If outputs get smaller while inputs get larger, a negative slope is created.* We say that a negatively sloped graph is *decreasing* as we move from left to right.

We make one further observation from Examples 2 and 3. The slope number appeared in the linear equation. The slope was $3.00 per mile for $C = 2.00 + 3.00x$, and the slope was $0.05 per copy for $y = 5.00 - 0.05x$. Thus the slope number multiplies the input, $x$. This leads us to the following conclusion.

♦ The slope of a linear equation is the same everywhere.

♦ The slope of a linear equation is the number that multiplies the input.

♦ If the input is $x$, the slope is the numerical coefficient of $x$.

CALCULATING SLOPE

The first three examples illustrated how to find slope from vertical and horizontal changes. We can find the slope directly from a table. In Example 4 we expand the input–output table to include columns for the change in inputs (horizontal change) and the change in outputs (vertical change) and show how to use these columns to calculate slope.

E X A M P L E   **4**

*Depreciation*   Immediately after a car or pick-up is purchased, its worth generally declines. This decline is called depreciation. Joe purchases a 1994 Ford V8 pick-up. Because of depreciation, its worth decreases for two years. In 1996 it is stolen and stripped. What's left of the pick-up is recovered a few days later, and in 1997 it is sold for scrap. Table 18 shows the pick-up's worth for these years. Figure 24 is the graph.

Calculate the slope for each segment of the graph using the data in Table 18. Recall that $\Delta x$ and $\Delta y$ are the changes in $x$ and $y$ found by subtraction.

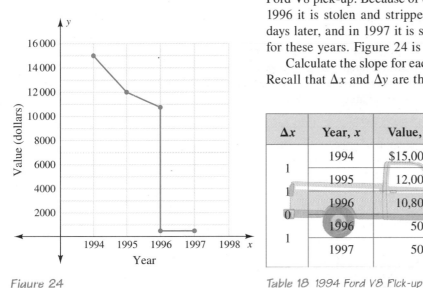

*Figure 24*

| $\Delta x$ | Year, $x$ | Value, $y$ | $\Delta y$ |
|---|---|---|---|
| | 1994 | $15,000 | |
| 1 | | | −3000 |
| 1 | 1995 | 12,000 | −1200 |
| | 1996 | 10,800 | |
| 0 | | | −10,300 |
| | 1996 | 500 | |
| 1 | | | 0 |
| | 1997 | 500 | |

*Table 18  1994 Ford V8 Pick-up*

SOLUTION

$$\text{slope} = \frac{\Delta y}{\Delta x} = \frac{-3000}{1} = -\$3000 \text{ per yr}$$

$$\text{slope} = \frac{\Delta y}{\Delta x} = \frac{-1200}{1} = -\$1200 \text{ per yr}$$

$$\text{slope} = \frac{\Delta y}{\Delta x} = \frac{-\$10,300}{0} = \text{undefined}$$

$$\text{slope} = \frac{\Delta y}{\Delta x} = \frac{0}{1} = 0$$

◆ ◆ ◆ ◆ ◆ ◆ ◆ ◆ ◆ ◆ ◆ ◆ ◆ ◆ ◆ ◆ ◆ ◆ ◆ ◆ ◆ ◆ ◆ ◆ ◆ ◆ ◆ ◆ ◆ ◆ ◆ ◆ ◆

Example 4 shows two new ideas about slope. We have assumed that the pick-up's worth dropped in 1996 from $10,300 to $500 instantaneously, creating a vertical line. It is unrealistic to assume an instantaneous loss; but because Joe felt the drop in value immediately, it is represented as vertical. *A vertical line with zero change in input has an* **undefined slope**. During the last year there is no change in value, and the slope is zero. A **zero slope** *occurs when there is a change in input but no change in output*. A line with zero slope is horizontal.

In the next two examples we assume the data fit a linear equation and calculate the slope. The choice of $x$ and $y$ in applications is important. Determine $x$ and $y$ by observing that output depends on input ($y$ depends on $x$). If no dependency is apparent, assign $x$ and $y$ alphabetically. In Example 5, involving Celsius and Fahrenheit, we assign $x$ to Celsius and $y$ to Fahrenheit, alphabetically.

EXAMPLE 5

*Temperature conversion*    Water freezes at 0° Celsius and 32° Fahrenheit. Water boils at 100° C and 212° F. Calculate the slope for these two data points, with Celsius as input and Fahrenheit as output.

SOLUTION

$$\text{slope} = \frac{\text{change in output}}{\text{change in input}} = \frac{212° \text{ F} - 32° \text{ F}}{100° \text{ C} - 0° \text{ C}}$$

$$= \frac{180° \text{ F}}{100° \text{ C}} = \frac{9° \text{ F}}{5° \text{ C}} = \frac{9}{5}° \text{ F per } ° \text{ C}$$

Would the slope remain the same if we used $(32° - 212°)$ divided by $(0° - 100°)$?

♦ ♦ ♦ ♦ ♦ ♦ ♦ ♦ ♦ ♦ ♦ ♦ ♦ ♦ ♦ ♦ ♦ ♦ ♦ ♦ ♦ ♦ ♦ ♦ ♦ ♦ ♦ ♦ ♦ ♦ ♦ ♦ ♦ ♦ ♦

If we identify any two points as $(x_1, y_1)$ and $(x_2, y_2)$, we can then calculate the slope of the line connecting the two points by finding the difference between the $y$-coordinates and dividing by the difference between the $x$-coordinates.

---

The slope between two points $(x_1, y_1)$ and $(x_2, y_2)$ is

$$\frac{\Delta y}{\Delta x} = \frac{y_2 - y_1}{x_2 - x_1}$$

---

Note the small numbers, or *subscripts*, to the right of $x$ and $y$. Here the subscripts are used to distinguish the two different points, $(x_1, y_1)$ and $(x_2, y_2)$. We could use $(a, b)$ and $(c, d)$ to write the formula for slope, but the subscripts emphasize both the coordinates and the relative positions of $x$ and $y$ in the formula.

In Example 6 we also assume the data fit a linear equation. We know the relationship of $y$ depends on $x$ and assign the variables accordingly.

EXAMPLE 6

| Size of Freezer: $x$ | Dry Ice: $y$ |
|:---:|:---:|
| 5 ft$^3$ | 20 lb |
| 15 ft$^3$ | 40 lb |

*Table 19* *Emergency Cooling*

*Keeping food frozen*    The owner's manual for a Kenmore freezer suggests leaving the freezer door closed if power goes off for less than 24 hours. If power is off longer, it recommends taking the food to a commercial frozen food locker or placing dry ice (frozen carbon dioxide) in the freezer. Recommended dry ice requirements per 24-hour period are 20 pounds for a 5-cubic-foot freezer and 40 pounds for a 15-cubic-foot freezer. Assume that the dry ice requirement is linear.

**a.** Which is input, and which is output?

**b.** Find the slope for these data.

SOLUTION

**a.** The number of pounds of dry ice needed depends on the cubic foot volume of the freezer, so input, $x$, is cubic feet and output, $y$, is pounds of dry ice (see Table 19).

**b.** $\text{slope} = \dfrac{y_2 - y_1}{x_2 - x_1} = \dfrac{(40 - 20) \text{ lb}}{(15 - 5) \text{ ft}^3} = \dfrac{20 \text{ lb}}{10 \text{ ft}^3} = 2 \text{ lb per ft}^3$

♦ ♦ ♦ ♦ ♦ ♦ ♦ ♦ ♦ ♦ ♦ ♦ ♦ ♦ ♦ ♦ ♦ ♦ ♦ ♦ ♦ ♦ ♦ ♦ ♦ ♦ ♦ ♦ ♦ ♦ ♦ ♦ ♦ ♦ ♦

---

HISTORICAL NOTE

The abbreviation for pound is lb and is from librae, the Roman unit of mass.

In Section 3.7 we will use the slope calculations from Examples 5 and 6 to obtain linear equations.

We can summarize what we have learned so far about slope as follows.

**Summary**

---

♦ Slope $= \dfrac{\text{vertical change}}{\text{horizontal change}} = \dfrac{\text{output change}}{\text{input change}} = \dfrac{\text{rise}}{\text{run}} = \dfrac{\Delta y}{\Delta x}$

♦ Slope formula: slope $= \dfrac{y_2 - y_1}{x_2 - x_1}$

♦ Lines with positive slope rise, or increase, from left to right.

♦ Lines with negative slope fall, or decrease, from left to right.

♦ Horizontal lines have zero slope.

♦ Vertical lines have undefined slope.

---

### VARIATIONS ON SLOPE

The following examples explore applications in which slope is given another name or is presented as a number in a different form.

*Slope and Percent.*    In the next two examples we look at slope used as *percent grade*. This is a common application in engineering, surveying, and geology.

**E X A M P L E    7**

**Highway grade**    Highway signs like the one in Figure 25 are placed near the top of mountain passes to warn truckers and motorists about the steepness of the downhill grade. A 6% grade indicates that there is a drop of 6 feet vertically for every 100 feet traveled horizontally.

**a.** What is the average slope of the road?

**b.** If this 6% grade is spread over 7 miles in horizontal distance, what is the change in elevation in feet?

Figure 25

**SOLUTION**    **a.** The average slope is $6\% = \frac{6}{100}$, which simplifies to $\frac{3}{50}$.

**b.** To find the drop in elevation, or vertical change, we multiply the percent slope times the horizontal distance in feet. Unit analysis is helpful in converting miles to feet.

$$\frac{6 \text{ ft}}{100 \text{ ft}} \cdot \frac{5280 \text{ ft}}{1 \text{ mi}} \cdot 7 \text{ mi} = 2217.6 \text{ ft}$$

♦ ♦ ♦ ♦ ♦ ♦ ♦ ♦ ♦ ♦ ♦ ♦ ♦ ♦ ♦ ♦ ♦ ♦ ♦ ♦ ♦ ♦ ♦ ♦ ♦ ♦ ♦ ♦ ♦ ♦ ♦ ♦ ♦ ♦ ♦

EXAMPLE **8**

*Geology*    Wizard Island in Crater Lake, Oregon, is a small cinder cone (see Figure 26). The sides of the cone form a 62.5% grade (see Figure 27).

**a.** What is the average slope of the cone?

**b.** If the average horizontal distance from the center of Wizard Island to the water line is 1220 feet, how high is the center of Wizard Island above the water line?

Figure 26

Figure 27

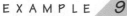

SOLUTION    **a.** The average slope is

$$62.5\% = \frac{62.5}{100} = \frac{625}{1000}$$

The slope is $\frac{5}{8}$ when simplified.

**b.** To find the vertical change we multiply the percent slope times the horizontal distance.

$$\frac{62.5 \text{ ft}}{100 \text{ ft}} \cdot 1220 \text{ ft} \approx 760 \text{ ft}$$

Thus the center of Wizard Island is about 760 feet above the water line.

♦ ♦ ♦ ♦ ♦ ♦ ♦ ♦ ♦ ♦ ♦ ♦ ♦ ♦ ♦ ♦ ♦ ♦ ♦ ♦ ♦ ♦ ♦ ♦ ♦ ♦ ♦ ♦ ♦ ♦ ♦ ♦ ♦ ♦

**Slope and Pitch.**    Slope is a logical descriptor of the steepness of a roof for someone familiar with mathematics, and it is closely related to the number used by carpenters—pitch. *The **pitch** of a roof is calculated by dividing the rise of the roof by the span of the roof.* The span is distance from the bottom of one side of the roof to the bottom of the other side of the roof (covering the full width of the house).

EXAMPLE **9**

*High-pitch roof*    The A-frame house is popular in climates with lots of snow. The pitch of the roof is quite large so that snow does not pile up on the roof.

**a.** What is the slope of the roof in Figure 28?

**b.** What is the pitch of the roof in Figure 28?

**c.** Compare the slope and the pitch.

Figure 28

SOLUTION **a.** $\text{slope} = \dfrac{\text{rise}}{\text{run}} = \dfrac{24 \text{ ft}}{12 \text{ ft}} = \dfrac{2}{1}$

**b.** $\text{pitch} = \dfrac{\text{rise}}{\text{span}} = \dfrac{24 \text{ ft}}{(12 + 12) \text{ ft}} = \dfrac{24}{24} = \dfrac{1}{1}$

**c.** The slope is twice as large as the pitch.

♦ ♦ ♦ ♦ ♦ ♦ ♦ ♦ ♦ ♦ ♦ ♦ ♦ ♦ ♦ ♦ ♦ ♦ ♦ ♦ ♦ ♦ ♦ ♦ ♦ ♦ ♦ ♦ ♦ ♦ ♦ ♦ ♦ ♦

E X A M P L E **10**    Low-pitch roof    In hot climates, where there is no snow, a low-pitch roof with a large overhang provides protection from rain and sun.

**a.** What is the slope of the roof in Figure 29?

**b.** What is the pitch of the roof in Figure 29?

**c.** Compare slope and pitch. Do slope and pitch always have this relation?

SOLUTION **a.** $\text{slope} = \dfrac{\text{rise}}{\text{run}} = \dfrac{2 \text{ ft}}{12 \text{ ft}} = \dfrac{1}{6}$

**b.** $\text{pitch} = \dfrac{\text{rise}}{\text{span}} = \dfrac{2 \text{ ft}}{(12 + 12) \text{ ft}} = \dfrac{2}{24} = \dfrac{1}{12}$

**c.** The slope, at $\frac{1}{6}$, is double the pitch, $\frac{1}{12}$. Yes; the relation of pitch to slope is

$$\text{pitch} = \frac{\text{rise}}{\text{span}} = \frac{\text{rise}}{2 \cdot \text{run}} = \frac{1}{2} \cdot \frac{\text{rise}}{\text{run}} = \frac{1}{2} \text{ slope}$$

♦ ♦ ♦ ♦ ♦ ♦ ♦ ♦ ♦ ♦ ♦ ♦ ♦ ♦ ♦ ♦ ♦ ♦ ♦ ♦ ♦ ♦ ♦ ♦ ♦ ♦ ♦ ♦ ♦ ♦ ♦ ♦ ♦ ♦

Figure 29

**Slope, Angle Measure, and Tangent.**    We close this section with a discussion of the relationship between slope and angle measure in a right triangle. As we look at a ramp, a roof, or even a ladder leaning against a wall, we ask what angle it makes with the horizontal. The relationship is described by the tangent equation,

$$\tan A = \frac{a}{b}$$

This equation may be described in words: For the right triangle $ABC$ (see Figure 30) the *tangent* of angle $A$ (measured in degrees) is the length of side $a$ divided by the length of side $b$.

Side $a$ divided by side $b$ is also the slope of the line connecting $A$ and $B$. The tangent function, abbreviated *tan*, is built into the scientific calculator. To change an angle measure, $A$, into a slope we use $A$ ⟨TAN⟩ or ⟨TAN⟩ $A$. We will look at changing slopes into angles in Examples 12 and 13.

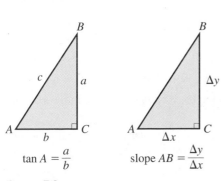

$$\tan A = \frac{a}{b} \qquad \text{slope } AB = \frac{\Delta y}{\Delta x}$$

Figure 30

EXAMPLE  **11**    What is the slope of lines making the following angles with the horizontal? Round all answers to the nearest tenth. *Set your calculator into degree measure.*

**a.** 22°    **b.** 45°    **c.** 76°

SOLUTION    **a.** [TAN] 22° ≈ 0.4    **b.** [TAN] 45° = 1.0    **c.** [TAN] 76° ≈ 4.0

♦ ♦ ♦ ♦ ♦ ♦ ♦ ♦ ♦ ♦ ♦ ♦ ♦ ♦ ♦ ♦ ♦ ♦ ♦ ♦ ♦ ♦ ♦ ♦ ♦ ♦ ♦ ♦ ♦ ♦ ♦ ♦ ♦ ♦

If your answers do not agree, make sure your calculator is set in degree measure and experiment with the order of entering the angle, *A*, and [TAN].

To change a slope, *m*, back to the angle measure, we use the inverse key: *m* [INV] [TAN], [INV] [TAN] *m*, or [2nd] [TAN] *m*. For practice start with the answers in Example 11, and use [INV] [TAN] to get back to the angle measures.

EXAMPLE  **12**    What is the angle formed by lines with these slopes? Round all answers to the nearest tenth.

**a.** $\frac{1}{2}$    **b.** 2

SOLUTION    **a.** [2nd] [TAN] .5 ≈ 26.6°    **b.** [2nd] [TAN] 2 ≈ 63.4°

♦ ♦ ♦ ♦ ♦ ♦ ♦ ♦ ♦ ♦ ♦ ♦ ♦ ♦ ♦ ♦ ♦ ♦ ♦ ♦ ♦ ♦ ♦ ♦ ♦ ♦ ♦ ♦ ♦ ♦ ♦ ♦ ♦ ♦

EXAMPLE  **13**    Highway grade    The 6% highway grade is much flatter than the 62.5% grade of Wizard Island, as you can see by comparing the scale drawings in Figure 31. Change the percent to a decimal, and use [2nd] [TAN] to find the angle off horizontal for each grade.

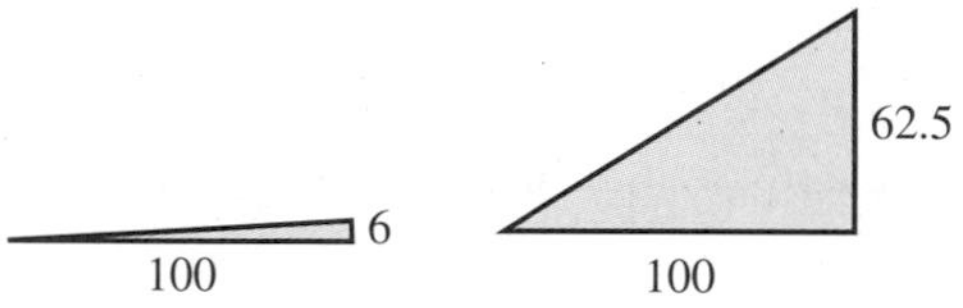

**a.** 6%    **b.** 62.5%

SOLUTION    **a.** [2nd] [TAN] .06 ≈ 3.4°    **b.** [2nd] [TAN] .625 ≈ 32°

♦ ♦ ♦ ♦ ♦ ♦ ♦ ♦ ♦ ♦ ♦ ♦ ♦ ♦ ♦ ♦ ♦ ♦ ♦ ♦ ♦ ♦ ♦ ♦ ♦ ♦ ♦ ♦ ♦ ♦ ♦ ♦ ♦ ♦

*Figure 31*

## EXERCISES 3.6

| Year | Men | Women |
|------|------|-------|
| 1890 | 26.1 | 22.0 |
| 1900 | 25.9 | 21.9 |
| 1910 | 25.1 | 21.6 |
| 1920 | 24.6 | 21.2 |
| 1930 | 24.3 | 21.3 |
| 1940 | 24.3 | 21.5 |
| 1950 | 22.8 | 20.3 |
| 1960 | 22.8 | 20.3 |
| 1970 | 23.2 | 20.8 |
| 1980 | 24.7 | 22.0 |
| 1990 | 26.1 | 23.9 |

*Median Age at First Marriage*

Reprinted with permission from *The World Almanac and Book of Facts 1992.* Copyright © 1991. All rights reserved. The World Almanac is an imprint of Funk & Wagnalls Corporation.

1. Graph both sets of data in the table on the same coordinate axes.

2. What is the meaning of the vertical distance between the graph of men's ages and the graph of women's ages? Where is it greatest? Where is it smallest?

3. What are some possible causes for the changes in age at first marriage?

4. In which decade was there zero change for women?

5. In which decades did the marriage age decrease for men?

6. In which decades did the marriage age increase for men?

7. In which decades did the marriage age remain the same for men?

8. For which decades does the women's age graph have a positive slope?

9. For which decades does the women's age graph have a negative slope?

10. For which decade does the women's age graph have the steepest slope?

11. For which decade does the men's age graph have a positive slope?

12. For which decade does the men's age graph have the steepest slope?

**In Exercises 13 to 16 what is the slope of each line or line segment?**

13.

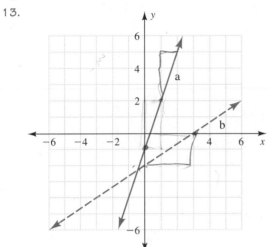

14.

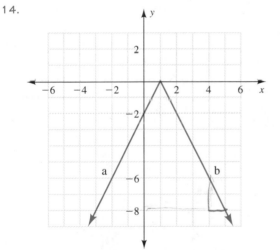

15.

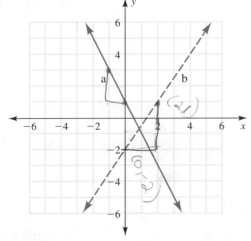

16.

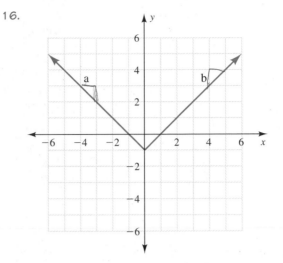

For Exercises 17 and 18 calculate the slope directly from the table. State the slope using units.

17.

| Input Change $\Delta x$ | Pounds $x$ | Cost $y$ | Output Change $\Delta y$ |
|---|---|---|---|
| $\Delta x =$ | 1 | $0.32 | $\Delta y =$ |
| $\Delta x =$ | 3 | 0.96 | $\Delta y =$ |
|  | 5 | 1.60 |  |

18.

| Input Change $\Delta x$ | Kilogram $x$ | Cost $y$ | Output Change $\Delta y$ |
|---|---|---|---|
| $\Delta x =$ | 1 | $0.50 | $\Delta y =$ |
| $\Delta x =$ | 2 | 1.00 | $\Delta y =$ |
|  | 3 | 1.50 |  |

For Exercises 19 to 26 determine whether the table represents a linear relationship. State the slope of the linear tables, using the given units.

19.

| Gallons | Cost |
|---|---|
| 2 | $3.00 |
| 3 | 4.50 |
| 4 | 6.00 |
| 5 | 7.50 |

20.

| Credit | Cost |
|---|---|
| 1 | $24.00 |
| 5 | 120.00 |
| 8 | 192.00 |
| 10 | 240.00 |

21.

| Hours | Earnings |
|---|---|
| 2 | $18 |
| 4 | 36 |
| 6 | 54 |
| 8 | 72 |

22.

| Cookies | Calories |
|---|---|
| 12 | 900 |
| 18 | 1350 |
| 24 | 1800 |
| 30 | 2250 |

**23.**

| Copies | Value |
|--------|-------|
| 0 | $15.00 |
| 10 | 12.50 |
| 20 | 10.00 |
| 25 | 8.75 |

**24.**

| Rides | Value |
|-------|-------|
| 0 | $20 |
| 2 | 16.50 |
| 4 | 13.00 |
| 10 | 2.50 |

**25.**

| Time (sec) | Distance (ft) |
|------------|---------------|
| 0 | 0 |
| 1 | .16 |
| 2 | 64 |
| 3 | 144 |

**26.**

| Length (ft) | Width (ft) |
|-------------|------------|
| 5 | 10 |
| 8 | 7 |
| 9 | 6 |
| 12 | 3 |

**In Exercises 27 to 34 plot and connect each pair of points.**

**a.** Show the vertical change and the horizontal change for each pair.

**b.** What is the slope of the line connecting each pair?

**27.** (0, 2) and (4, 3)       **28.** (2, 0) and (4, 3)

**29.** (−2, 3) and (0, −4)     **30.** (−3, 2) and (4, 0)

**31.** (4, 3) and (4, 4)       **32.** (4, 3) and (3, 3)

**33.** (0, 2) and (−2, 2)      **34.** (4, 4) and (−4, 4)

**Find the slopes connecting the pairs of points in Exercises 35 to 38.**

**35.** (a, b) and (c, d)       **36.** (0, 0) and (m, n)

**37.** (a, 0) and (0, b)       **38.** (m, n) and (p, q)

**39.** In calculating the slope between (8, 2) and (4, 5), one student started with (5 − 2) divided by (4 − 8). Another student started with (2 − 5) divided by (8 − 4). Will they both obtain the correct slope? Explain why or why not.

**40.** Describe in complete sentences how to determine from a table whether its graph is curved or straight.

**41.** Describe in words how to give meaning to the slope of a line.

**42.** Describe in words how to determine the slope of a line from its graph.

**In Exercises 43 to 46 what is the slope of each linear equation? What are the units for the slope?**

**43.** $T = 1.35g$; $g$ = number of gallons, $T$ = total cost in dollars

**44.** $C = 65x$; $x$ = number of cookies, $C$ = total calories

**45.** $E = 6.25h$; $h$ = number of hours, $E$ = earnings in dollars

**46.** $C = 1.09(m - 1) + 1.58$; $m$ = number of minutes, $C$ = total cost in dollars

The figure in Exercises 47 and 48 show wave forms in electronics. For what intervals does each graph have a positive slope, a negative slope, a zero slope? At what inputs is the slope of a segment not defined?

**47.**

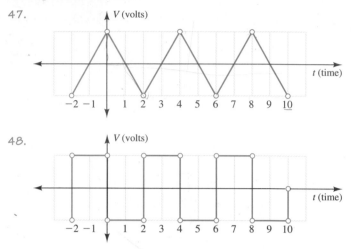

**48.**

*VARIATIONS ON SLOPE*

**49.** A narrow mountain road has a 9% grade. What is the average slope of the road? If this grade covers a 2-mile horizontal distance, what is the rise in elevation, in feet?

**50.** A ski lift goes up a 50% grade. What is the average slope of the lift? If the ski lift covers a 3000-foot horizontal distance, what is the change in elevation?

**51.** The slope of a staircase is the fraction formed by the riser divided by the tread (see the figure). What is the slope fraction for these stair measurements? (Measurements are in inches.)

**a.** Riser 7.5, tread 10.5       **b.** Riser 6.75, tread 10.25

**c.** Riser 7.25, tread 10.75      **d.** Riser 5, tread 12

**52.** What is the percent grade for each of the stair measurements in Exercise 51?

**53.** Is the meaning of slope on a set of stairs the same as in mathematics? Explain, using complete sentences.

**54.** An access ramp has a slope of $\frac{1}{12}$. What percent grade is this?

**55.** A loading ramp has a slope of $\frac{1}{4}$. What percent grade is this?

**56.** An access ramp has a slope of $\frac{1}{8}$. What percent grade is this?

**For Exercises 57 to 60 find the slope and the pitch.**

57.

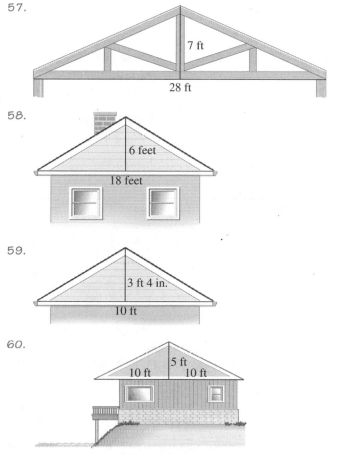

7 ft

28 ft

58.

6 feet

18 feet

59.

3 ft 4 in.

10 ft

60.

5 ft

10 ft        10 ft

**In Exercises 61 and 62 use a ruler to connect *A* and the appropriate vertical mark in order to determine the angle of the line with the indicated slope. Check your answer with a calculator by applying** $\boxed{\text{INV}}$ $\boxed{\text{TAN}}$ **to the slope number.**

61.  a. $\frac{1}{10} = 0.1$    b. $\frac{2}{10} = \frac{1}{5} = 0.2$    c. $\frac{3}{10} = 0.3$

d. $\frac{3}{4} = 0.75$    e. $\frac{5}{100} = 0.05$

62.  a. $\frac{4}{10} = 0.4$    b. $\frac{6}{10} = 0.6$    c. $\frac{85}{100} = 0.85$

d. $\frac{10}{10} = 1.0$    e. $\frac{15}{100} = 0.15$

**In Exercises 63 and 64 use a ruler to find the slope of the line passing from *A* through the indicated angle mark (see the figure).**

63.  a. 10°       b. 20°       c. 30°

64.  a. 35°       b. 40°       c. 45°

65. Use $\boxed{\text{2nd}}$ $\boxed{\text{TAN}}$ or $\boxed{\text{INV}}$ $\boxed{\text{TAN}}$ to change the staircase slopes to angles in Exercise 51.

66. Use $\boxed{\text{2nd}}$ $\boxed{\text{TAN}}$ or $\boxed{\text{INV}}$ $\boxed{\text{TAN}}$ to convert the slopes in Exercises 49 and 50 to angles. Compare your results with those from the protractor.

*PROJECTS*

67. ***Circumference and Diameter.*** Measure and record in a table the diameter and circumference of ten circular objects.

a. Graph your results, with diameter as input and circumference as output.

b. Draw a straight line that seems to pass through most of your data points.

c. Where should your line cross the *y*-axis? the *x*-axis?

d. What is the slope of the line?

68. ***Staircase Slope.*** Measure the riser and tread for six sets of stairs. Try to find different locations such as an office building, a tourist attraction, and a home.

a. Calculate the slope of each set of stairs.

b. Calculate the percent grade for each set of stairs.

c. Calculate the angle in degrees for each set of stairs.

d. What might happen if one step in a flight of stairs had a different size riser?

e. Discuss the relationship between the steepness of the stairs and the purpose for the stairs.

**The figure shows a protractor on a graph with both horizontal and vertical scales marked in centimeters. By using a ruler to connect *A* and points on the vertical scale, we can read the angle measures for many slopes. The line connecting *A* and *B* has a slope of $\frac{5}{10} = \frac{1}{2} = 0.5$. The lines passes between 26° and 27° on the protractor.**

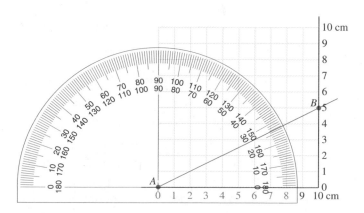

## 3.7   EQUATIONS OF A STRAIGHT LINE

**OBJECTIVES**   Identify the x- and y-intercepts of a straight line. ◆ Identify the slope and y-intercepts from an equation in the form $y = mx + b$. ◆ Interpret the meaning of the intercepts in a problem situation. ◆ Build an equation in the form $y = mx + b$ from two data points. ◆ Draw the graph of a line with a given slope through the y-intercept. ◆ Draw the graph of a line with a given slope through a given point. ◆ Identify problem situations that create parallel lines.

**WARM-UP**   Find the slope for each set of data points.

$y = mx + b$

**1.** (4, 16.80), (20, 4)          **2.** (0, 5.15), (1, 5.80)          **3.** (0, 5), (20, 4)

**4.** (0, 5), (2, 4)             **5.** (0, 32), (100, 212)          **6.** (5, 20), (15, 40)

**7.** $(x_1, y_1), (x_2, y_2)$                                              ◆

I n the last section we examined the rate of change, or slope, of a graph. In this section we look at the intercepts—where a graph crosses the x- and y-axes—and the meanings of these points. We will use the slope and y-intercept to build the equation of a straight line from its graph or from data. Finally, we will emphasize the significance of slope in the equation of a line by exploring the concept of parallel lines.

### EQUATIONS AND INTERCEPTS

In the last section the graph for the photocopy card crossed the y-axis at $5.00, the initial cost of the photocopy card. The **y-intercept** *is the point where the graph crosses the y-axis.*

**EXAMPLE   1**   **Commuter train tickets**   A regional mass transit "Pay in Advance" ticket costs $20. A commuter regularly makes trips with an $0.80 fare. The gates automatically deduct the fare.

**a.** Build a table and graph to show the worth of the ticket (output) after x trips (input).

**b.** Is the relationship linear?

**c.** What is the slope? What is its meaning?

**d.** What is the y-intercept? What is its meaning?

**e.** What is the equation of the line?

$M = \dfrac{16.80 - 4.00}{4.00 - 20} = \dfrac{12.80}{-16} = -.80$

mis Remaining
value decreases
by .80 per
trip

**SOLUTION**   **a.** We first construct Table 21.

**b.** The graph of the data points in Table 21 is shown in Figure 32. The points lie on a straight line, so the relationship is linear.

**c.** The slope is calculated from any two data points—for example, the first and last points in Table 21: (0, 20) and (20, 4).

$$\text{slope} = \frac{\$4 - \$20}{20 - 0} = \frac{-\$16}{20} = -\$0.80$$

The slope is the fare for one trip, the rate per ride. It is negative because the fare is being subtracted from the ticket's initial cost.

**d.** The y-intercept is $20, the initial cost of the ticket.

| Trips x | Remaining Value: y |
|---------|--------------------|
| 0       | $20.00             |
| 4       | 16.80              |
| 8       | 13.60              |
| 12      | 10.40              |
| 16      | 7.20               |
| 20      | 4.00               |

*Table 21* Prepaid Transit Ticket

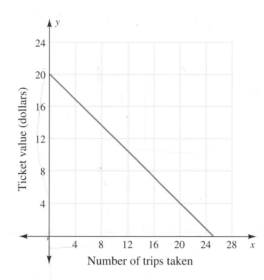

*Figure 32*

**e.** The equation of the line is obtained from the initial description, a $20 ticket less $0.80 for each trip:

$$y = 20 - 0.80x$$

◆ ◆ ◆ ◆ ◆ ◆ ◆ ◆ ◆ ◆ ◆ ◆ ◆ ◆ ◆ ◆ ◆ ◆ ◆ ◆ ◆ ◆ ◆ ◆ ◆ ◆ ◆ ◆ ◆ ◆ ◆ ◆

In Example 1 we wrote the equation of the straight line from the description, $y = 20 - 0.80x$. Observe that the $y$-intercept and the slope both appear in the equation. The $y$-intercept is $20, and the slope is $-$0.80.

---

A linear equation can be written in the form

$$y - b + mx \qquad \text{or} \qquad y = mx + b$$

where $b$ is the $y$-intercept and $m$ is the slope.

---

The letter $m$ is traditionally chosen for slope; it may refer to the word *modus*, which is Latin for "measure." Think of $m$ as *m*ultiplying $x$, if that is helpful.

**E X A M P L E   2**

**Water bills**   The city water bill is made up of a $5.15 basic charge plus $0.65 per thousand gallons (kgal) used.

**a.** Build an input–output table and graph for the cost of $x$ thousand gallons.
**b.** If the relationship is linear, find its slope and give the meaning of the slope.
**c.** Identify the $y$-intercept and its meaning.
**d.** Write the equation of the line in the form $y = mx + b$.

**SOLUTION**   **a.** We first construct Table 22.

| Water: $x$ (kgal) | Cost: $y$ |
|:---:|:---:|
| 0 | $5.15 |
| 1 | 5.80 |
| 3 | 7.10 |
| 5 | 8.40 |
| 7 | 9.70 |

*Table 22 City Water Cost*

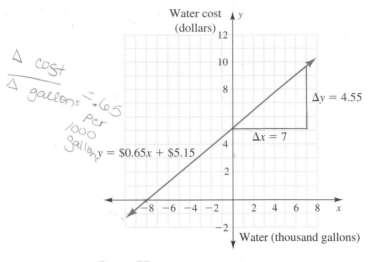

*Figure 33*

**b.** The graph in Figure 33 is a straight line. (In Example 4 we will discuss why it is extended into other quadrants.) Because the graph is linear, any pair of coordinates will give the slope. Using the first and last data points in Table 22 gives

$$\text{slope} = m = \frac{\$9.70 - \$5.15}{(7 - 0) \text{ kgal}} = \frac{\$4.55}{7 \text{ kgal}} = \$0.65 \text{ per thousand gal}$$

The slope is the rate, or cost, per thousand gallons.

**c.** The graph crosses the *y*-axis at $5.15, which is the basic cost of the service and is the amount paid even if no water is used. Thus the *y*-intercept is (0, $5.15).

**d.** Using parts b and c, we obtain the equation $y = mx + b = \$0.65x + \$5.15$. The equation describes the total cost of *x* thousand gallons.

◆ ◆ ◆ ◆ ◆ ◆ ◆ ◆ ◆ ◆ ◆ ◆ ◆ ◆ ◆ ◆ ◆ ◆ ◆ ◆ ◆ ◆ ◆ ◆ ◆ ◆ ◆ ◆ ◆ ◆ ◆ ◆ ◆ ◆ ◆ ◆ ◆ ◆ ◆

(*a*, 0)
*x*-intercept    (0, *b*)
*y*-intercept

*Figure 34*

The *y*-intercept is the point [(0, *b*) in Figure 34] where a graph crosses the *y*-axis. The *y*-intercept is where *x* is zero. The **x-intercept** *is the point* [(*a*, 0) in Figure 34] *where a graph crosses the x-axis*. The *x*-intercept is where *y* is zero.

E X A M P L E    **3**    Find the *x*-intercept for the mass transit equation, $y = 20 - 0.80x$, from Example 1. What does the *x*-intercept mean?

SOLUTION    The graph of $y = 20 - 0.80x$ crosses the *x*-axis at $y = 0$, so the equation becomes

$$0 = 20 - 0.80x$$

$$-20 + 0 = 20 - 0.80x - 20 \qquad \text{Subtract 20.}$$

$$-20 = -0.80x$$

$$-20 \div -0.80 = -0.80x \div -0.80 \qquad \text{Divide by } -0.80.$$

$$25 = x$$

The line crosses the *x*-axis at $x = 25$ and $y = 0$. The $y = 0$ means the ticket has no remaining value; 25 is the number of $0.80 trips purchased for $20.

◆ ◆ ◆ ◆ ◆ ◆ ◆ ◆ ◆ ◆ ◆ ◆ ◆ ◆ ◆ ◆ ◆ ◆ ◆ ◆ ◆ ◆ ◆ ◆ ◆ ◆ ◆ ◆ ◆ ◆ ◆ ◆ ◆ ◆ ◆ ◆ ◆ ◆ ◆

EXAMPLE 4

**Water bills, continued**  Find the *x*-intercept for the city water cost equation, $y = \$0.65x + \$5.15$, from Example 2. Is there any meaning to the *x*-intercept? (*Hint*: Look at Figure 33.)

SOLUTION  The *x*-intercept is where $y = 0$.

$$0 = 0.65x + 5.15$$

$$0 - 5.15 = 0.65x + 5.15 - 5.15 \qquad \text{Subtract 5.15.}$$

$$-5.15 = 0.65x$$

$$-5.15 \div 0.65 = 0.65x \div 0.65 \qquad\qquad \text{Divide by 0.65.}$$

$$-7.92 \approx x$$

In Figure 33 the units on the *x*-axis are thousands of gallons. Unless we are pumping water back into the city water system, a negative number of gallons has no meaning; therefore the *x*-intercept and portions of the graph in the second and third quadrants have no meaning in this situation.

♦ ♦ ♦ ♦ ♦ ♦ ♦ ♦ ♦ ♦ ♦ ♦ ♦ ♦ ♦ ♦ ♦ ♦ ♦ ♦ ♦ ♦ ♦ ♦ ♦ ♦ ♦ ♦ ♦ ♦ ♦ ♦ ♦ ♦ ♦ ♦ ♦ ♦

---

To find the *x*-intercept, let $y = 0$ and solve for *x*.

---

## EQUATIONS FROM COORDINATES

In Section 3.6 we found the slope for a Celsius and Fahrenheit temperature problem and for dry ice requirements in freezers during power failures. We now use those slopes and data to find linear equations.

EXAMPLE 5

**Temperature conversion revisited**  Table 23 gives the two coordinate points relating Celsius and Fahrenheit temperatures. The slope of the line containing these data is

$$\text{slope} = m = \frac{212 - 32}{100 - 0} = \frac{180}{100} = \frac{9}{5}$$

Find the equation of the line, using $y = mx + b$.

|  | Input: °C | Output: °F |
|---|---|---|
| Water freezes | 0 | 32 |
| Water boils | 100 | 212 |

*Table 23 Celsius and Fahrenheit*

SOLUTION  We substitute slope $m = \frac{9}{5}$ and *y*-intercept $b = 32$ into the equation $y = mx + b$ to obtain $y = \frac{9}{5}x + 32$. Using letters that represent Celsius and Fahrenheit, we obtain the equation

$$F = \frac{9}{5}C + 32$$

♦ ♦ ♦ ♦ ♦ ♦ ♦ ♦ ♦ ♦ ♦ ♦ ♦ ♦ ♦ ♦ ♦ ♦ ♦ ♦ ♦ ♦ ♦ ♦ ♦ ♦ ♦ ♦ ♦ ♦ ♦ ♦ ♦ ♦ ♦ ♦ ♦ ♦

In the examples thus far we have known or have easily determined the *y*-intercept, $(0, b)$, and have then built the equation $y = mx + b$. In Example 6 we need equation solving to obtain the *y*-intercept, *b*. Example 6 returns to the Kenmore freezer data.

EXAMPLE 6

| Size of Freezer: $x$ | Dry Ice: $y$ |
|---|---|
| 5 ft³ | 20 lb |
| 15 ft³ | 40 lb |

Table 24 Emergency Cooling

**Keeping food frozen**   The owner's manual for a Kenmore freezer recommends that dry ice be placed in the freezer if the power is off for longer than 24 hours. The data are shown in Table 24. In Section 3.6 we found the slope:

$$\text{slope} = m = \frac{(40 - 20)\text{ lb}}{(15 - 5)\text{ ft}^3} = \frac{20\text{ lb}}{10\text{ ft}^3} = 2 \text{ lb per ft}^3$$

Find an equation to calculate the amount of dry ice needed for freezers of this general size, and confirm that the equation agrees with the graph of the data.

SOLUTION   We place one data point and the slope in the equation, $y = mx + b$. Substitute $m = 2$ and $(5, 20)$ into the equation:

$$20 = 2(5) + b$$
$$b = 10$$

Substitute $b = 10$ into $y = 2x + b$ to obtain the equation relating $x$ cubic feet to $y$ pounds of dry ice needed in each 24-hour period: $y = 2x + 10$. Figure 35 shows that the graph of the line connecting $(5, 20)$ and $(15, 40)$ passes through the $y$-intercept $(0, 10)$, which confirms our symbolic work.

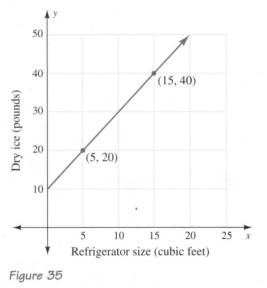

Figure 35

♦ ♦ ♦ ♦ ♦ ♦ ♦ ♦ ♦ ♦ ♦ ♦ ♦ ♦ ♦ ♦ ♦ ♦ ♦ ♦ ♦ ♦ ♦ ♦ ♦ ♦ ♦ ♦ ♦ ♦ ♦ ♦ ♦

In Example 6, because each data point makes the equation true, we may substitute either data point into the equation to solve for $b$. If we substitute the other data point, $(15, 40)$, for $x$ and $y$ in $y = 2x + b$, we obtain $40 = 2(15) + b$, and again $b = 10$, the $y$-intercept.

**Finding the y-intercept from Two Data Points**

1. Find the slope, and substitute it into $y = mx + b$.
2. Select one data point and substitute it into the equation from step 1, and solve for $b$.
3. Place $b$ into the equation from step 1.

PARALLEL LINES

In a number of problem situations in Chapter 1 we suggested the question "How does new data or information change the graph?" We now return to the changes that shift a graph up or down. When the graph is linear, a shift up or down makes **parallel lines**.

EXAMPLE **7**    *Photocopy costs revisited*    In Section 3.6 our prepaid photocopy machine card cost $5.00, and each photocopy cost $0.05. Figure 36 shows the graph for the card's value after photocopies have been made. Suppose the card costs $10.00 instead of $5.00 and the copies remain the same price.

**a.** What is the new equation for the card's value?

**b.** How does the graph change?

SOLUTION    **a.** The slope, or the rate, remains $0.05 per copy. The initial cost is $10.00 instead of $5.00. The equation is therefore $y = 10.00 - 0.05x$.

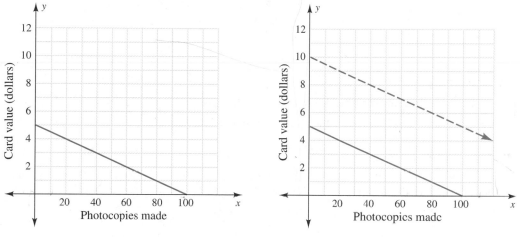

Figure 36                                         Figure 37

**b.** The graph, in Figure 37, shifts up to a $y$-intercept of $10.00. The new graph is parallel to the old graph.

◆ ◆ ◆ ◆ ◆ ◆ ◆ ◆ ◆ ◆ ◆ ◆ ◆ ◆ ◆ ◆ ◆ ◆ ◆ ◆ ◆ ◆ ◆ ◆ ◆ ◆ ◆ ◆ ◆ ◆ ◆ ◆ ◆ ◆ ◆ ◆

Parallel lines have the same slope but different $y$ intercepts.

**Graphing Calculator Technique**

> **Obtaining an Equation from Sets of Data**
>
> The graphing calculator determines the equation of a line from two data points with the STAT function. The STAT function eliminates the need to use symbols to find $m$ and $b$ in writing the equation of a line.
>
> ***TI 81:***
>
> Clear prior data:    [2nd] [STAT] [▶] [▶] 2 [ENTER]
>
> Enter new data:    [2nd] [STAT] [▶] [▶] 1, then enter input–output pairs.
>
> Find equation:    [2nd] [STAT] 2 [ENTER]
> The calculator gives $a$ as the $y$-intercept and $b$ as the slope, so write the equation as $y = a + bx$. The $r$ indicates how closely the data fit a straight line. If $r = 1$ or $r = -1$, the data are linear.
>
> ***TI 82:***
>
> Clear prior lists:    [STAT] 4 [2nd] [L₁] [,] [2nd] [L₂] [ENTER]
>
> Enter new data:    [STAT] 1, then enter inputs under $L_1$ and outputs under $L_2$.
>
> Find equation:    [STAT] [▶] 5 [ENTER]
> Write the equation as $y = ax + b$. The $r$ indicates how closely the data fit a straight line. If $r = 1$ or $r = -1$, the data are linear.

## EXERCISES 3.7

**Refer to Example 1 to answer Exercises 1 to 5.**

1. Why is the graph in the first quadrant only?

2. How do we estimate the scales to be placed on the axes?

3. Do fraction or decimal inputs make sense? Why or why not?

4. How would the graph change if the original ticket price increased from $20 to $25?

5. How would the graph change if the cost of a trip changed from $0.80 to $0.90?

**Refer to Example 2 to answer Exercises 6 to 8.**

6. How would the graph change if the cost per thousand gallons increased to $0.75?

7. How would the graph change if the basic charge changed to $6.00?

8. Do fraction or decimal inputs make sense? Why or why not?

9. Use the equation in Example 5 to find the $x$-intercept. What is its meaning?

10. Find the Celsius temperature for normal body temperature, 98.6° Fahrenheit.

11. Find the Celsius temperature for a nice summer day, 72° Fahrenheit.

12. If the Celsius temperature is $-40°$, what is the Fahrenheit temperature?

13. In Example 6, what is the $x$-intercept? Does it have any meaning?

14. In Example 6, what is the $y$-intercept? Does it have any meaning?

15. What is the $x$-intercept of each line in Figure 37 (Example 7)?

16. What is the $x$-intercept of each line, $y = \frac{3}{4}x + 1$ and $y = \frac{3}{4}x - 2$?

17. What are the slopes for $AD$ and $BC$ in the figure? What are the equations of lines $AD$ and $BC$?

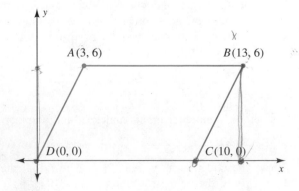

18. What is the $x$-intercept of each line, $y = 5 - \frac{1}{2}x$ and $y = 2 - \frac{1}{2}x$?

**Identify the slope, $m$, and the $y$-intercept, $b$, in the equations in Exercises 19 to 26. Assume that the letter on the left is the output variable.**

19. $D = 55t$

20. $C = 2\pi r$

21. $C = 8 + 2\pi r$

22. $D = 225 + 45t$

23. $P = 2.98n + 0.50$

24. $R = 0.08m + 25$

25. $V = 50 - 0.29n$

26. $V = 20 - 0.19n$

**In Exercises 27 to 32 find the equation of the line containing the two data points. First find the slope, and then use one data point with $y = mx + b$ to find the $y$-intercept, $b$.**

27. $(2, 3), (7, 1)$

28. $(5, 2), (3, 3)$

29. $(-5, 6), (-4, -2)$

30. $(3, -4), (-1, 5)$

31. $(13, 6), (10, 0)$

32. $(3, 6), (0, 0)$

**For Exercises 33 to 38 sketch a line with the given slope that passes through the indicated point.**

33. slope $= \frac{1}{2}$; point $= (2, 3)$

34. slope $= \frac{3}{4}$; point $= (0, 2)$

35. slope $= \frac{2}{3}$; point $= (4, 0)$

36. slope $= -\frac{1}{2}$; point $= (1, 2)$

37. slope $= -\frac{2}{1}$; point $= (-2, 1)$

38. slope $= -\frac{3}{4}$; point $= (-2, 0)$

**In Exercises 39 to 42, answer the following questions.**

a. Which fact gives the slope?

b. Which fact gives the $y$-intercept?

c. Write the equation using $y = mx + b$.

39. Hwang prepays $50 on racquetball court rental of $2 per hour. The equation describes the prepaid amount that remains after x hours of rental time.

40. Yolanda's $500 monthly expense account is set up through an automatic teller machine (ATM). She withdraws funds, using the $40 Fast Cash Option. The equation describes the amount that remains in her account after $x$ withdrawals during the month.

41. Carmen rents a Cessna 152 for $42 per hour plus a $28 insurance fee. The equation describes the total cost of $x$ hours flying time.

42. Alberto earns a weekly salary of $250 plus 10% of his sales volume. The equation describes the total weekly earnings for $x$ dollars in sales.

43. Describe, using full sentences, what words identify the slope in Exercises 39 to 42.

44. Describe, using full sentences, what words identify the $y$-intercept in Exercises 39 to 42.

45. In the late afternoon a 7-minute call to Sweden costs $6.01. On another afternoon an 8-minute call costs $6.79. Assume that the cost of a call is linear and $x$ is the time in minutes. Find a linear equation that gives the cost of an afternoon call.

46. One Sunday a 22-minute call to Sweden cost $23.71. On another Sunday a 31-minute call cost $33.16. Assume that the cost of a call is linear and $x$ is the time in minutes. Find a linear equation that gives the cost of a Sunday call.

47. Experiments show that a heat pump has an output of 36,000 Btu/hr at 48° F outside temperature. The same heat pump has an output of 15,000 Btu/hr at 18° F outside temperature. Assume that the output in Btu's is linear for input temperatures between 18° F and 48 °F.

  a. Find an equation that gives the Btu output at any temperature input.

  b. Why might there be limitations on the inputs for this equation?

  c. What is the $y$-intercept, and does it have any meaning?

  d. What is the $x$-intercept, and does it have any meaning?

48. West Coast Electric charges a $5.00 basic fee and $0.03715 per kilowatt hour used.

  a. What is the linear equation describing the total cost, in terms of kilowatt hours used?

  b. If the basic fee rises to $17.00, what is the new equation? What is the effect on the graph?

**Which of the changes described in Exercises 49 to 54 will create a graph parallel to the graph of the original equation, given in parentheses?**

49. The daily rental on a car goes up $10, but the cost per mile is the same ($y = \$35 + 0.08x$).

50. The basic charge for water service rises $2, but the cost per thousand gallons remains the same ($y = 5.15 + 0.65x$).

51. The cost per gallon of gas goes up $0.20 per gallon, but the tank still holds the same amount of gas ($C = 1.50g$).

52. Rosa buys a $30 mass transit ticket instead of a $20 ticket, but the cost of her fare remains $0.80 per ride ($V = 20 - 0.80x$).

53. Duane increases his speed by 10 miles an hour and drives the same route ($D = 40t$).

54. The bank raises its monthly service charge from $3 to $5 on a 12% loan ($C = 3 + 0.01x$). (*Note*: A 12% loan costs 1% each month.)

*GRAPHING CALCULATOR*

**Set a viewing window with $x$ and $y$ in the interval $[-10, 10]$. Graph the equations in Exercises 55 to 58 to verify the role of $m$ and $b$ in controlling the graph of an equation in the form $y = mx + b$. For each exercise tell what is the same in the graphs and why.**

55. $y = 2x - 4$, $y = 2x$, $y = 2x + 2$

56. $y = 3x + 4$, $y = 2x + 4$, $y = -1x + 4$

57. $y = -2x + 3$, $y = -5x + 3$, $y = 3x + 3$

58. $y = 3x + 4$, $y = 3x$, $y = 3x - 2$

*PROJECTS*

59. **Lines of Symmetry.** The **line of symmetry** *is a line over which a figure folds onto itself.*

  a. Trace the figure in Exercise 17 and the two figures below onto a plain sheet of paper. Fold the paper to find the line or lines of symmetry, if they exist, for each figure. Write the equation of each figure's line or lines of symmetry.

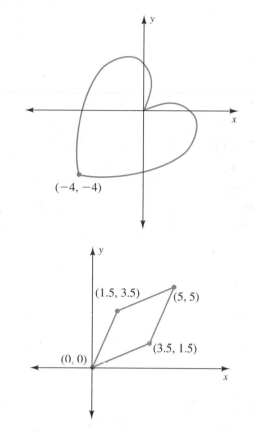

  b. We may place a geometric figure at any convenient location on the coordinate axes. Suppose a square has one corner at the origin and the other corners at (6, 0), (6, 6), and (0, 6). One of the four lines of symmetry is a vertical line with an undefined slope. Its equation does not fit the $y = b + mx$ form but, rather, is the line $x = 3$. Find the equations of the other three lines of symmetry.

60. **Parallelograms.** Three of the four vertices (corners) of a parallelogram are given. Plot them. Find a fourth vertex. Remember that the opposite sides of a parallelogram are parallel and have the same slope.

  a. (0, 0) (3, 4) (5, 0)        b. (1, 1) (5, 3) (1, 8)

  c. There are three different solutions to part a and to part b. Find all three, plot them, and compare the area of this large triangle with the area of the triangle formed by the original three vertices.

  d. Choose any three points on the coordinate plane. Find three other points that will each create a parallelogram with the first three points as vertices. Compare the area of the triangle formed by your original three vertices with the area of the triangle formed by the three possible answers.

## CHAPTER 3 SUMMARY

### Vocabulary

*For definitions and page references, see the Glossary/Index.*

addition property
of equations

common monomial factors

dependent variable

distributive property

empty set

equation

equivalent equations

factoring

independent variable

in terms of

like terms

linear equation

line of symmetry

monomial

multiplication property
of equations

nonlinear equations

numerical coefficient

parallel lines

slope

solution

solution set

solving an equation

subscripts

term

undefined slope

*x*-intercept

*y*-intercept

zero slope

### Concepts

Ordered pairs (*x*, *y*) from an input–output table or on a graph make the equation true.

Solutions to equations may be found by guessing or observation, from an input–output table, from a graph, or with step-by-step symbolic procedures.

Symbolic solutions to equations use any or all of these steps:

♦ The reverse order of operations.

♦ Addition property of equations: Adding the same number to both sides of an equation produces an equivalent equation.

♦ Multiplication property of equations: Multiplying both sides of an equation by the same nonzero number produces an equivalent equation.

♦ The distributive property and adding like terms.

Solving formulas results in answers in expression form. Solving equations often results in numerical answers.

Because the scales on the axes affect the appearance of steepness of a line, we use a number, *slope*, to describe the steepness. The slope describes a rate of change.

$$\text{slope} = \frac{\text{vertical change}}{\text{horizontal change}} = \frac{\text{rise}}{\text{run}} = \frac{\text{change in output}}{\text{change in input}} = \frac{\Delta y}{\Delta x}$$

$$m = \frac{y_2 - y_1}{x_2 - x_1} \quad \text{Slope formula}$$

Straight lines have constant slope; that is, the slope between any two points on a straight line is the same as that between any other two points on the same line.

Curved graphs have slopes that change as we move along the graph. A vertical line has *undefined slope*, because the change in *x* is zero.

A horizontal line has *zero slope*, because the change in *y* is zero.

Lines that drop from left to right have *negative slope*.

Lines that rise from left to right have *positive slope*.

All *linear equations* may be written in the form $y = mx + b$. The letter *m* is the coefficient of the input variable and represents the slope of the line. The letter *b* indicates the intersection of the line with the *y*-axis, the *y*-intercept.

## CHAPTER 3 REVIEW EXERCISES

1. The Ajax Car Rental cost schedule for a compact car is shown in the table. The cost is also shown in the figure.

   a. What is the input, or independent variable?

   b. What is the output, or dependent variable?

   c. What is the cost of driving 125 miles?

   d. How many miles may be driven for $57?

   e. What is the *y*-intercept? What does it mean?

   f. What is the slope of the line? What does it mean?

   g. What is the equation of the line?

   h. What is the *x*-intercept? What does it mean?

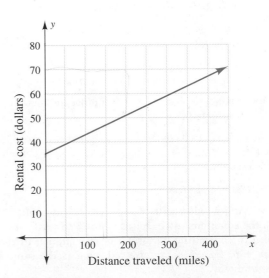

| Miles: *x* | Cost: *y* |
|------------|-----------|
| 50 | $39 |
| 100 | 43 |
| 150 | 47 |
| 200 | 51 |
| 250 | 55 |

2. Simplify the following.

   a. $-2(x - 4)$     b. $3(x + y - 5)$

   c. $3 - (x - 4)$     d. $5 - 3(x - 4)$

e.

| Multiply | $x^2$ | $+2x$ | $-1$ |
|---|---|---|---|
| $2x$ | | | |

3. Name the common factor or units in each of the following, and then use the distributive property to change to factored form. Complete all additions and subtractions.

a. $15x - 3y + 6z$

b. $8x^2 - 4x - 6$

c. $10 \text{ cm} - 4.8 \text{ cm} + 6.2 \text{ cm}$

d. $1\frac{1}{2} \text{ ft} + 6\frac{1}{2} \text{ ft} - 4\frac{1}{4} \text{ ft}$

e.

| Factor | | |
|---|---|---|
| | $6xy^2$ | $+3xy$ | $+9x^2y^2$ |

4. Add like terms. Remove parentheses only as necessary.

a. $3x + 2y - 2x - 3y + 4x - 4y$

b. $x^3 - 3x^2 + x - 2x^2 + 6x - 2$

c. $2(x - 2) + 3(x - 1)$

d. $12(x - 1) - 5(x - 1)$

5. Change to like units and add. (*Metric note*: 10 mm = 1 cm, 100 cm = 1 m.)

a. $3 \text{ mm} + 4.5 \text{ cm} + 8.4 \text{ mm}$

b. $1.8 \text{ m} + 3.6 \text{ m} + 18 \text{ cm}$

6. Use the table to solve the following equations.

a. $3x - 2 = 4$          b. $3x - 2 = 13$

c. $3x - 2 = 8$          d. $3x - 2 = 16$

| $x$ | $y = 3x - 2$ |
|---|---|
| 2 | 4 |
| 3 | 7 |
| 4 | 10 |
| 5 | 13 |

7. Solve these equations using the figure.

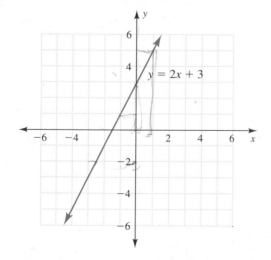

a. $2x + 3 = 5$          b. $2x + 3 = 1$

c. $2x + 3 = -2$

8. Solve these equations using the figure.

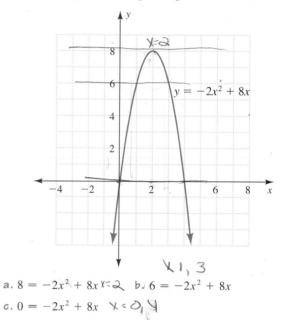

a. $8 = -2x^2 + 8x$          b. $6 = -2x^2 + 8x$

c. $0 = -2x^2 + 8x$

9. Use the figure and the equation $x + 1 = 3 - x$ to do the following.

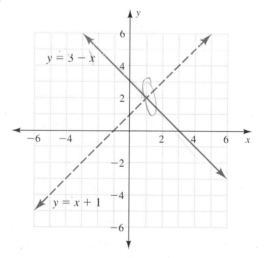

a. Find the point of intersection of the two lines.

b. Substitute the intersection point into each equation shown on the graph.

c. Solve the equation $x + 1 = 3 - x$.

d. Describe how the equation $x + 1 = 3 - x$ relates to the graph.

**In Exercises 10 to 13 solve for $x$ in each of the equations.**

10. $3x - 1 = x + 1$          11. $-2(x + 4) = 10$

12. $5 - 3(x - 4) = x + 9$          13. $-2(x - 3) = -0.5x + 3$

**Solve for $b$ in Exercises 14 and 15.**

14. $3 = 4(-2) + b$          15. $-4 = \frac{2}{3}(-6) + b$

Solve each of the formulas in Exercises 16 to 25 for the indicated variable.

16. $C = K - 273$ for $K$

17. $D = rt$ for $t$

18. $PV = nRT$ for $T$

19. $ax + by = c$ for $x$

20. $A = 2l + 2w$ for $l$

21. $P_1V_1 = P_2V_2$ for $P_2$

22. $C = 35 + 5(x - 100)$ for $x$

23. $A = 0.5h(a + b)$ for $a$

24. $\dfrac{P_1V_1}{T_1} = \dfrac{P_2V_2}{T_2}$ for $V_2$

25. $y_2 - y_1 = m(x_2 - x_1)$ for $m$

The two students in Exercises 26 and 27 have the same total test scores but different homework points. If the final exam is worth 150 points, is it possible for each student to earn 80% or greater?

26. Student 1 has
 tests (100 each): 82, 72
 quizzes (20 each): 20, 0, 20, 20, 18
 homework (70 pts): 12

27. Student 2 has
 tests (100 each): 82, 72
 quizzes (20 each): 15, 15, 15, 16, 16
 homework (70 pts): 70

28. Graph $y = |x - 1|$. From your graph solve these equations.

 a. $|x - 1| = 4$

 b. $|x - 1| = 2$

 c. $|x - 1| = 0$

 d. $|x - 1| = -2$

29. Graph the conditional equations

$$y = 1 - x \text{ for } x < 1$$

 and

$$y = x - 1 \text{ for } x \geq 1$$

For Exercises 30 to 33 determine whether the data in the table are linear. If they are linear, give the slope.

30.

| $x$ | $y$ |
|---|---|
| 1 | 5 |
| 3 | 2 |
| 5 | −1 |

31.

| $x$ | $y$ |
|---|---|
| 1 | 4 |
| 3 | 7 |
| 5 | 11 |

32.

| $x$ | $y$ |
|---|---|
| 1 | 5 |
| 3 | 8 |
| 5 | 13 |

33.

| $x$ | $y$ |
|---|---|
| 1 | 6 |
| 3 | 1 |
| 5 | −4 |

Find the slope for each graph in Exercises 34 to 37.

34.

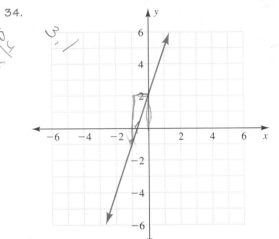

35.

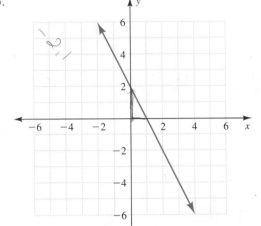

36.

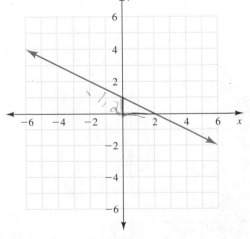

37.

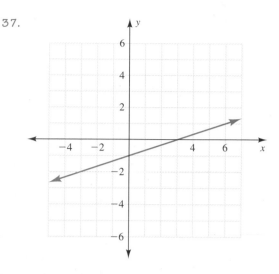

**For Exercises 38 to 41 answer the following questions.**

a. **Write each problem situation using data points. Select the inputs appropriately.**

b. **What is the slope of the line through these points?**

c. **What is the meaning of the slope?**

38. 14 minutes at $4.44 and 5 minutes at $2.10 (long distance phone calls)

39. 2 minutes at $0.26 and 7 minutes at $0.91 (local toll phone calls)

40. 12 feet takes 12 hours and 32 feet takes 17 hours (sidewalk repair)

41. 6 dozen in 2 hours and 12 dozen in 3 hours (cookie baking)

42. Sketch a line with slope $\frac{3}{4}$ that passes through $(1, -2)$.

43. Sketch a line with slope $-\frac{3}{4}$ that passes through $(-2, 1)$.

44. Sketch a line with slope $\frac{1}{3}$ that passes through $(1, -2)$.

45. Sketch a line with slope $-\frac{1}{3}$ that passes through $(2, 0)$.

**For what intervals do the electronics graphs in Exercises 46 and 47 have a positive slope? a negative slope? a zero slope? At what input is the slope of a line segment not defined?**

46.

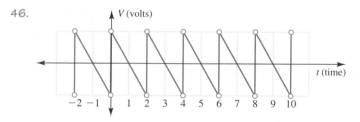

47.

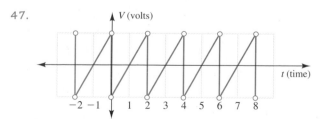

48. A long distance call from Eugene, Oregon, to Boston, Massachusetts, costs $2.10 for 5 minutes and $4.44 for 14 minutes.

a. What linear equation describes the data?

b. From another source the rate is quoted as $1.06 for the first minute and $0.26 per minute thereafter. Write this as an equation. (*Hint*: Define $x$ as the total number of minutes talked.)

c. Are the answers to parts a and b equivalent?

49. A 12-foot sidewalk repair takes 12 hours. A 32-foot sidewalk repair takes 17 hours.

a. Find the linear equation for the time of repair in terms of length of sidewalk.

b. What is the meaning of the slope in your equation?

c. What is the meaning of the $y$-intercept in the equation?

50. Ajax Car Rental charges a $35 fee plus $0.08 per mile. A competing firm, Help-U Company, charges a $30 fee and $0.10 per mile.

a. Write an equation for each company that gives the total rental cost for $x$ miles in one day.

b. For what mileage, $x$, will the costs of the two companies' cars be the same?

c. Which firm, Help-U or Ajax, has a steeper slope if graphed on the same axes? Why?

**Tell whether the pairs of lines in Exercises 51 to 54 are parallel.**

51. $y = 3x + 2$ and $y = -3x + 2$

52. $y = 2x + 3$ and $y = 2x - 3$

53. $y = 5x + 3$ and $y = \frac{1}{5}x + 3$

54. $y = 5x + 3$ and $y = -5x + 3$

**In Exercises 55 to 58 determine whether the changes will create a graph parallel to that of the original equation given in parentheses.**

55. The cost of the first minute of the phone call rises, but the cost per minute stays the same ($C = \$1.58 + \$0.65m$).

56. It takes longer to heat up the stove, but the minutes per batch of cookies remains the same ($T = 10 + 8m$).

57. The cost per mile for a rental car increases, but the cost for one day remains fixed ($C = \$45.00 + \$0.10m$).

58. The phone company keeps the installation charge fixed and increases the basic charge per month ($C = \$14.97m + \$50.00$).

## CHAPTER 3 TEST

1. Translate these sentences into equations. Define the variables.

    a. The output is 25% of the input.

    b. The output is twice the input.

    c. The monthly cost of a Value Plan checking account is $3.00 plus $0.50 for each check over the first 15 checks.

**In Exercises 2 and 3 multiply each expression to remove the parentheses and then add like terms.**

2. $3(x + 2) - 2(x - 3)$

3. $x(2y - 5z) + y(3x + 8z) - 4xy$

4. A rectangle has the area shown in the figure. What are the length and width?

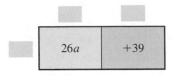

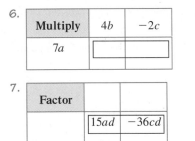

| | 26a | +39 |
|---|---|---|

5. Factor $14x - 7y$.

**For Exercises 6 and 7 multiply or factor the expressions in table form.**

6.

| Multiply | 4b | −2c |
|---|---|---|
| 7a | | |

7.

| Factor | | |
|---|---|---|
| | 15ad | −36cd |

8. Circle the solutions to the equation $5 - 2x = 3$ on the table and on the graph.

| Input: x | Output: $y = 5 - 2x$ |
|---|---|
| −1 | 7 |
| 0 | 5 |
| 1 | 3 |
| 2 | 1 |
| 3 | −1 |

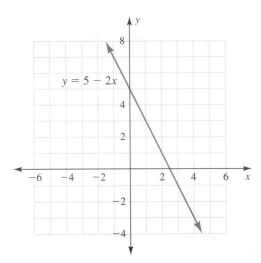

9. Explain how to estimate the solution to $5 - 2x = 0$ from the table in Exercise 8.

10. Explain how to estimate the solution to $5 - 2x = 0$ using the graph in Exercise 8.

11. What is the slope of the line $y = 5 - 2x$ graphed in Exercise 8?

12. In the figure, which line—(1), (2), (3), or (4)—shows the distance traveled by a bicyclist averaging 10 miles per hour?

**For Exercises 13 to 16 use the figure to find the slope of each line.**

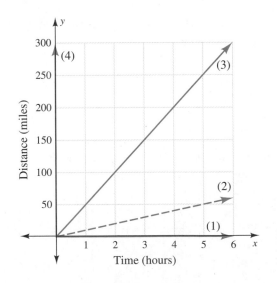

13. Line (1)     14. Line (2)     15. Line (3)     16. Line (4)

17. Explain why $17 - 3(x + 4)$ is not equal to $14(x + 4)$.

**In Exercises 18 to 20 use the reverse order of operations to solve each equation for $x$.**

18. $2x + 3 = 17$     19. $5 - 2x = 3$     20. $\frac{1}{2}x + 5 = -3$

21. Solve for $b$ in $y = mx + b$ if $m = -\frac{2}{3}$ and $(x, y) = (6, 3)$.

**In Exercises 22 and 23 solve for $x$.**

22. $3(x - 5) = x + 9$     23. $3(x + 4) = 2(1 - x)$

24. Solve $P_1 V_1 = P_2 V_2$ for $V_2$.

25. Solve $A = \dfrac{bh}{2}$ for $h$.

**Use the table to answer the questions in Exercises 26 to 29.**

| Year | Population: Ages 5 to 17 |
|------|-------------------------|
| 1950 | 30 million |
| 1960 | 44 million |
| 1970 | 52 million |
| 1980 | 48 million |
| 1990 | 45 million |

26. What is the rate of change of school age population between 1950 and 1960?

27. What is the rate of change of school age population between 1970 and 1980?

28. Estimate the school age population for 1965.

29. Predict the school age population for 2000. Explain your prediction.

30. The Help-U Company rents a compact car for the costs shown in the table. The costs are graphed in the figure.

| Miles: $x$ | Cost: $y$ |
|------------|-----------|
| 50 | $35 |
| 100 | 40 |
| 150 | 45 |
| 200 | 50 |
| 250 | 55 |

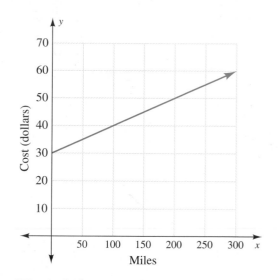

a. What is the input, or independent variable?

b. What is the output, or dependent variable?

c. What is the cost for driving 125 miles?

d. How many miles can be driven for $57?

e. What is the $y$-intercept? What does it mean?

f. What is the slope of the line? What does it mean?

g. What is the equation of the line?

h. What is the $x$-intercept? What does it mean?

## Ratio,

## Rates, and

## Proportional Reasoning

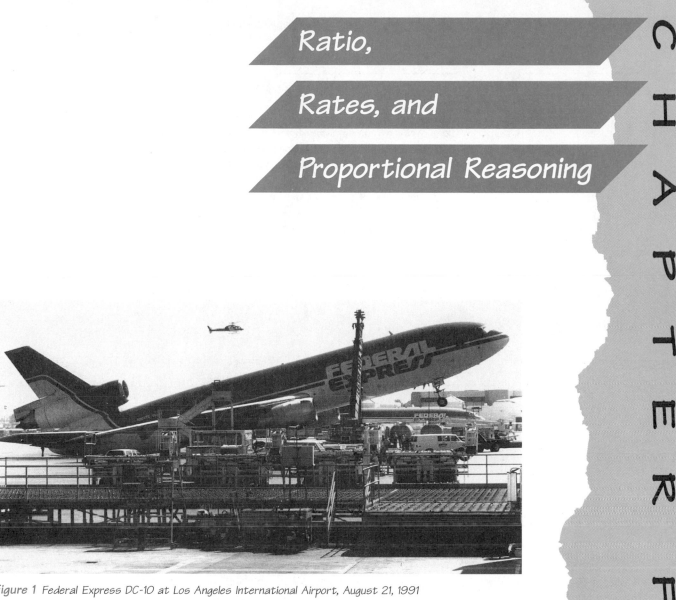

*Figure 1 Federal Express DC-10 at Los Angeles International Airport, August 21, 1991*

**THE AIRPLANE IN** Figure 1 was out of balance and became an

embarrassment when it tipped. The balance point in an airplane is called

the *center of gravity*. In this case, loading or unloading the plane caused

the center of gravity to shift behind the main landing gear, making the

plane tilt. In geometric figures of uniform thickness, the center of gravity

is called the *centroid*. Calculating centroids is discussed in Section 4.5.

## 4.1    RATIOS, RATES, AND PERCENTS

**OBJECTIVES**    Write a ratio in three different forms. ◆ Identify equivalent ratios. ◆ Simplify ratios containing expressions. ◆ Simplify ratios containing units. ◆ Determine quantities allocated under a continued ratio distribution. ◆ Determine ratios of atoms in chemical formulas. ◆ Write rates as ratios. ◆ Write percents as ratios and vice versa. ◆ Use unit analysis to convert rates and solve problems.

**WARM-UP**    To facilitate comparing the operations, change mixed numbers to improper fractions before computation.

*Example*: Add, subtract, multiply, and divide $2\frac{2}{3}$ and $1\frac{1}{4}$.

$$2\frac{2}{3} + 1\frac{1}{4} = \frac{8}{3} + \frac{5}{4} = \frac{32}{12} + \frac{15}{12} = \frac{47}{12} = 3\frac{11}{12}$$

$$2\frac{2}{3} - 1\frac{1}{4} = \frac{8}{3} - \frac{5}{4} = \frac{32}{12} - \frac{15}{12} = \frac{17}{12} = 1\frac{5}{12}$$

$$2\frac{2}{3} \cdot 1\frac{1}{4} = \frac{8}{3} \cdot \frac{5}{4} = \frac{2}{3} \cdot \frac{5}{1} = \frac{10}{3} = 3\frac{1}{3}$$

$$2\frac{2}{3} \div 1\frac{1}{4} = \frac{8}{3} \div \frac{5}{4} = \frac{8}{3} \cdot \frac{4}{5} = \frac{32}{15} = 2\frac{2}{15}$$

1. Add, subtract, multiply, and divide $\frac{1}{3}$ and $\frac{3}{7}$.

2. Add, subtract, multiply, and divide $3\frac{1}{5}$ and $1\frac{3}{4}$.

Compare answers with another student to check your work.    ◆

In this section we examine ratios, rates, and percents. Although many ratio examples are numerical, we return to algebraic forms in determining slopes of linear equations and in simplifying ratios. Because rates and percents play an important role in many algebraic applications, this section offers practice in writing rates as fractions and in unit analysis.

### RATIOS

If an adult dosage of tetracycline hydrochloride is 250 milligrams, how much should a child receive? Medication dosages for infants and children are always less than the adult dosage. Proper dosage may be calculated by comparing the weight, age, or body surface area of the child with the adult standards. (Body surface area charts are found in textbooks on medical dosage calculation for prenursing programs. Body surface area estimates are also important in burn treatment.) Because a **ratio** *is a comparison of two like or unlike quantities*, ratios are common in dosage calculations. There are three common ways to write a ratio: as a fraction, with the word *to*, or with a colon.

**EXAMPLE 1**    *Medication dosages*    Adult dosage refers to a person who weighs 150 pounds, is at least 150 months of age, or has a body surface area of 1.7 square meters. Suppose a 60-pound, 8-year-old child has a body surface area of 1 square meter. Write the ratios in all three ways—as fractions, with the word *to*, and with a colon. Use a calculator to change the fractions to decimal form.

**a.** What is the ratio of the child's weight to adult weight?

**b.** What is the ratio of the child's age to adult age?

**c.** What is the ratio of the child's body surface area to adult body surface area?

**d.** Compare the ratios provided by the three measures.

SOLUTION   **a.** $\dfrac{60 \text{ lb}}{150 \text{ lb}} = \dfrac{2}{5} = \dfrac{0.4}{1}$

The other ratios are 0.4 to 1 and 0.4 : 1.

**b.** $8 \text{ yr} \cdot \dfrac{12 \text{ mo}}{1 \text{ yr}} = 96 \text{ mo}$

$\dfrac{96 \text{ mo}}{150 \text{ mo}} = \dfrac{6 \cdot 16}{6 \cdot 25} = \dfrac{16}{25} = \dfrac{0.64}{1}$

The other ratios are 0.64 to 1 and 0.64 : 1.

**c.** $\dfrac{1 \text{ m}^2 \text{ body surface area}}{1.7 \text{ m}^2 \text{ body surface area}} = \dfrac{1}{1.7} = \dfrac{0.59}{1}$

The other ratios are 0.59 to 1 and 0.59 : 1.

**d.** The weight ratio would give the lowest dosage, and the age ratio would give the highest dosage.

♦ ♦ ♦ ♦ ♦ ♦ ♦ ♦ ♦ ♦ ♦ ♦ ♦ ♦ ♦ ♦ ♦ ♦ ♦ ♦ ♦ ♦ ♦ ♦ ♦ ♦ ♦ ♦ ♦ ♦ ♦ ♦ ♦ ♦ ♦

Slope is another example of a ratio. Recall that

$$\text{slope} = \frac{\text{rise}}{\text{run}} = \frac{\text{change in output}}{\text{change in input}} = \frac{\Delta y}{\Delta x}$$

EXAMPLE   **2**   Find the slope of the graph of $y = -2x$, and interpret the slope as a ratio.

SOLUTION   *Note*: We could read the slope directly, as the coefficient of $x$ in the equation, but instead we will review finding the slope from Table 1 and from coordinates on the graph in Figure 2.

| $\Delta x$ | $x$ | $y$ | $\Delta y$ |
|:---:|:---:|:---:|:---:|
| | $-2$ | $4$ | |
| $+1$ | | | $-2$ |
| | $-1$ | $2$ | |
| $+1$ | | | $-2$ |
| | $0$ | $0$ | |
| $+1$ | | | $-2$ |
| | $1$ | $-2$ | |
| $+1$ | | | $-2$ |
| | $2$ | $-4$ | |

Table 1

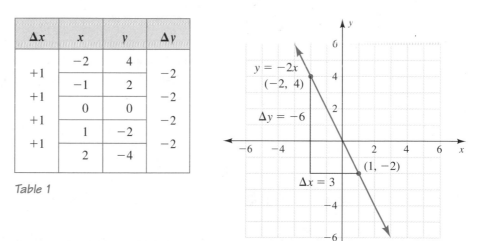

Figure 2

Table 1 shows a 2-unit decrease in output for a 1-unit increase in input. Thus $y = -2x$ has a slope of $\frac{-2}{1}$, or $-2$. The slope between two arbitrarily chosen coordinates, $(-2, 4)$ and $(1, -2)$, is

$$\frac{y_2 - y_1}{x_2 - x_1} = \frac{-2 - 4}{1 - (-2)} = \frac{-6}{3} = \frac{-2}{1}$$

The slope is the ratio $-2$ to 1. The line drops 2 units for each unit change to the right.

 ♦ ♦ ♦ ♦ ♦ ♦ ♦ ♦ ♦ ♦ ♦ ♦ ♦ ♦ ♦ ♦ ♦ ♦ ♦ ♦ ♦ ♦ ♦ ♦ ♦ ♦ ♦ ♦ ♦ ♦ ♦ ♦ ♦ **♦ ♦ ♦**

A baseball batting average is the ratio of hits to times at bat. Many ratios, such as batting averages, are changed from fraction to decimal form to ease comparison.

$$\text{batting average} = \frac{\text{number of hits}}{\text{number of times at bat}}$$

**E X A M P L E    3**

*Batting averages*    Kirby Puckett of the Minnesota Twins had 195 hits in 611 times at bat, while Tony Gwynn of the San Diego Padres had 168 hits in 530 times at bat. Write each batting average as a ratio in fraction form, and find the equivalent decimal, rounded to nearest thousandth. Who had the higher batting average?

**SOLUTION**     Puckett had an average of $\frac{195}{611} \approx .319$, while Gwynn's average was $\frac{168}{530} \approx .317$. Puckett's was slightly higher.

♦ ♦ ♦ ♦ ♦ ♦ ♦ ♦ ♦ ♦ ♦ ♦ ♦ ♦ ♦ ♦ ♦ ♦ ♦ ♦ ♦ ♦ ♦ ♦ ♦ ♦ ♦ ♦ ♦ ♦ ♦ ♦ ♦ ♦ ♦ ♦

Batting averages are generally written as ratios out of a thousand, so these two players have a .319 and a .317 average, respectively. (*Respectively* means "in the same order as listed" or "as given." The use of *respectively* saves repeating information in a paragraph.) If ratios have the same decimal or simplify to the same fraction, they are equivalent.

---

Two ratios are *equivalent* if they simplify to the same number.

---

**E X A M P L E    4**

Show that $\frac{4}{6}$ and $\frac{6}{9}$ are equivalent ratios.

**SOLUTION**

$$\frac{4}{6} = \frac{2 \cdot 2}{3 \cdot 2} = \frac{2}{3} \quad \text{and} \quad \frac{6}{9} = \frac{2 \cdot 3}{3 \cdot 3} = \frac{2}{3}$$

Each ratio contains a factor common to the numerator and denominator. These common factors simplify to $\frac{1}{1}$ and cancel out, leaving the same ratio: $\frac{2}{3}$.

♦ ♦ ♦ ♦ ♦ ♦ ♦ ♦ ♦ ♦ ♦ ♦ ♦ ♦ ♦ ♦ ♦ ♦ ♦ ♦ ♦ ♦ ♦ ♦ ♦ ♦ ♦ ♦ ♦ ♦ ♦ ♦ ♦ ♦ ♦ ♦

Simplifying ratios that contain only numbers, expressions, or units is exactly like simplifying fractions: We look for common factors in the numerator and denominator and cancel those factors.

**E X A M P L E    5**

Write each ratio in fraction form, identify the common factors, and simplify.

**a.** $3x^2 : 15x$     **b.** $(a + ab)$ to $a$     **c.** $2\sqrt{3}$ to $4\sqrt{3}$

**d.** $(a + b)(a - b)$ to $(a - b)$

**SOLUTION**     **a.** $3x$ is a common factor;

$$\frac{3x^2}{15x} = \frac{3x \cdot x}{3x \cdot 5} = \frac{x}{5}$$

**b.** $a$ is a common factor;

$$\frac{a + ab}{a} = \frac{a(1 + b)}{a} = \frac{(1 + b)}{1}$$

**c.** $2\sqrt{3}$ is a common factor;

$$\frac{2\sqrt{3}}{4\sqrt{3}} = \frac{2\sqrt{3}}{2 \cdot 2\sqrt{3}} = \frac{1}{2}$$

**d.** $(a - b)$ is a common factor;

$$\frac{(a + b)(a - b)}{(a - b)} = \frac{(a + b)}{1}$$

♦ ♦ ♦ ♦ ♦ ♦ ♦ ♦ ♦ ♦ ♦ ♦ ♦ ♦ ♦ ♦ ♦ ♦ ♦ ♦ ♦ ♦ ♦ ♦ ♦ ♦ ♦ ♦ ♦ ♦ ♦ ♦

If the ratios contain units of measurement of the same type (length, mass, capacity), the units should be made the same before the ratios are simplified or are compared with another ratio. See Table 2 for some equivalent units of measurement.

| |
|---|
| 1000 milliliters = 1 liter |
| 1000 grams = 1 kilogram |
| 1000 meters = 1 kilometer |
| 100 centimeters = 1 meter |
| 16 ounces = 1 pound |
| 1 yard = 36 inches |
| 1 yard = 3 feet |
| 1 mile = 5280 feet |
| 4 quarts = 1 gallon |

*Table 2* Measurement Facts

**E X A M P L E   6**

Change the units to common units, and simplify the ratios.

**a.** 100 centimeters:2 meters        **b.** 1500 grams to 1 kilogram

**c.** 15 feet to 20 miles        **d.** $k$ inches to $m$ yards

**e.** $b$ quarts to $c$ gallons

**SOLUTION**   **a.** $\dfrac{100 \text{ cm}}{2 \text{ m}} = \dfrac{1 \text{ m}}{2 \text{ m}} = \dfrac{1}{2}$  or  $\dfrac{100 \text{ cm}}{200 \text{ cm}} = \dfrac{1}{2}$

**b.** $\dfrac{1500 \text{ g}}{1 \text{ kg}} = \dfrac{1.5 \text{ kg}}{1 \text{ kg}} = \dfrac{3}{2}$  or  $\dfrac{1500 \text{ g}}{1000 \text{ g}} = \dfrac{3}{2}$

**c.** $\dfrac{15 \text{ ft}}{20 \text{ mi}} = \dfrac{15 \text{ ft}}{105,600 \text{ ft}} = \dfrac{1}{7040}$

**d.** $\dfrac{k \text{ in.}}{m \text{ yd}} = \dfrac{k \text{ in.}}{36m \text{ in.}} = \dfrac{k}{36m}$

**e.** $\dfrac{b \text{ qt}}{c \text{ gal}} = \dfrac{b \text{ qt}}{4c \text{ qt}} = \dfrac{b}{4c}$

♦ ♦ ♦ ♦ ♦ ♦ ♦ ♦ ♦ ♦ ♦ ♦ ♦ ♦ ♦ ♦ ♦ ♦ ♦ ♦ ♦ ♦ ♦ ♦ ♦ ♦ ♦ ♦ ♦ ♦ ♦ ♦

In addition to the colon, "to," and fraction forms of ratios, there are three other ratios of importance: continued ratios, rates, and percents.

**CONTINUED RATIOS**

A **continued ratio** *is a ratio of three or more quantities.* In business a continued ratio might represent an individual's interest in an estate or in profit sharing. Such a ratio might be used to describe the quantities of ingredients in food manufacturing or to allocate time spent by a specialist on a number of different projects.

EXAMPLE 7

**Profit sharing**    The owners of a business have contributed funds in the *continued ratio* 5:3:1. If there is a $90,000 profit, how much will each receive?

SOLUTION

The ratio 5:3:1 means that for $5 + 3 + 1 = 9$ total parts, 5 go to the first owner, 3 to the second owner, and 1 to the third owner. If the profit is $90,000, one part will be

$$\$90,000 \div 9 = \$10,000$$

The three owners will receive $10,000 times their shares, or $50,000, $30,000, and $10,000, respectively.

♦ ♦ ♦ ♦ ♦ ♦ ♦ ♦ ♦ ♦ ♦ ♦ ♦ ♦ ♦ ♦ ♦ ♦ ♦ ♦ ♦ ♦ ♦ ♦ ♦ ♦ ♦ ♦ ♦ ♦ ♦ ♦ ♦ ♦ ♦ ♦ ♦ ♦

EXAMPLE 8

**Time on the job**    One day an accounting consultant spent 4 hours on project A, 3 hours on project B, and 1 hour on project C.

**a.** Write a continued ratio to describe the time spent on the three projects.

**b.** Determine the amount charged to each project if the consultant's daily salary is $400.

SOLUTION

**a.** The ratio 4:3:1 describes the time spent on projects A, B, and C.

**b.** The total parts are $4 + 3 + 1 = 8$. The daily salary is divided by the number of hours worked, so each hour is worth $400 divided by 8, or $50. Multiply the hourly rate by the time spent on each project: $4 \cdot \$50$ is charged to project A; $3 \cdot \$50$ is charged to project B; and $1 \cdot \$50$ is charged to project C. Thus projects A, B, and C are charged $200, $150, and $50, respectively.

♦ ♦ ♦ ♦ ♦ ♦ ♦ ♦ ♦ ♦ ♦ ♦ ♦ ♦ ♦ ♦ ♦ ♦ ♦ ♦ ♦ ♦ ♦ ♦ ♦ ♦ ♦ ♦ ♦ ♦ ♦ ♦ ♦ ♦ ♦ ♦ ♦

Chemical formulas illustrate another continued ratio—the ratio of atoms in a molecule. The water molecule, with formula $H_2O$, has 2 atoms of hydrogen for each atom of oxygen. A subscript 1 is assumed on oxygen: $O = O_1$. Nicotine, $C_{10}H_{14}N_2$, is composed of 10 atoms of carbon, 14 atoms of hydrogen, and 2 atoms of nitrogen.

Three molecules of water would be written $3H_2O$, to represent 6 atoms of hydrogen combined with 3 of oxygen. The 3 in front is a coefficient of the entire formula and is applied to each chemical element. It does not change the ratio of the elements in the formula.

EXAMPLE 9

**Chemical formulas**    Two molecules of caffeine is written $2C_8H_{10}N_4O_2$. How many atoms of each element are contained in these two molecules?

SOLUTION

The formula $2C_8H_{10}N_4O_2$ represents 16 atoms of carbon, 20 atoms of hydrogen, 8 atoms of nitrogen, and 4 atoms of oxygen.

♦ ♦ ♦ ♦ ♦ ♦ ♦ ♦ ♦ ♦ ♦ ♦ ♦ ♦ ♦ ♦ ♦ ♦ ♦ ♦ ♦ ♦ ♦ ♦ ♦ ♦ ♦ ♦ ♦ ♦ ♦ ♦ ♦ ♦ ♦ ♦ ♦

## RATES

Rates as ratios are important in applications. A **rate** often contains the word *per*, which means "for each" or "for every." Some rates are abbreviated with just three letters, such as mpg and rpm in Example 10. When we change a rate into the fraction form of a ratio, we place the word that follows *per* in the denominator.

EXAMPLE **10**   Write these rates in fraction form.

**a.** mph (miles per hour)          **b.** liters per minute

**c.** mpg (miles per gallon)        **d.** rpm (revolutions per minute)

**e.** dollars per hour

SOLUTION   Observe the abbreviations used within the fraction form.

**a.** $\dfrac{\text{mi}}{\text{hr}}$    **b.** $\dfrac{\text{L}}{\text{min}}$    **c.** $\dfrac{\text{mi}}{\text{gal}}$    **d.** $\dfrac{\text{rev}}{\text{min}}$    **e.** $\dfrac{\text{dollars}}{\text{hr}}$

◆ ◆ ◆ ◆ ◆ ◆ ◆ ◆ ◆ ◆ ◆ ◆ ◆ ◆ ◆ ◆ ◆ ◆ ◆ ◆ ◆ ◆ ◆ ◆ ◆ ◆ ◆ ◆ ◆ ◆ ◆ ◆ ◆ ◆ ◆

To change from one rate to another we use unit analysis. Recall that this process combines simplifying fractions with multiplying by ratios of equivalent units.

EXAMPLE **11**   Leaky faucets   Suppose a faucet leaks 1 drop per second. Estimate how many gallons per month are wasted.

SOLUTION   To find how many gallons of water are wasted during a 30-day month, we need to convert drops to gallons and seconds to months. An experimental estimate gives 16 drops per teaspoon. Standard measure gives 3 teaspoons per tablespoon, 16 tablespoons per cup, 4 cups per quart, 4 quarts per gallon. We set up ratios so that the original units cancel out and the desired units are in the proper position: gallons divided by months.

$$\frac{1\ \text{drop}}{1\ \text{sec}} \cdot \frac{1\ \text{tsp}}{16\ \text{drops}} \cdot \frac{1\ \text{tbsp}}{3\ \text{tsp}} \cdot \frac{1\ \text{C}}{16\ \text{tbsp}} \cdot \frac{1\ \text{qt}}{4\ \text{C}} \cdot \frac{1\ \text{gal}}{4\ \text{qt}} \cdot \frac{60\ \text{sec}}{1\ \text{min}} \cdot \frac{60\ \text{min}}{1\ \text{hr}}$$

$$\cdot \frac{24\ \text{hr}}{1\ \text{day}} \cdot \frac{30\ \text{days}}{1\ \text{mo}}$$

$$= \frac{60 \cdot 60 \cdot 24 \cdot 30\ \text{gal}}{16 \cdot 3 \cdot 16 \cdot 4 \cdot 4\ \text{mo}}$$

$$\approx 211\ \text{gallons per month}$$

◆ ◆ ◆ ◆ ◆ ◆ ◆ ◆ ◆ ◆ ◆ ◆ ◆ ◆ ◆ ◆ ◆ ◆ ◆ ◆ ◆ ◆ ◆ ◆ ◆ ◆ ◆ ◆ ◆ ◆ ◆ ◆ ◆ ◆ ◆

If you do Example 11 on a calculator, you need to be careful about grouping. Use these keystrokes, and compare answers. Which is correct?

**a.** 60 $\boxed{\times}$ 60 $\boxed{\times}$ 24 $\boxed{\times}$ 30 $\boxed{\div}$ 16 $\boxed{\times}$ 3 $\boxed{\times}$ 16 $\boxed{\times}$ 4 $\boxed{\times}$ 4

**b.** 60 $\boxed{\times}$ 60 $\boxed{\times}$ 24 $\boxed{\times}$ 30 $\boxed{\div}$ 16 $\boxed{\div}$ 3 $\boxed{\div}$ 16 $\boxed{\div}$ 4 $\boxed{\div}$ 4

**c.** 60 $\boxed{\times}$ 60 $\boxed{\times}$ 24 $\boxed{\times}$ 30 $\boxed{\div}$ $\boxed{(}$ 16 $\boxed{\times}$ 3 $\boxed{\times}$ 16 $\boxed{\times}$ 4 $\boxed{\times}$ 4 $\boxed{)}$

Recall that the fraction bar is a grouping symbol. Option a illustrates why the numerator and denominator must often be entered within parentheses. Because of the order of operations built into the scientific calculator, the keystrokes in option a will not give the correct solution. Both option b and option c are correct.

## PERCENTS

A **percent** is also a rate, meaning "per hundred." The quantity *15 percent* means 15 per hundred and is written $\frac{15}{100}$ as a fraction. The ratio *one part to the total* may be divided to find a percent.

E X A M P L E    **12**    Here is how Monique spent her time during one week. What percent of her week was spent on each activity?

| | |
|---|---|
| Classes: | 12 hours |
| Study: | 30 hours |
| Work: | 20 hours |
| Sleep: | 56 hours (8 hours each night) |
| Transportation: | 7 hours |
| Other: | remaining time (at school between classes, with family, eating, etc.) |

SOLUTION    24 hours per day $\times$ 7 days per week = 168 hours per week

| | |
|---|---|
| Classes: | $12 \div 168 = 0.071 = 7.1\%$ |
| Study: | $30 \div 168 = 0.179 = 17.9\%$ |
| Work: | $20 \div 168 = 0.119 = 11.9\%$ |
| Sleep: | $56 \div 168 = 0.333 = 33.3\%$ |
| Transportation: | $7 \div 168 = 0.042 = 4.2\%$ |
| Other: | $168 - (12 + 30 + 20 + 56 + 7) = 43$ hr remaining |
| | $43 \div 168 = 0.256 = 25.6\%$ |

◆ ◆ ◆ ◆ ◆ ◆ ◆ ◆ ◆ ◆ ◆ ◆ ◆ ◆ ◆ ◆ ◆ ◆ ◆ ◆ ◆ ◆ ◆ ◆ ◆ ◆ ◆ ◆ ◆ ◆ ◆ ◆ ◆ ◆ ◆ ◆ ◆ ◆ ◆

## EXERCISES 4.1

**For Exercises 1 to 10 identify the slope in each linear equation.**

1. $y = 3x$
2. $y = 4 + 2x$
3. $y = 2 - 4x$
4. $y = -5x$
5. $y = \frac{1}{2}x$
6. $y = -\frac{1}{4}x$
7. $y = 2.98x + 5.00$
8. $y = 20 - 1.50x$
9. $y = 5 - 0.05x$
10. $y = 0.65x + 5.15$

**In Exercises 11 and 12 identify the common factors, and then simplify the ratios.**

11. a. 15 to 35    b. 48 to 16    c. 5280 to 3600
    d. $12x : 4x^2$    e. $2xy$ to $6x^2y$

12. a. 84 to 32    b. 52 to 26    c. 1024 to 288
    d. $18n : 6n^3$    e. $6x^2y$ to $2xy$

**In Exercises 13 and 14 write the ratios as fractions, identify the common factors, and then simplify.**

13. a. $a(b + c)$ to $a$    b. $(a + b)(b + c)$ to $(a + b)$
    c. $abc$ to $ace$    d. $\dfrac{2(1 + \sqrt{3})}{2}$

14. a. $a(b + c)$ to $(b + c)$    b. $b(a + b)$ to $b$
    c. $def$ to $dig$    d. $\dfrac{2}{2(1 + \sqrt{3})}$

**In Exercises 15 and 16 change the units to like units, then simplify the ratios.**

15. a. $\frac{1}{2}$ foot to 1 inch
    b. 3000 grams to 1 kilogram
    c. 32 ounces to 5 pounds

d. $2\frac{2}{3}$ yards to 2 feet
e. 200 milliliters to 20 liters
f. 2 years to 150 months
g. 20 minutes to $\frac{1}{4}$ hour
h. $m$ minutes to $h$ hours

16. a. 1 foot to 3 inches
    b. 2 meters to 1200 centimeters
    c. 1500 meters to 1 kilometer
    d. 2 quarts to 4 gallons
    e. 2 liters to 2000 milliliters
    f. 4 years to 150 months
    g. 120 minutes to $\frac{1}{2}$ hour
    h. $d$ days to $w$ weeks

17. In several parts of Exercises 15 and 16 there were two ways to change to common units. Did the choice of unit changed make any difference in the simplified answer? Explain why.

18. Is it possible to change the ratio *1 gallon to 300 square feet* to common units? Where might such a ratio be used?

**For Exercises 19 to 26 write a ratio from the sentence and simplify it, if possible.**

19. Sergei adds 3 cans of water to each can of frozen orange juice.

20. Frantisek adds 2 cans of milk to a can of soup.

21. Stephanie pays $480 tuition for 12 credits.

22. Jocelyn earns $250 in a 40-hour week.

23. Mikael adds a half pint of oil to 1 gallon of gasoline. (*Hint*: 2 pints equals 1 quart, and 1 gallon equals 4 quarts.)

24. Sylvia adds 1 tablespoon of plant fertilizer to a gallon of water.

25. Betty orders an IV of 240 cc in 16 hours.

26. Johann fills an IV order for 250 mL in 12 hours.

**Batting average is hits divided by number of times at bat. Calculate a decimal batting average for each of the players in Exercises 27 and 28. Round to the nearest thousandth.**

27. a. Roberto Clemente: 209 hits in 585 times at bat

    b. Juan Gonzalez: 152 hits in 584 times at bat (The hits included 43 home runs.)

28. a. Ken Griffey, Jr.: 174 hits in 565 times at bat

    b. Ivan Calderon: 141 hits in 470 times at bat

**What is the ratio of atoms in the chemical formulas in Exercises 29 to 32? (Na is sodium and Ca is calcium; the other elements were listed in the text.)**

29. $C_9H_8O_4$ (the active ingredient in aspirin)

30. $NaHCO_3$ (baking soda)

31. $CaCO_3$ (chalk)

32. $C_7H_5N_3O_6$ (T.N.T.)

33. A $1 million estate is to be divided among Goodwill, Habitat for Humanity, and United Way in the ratio $10:10:5$. What is each share worth?

34. Cosmo, Timo, and Sven earn a profit of 2500 crowns (Swedish currency) on a dance. They split the profits $4:3:3$. How much will each receive?

35. Virginia, James, and Ursula receive a $5000 advance on royalties. All three authors are writing the book, but Ursula contributes another full share by keying the book into the computer. They agree to split the $5000 in a $1:1:2$ ratio. How much should each earn?

36. A rope of red licorice is 36 inches long. Each of three children is to get a share of the rope proportional to his or her age. Lee is 12, Terry is 9, and Alice is 6. Write a continued ratio of the ages. How long a piece does each get?

**In Exercises 37 and 38 write a sentence using each of the rates in an appropriate setting.**

37. a. stitches per inch      b. feet per second

    c. miles per gallon      d. revolutions per hour

38. a. revolutions per minute      b. gallons per hour

    c. kilometers per hour      d. calories per day

**In Exercises 39 to 46 arrange the facts listed after each problem into a unit analysis that solves the problem. Round to the nearest hundredth.**

39. 140 pounds is how many kilograms (kg)?
2.2 lb is 1 kg

40. 200 milliliters (mL) is how many liters (L)?
1 L is 1000 mL

41. 300 milliliters of water is how many grams (g)?
1 kg is 1000 mL of water
1000 g is 1 kg

42. 15 feet is how many meters?
1 ft is 12 in.
1 m is 39.37 in.

43. 1 gallon for five miles is how many dollars per day of driving?
1 hr to travel 55 mi
1 gal is $1.25
1 driving day is 10 hr

44. 60 miles per hour is how many feet per second?
1 mi is 5280 ft
1 min is 60 sec
1 hr is 60 min

45. 250 mL in 12 hours is how many microdrops per minute?
60 microdrops is 1 mL
1 hr is 60 min

46. An infant of 1 year should receive how many milligrams (mg)?
Adult dosage is 500 mg
150 mo = 1 adult dosage
1 yr = 12 mo

*PROBLEM SOLVING*

47. Compare alcohol consumption by finding the alcohol content of each serving. What may be concluded about the results?

    a. Three 6-ounce glasses of wine at 12% alcohol

    b. Three 12-ounce cans of beer at 6% alcohol

    c. Three 2-ounce glasses of 72-proof whiskey at 36% alcohol

48. *A True Scenario from the 1970s.* The 12 500-watt light fixtures in a classroom are connected to a 4-light switch panel.

    a. If all the window curtains are opened and one switch is turned off, what percent of the electricity is saved?

    b. How many watts are saved if the bulbs are changed to 300 watts and all lights are turned on? What percent savings is this from the original lighting set-up?

    c. Which yields the greater savings?

**In Exercises 49 and 50 a *mole* is a chemical unit of measure, not a small, furry animal. Round answers to the nearest tenth.**

49. Apply unit analysis to 25.0 grams $CH_4$ to find grams HCN. (It is assumed that the environment has an ample source of nitrogen, N.)
1 mole $CH_4$ is 16.0 grams $CH_4$
1 mole HCN is 27.0 grams HCN
2 moles $CH_4$ is equated with 2 moles HCN

50. Apply unit analysis to 7.0 grams Na to find grams NaCl. (It is assumed that the environment has an ample source of chlorine, Cl.)
    1 mole NaCl is 58.5 grams NaCl
    2 moles Na is equated with 2 moles NaCl
    1 mole Na is 23.0 grams Na

51. A lawn mower cuts 100 feet in 30 seconds. How many miles per hour is this?

52. The oil in a lawn mower needs to be changed every 25 hours. If the mower travels 100 feet in 30 seconds and shuts off when stopped, how many miles will it travel between oil changes?

53. How far is it possible to bicycle in 10 minutes at 20 miles per hour?

54. How far is it possible to walk in 30 minutes at 6 miles per hour?

55. A circle contains 360°. On an old-fashioned clockface watch, what is the rate in degrees per hour for

    a. the minute hand?    b. the hour hand?

    c. the second hand?

PROJECTS

56. **Road Use Taxes.** A sign on a freight truck reads: *This truck paid $8025 in road use taxes last year.*

    a. If the truck owner paid $0.07 per gallon in road taxes and gets 5 miles per gallon, how many miles did the truck travel?

    b. At 50 miles per hour average and 10 hours of travel per day, could one driver possibly have driven the truck that far?

    c. Are there other road use taxes that could explain any discrepancy?

    d. At 50 miles per hour average and 10 hours of travel per day, how far could one truck driver travel in a year? State your assumptions about the number of days traveled.

    e. At $0.07 per gallon and 5 miles per gallon, what amount of fuel taxes would have been paid? If the truck owner did pay $8025 in road use taxes, what percent was from fuel taxes and what percent was from other taxes?

57. **Shoelace Length.** Is there a relationship between the length of a shoelace and the number of eyelets (holes) in a shoe?  Gather data from at least ten different styles of shoes. Make a table showing information about the shoe (dress shoe, work boot, baby shoe, low-cut gym shoe, high-top basketball shoe, etc.) as well as the number of eyelets and the length of the shoelaces. Plot the data on a graph, with number of eyelets as inputs. Describe your conclusions in full sentences. Include any ratios that seem appropriate.

58. **Geometry.** Investigate the following questions. First use ratios in numerical examples, and then use ratios in formulas. Illustrate your findings with drawings, and summarize your findings in complete sentences.

    a. When we double the dimensions of a square and a circle, how do the perimeter and circumference change?

    b. When we double the dimensions of a square and a circle, how does the area change?

    c. When we double the dimensions of a square and a circle, how does the volume change?

    d. What do the results say about medication dosages for a 3-foot-tall child compared with those for a 6-foot-tall adult? Why would weight be a better measure for dosage ratios than height?

---

## 4.2   PROPORTIONS AND PROPORTIONAL REASONING

**OBJECTIVES**   Investigate division with numbers approaching zero.   ◆   Determine input that makes an expression undefined.   ◆   Write equivalent ratios as proportions.   ◆   Verify a proportion using cross multiplication.   ◆   Form proportions containing units.   ◆   Use cross multiplication to solve for a variable.   ◆   Set up proportions to solve application problems.

**WARM-UP**   Multiply each of these using the distributive property.

1. $6(x + 4)$     2. $4(x + 5)$     3. $8(x + 8)$

4. $20(x - 1)$     5. $6(x - 6)$     6. $-4(x - 3)$

7. $-4(x + 1)$     8. $-5(x - 1)$     9. $-2(x + 3)$

10. $-3(x - 3)$

Solve each equation for *x*.

**11.**  $x - 9 = 0$          **12.** $2x + 4 = 0$          **13.** $2x - 24 = 0$

**14.**  $12x - 84 = 0$

**15.** Multiply both sides of the equation by $bd$, then simplify each side.

$$\frac{a}{b} = \frac{c}{d}$$

What do you observe?                                                              ♦

## PROPORTIONS AND PROPORTIONAL STATEMENTS

Safety guidelines for home and professional ladders warn people to set the top of a straight ladder four times higher than the distance the foot of the ladder is from the wall. The height-to-base ratio is 4 to 1. This 4:1 ratio is also the slope of the ladder.

EXAMPLE   *1*     *Safe ladder ratio*     Each ladder below reaches 12 feet up a wall. Write a simplified height-to-base ratio for each ladder position. Compare the simplified ratio with 4:1. How is an accident likely to happen if the ratio is not safe?

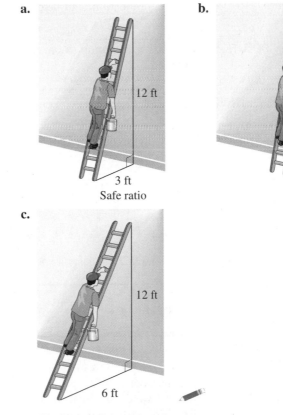

a.

12 ft

3 ft

Safe ratio

b.

12 ft

2 ft

c.

12 ft

6 ft

SOLUTION   **a.** 12:3 = 4:1, a safe ratio

**b.** 12:2 = 6:1, not a safe ratio; the ladder may tip over backwards because the ratio is larger than 4:1.

**c.** 12:6 = 2:1, not a safe ratio; the bottom of the ladder may slide away from the wall because the ratio is smaller than 4:1.

♦ ♦ ♦ ♦ ♦ ♦ ♦ ♦ ♦ ♦ ♦ ♦ ♦ ♦ ♦ ♦ ♦ ♦ ♦ ♦ ♦ ♦ ♦ ♦ ♦ ♦ ♦ ♦ ♦ ♦ ♦ ♦ ♦ ♦ ♦ ♦ ♦

In part a of Example 1 the safe ladder ratios form a proportion:

$$12:3 = 4:1 \quad \text{or, as a fraction,} \quad \frac{12}{3} = \frac{4}{1}$$

This leads us to the definition of a proportion:

> Two equal ratios form a **proportion**:
>
> $$\frac{a}{b} = \frac{c}{d}$$
>
> where $b \neq 0$ and $d \neq 0$.

A ladder reaching 20 feet up a wall and set 5 feet from the base of the wall would also be safe, because the 20 to 5 ratio simplifies to 4 to 1. This ladder's proportion is

$$20:5 = 4:1 \quad \text{or} \quad \frac{20}{5} = \frac{4}{1}$$

The numbers on the diagonal in the fraction proportion (5 and 4, 1 and 20) multiply to the same number: 20. This diagonal property is true for all proportions.

**E X A M P L E   2**

**SOLUTION**

Multiply the numbers on the diagonal to determine whether each of the following is a proportion, and determine whether each simplifies to the same ratio.

**a.** $\frac{9}{12} \overset{?}{=} \frac{6}{8}$   **b.** $\frac{1}{3} \overset{?}{=} \frac{30}{100}$

**a.** Yes, this is a proportion. Both diagonals multiply to 72, and each ratio simplifies to $\frac{3}{4}$.

$$\overset{72}{\underset{}{\phantom{9}}} \quad \overset{72}{\phantom{6}}$$
$$\frac{9}{12} \times \frac{6}{8}$$

**b.** No, this is not a proportion. The diagonals multiply to different numbers, and the ratios simplify to different numbers.

$$\overset{100}{\phantom{1}} \quad \overset{90}{\phantom{30}}$$
$$\frac{1}{3} \times \frac{30}{100}$$

◆ ◆ ◆ ◆ ◆ ◆ ◆ ◆ ◆ ◆ ◆ ◆ ◆ ◆ ◆ ◆ ◆ ◆ ◆ ◆ ◆ ◆ ◆ ◆ ◆ ◆ ◆ ◆ ◆ ◆ ◆

Example 2 suggests a special multiplication:

**The Cross Multiplication Property**

> The proportion $\dfrac{a}{b} = \dfrac{c}{d}$ implies $a \cdot d = b \cdot c$.

*The equation $a \cdot d = b \cdot c$ is called* **cross multiplication** *because drawing lines between the parts being multiplied forms an X-shaped cross. In Section 4.1 we determined whether two ratios were equal by simplifying them or by dividing them to the same decimal. Cross multiplication gives a third way to compare ratios.

Proportions written in the form $a:b = c:d$ also can be written $a \cdot d = b \cdot c$; this form emphasizes the fact that equal products come from the multiplication of the two inside numbers and the multiplication of the two outside numbers. We call this way of writing a proportion the *inner–outer product method*. The inner–outer product form is an optional alternative to writing a proportion with fractions.

EXAMPLE **3**

Verify that $8:12 = 12:18$ using the inner–outer product and cross-multiplication methods.

SOLUTION

*Inner–outer product method*:

$$\text{Inner product} = 144$$
$$8:12 = 12:18$$
$$\text{Outer product} = 144$$

*Cross-multiplication method*:

$$\overset{144}{\dfrac{8}{12}}\diagdown\diagup\overset{144}{\dfrac{12}{18}}$$

♦ ♦ ♦ ♦ ♦ ♦ ♦ ♦ ♦ ♦ ♦ ♦ ♦ ♦ ♦ ♦ ♦ ♦ ♦ ♦ ♦ ♦ ♦ ♦ ♦ ♦ ♦ ♦ ♦ ♦ ♦ ♦ ♦ ♦ ♦ ♦ ♦

Examples 4 and 5 illustrate that a selection of four numbers or expressions may be arranged into proportions in several ways.

EXAMPLE **4**

Place the numbers 2, 4, 5, and 10 in pairs of fractions. Which pairs of fractions make proportions? Which are not proportions?

SOLUTION

Some pairs are

$$\frac{2}{4} = \frac{5}{10}, \quad 2 \cdot 10 = 4 \cdot 5 \qquad \text{A proportion}$$
$$\frac{4}{10} = \frac{2}{5}, \quad 4 \cdot 5 = 2 \cdot 10 \qquad \text{A proportion}$$
$$\frac{10}{5} = \frac{4}{2}, \quad 10 \cdot 2 = 5 \cdot 4 \qquad \text{A proportion}$$
$$\frac{2}{5} \neq \frac{10}{4}, \quad 2 \cdot 4 \neq 5 \cdot 10 \qquad \text{Not a proportion}$$

♦ ♦ ♦ ♦ ♦ ♦ ♦ ♦ ♦ ♦ ♦ ♦ ♦ ♦ ♦ ♦ ♦ ♦ ♦ ♦ ♦ ♦ ♦ ♦ ♦ ♦ ♦ ♦ ♦ ♦ ♦ ♦ ♦ ♦ ♦ ♦ ♦

EXAMPLE **5**

*Medication dosages* Form proportions showing that a child of age 96 months should receive a dosage of 160 milligrams if 250 milligrams is the dosage for persons 150 months or older.

SOLUTION

Some proportions are

$$\frac{96 \text{ mo}}{150 \text{ mo}} = \frac{160 \text{ mg}}{250 \text{ mg}}$$

$$\frac{96 \text{ mo}}{160 \text{ mg}} = \frac{150 \text{ mo}}{250 \text{ mg}}$$

$$\frac{250 \text{ mg}}{150 \text{ mo}} = \frac{160 \text{ mg}}{96 \text{ mo}}$$

♦ ♦ ♦ ♦ ♦ ♦ ♦ ♦ ♦ ♦ ♦ ♦ ♦ ♦ ♦ ♦ ♦ ♦ ♦ ♦ ♦ ♦ ♦ ♦ ♦ ♦ ♦ ♦ ♦ ♦ ♦ ♦ ♦ ♦ ♦ ♦ ♦

When proportions contain units, the units must match either side to side or within numerator and denominator pairs. In either case, when we cross multiply, the product of the units is identical for each multiplication.

## SOLVE PROPORTIONS

We can cross multiply to solve for a variable.

E X A M P L E   **6**

**Purchasing ribbon**    Samantha needs more ribbon for a sewing project. She has \$3 remaining in the project's budget. If 7 yards of ribbon costs \$8, how many yards is it possible for her to buy for \$3?

SOLUTION

$$\frac{7}{8} = \frac{x}{3} \qquad \text{Set up the proportion.}$$

$$21 = 8x \qquad \text{Cross multiply.}$$

$$\frac{21}{8} = \frac{8x}{8} \qquad \text{Divide by 8.}$$

$$x = 2\tfrac{5}{8} \text{ yd}$$

***Check***: $7 \cdot 3 \overset{?}{=} 8\left(2\tfrac{5}{8}\right)$  ✔

♦ ♦ ♦ ♦ ♦ ♦ ♦ ♦ ♦ ♦ ♦ ♦ ♦ ♦ ♦ ♦ ♦ ♦ ♦ ♦ ♦ ♦ ♦ ♦ ♦ ♦ ♦ ♦ ♦ ♦ ♦ ♦ ♦ ♦ ♦ ♦ ♦ ♦

In the next two examples we use proportions to estimate wildlife populations.

E X A M P L E   **7**

**Fish population**    At random locations in a certain lake, fish management personnel catch and tag 50 fish. They return the fish unharmed to the lake. They come back to the same locations 2 weeks later and catch 100 fish. If 4 of the fish are tagged and 96 are not tagged, estimate the total population of fish. What assumptions are made?

SOLUTION

$$\frac{50 \text{ initial tagged}}{x \text{ total population}} = \frac{4 \text{ caught with tag}}{100 \text{ total caught}}$$

$$50 \cdot 100 = 4 \cdot x$$

$$x = 1250$$

***Check***: $50(100) \overset{?}{=} 4(1250)$  ✔

This solution uses a ratio of tagged fish to total population. Other solutions are possible, using tagged to untagged or untagged to total population.

We assume that no tagged fish died and that the tagged fish returned to the general fish population and mixed sufficiently so that they and the other fish were equally likely to be caught the second time.

♦ ♦ ♦ ♦ ♦ ♦ ♦ ♦ ♦ ♦ ♦ ♦ ♦ ♦ ♦ ♦ ♦ ♦ ♦ ♦ ♦ ♦ ♦ ♦ ♦ ♦ ♦ ♦ ♦ ♦ ♦ ♦ ♦ ♦ ♦ ♦ ♦ ♦

E X A M P L E   **8**

**Bat and mosquito population**    Researchers sample several locations on the ceiling of a cave and find an average of 18 adult bats per square foot. They estimate the dimensions of the ceiling to be 30 feet by 75 feet. Use a proportion to estimate the number of adult bats, $x$, in the cave. Then estimate the number of mosquitoes, $m$, eaten in 30 nights if each bat consumes 500 mosquitoes per night.

SOLUTION

$$\frac{18 \text{ bats}}{1 \text{ ft}^2} = \frac{x \text{ bats}}{(30 \text{ ft})(75 \text{ ft})}$$

$$x = 40{,}500 \text{ bats}$$

$$m = \frac{40{,}500 \text{ bats}}{1 \text{ night}} \cdot \frac{500 \text{ mosquitoes}}{1 \text{ bat}} \cdot 30 \text{ nights}$$

$$m = 607{,}500{,}000 \text{ mosquitoes}$$

♦ ♦ ♦ ♦ ♦ ♦ ♦ ♦ ♦ ♦ ♦ ♦ ♦ ♦ ♦ ♦ ♦ ♦ ♦ ♦ ♦ ♦ ♦ ♦ ♦ ♦ ♦ ♦ ♦ ♦ ♦ ♦ ♦ ♦ ♦ ♦ ♦ ♦

### DIVISION BY NUMBERS NEAR ZERO AND UNDEFINED EXPRESSIONS

We close this section by examining the effect on a graph of division by numbers near zero and its implication for expressions and equations.

E X A M P L E   9   *Walking time*   Suppose that we have 2 miles to walk. Let $x$ be the speed at which we walk. Because time traveled is the distance divided by rate, the time, $y$, needed to walk 2 miles is $y = 2/x$. What happens to the graph of $y = 2/x$ as $x$ gets close to zero?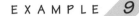

SOLUTION

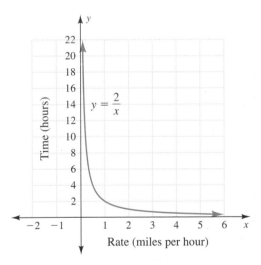

| Rate (mph) $x$ | Time (hours) $y = 2/x$ |
|---|---|
| 5 | $2/5 = 0.4$ |
| 4 | $2/4 = 0.5$ |
| 2 | $2/2 = 1.0$ |
| 1 | $2/1 = 2.0$ |
| 0.5 | $2/0.5 = 4.0$ |
| 0.25 | $2/0.25 = 8.0$ |
| 0.1 | $2/0.1 = 20.0$ |

*Table 3*  $y = 2/x$ *as x approaches zero*

Table 3 shows outputs for $y = 2/x$ if $x$ is close to zero. If we walk 5 miles per hour, it will take us 0.4 hour to walk 2 miles. (How many minutes is 0.4 hour?) If we walk 0.1 mile (about 1 city block) per hour, it will take us 20 hours to walk the 2 miles. We use dots to plot the input–output data from Table 3 and then connect the dots, because the rate, $x$, could be any fraction or decimal.

The graph of $y = 2/x$ forms a curve. The curve in Figure 3 is one portion of a *rectangular hyperbola*. A second portion, not shown and not connected to the first, is in the third quadrant. It is formed by negative inputs. Because rate cannot be negative, we disregard the second portion for this discussion.

If we trace the graph from right to left, the rate gets smaller and the time increases. The curve rises as it approaches the $y$-axis. Although it gets close, the curve never touches the $y$-axis. A point on the $y$-axis would have a rate equal to 0 miles per hour, and at that rate there is no motion. It is impossible to travel 2 miles if we do not move.

We describe the behavior of the curve as *y approaches positive infinity as x approaches zero*. Because the equation $y = 2/x$ has no output for $x = 0$, we say *the equation is undefined at x = 0*.

*Figure 3*

---

Outputs approach infinity whenever the denominator of a simplified fraction approaches zero.

---

We must always watch for inputs that create division by zero and undefined expressions.

EXAMPLE 10    What input will make each expression undefined? If a graphing calculator is available, graph each expression as an equation ($y = $ expression), and observe the behavior near the identified input.

a. $\dfrac{x}{x-2}$    b. $\dfrac{-2}{x+3}$    c. $\dfrac{x+3}{x+4}$    d. $\dfrac{1}{\sqrt{x}}$

SOLUTION    a. $x = 2$    b. $x = -3$    c. $x = -4$    d. $x = 0$

The expression in part d will also be undefined if $x$ is any negative number, because the square root of a negative number has no real number solution. When tracing a graph, the graphing calculator gives a blank for the $y$-coordinate of an undefined expression.

♦ ♦ ♦ ♦ ♦ ♦ ♦ ♦ ♦ ♦ ♦ ♦ ♦ ♦ ♦ ♦ ♦ ♦ ♦ ♦ ♦ ♦ ♦ ♦ ♦ ♦ ♦ ♦ ♦ ♦ ♦ ♦ ♦ ♦ ♦ ♦ ♦

Calculators are designed to reject an attempt to divide by zero. Try $2 \div 0$ on a calculator to see how the calculator indicates that this operation is in error. It is important to recognize this "error" message at all times.

Example 11 reminds us of two important concepts. First, when we are working with fractional equations, we need to remember to exclude inputs that cause a zero denominator. Second, if any numerator or denominator in a proportion contains two or more terms, it is essential to apply the distributive property within the cross multiplication.

EXAMPLE 11    Identify all numbers that cause zero denominators. Solve each proportion for $x$.

a. $\dfrac{4}{x-1} = \dfrac{5}{x+1}$    b. $\dfrac{9}{2x} = \dfrac{12}{2x+4}$

SOLUTION    a. $\dfrac{4}{x-1} = \dfrac{5}{x+1}$,    $x \neq 1$ and $x \neq -1$

$4(x+1) = 5(x-1)$

$4x + 4 = 5x - 5$

$4 + 5 = 5x - 4x$

$9 = x$

*Check*: $\dfrac{4}{9-1} \stackrel{?}{=} \dfrac{5}{9+1}$ ✔

b. $\dfrac{9}{2x} = \dfrac{12}{2x+4}$,    $x \neq 0$ and $x \neq -2$

$9(2x+4) = 12(2x)$

$18x + 36 = 24x$

$36 = 6x$

$6 = x$

*Check*: $\dfrac{9}{2 \cdot 6} \stackrel{?}{=} \dfrac{12}{2 \cdot 6 + 4}$ ✔

The parentheses were used in the second step of each solution to assure proper application of the distributive property.

♦ ♦ ♦ ♦ ♦ ♦ ♦ ♦ ♦ ♦ ♦ ♦ ♦ ♦ ♦ ♦ ♦ ♦ ♦ ♦ ♦ ♦ ♦ ♦ ♦ ♦ ♦ ♦ ♦ ♦ ♦ ♦ ♦ ♦ ♦ ♦ ♦

## EXERCISES 4.2

In Exercises 1 to 6 simplify each ladder ratio, and indicate whether each is a safe ratio. If the ratio represents an unsafe ladder, guess whether the ladder will slip or tip.

1. 15 feet to 5 feet
2. 18 feet to 5 feet
3. 18 feet to 4 feet
4. 20 feet to 4 feet
5. 20 feet to 5 feet
6. 18 feet to 4.5 feet

7. Refer to Example 4. Write the fourth proportion that may be built with the numbers 2, 4, 5, and 10. Use cross multiplication to check. Write another false statement.

8. Make four proportions with the numbers 2, 3, 4, and 6. Use cross multiplication to eliminate any false statements.

For Exercises 9 to 14 show the cross multiplication used to determine which are proportions and which are false statements. You may prefer to write each in fraction form first.

9. $6:8 = 15:20$
10. $8:10 = 12:15$
11. $4:6 = 6:9$
12. $9:12 = 15:18$
13. $9:21 = 21:35$
14. $9:6 = 15:10$

Solve the proportions in Exercises 15 to 22 by cross multiplication.

15. $\dfrac{3}{4} = \dfrac{x}{15}$
16. $\dfrac{5}{3} = \dfrac{27}{x}$
17. $\dfrac{x}{5} = \dfrac{2}{3}$

18. $\dfrac{5}{x} = \dfrac{3}{8}$
19. $\dfrac{7}{3} = \dfrac{5}{x}$
20. $\dfrac{2}{5} = \dfrac{x}{12}$

21. $\dfrac{4}{x} = \dfrac{3}{7}$
22. $\dfrac{x}{6} = \dfrac{7}{5}$

Solve the unit problems in Exercises 23 to 28 using the conversion information in the table. Use either a proportion or unit analysis. Round to the nearest hundredth.

| |
|---|
| 1 acre = 43,560 square feet |
| 1 liter = 1.0567 quarts |
| 1 kilogram = 2.2 pounds |
| 1 kilometer = 1000 meters |
| 1 square mile = 640 acres |
| 1 meter = 39.37 inches |
| 1 kilometer = 0.621 miles |

23. Change 65 inches to meters.
24. Change 160 pounds to kilograms.
25. Change 15 miles to kilometers.
26. Change 16 quarts to liters.
27. Change 640 acres to square feet.
28. Change 10,000 acres to square miles.

For what input will each of the fractional expressions in Exercises 29 to 34 be undefined? If a graphing calculator is available, graph each as an equation ($y =$ expression), and observe the behavior near the identified input.

29. $\dfrac{5}{x - 3}$
30. $\dfrac{3}{x + 4}$
31. $\dfrac{x + 2}{x + 5}$

32. $\dfrac{x - 3}{x}$
33. $\dfrac{2x + 3}{\sqrt{x}}$
34. $\dfrac{3x - 2}{\sqrt{x}}$

In Exercises 35 to 42 what inputs must be excluded from the solution set for each equation because they create an undefined expression? Solve for $x$. Use the distributive property as appropriate.

35. $\dfrac{x - 1}{2} = \dfrac{x + 3}{3}$
36. $\dfrac{x - 4}{3} = \dfrac{x - 1}{4}$

37. $\dfrac{x + 4}{15} = \dfrac{x + 2}{10}$
38. $\dfrac{x - 6}{10} = \dfrac{x - 4}{6}$

39. $\dfrac{9}{x + 7} = \dfrac{6}{x + 3}$
40. $\dfrac{8}{x - 1} = \dfrac{12}{x + 2}$

41. $\dfrac{2}{3x} = \dfrac{5}{6x + 3}$
42. $\dfrac{3}{4x} = \dfrac{2}{3x - 1}$

43. How far from a wall must the base of an extension (flexible-length) ladder be set in order to safely reach 21 feet up a wall? (*Hint:* What is the safe ladder ratio?)

44. Tuesday afternoon Sean traveled 252 miles on 9 gallons of gas. If the same highway and driving conditions continue, how far can he expect to travel on a full 13-gallon tank?

45. An access ramp needs to rise 54 inches. If the ramp is to have a slope ratio of 1:8, what horizontal distance is needed to build the ramp? Give the answer in feet.

46. An access ramp needs to rise 18 inches. If the ramp is to have a slope ratio of 1:8, what horizontal distance is needed to build the ramp? Give the answer in feet.

47. A staircase has a slope of 7 to 11. What horizontal distance is needed for an 8-foot vertical distance?

48. A staircase has a slope of 6.5 to 11.5. What horizontal distance is needed for an 8.5-foot vertical distance?

49. If a 10-foot storm surge (high wave of water) from Hurricane Andrew hits the Louisiana coast and comes inland 13 miles, what is the average slope of the coastal region at that location? (*Hint:* The units must be the same for a sensible answer.)

50. If a 10-foot storm surge hits the Oregon coast and comes inland $\frac{1}{8}$ mile, what is the average slope of the coastline at that location? (*Hint:* The units must be the same.)

51. A 40-pound child should receive how many units of penicillin if the dosage is 500,000 units for a 150-pound adult?

52. How many milligrams (mg) of atrophine sulfate should be given to a 6-month infant if the dosage is 0.4 mg for a patient of 150 months?

53. **Bird Population.** A wildlife management team traps pheasants in nets and tags them at randomly located areas in a fire-damaged setting. They tag 75 birds altogether. Two weeks later they trap again, and they capture 10 tagged birds and 55 untagged birds. Use a proportion to estimate the population. Assume that the birds didn't learn to avoid the nets after being caught the first time.

54. **Fish Population.** Suppose 15,000 hatchery fish are released in a river. At maturity these fish return to the river along with the native fish. Assume that the hatchery fish have a 5% survival and return rate. Of 85 mature fish caught by people fishing along the river, 82 have the clipped fin of the hatchery fish. Use a proportion to estimate the number of native fish in the river.

Cross multiplication may be used to solve more complicated equations if we first change the equation into a proportion. Solve the equations in Exercises 55 and 56 by first adding the fractions on the left side using a common denominator and then cross multiplying to solve the resulting proportion.

55. a. $\dfrac{1}{5} + \dfrac{1}{3} = \dfrac{1}{t}$    b. $\dfrac{1}{4} + \dfrac{1}{6} = \dfrac{2}{t}$

56. a. $\dfrac{1}{3} + \dfrac{1}{6} = \dfrac{1}{t}$    b. $\dfrac{1}{3} + \dfrac{1}{7} = \dfrac{2}{t}$

Chemistry and physics have several laws relating pressure, volume, and temperature of gases, as well as concentration and volume. All contain subscripted variables. We solved several of these formulas in Section 3.5 for certain variables. In Exercises 57 to 60 use cross multiplication to transform the proportions to a formula without denominators.

57. $\dfrac{V_1}{T_1} = \dfrac{V_2}{T_2}$    58. $\dfrac{V_1}{C_2} = \dfrac{V_2}{C_1}$

59. $\dfrac{P_1}{V_2} = \dfrac{P_2}{V_1}$    60. $\dfrac{P_1 V_1}{T_1} = \dfrac{P_2 V_2}{T_2}$

In Exercises 61 to 64 "undo" the cross multiplication, and change these chemistry formulas back to a proportion form.

61. $C_1 V_1 = C_2 V_2$; make a proportion with the Vs on top.

62. $P_1 V_1 = P_2 V_2$; make a proportion with the Ps on top.

63. $V_1 T_2 = V_2 T_1$; make a proportion with the Ts on top.

64. $P_1 V_1 = P_2 V_2$; make a proportion with the Vs on top.

### PROBLEM SOLVING

65. The ratio of width to length of a rectangle is 1:2. The length is 6 units less than three times the width. Set up a proportion for width to length, and use it in finding both dimensions. What is the area of the rectangle?

66. The base and height of a triangle are in the ratio 3:4. The height is 3 units more than the base. Set up a proportion, and use it in finding both dimensions. What is the area of the triangle?

### PROJECTS

67. **Line Segment Ratios.** The figure shows three line segments that have been formed into a triangle. The sides are labeled with lengths $a$, $b$, and $c$.

    a. What are the six possible ratios of the three line segments?

    b. For the triangle shown, which ratios are greater than 1?

    c. Which ratios are smaller than 1?

    d. Which ratio is the largest of the six?

    e. Which ratio is the smallest of the six?

    f. If segments $a$, $b$, and $c$ were redrawn with $a + b = c$, what would the resulting figure look like?

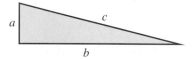

68. **Population Estimates.** Place a large number (at least 100, but uncounted) of like objects (dry beans, marbles, coins) in a paper bag.

    a. Remove a handful, count, and "tag" them with some sort of label. Return the tagged objects to the bag and mix thoroughly.

    b. Remove another handful, and count the tagged objects.

    c. Set up a proportion to estimate the total number of objects in the bag. State your assumptions.

    d. How could you improve your estimate? What other ways are there to count the objects? Count the objects or use another method to estimate the total objects.

69. **Clock Angles.** Examine the clock faces below. In answering the questions, explain your reasoning carefully and completely.

    a. What is the measure of the angle formed by the hands of a clock at 7:30?

    b. What is the measure of the angle at 4:35?

## SIMILAR FIGURES AND SIMILAR TRIANGLES

**OBJECTIVES**  Identify similar figures. ◆ Identify corresponding parts of similar figures. ◆ Write proportions to determine unknown lengths in similar triangles. ◆ Write expressions for the length of one segment in terms of the length of another. ◆ Explore common ratios of sides of similar right triangles and their relationship to trigonometry. ◆ Determine ratios of sides of similar right triangles using a scientific calculator.

**WARM-UP**  Solve for $x$. Identify any inputs making undefined expressions.

**1.** $\dfrac{20}{12} = \dfrac{x}{7.5}$  **2.** $\dfrac{5}{3} = \dfrac{x}{100}$  **3.** $\dfrac{5.5}{3} = \dfrac{x}{34}$

**4.** $\dfrac{4}{10 - x} = \dfrac{9}{10}$  **5.** $\dfrac{4}{3} = \dfrac{x + 4}{5}$                            ◆

### SIMILAR FIGURES

A magazine layout designer wishes to enlarge a photograph to fit the bottom of a page. The photograph's size is 12 cm (base) by 7.5 cm (height). The enlargement is to have a base of 20 cm. To finish the rest of the page layout, the designer needs to know the height of the photograph after the enlargement.

Initially, the designer might estimate 15 centimeters, which is double 7.5 centimeters, because 12 to 20 is almost doubled. To be more exact, the designer uses the concept of similarity. In Figure 4 a diagonal is drawn from the lower left to the upper right corner of a rectangle with the same shape as the photograph. The diagonal is then extended until it crosses a vertical line at width 20 centimeters. The height of the vertical line at width 20 is then the height of the enlarged photograph. The approximate height is 13 centimeters.

An enlargement (or reduction) of a photograph creates a rectangle of the same shape as the original but of a different size. The right angles forming the corners of the rectangle remain square corners during the enlargement. This gives us the basic idea of similar figures.

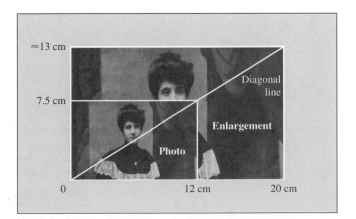

*Figure 4*

> **Similar figures** have corresponding sides that are proportional and corresponding angles that are equal.

The proportionality of similar figures applies to diagonals, interior heights, diameters, radii (plural of radius), or any other straight line associated with the figures. **Corresponding** means *in the same relative position.* Most of the similar figure work in this course will be with similar triangles.

SIMILAR TRIANGLES

> ◆ Similar triangles have corresponding angles that are equal.
> ◆ Similar triangles have corresponding sides that are proportional.

EXAMPLE 1

**a.** Name the corresponding parts of the triangles. Give the ratios of the corresponding sides. Indicate why the triangles are similar.

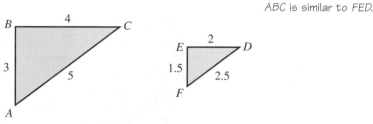

*ABC is similar to FED.*

**b.** Name the corresponding parts of the triangles, and indicate why the triangles are similar.

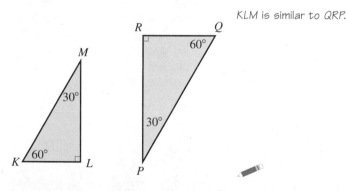

*KLM is similar to QRP.*

SOLUTION

**a.** The longest sides in both triangles are corresponding, so *AC* corresponds with *FD*. The middle-length sides are the next correspondence: *BC* and *ED*. Finally, the shortest sides give the third correspondence: *AB* and *FE*.

The ratio *AC* to *FD* is 5 to 2.5; the ratio simplifies to 2 to 1. The ratio *BC* to *ED* is 4 to 2; the ratio simplifies to 2 to 1. The ratio *AB* to *FE* is 3 to 1.5; the ratio simplifies to 2 to 1.

The triangles are similar because the ratios of corresponding sides are the same: 2 to 1.

**b.** Angle *L* corresponds with angle *R* because they are both right angles. Angle *M* corresponds with angle *P* because they are both 30°. Angle *K* corresponds with angle *Q* because they are both 60°.

The two triangles are similar because their corresponding angles are equal.

◆ ◆ ◆ ◆ ◆ ◆ ◆ ◆ ◆ ◆ ◆ ◆ ◆ ◆ ◆ ◆ ◆ ◆ ◆ ◆ ◆ ◆ ◆ ◆ ◆ ◆ ◆ ◆ ◆ ◆ ◆ ◆

EXAMPLE 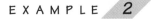 2    Explain why the pairs of triangles are not similar.

**a.**

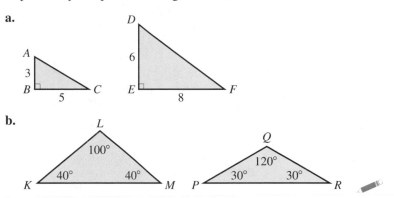

**b.**

SOLUTION    **a.** The ratio of the shortest sides is $AB:DE = 3:6$, which simplifies to $1:2$. The ratio of the middle sides is $BC:EF = 5:8$. The ratios of corresponding sides are different, so the triangles are not similar.

**b.** Angle $L$ corresponds to angle $Q$ because they are the largest angles in the two triangles; but they are not equal, so the triangles are not similar.

♦ ♦ ♦ ♦ ♦ ♦ ♦ ♦ ♦ ♦ ♦ ♦ ♦ ♦ ♦ ♦ ♦ ♦ ♦ ♦ ♦ ♦ ♦ ♦ ♦ ♦ ♦ ♦ ♦ ♦ ♦ ♦ ♦

In our opening example the figures are similar because the angles remained the same as we extended the lines. Thus we apply proportions to find the lengths. Our proportion is

$$\frac{\text{base of enlargement}}{\text{base of original}} = \frac{\text{height of enlargement}}{\text{height of original}}$$

$$\frac{20 \text{ cm}}{12 \text{ cm}} = \frac{x}{7.5 \text{ cm}}$$

$$(20 \text{ cm}) \cdot (7.5 \text{ cm}) = x(12 \text{ cm})$$

$$x = 12.5 \text{ cm}$$

The ratio of the enlargement to the original is 20 to 12, or 5 to 3. The ratio of the new figure to the original figure is called the *scale ratio*. The settings on a photocopy machine—*reduce to 67%* or *enlarge to 150%*—are scale ratios written as percents.

EXAMPLE 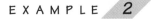 3    *Photo enlargements*    What percent is a photo enlargement with a scale ratio 5 to 3?

SOLUTION    *Percent* means per hundred, so we write a proportion with the scale ratio:

$$\frac{5}{3} = \frac{x}{100}$$

$$500 = 3x$$

$$\frac{500}{3} = x$$

$$x \approx 167$$

The enlargement is 167% of the original.

♦ ♦ ♦ ♦ ♦ ♦ ♦ ♦ ♦ ♦ ♦ ♦ ♦ ♦ ♦ ♦ ♦ ♦ ♦ ♦ ♦ ♦ ♦ ♦ ♦ ♦ ♦ ♦ ♦ ♦ ♦ ♦ ♦

As implied in our original definition of similar figures, similarity is not limited to triangles; however, the corresponding parts may be less obvious on other figures.

EXAMPLE 4    Use a ruler marked in millimeters to measure corresponding parts of the shapes. Write ratios of corresponding lengths, and determine similarity.

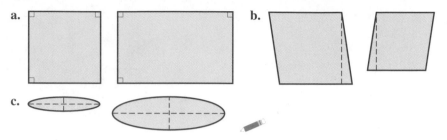

SOLUTION   **a.** These two figures are rectangles. The ratio of the bases is not the same as the ratio of the heights, so the rectangles are not similar.

**b.** These two figures are diamonds, or rhombuses, with four equal sides. We compare corresponding heights and corresponding sides. The ratios are reasonably close, so we assume the figures are similar.

**c.** These two figures are ellipses. The ratio of vertical distances is not the same as the ratio of horizontal distances. Therefore the ellipses are not similar.

♦ ♦ ♦ ♦ ♦ ♦ ♦ ♦ ♦ ♦ ♦ ♦ ♦ ♦ ♦ ♦ ♦ ♦ ♦ ♦ ♦ ♦ ♦ ♦ ♦ ♦ ♦ ♦ ♦ ♦ ♦ ♦ ♦ ♦ ♦ ♦ ♦ ♦

For the rest of this section we work with similar triangles. We first examine indirect measurement and then close the section with similar triangles and trigonometry.

### INDIRECT MEASUREMENT

The proportionality of similar triangles gives us a powerful tool for finding unknown lengths. The following examples involve *indirect measurement*— finding the measure of objects we are able to see but not actually measure.

In these examples note how the objects are viewed in the drawings. In some examples we draw a *side view*. In others we pretend we are in the sky looking down, and we draw a *top view*. Looking at a scene from one position often reveals a mathematical relationship that does not appear from any other. The proportional relationship between heights and shadows in Example 5 was known to the ancient Greeks.

EXAMPLE 5    Trees and shadows    At 10:00 one morning a tree casts a 34-foot shadow along the ground, while a $5\frac{1}{2}$-foot-tall person casts a shadow 3 feet long. How tall is the tree? (Because of space limitations, the triangles shown in Figure 5 are similar but are not to scale with each other.)

*Geometry fact*: The sun's rays form equal angles at the top of each triangle. It is assumed that the tree and the person are at right angles to the ground. This information is sufficient to create similar triangles.

SOLUTION   Form a proportion with the heights and shadows:

$$\frac{\text{person height}}{\text{person shadow}} = \frac{\text{tree height}}{\text{tree shadow}}$$

$$\frac{5.5 \text{ ft}}{3 \text{ ft}} = \frac{x}{34 \text{ ft}}$$

$$(34 \text{ ft})(5.5 \text{ ft}) = x(3 \text{ ft})$$

$$x \approx 62.3 \text{ ft}$$

The tree is approximately 62.3 feet tall.

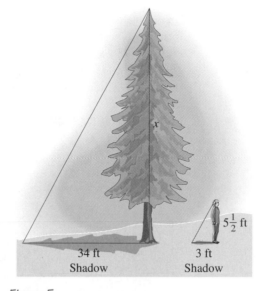

*Figure 5*

♦ ♦ ♦ ♦ ♦ ♦ ♦ ♦ ♦ ♦ ♦ ♦ ♦ ♦ ♦ ♦ ♦ ♦ ♦ ♦ ♦ ♦ ♦ ♦ ♦ ♦ ♦ ♦ ♦ ♦ ♦ ♦ ♦ ♦ ♦ ♦ ♦

EXAMPLE  6

**River width**   Use the similar triangles shown in Figure 6 to find the distance across the river. This figure is drawn as a top view. There is a tree, *E*, on the river bank that is used as a sighting point. Length *AB* is 24 inches, *BC* is 6 inches, and *CD* is 30 feet.

***Geometry fact***: The two angles at point *C*, inside the triangles, are equal. We say, "Angle *BCA* equals angle *DCE*." The angles at *B* and *D* are right angles, so angle *E* and angle *A* are equal. Thus the triangles have equal corresponding angles and are similar.

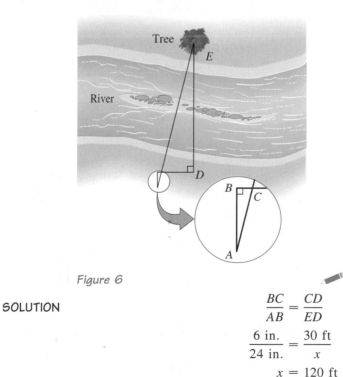

*Figure 6*

SOLUTION

$$\frac{BC}{AB} = \frac{CD}{ED}$$

$$\frac{6 \text{ in.}}{24 \text{ in.}} = \frac{30 \text{ ft}}{x}$$

$$x = 120 \text{ ft}$$

The distance across the river is about 120 feet.

♦ ♦ ♦ ♦ ♦ ♦ ♦ ♦ ♦ ♦ ♦ ♦ ♦ ♦ ♦ ♦ ♦ ♦ ♦ ♦ ♦ ♦ ♦ ♦ ♦ ♦ ♦ ♦ ♦ ♦ ♦ ♦ ♦ ♦ ♦ ♦

**E X A M P L E   7**   Street light and shadows   A 6-foot-tall person has an 8-foot shadow formed by a streetlight. The person is standing 12 feet from the streetlight. Figure 6 shows a side view of the situation. How high is the streetlight?

***Geometry fact***: The triangles in this set overlap. The angle at the tip of the shadow is in both the large and the small triangle. This angle and the right angles formed by the streetlight and by the person give sufficient information to identify similar triangles.

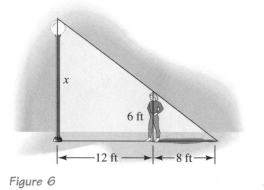

*Figure 6*

**SOLUTION**   The length of the base of the larger triangle is the sum of the 12 feet between the lamppost and the person and the 8-foot shadow.

$$\frac{\text{person height}}{\text{person shadow}} = \frac{\text{height of light}}{\text{base of triangle}}$$

$$\frac{6 \text{ ft}}{8 \text{ ft}} = \frac{x}{(12 + 8) \text{ ft}}$$

$$(6 \text{ ft})(20 \text{ ft}) = x(8 \text{ ft})$$

$$x = 15 \text{ ft}$$

The streetlight is 15 feet high.

♦ ♦ ♦ ♦ ♦ ♦ ♦ ♦ ♦ ♦ ♦ ♦ ♦ ♦ ♦ ♦ ♦ ♦ ♦ ♦ ♦ ♦ ♦ ♦ ♦ ♦ ♦ ♦ ♦ ♦ ♦ ♦ ♦

When we are working with similar triangles, it is sometimes necessary to write the length of one line segment in terms of the lengths of the other line segments. This is a geometric application of the phrase *in terms of.*

**E X A M P L E   8**   Write an expression describing *d* in each drawing below; that is, write *d in terms of* the lengths of the other line segments.

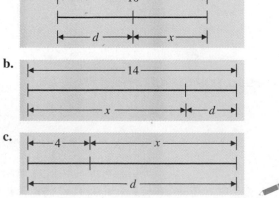

SOLUTION    **a.** The whole is 10, and one part is $x$; $d = 10 - x$.

**b.** The whole is 14, and one part is $x$; $d = 14 - x$.

**c.** One part is $x$, and the other is 4; the whole is $d = 4 + x$.

♦ ♦ ♦ ♦ ♦ ♦ ♦ ♦ ♦ ♦ ♦ ♦ ♦ ♦ ♦ ♦ ♦ ♦ ♦ ♦ ♦ ♦ ♦ ♦ ♦ ♦ ♦ ♦ ♦ ♦ ♦ ♦ ♦ ♦

E X A M P L E    *9*    Find an expression for $d$ in each drawing below. Use similar triangles and pro-portions to determine $x$. Round $x$ to the nearest hundredth.

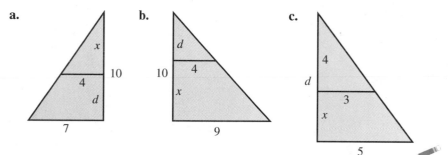

SOLUTION    **a.** Line segment $d$ is $10 - x$. Solve for $x$:

$$\frac{4}{x} = \frac{7}{10}$$

$$4 \cdot 10 = 7x$$

$$x \approx 5.71$$

**b.** Line segment $d$ is $10 - x$. Solve for $x$:

$$\frac{4}{10 - x} = \frac{9}{10}$$

$$4 \cdot 10 = 9(10 - x)$$

$$40 = 90 - 9x$$

$$9x = 50$$

$$x \approx 5.56$$

**c.** Line segment $d$ is $x + 4$. Solve for $x$:

$$\frac{3}{4} = \frac{5}{4 + x}$$

$$3(4 + x) = 5 \cdot 4$$

$$12 + 3x = 20$$

$$3x = 8$$

$$x \approx 2.67$$

There are several ways to write the proportions. These solutions have the ratio of base to height in each triangle; height to base would be equally correct.

♦ ♦ ♦ ♦ ♦ ♦ ♦ ♦ ♦ ♦ ♦ ♦ ♦ ♦ ♦ ♦ ♦ ♦ ♦ ♦ ♦ ♦ ♦ ♦ ♦ ♦ ♦ ♦ ♦ ♦ ♦ ♦ ♦ ♦

### RATIOS, PROPORTIONS, AND TRIGONOMETRY

The proportionality of similar triangles and the need to solve problems with a minimum of information led mathematicians to the discovery that knowing one other angle of a right triangle is sufficient to find the ratio of the sides of the triangle. In part a of Example 10, angle $A$ is 76°. The ratio of side $BC$ to side $AC$ for a 76° right triangle is 4 : 1. Confirm this by measuring the figure in millimeters.

Angles of other sizes in right triangles create different ratios of side *BC* to side *AC*. The ratios have been known for a long time and are available from the *tangent* ( TAN ) key on any scientific calculator. Two other keys, sine ( SIN ) and cosine ( COS ), give ratios of other pairs of sides in a right triangle. These ratios are described in the exercises and are part of a course called *trigonometry*, literally translated as "*triangle measurement*."

We focus on tangent not because it is more important than sine or cosine but because it gives the slope of a line, as discussed in Section 3.6, and hence is more closely related to our current topic.

EXAMPLE 10

Use TAN *A* on a scientific calculator to obtain the tangent ratios for these angles. Round the answers to the nearest tenth. Measure the height and base of each triangle and calculate the slope to see that the calculator tangent ratio is reasonable.

**a.** $A = 76°$        **b.** $A = 45°$        **c.** $A = 35°$

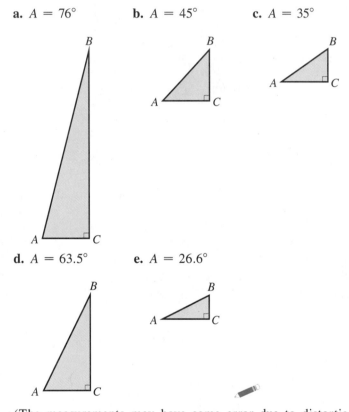

**d.** $A = 63.5°$        **e.** $A = 26.6°$

SOLUTION

(The measurements may have some error due to distortion in transferring from computer graphics to a printed textbook.)

**a.** tan 76° ≈ 4.0 to 1        **b.** tan 45° = 1 to 1        **c.** tan 35° ≈ .70 to 1
**d.** tan 63.5° ≈ 2.0 to 1      **e.** tan 26.6° ≈ 0.50 to 1

Tangents, like square roots, are generally irrational numbers. The angles in Example 10 were chosen because they give simple tangent ratios.

♦ ♦ ♦ ♦ ♦ ♦ ♦ ♦ ♦ ♦ ♦ ♦ ♦ ♦ ♦ ♦ ♦ ♦ ♦ ♦ ♦ ♦ ♦ ♦ ♦ ♦ ♦ ♦ ♦ ♦ ♦ ♦ ♦ ♦ ♦ ♦ ♦

As mentioned in Section 3.6, on most scientific calculators the angle in degrees is entered first and then TAN . On graphing calculators the order is usually TAN followed by the angle. No keystroke is needed for the degrees; however, there are other forms of angle measure. If your answers do not agree with those given, it may be necessary to change your calculator into degree mode. Check your calculator operating guide for instructions on how to change into degree mode. This change is usually done with a DRG key or a MODE key.

## EXERCISES 4.3

1. If the designer in the introduction to this section wanted to reduce the photograph to an 8-centimeter base, what would the new height be? Estimate as well as calculate, using Figure 4 and a proportion.

2. In the introductory example, what is the slope of the diagonal line
   a. in the original photograph?   b. in the illustration?
   Write the slope as a ratio of two integers.

**For Exercises 3 and 4 trace the rectangles, and use the designer's diagonal method to enlarge them to the specified base.**

3. Enlarge to base 2.5 inches.

4. Enlarge to base 3 inches.

**In Exercises 5 to 8 name the corresponding parts of the triangles. Measure the triangles, and determine whether they are similar.**

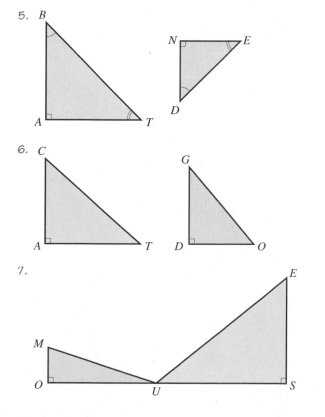

8.

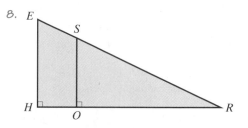

**Measure and compare ratios of corresponding parts to determine whether the figures in Exercises 9 to 12 are similar. Use millimeters.**

9.

10.

11.

12.

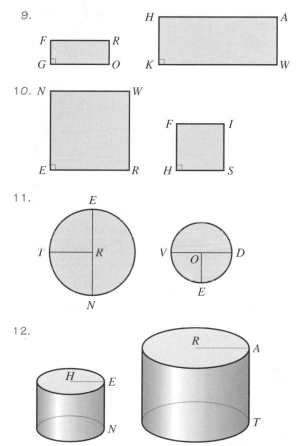

13. Name three geometric figures that are always similar.

14. Why are certain shapes always similar and other shapes only sometimes similar?

15. A rectangle is 10 centimeters by 12 centimeters. What is its new size after a photocopy machine
    a. reduces it to 67%?   b. enlarges it to 150%?

16. A rectangle is 8 centimeters by 14 centimeters. What is its new size after a photocopy machine
    a. reduces it to 67%?   b. enlarges it to 150%?

17. A model car has scale 1 to 25 in comparison with the full-size car. If a sun visor is 5 inches by 14 inches in the full-size car, what are its dimensions in the scale model?

18. A model plane has scale 1 to 20 with the full-size plane. If a window is 10 inches by 12 inches in the full-size plane, what are the window dimensions in the scale model?

19. At 3:00 in the afternoon a 30-foot tree casts a 35-foot shadow. A person 4 feet tall will cast how long a shadow?

20. At 5:00 in the afternoon a 30-foot tree casts a 125-foot shadow. How long is the shadow of a person 5.5 feet tall?

21. The person who estimates the amount of wood in a tree is called a timber cruiser. A timber cruiser holds her arm parallel to the ground. In her hand she holds a stick, vertically, 27 inches from her eye (see the figure). A 14-inch length on the stick lines up with the top and bottom of a tree. The distance from the cruiser to the tree is 78 feet. How tall is the tree?

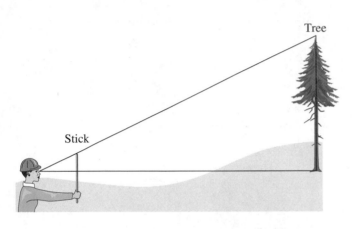

22. The timber cruiser in Exercise 21 lines up another tree in the same way. The second tree matches up with 30 inches on the stick when she is 60 feet from the base of the tree. How tall is the second tree?

**Determine the coordinates of *A* and *B* (and *C* where appropriate) in Exercises 23 to 26. Use slope, proportions, the slope triangles, or counting, as necessary.**

23.

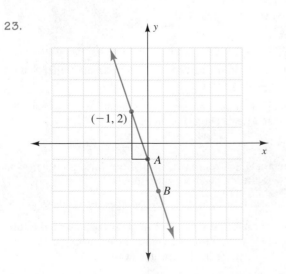

24.

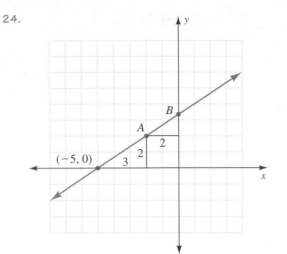

25.

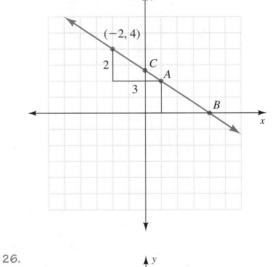

26.

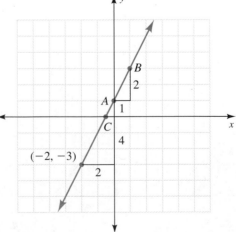

**Identify the coordinates labeled *A* and *B* in Exercises 27 to 30. Use properties of similar triangles as needed.**

27.

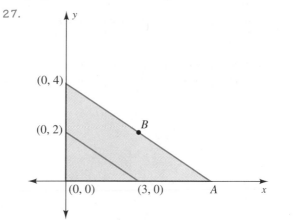

28.    (0, 0)

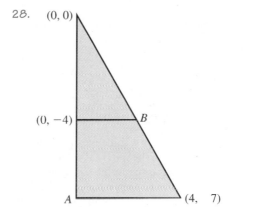

29.    *B*

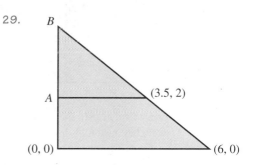

30.

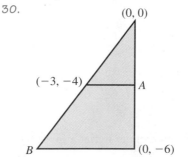

**In Exercises 31 to 34 draw line segments to illustrate the situations, and label them.**

31. The whole segment is 8. One part is *x*. A second part is $8 - x$.

32. The whole segment is $8 + x$. One part is *x*. The other part is 8.

33. The whole segment is 12. One part is *x*.

34. The whole segment is 5. One part is $5 - x$.

**Use proportions to find *x* in Exercises 35 to 38.**

35.                        36.

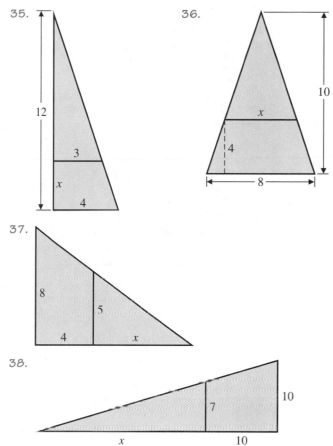

37.

38.

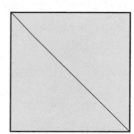

**PROBLEM SOLVING**

For Exercises 39 to 42 refer to the squares shown in the figure, cut into two *isosceles right triangles* by diagonals. An *isosceles triangle is a triangle with two equal sides.* (*Iso* means "same"; *sceles* refers to "leg," as in *scelalgia*, which is a medical term for pain in the leg.)

39. Using a millimeter ruler, find the ratio of the length of the diagonal to the length of one side for each square. Write the ratio in decimal form. Is this ratio true for all squares?

40. What is the angle measure of each of the three angles in the triangles? Do the three angles of each triangle add up to 180°?

41. The infield portion of a baseball playing field is called a diamond. It is actually a square, as in the figure, with 90 feet on each side. The ratio of the diagonal to the side of a square is 1.4142 when rounded to four decimal places. Find the distance diagonally across a baseball diamond.

42. Use your measurements to find the area of each square in the figure. What is the ratio of the areas, in decimal form? Is the ratio of the areas the same as the ratio of the sides of the squares?

**For Exercises 43 to 46 refer to the *equilateral triangles* in the figure. An *equilateral triangle has three equal sides and three equal angles*. (Use a millimeter ruler.)**

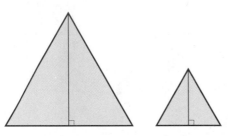

43. What is the ratio of the height to the side of each triangle? Write the ratio in decimal form. (*Hint:* Measure to the nearest millimeter.)

44. What is the measure of each angle of an equilateral triangle? (*Hint:* What is the sum of the measures of the angles of a triangle?)

45. Measure the bases and heights, and calculate the area of each triangle in the figure. Is the ratio of areas the same as the ratio of sides? Explain.

46. If we double the side of an equilateral triangle, what happens to the area?

47. Is tangent a linear relationship? (*Hint:* Use the data from Example 10 to make a table with angle *A* as input and tan *A* as output. If tangent is linear, then the slope between any two data points should simplify to the same number.)

PROJECT

48. ***Sine and Cosine Ratios from Measurement***

 a. Trace the triangles in Example 10. Measure and label all three sides.

 b. Copy the table below, and extend it to hold data for all five triangles.

| Triangle | Angle *A* | Sine Ratio | Cosine Ratio |
|---|---|---|---|
| a | | *BC/AB* = | *AC/AB* = |
| b | | *BC/AB* = | *AC/AB* = |
| c | | *BC/AB* = | *AC/AB* = |
| | | | |
| | | | |

 c. The ladder safety rule is $\tan A = \frac{4}{1}$. Use $\boxed{\text{INV}}$ $\boxed{\text{TAN}}$ 4 to find angle *A*. Round to the nearest degree. What is $\boxed{\text{COS}}$ *A*? What is $\boxed{\text{SIN}}$ *A*?

 d. Some forms of the ladder safety guidelines indicate that the ratio of the distance from the wall to the length of the ladder should be 1 to 4. This is the cosine ratio, $\cos B = \frac{1}{4}$, instead of the tangent ratio. Use $\boxed{\text{INV}}$ $\boxed{\text{COS}}$ .25 to find angle *B*. What is $\boxed{\text{TAN}}$ *B*? Are the two rules close enough to be interchangeable?

---

## MID-CHAPTER 4 TEST

**For Exercises 1 and 2 write the ratio in simplified form. Identify the inputs for which the original expression will be undefined.**

1. $12xy$ to $15y^2z$

2. $\dfrac{(x-2)(x+2)}{x(x+2)}$

**Change the ratios in Exercises 3 to 5 to like units, and simplify.**

3. 18 inches : 6 feet

4. 3 meters to 75 centimeters

5. *x* hours to *y* minutes

**In Exercises 6 and 7 describe in complete sentences an appropriate situation for each rate.**

6. gallons per mile

7. pounds per week

8. Solve for *x*: $\dfrac{5}{x} = 500{,}000$.

9. Calories in a diet are distributed in a ratio of 2 : 3 : 5 for fat, carbohydrate, and protein. If 1500 calories are to be consumed, how many may be allocated to each source?

10. The angle measures in a triangle are in the ratio 1 : 1 : 2. What is the measure of each angle?

11. An access ramp needs to rise 48 inches. If the ramp is to have a slope ratio of 1 : 12, what horizontal distance is needed to build the ramp?

**Solve the equations in Exercises 12 to 15.**

12. $\dfrac{3}{5} = \dfrac{16}{x}$

13. $\dfrac{x}{12} = \dfrac{10}{8}$

14. $\dfrac{x+5}{8} = \dfrac{2x-1}{12}$

15. $\dfrac{3x}{5x-2} = \dfrac{3}{4}$

16. Solve for *b*: $\dfrac{a}{b} = \dfrac{c}{d}$.

17. Use similar triangles and proportions to find *x* in the figure.

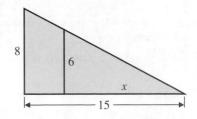

18. Use similar triangles and proportions as needed to find coordinates *A* and *B* in the figure.

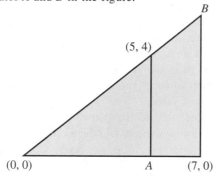

19. The model for an experimental airplane is $\frac{3}{4}$ full size. The main body of the model is 18 feet. How large is the main body of the full-size plane?

20. Ten thousand hatchery trout, each with a clipped fin, are released in a lake. Two weeks later fishing season starts. Within two days 260 trout are caught, and 250 have a clipped fin. Estimate the number of native trout (not tagged) originally in the lake.

## 4.4    QUANTITY AND VALUE TABLES

**OBJECTIVES**    Distinguish between quantity and value in problem situations. ♦ Summarize quantity and value facts in tables. ♦ Determine when the sum of the quantities is meaningful. ♦ Determine when the product of quantity and value is meaningful. ♦ Locate the appropriate position for an average rate (value) in a quantity-value table. ♦ Use quantity and rate (value) tables with variables.

**WARM-UP**    Solve for $x$.

1. $24(1.875) + 3x = (24 + x)(2.50)$      2. $9(35) + 140x = (9 + x)(103)$

3. $250 - 0.40(2500 - x)$                                            ♦

*SMART*
*Shel Silverstein*

My dad gave me one dollar bill
'Cause I'm his smartest son,
And I swapped it for two shiny quarters
'Cause two is more than one!

And then I took the quarters
And traded them to Lou
For three dimes—I guess he don't know
That three is more than two!

Just then, along came old blind Bates
And just 'cause he can't see
He gave me four nickels for my three dimes,
And four is more than three!

And I took the nickels to Hiram Coombs
Down at the seed-feed store,
And the fool gave me five pennies for them,
And five is more than four!

And then I went and showed my dad,
And he got red in the cheeks
And closed his eyes and shook his head—
Too proud of me to speak!

## QUANTITY AND VALUE

*SMART* illustrates the fact that quantity and value are two different concepts. A young child learns quantity—how many—with counting. Value—what something is worth—is learned much later.

In this section and the next we examine quantity and value in a variety of applications. In each application it will be important to clearly distinguish quantities from values. *Quantity* answers the question "How many?" or "How much?" There are many types of value.

*Value*, as the poem illustrates, may be *monetary worth*. Value is most commonly given as a *rate*. As we saw in Section 4.1, the word *per* identifies a rate: cents per dime, cents per nickel, percent, miles per hour, cost per item, pounds per square inch, and so forth.

Values may also be *numbers assigned to specific outcomes or activities*. Finishing positions (first, second, third) in a sporting event are given points to determine a winning team. In basketball, the points per basket depends on the position from which the ball is thrown. Letter grades are given points for the purpose of calculating a grade point average.

## QUANTITY-VALUE TABLES

In the poem *SMART*, confusing value and quantity created a significant decrease in what the child's present from his dad was worth. Calculation of total worth is a central idea of this section. Table 4 illustrates the pattern of *multiplying the quantity of each item times the value of each item (or the rate) to obtain the item total value*.

| Quantity | Value | Item Total |
|----------|-------|------------|
| 1 dollar | $1.00 | $1.00 |
| 2 quarters | 0.25 | 0.50 |
| 3 dimes | 0.10 | 0.30 |
| 4 nickels | 0.05 | 0.20 |
| 5 pennies | 0.01 | 0.05 |

*Table 4* SMART

---

*Item total value* is the product of the quantity of an item and the item's value (rate).

---

The sum of the item total column has no meaning in the *SMART* table. Examples 1 through 4 also show the pattern of multiplying quantity times value or rate to get item total value. Unlike the *SMART* table, however, these tables provide a meaningful overall total when the item totals are then added.

EXAMPLE 1

*Grocery bill*    Cash register receipts from many large grocery stores give the name of the item, the quantity purchased, the unit item value, and the item total value, as well as the total cost. Make a table showing the quantity and value for each item in the purchase, and calculate the item total value and the total cost. The purchase is 0.13 pound of garlic at $2.88 per pound, 1.98 pounds of bananas at $0.58 per pound, 4 cans of pears at $0.78 per can, and 2 cans of peaches at $1.05 per can.

SOLUTION

| Item | Quantity | Value (rate) | Item Total |
|------|----------|--------------|------------|
| Garlic | 0.13 lb | $2.88/lb | $0.38 |
| Bananas | 1.98 lb | 0.58/lb | 1.15 |
| Pears | 4 cans | 0.78/can | 3.12 |
| Peaches | 2 cans | 1.05/can | 2.10 |
| Total | | | $6.75 |

◆ ◆ ◆ ◆ ◆ ◆ ◆ ◆ ◆ ◆ ◆ ◆ ◆ ◆ ◆ ◆ ◆ ◆ ◆ ◆ ◆ ◆ ◆ ◆ ◆ ◆ ◆ ◆ ◆

Adding the item total column gives the total cost of the purchase. In some circumstances the sum of the quantity column is sensible.

EXAMPLE  2

**Lunch for four**   Make a table for this meal at Jaime's Hamburgers, identifying quantity and value for each item ordered: 4 hamburgers at $3.85 each, 2 orders of fries at ($1.65 each), coffee ($0.95), and milk ($1.10). Calculate the item total and the total cost of the meal.

SOLUTION

| Item | Quantity | Value (rate) | Item Total |
|------|----------|--------------|------------|
| Hamburger | 4 | $3.85 | $15.40 |
| Fries | 2 | 1.65 | 3.30 |
| Coffee | 2 | 0.95 | 1.90 |
| Milk | 2 | 1.10 | 2.20 |
| Total | 10 items | | $22.80 |

♦ ♦ ♦ ♦ ♦ ♦ ♦ ♦ ♦ ♦ ♦ ♦ ♦ ♦ ♦ ♦ ♦ ♦ ♦ ♦ ♦ ♦ ♦ ♦ ♦ ♦ ♦ ♦ ♦ ♦ ♦ ♦ ♦ ♦ ♦

The sum of the quantity column in the table in Example 2 gives the total number of items purchased. In contrast, the sum of the quantity column in the table in Example 1 is a sum of pounds and cans and is therefore not meaningful.

In Example 2 dividing $22.80 by 10 gives the average cost, $2.28, of the 10 items. The average cost is meaningful but not particularly useful. In Examples 3 and 4 a division of the total value by total quantity gives an average that is both meaningful *and* useful. Such an average is frequently the desired outcome of a problem situation.

EXAMPLE 3

**Calculating a grade point average (GPA)**   To calculate grade point averages, some schools set these rates: 4 points per A, 3 points per B, 2 points per C, and 1 point per D. Make a quantity and value (rate) table for the following grade report. Calculate the grade point average.

French:     4 credit hours     C
Algebra:    4 credit hours     A
Writing:    3 credit hours     B
Tennis:     1 credit hour      A

SOLUTION

| Subject and Grade | Quantity (credit hours) | Value or Rate (grade points) | Item Total |
|-------------------|-------------------------|------------------------------|------------|
| French, C | 4 hr | 2 pts per hr | 8 pts |
| Algebra, A | 4 hr | 4 pts per hr | 16 pts |
| Writing, B | 3 hr | 3 pts per hr | 9 pts |
| Tennis, A | 1 hr | 4 pts per hr | 4 pts |
| Total | 12 hours | | 37 points |

The grade point average is 37 points divided by 12 hours, or approximately 3.08. The average, 3.08, is placed in the last row of the value or rate column.

♦ ♦ ♦ ♦ ♦ ♦ ♦ ♦ ♦ ♦ ♦ ♦ ♦ ♦ ♦ ♦ ♦ ♦ ♦ ♦ ♦ ♦ ♦ ♦ ♦ ♦ ♦ ♦ ♦ ♦ ♦ ♦ ♦ ♦ ♦

The last number in the value column is an average. The average is calculated by dividing the sum of the individual item totals by the total quantity. The value or rate column is not added.

The four-step problem-solving process may be helpful in organizing our thinking.

EXAMPLE 4

Investment earnings    Silvia invests $2000 at 8% interest and $2500 at 6% interest. Use the four-step process to determine the total earnings from her investments.

SOLUTION    *Understand*: What formulas are relevant? Interest earned in 1 year is the product of the investment and the interest rate. Interest rates are changed from percents to decimals.

*Plan*: Set up a quantity and value table, as shown in Table 5.

| Quantity (money invested) | Value (interest rate) | Item Total (interest earned) |
|---|---|---|
| $2000 | 0.08 | $160.00 |
| $2500 | 0.06 | $150.00 |
| Total:    $4500 | | $310.00 |

Table 5 *Investment*

*Carry out the plan*: The total earnings are $310.

*Check and extend*: Is the sum of the quantity column meaningful? Is dividing the total interest earned by the sum of the quantity column meaningful? The quantity column sum gives the total money invested, $4500. The quotient of the total interest earned and the total money invested ($310 ÷ $4500) gives the average interest rate earned on the total invested, approximately 6.9%. Thus both calculations are meaningful.

♦ ♦ ♦ ♦ ♦ ♦ ♦ ♦ ♦ ♦ ♦ ♦ ♦ ♦ ♦ ♦ ♦ ♦ ♦ ♦ ♦ ♦ ♦ ♦ ♦ ♦ ♦ ♦ ♦ ♦ ♦ ♦

What makes the pattern *quantity times value equals item total value* so powerful is that it is not limited to purchases, investments, or grade calculation. It is a tool in medicine, business, science, transportation, and countless other fields.

QUANTITY AND VALUE TABLES WITH VARIABLES

We now explore quantity and value settings requiring variables. Look carefully at how two expressions are built for the bottom right corner of each table.

EXAMPLE 5

GPAs, continued    A student with a 1.875 GPA has 24 credit hours. Using the grade rates from Example 3, determine how many credit hours of Bs are needed to raise his GPA to a 2.5. (*Hint*: Let $x$ be the number of credit hours of Bs.)

SOLUTION    *Plan*: The desired GPA is 2.50. Because it is an overall *average*, it is placed in the last row of the value column, but it is not the total of that column. The sum of the last column (the total points from current and future grades) gives one expression for the lower right corner. The product of the total quantity of credit hours and the desired GPA gives a second expression for the lower right corner.

| Item | Quantity (credit hours) | Value (GPA) | Item Total (grade points) |
|---|---|---|---|
| Current GPA status | 24 | 1.875 points | 24(1.875) |
| Future B grades | $x$ | 3.00 points | $3x$ |
| Total | $24 + x$ | 2.50 average | (Two expressions: the sum down and the product across) |

Table 6 Grade Point Average

*Carry out the plan*: The two expressions for the lower right corner of Table 6 are *the total down*, $24(1.875) + 3x$, and *the product across*, $(24 + x)(2.50)$. We set them equal and solve for $x$:

$$24(1.875) + 3x = (24 + x)(2.50)$$
$$45 + 3x = 60 + 2.50x$$
$$0.5x = 15$$
$$x = 30 \text{ hr}$$

*Check*: $24(1.875) + 3(30) \overset{?}{=} (24 + 30)(2.50)$ ✔

◆ ◆ ◆ ◆ ◆ ◆ ◆ ◆ ◆ ◆ ◆ ◆ ◆ ◆ ◆ ◆ ◆ ◆ ◆ ◆ ◆ ◆ ◆ ◆ ◆ ◆ ◆ ◆ ◆ ◆ ◆ ◆ ◆

Finding two expressions for the lower right corner is the key to building an equation.

EXAMPLE 6

More GPAs    Suppose the student in Example 5 wants to raise his GPA to a 3.00. How many credit hours of Bs are needed to raise his GPA to a 3.00? Use Table 7 to write the two expressions that describe the lower right corner.

| Item | Quantity (credit hours) | Value (GPA) | Item Total |
|---|---|---|---|
| Current GPA | 24 | 1.875 points | 24(1.875) |
| Desired grades | $x$ | 3.00 points | $3x$ |
| Total | $24 + x$ | 3.00 average | (Write two expressions) |

Table 7 Grade Point Average

SOLUTION    The two expressions for the lower right corner are *the total down*, $24(1.875) + 3x$, and *the product across*, $(24 + x)(3.00)$. We set them equal and solve for $x$:

$$24(1.875) + 3x = (24 + x)(3.00)$$
$$45 + 3x = 72 + 3.00x$$
$$0 = 27 \quad \text{No real number solution}$$

There is no real number $x$ that makes the equation true, so the solution set is empty, { }. What does *no real number solution* mean in terms of the problem situation?

◆ ◆ ◆ ◆ ◆ ◆ ◆ ◆ ◆ ◆ ◆ ◆ ◆ ◆ ◆ ◆ ◆ ◆ ◆ ◆ ◆ ◆ ◆ ◆ ◆ ◆ ◆ ◆ ◆ ◆ ◆ ◆ ◆ ◆ ◆

EXAMPLE 7

**A hot bath**    One cold winter evening you plan to take a leisurely bath, but you get called away after you turn the water on. When you get back, you discover that only the cold water was turned on. The tub is one-fourth full (9 gallons). The cold water temperature in winter is about 35°. The hot water heater is set at 140°. You would rather not waste water by draining the tub and starting over. Use a quantity-value table to determine how much hot water must be added to correct the temperature to the desired 103°.

SOLUTION

| Item | Quantity | Value (temperature) | Item Total |
|------|----------|---------------------|------------|
| Cold water | 9 gal | 35° | 9(35) |
| Hot water | $x$ gal | 140° | 140$x$ |
| Total | 9 + $x$ | 103° | (Two expressions) |

*Table 8* Bath Water

The two expressions for the lower right corner of Table 8 are *the total down*, $9(35) + 140x$, and *the product across*, $(9 + x)(103°)$. We set them equal and solve for $x$:

$$9(35) + 140x = (9 + x)(103)$$

$$315 + 140x = 927 + 103x$$

$$37x = 612$$

$$x \approx 16.5 \text{ gal}$$

You need to add approximately 16.5 gallons of hot (140°) water. Will the added water raise the water level too high (either before or after you step into the tub)?

♦ ♦ ♦ ♦ ♦ ♦ ♦ ♦ ♦ ♦ ♦ ♦ ♦ ♦ ♦ ♦ ♦ ♦ ♦ ♦ ♦ ♦ ♦ ♦ ♦ ♦ ♦ ♦ ♦ ♦ ♦ ♦ ♦ ♦ ♦ ♦

In Example 8 the total quantity is known, but the two individual quantities are both unknown. Look carefully at how they are described.

EXAMPLE 8

**An antifungal footwash**    Jasmine orders a 10% boric acid solution for an antifungal footwash. How much water must be added to a 40% boric acid stockroom solution to make 2500 milliliters of a 10% solution? How much boric acid solution will be needed?

SOLUTION

**Understand the problem**: Work through a table with a guess of half water and half stockroom solution. Table 9 shows 1250 milliliters of water added to 1250 milliliters of boric acid solution. Observe that the water has a 0 in the value column because it contains no boric acid. The resulting mixture is $\frac{500}{2500} = 0.20$, or a 20% solution.

| Item | Quantity | Value | Item Total |
|------|----------|-------|------------|
| Water | 1250 mL | 0 | 0 |
| Boric acid | 1250 mL | 0.40 | 500 |
| Total | 2500 mL | | 500 |

*Table 9*

*Plan*: Because we want a 10% solution, we estimate that we need considerably more water, almost 2000 milliliters. This time we enter the 10% in the average position and use it to write one of the two expressions for the corner of Table 10.

*Carry out the plan*: We complete Table 10.

| Item | Quantity | Value | Item Total |
|------|----------|-------|------------|
| Water | $x$ | 0 | 0 |
| Boric acid | $2500 - x$ | 0.40 | $0.40(2500 - x)$ |
| Total | 2500 mL | 0.10 | (Two expressions) |

*Table 10*

The sum of the item totals is $0.40(2500 - x)$. The product across the last row is $2500(0.10) = 250$. These two expressions are equal and give the equation:

$$250 = 0.40(2500 - x)$$
$$250 = 1000 - 0.40x$$
$$0.40x = 750$$
$$x = 1875 \text{ mL water}$$
$$2500 - x = 625 \text{ mL boric acid solution}$$

*Check*: $1875(0) + (625)(0.40) \stackrel{?}{=} 2500(0.10)$ ✔

◆ ◆ ◆ ◆ ◆ ◆ ◆ ◆ ◆ ◆ ◆ ◆ ◆ ◆ ◆ ◆ ◆ ◆ ◆ ◆ ◆ ◆ ◆ ◆ ◆ ◆ ◆ ◆ ◆ ◆ ◆ ◆ ◆ ◆ ◆ ◆

In Example 8 we used the variable $x$ in the description of both unknown quantities: $x$ for water and $2500 - x$ for boric acid. Example 9 practices this notation in other settings. The same technique was used in Section 4.3 to describe two line segments adding to a given length.

EXAMPLE 9

A total quantity is given. If $x$ is one part of the quantity, what expression describes the other part?

**a.** 30 pounds    **b.** 45 coins    **c.** $2000    **d.** 15 liters

SOLUTION    **a.** $30 - x$ pounds    **b.** $45 - x$ coins    **c.** $\$2000 - x$

**d.** $15 - x$ liters

◆ ◆ ◆ ◆ ◆ ◆ ◆ ◆ ◆ ◆ ◆ ◆ ◆ ◆ ◆ ◆ ◆ ◆ ◆ ◆ ◆ ◆ ◆ ◆ ◆ ◆ ◆ ◆ ◆ ◆ ◆ ◆ ◆ ◆ ◆ ◆

## EXERCISES 4.4

**Identify the quantity and value (or rate) in each of the problem settings in Exercises 1 to 10.**

1. Amel, a veterinarian, has 100 pounds of dog food containing 12% protein and 50 pounds of dog food containing 15% protein.

2. La Deane, a veterinarian, has 30 kilograms of cat food containing 8% fat and 40 kilograms of cat food containing 14% fat.

3. Demi's snack shop has 5 kilograms of peanuts at $8.80 per kilogram and 2 kilograms of cashews at $24.20 per kilogram.

4. Reuel's café has 100 pounds of coffee at $9.00 per pound and 100 pounds of coffee at $10.80 per pound.

5. Li, a chemist, has 150 milliliters of 18-molar sulfuric acid and 100 milliliters of 3-molar sulfuric acid. (*Molar* is a chemical term; the larger the molarity number the more concentrated the acid.)

6. Ingrid, a chemist, has 0.2 liter of 16-molar nitric acid and 0.5 liter of 6-molar nitric acid.

7. Serena has 15 dimes and 20 quarters.

8. Andrzej has 12 half-dollars and 30 nickels.

9. Loki drives 3 hours at 80 kilometers per hour and 2 hours at 30 kilometers per hour.

10. Kana drives 4 hours at 50 miles per hour and 3 hours at 18 miles per hour.

For Exercises 11 to 14 do the following.

    a. Set up a quantity-value table for each problem situation.

    b. Is the sum of the quantity column meaningful?

    c. Is dividing the sum of the item total column by the sum of the quantity column meaningful? State your conclusion in terms of the problem setting.

    d. Answer the question in the problem.

11. Abraham buys 3 pounds of grapes at $0.98 per pound, 5 pounds of potatoes at $0.49 per pound, and 2 pounds of broccoli at $0.89 per pound. What is the total cost of his purchase?

12. Hwang works 30 hours at one job for $4.50 per hour and then 15 hours at another job for $6.25 per hour. What are his total earnings?

13. Zohreh blends 0.2 liter of 16-molar nitric acid with 0.3 liter of 6-molar nitric acid. What is the molar value of 1 liter of her mixture?

14. Ludvina buys 15 airletters at $0.45, 50 first-class postage stamps at $0.32, and 10 postcards at $0.20. What is the total cost of her purchase?

Solve Exercises 15 to 24 by setting up a quantity-value table and building an equation from the table.

15. Suppose only the hot water had been turned on in Example 7 and the bathtub contained 12 gallons of water at 140°. How many gallons of summer cold water (60°) would be needed to drop the temperature to 103°?

16. Suppose only the hot water had been turned on in Example 7 and the bathtub contained 12 gallons of water at 140°. How many gallons of winter cold water (35°) would be needed to drop the temperature to 103°?

17. If equal parts of hot (140°) and winter cold (35°) water are used, what is the temperature of the mixture in the tub?

18. If equal parts of hot (140°) and summer cold (60°) water are used, what is the temperature of the mixture in the tub?

19. If 15 gallons of summer cold water (60°) are used, how many gallons of hot water (140°) need to be added to raise the temperature to 104°?

20. If 15 gallons of winter cold water (35°) are used, how many gallons of hot water (140°) need to be added to raise the temperature to 104°?

21. Georgia purchases 200 shares of Boeing stock at $54 per share. How many shares of Nike stock may be purchased at $71 per share if she has a total of $25,000 to invest?

22. Mikhail has $22,000 to invest. How many shares of Ford stock at $37 per share may he buy if he also buys 300 shares of General Motors at $24 per share?

23. Abdulla needs to produce a 12-molar sulfuric acid solution by blending an unknown quantity of 3-molar sulfuric acid with 0.3 liter of 18-molar sulfuric acid. How many liters of the 3-molar acid will he need?

24. Tsuki needs to make an 8-molar hydrochloric acid. She wants to add an unknown quantity of 12-molar hydrochloric acid to 0.5 liter of 6-molar hydrochloric acid. How many liters of the 12-molar acid will she need?

Each quantity listed in Exercises 25 to 30 is the sum of two numbers. The variable $x$ is one of the two numbers. What is the other number?

25. 300 pounds

26. 50 kilograms

27. $15,000

28. 18 hours

29. 16 liters

30. 24 ounces

In Exercises 31 and 32 solve by setting up a quantity-value table and building an equation from the table.

31. Alan blends two types of coffee beans to sell at a price of $8.35 per pound. Colombian coffee beans sell for $7.25 per pound, and Sumatran beans sell for $10.00 per pound. If he wishes to make 300 pounds of blend, how many pounds of each are needed? What is the ratio of the two types of coffee, Colombian to Sumatran?

32. Bridget's blend of coffee combines Colombian beans at $7.25 per pound with Sumatran beans at $10.00 per pound. She wishes to blend 300 pounds to sell at $9.45 per pound. How many pounds of each does she need? What is the ratio of the two types of coffee, Colombian to Sumatran?

33. Two terms of straight-A grades will help which student's GPA more: a second-year full-time student or a fourth-year full-time student? Explain why.

34. Explain how one might predict whether an outcome will have a large or a small effect on an average?

*PROBLEM SOLVING*

Bill and Pat's Nut Shop mixes cashews and peanuts. Cashews sell for $24 a kilogram, and peanuts sell for $10 a kilogram. How many kilograms of each are needed to blend 50 kilograms to sell for the prices listed in Exercises 35 and 36? (Do a quantity-value table for each of the prices.) What is the ratio of cashews to peanuts in the resulting blends?

35. a. $12 per kilogram    b. $16 per kilogram

36. a. $18 per kilogram    b. $20 per kilogram

Maria has $15,000 to invest for one year. Investment A offers 5%, and investment B offers 8%. She may withdraw her money from investment A quickly if she needs it. Set up a quantity-value table for each part of Exercises 37 and 38.

37. a. If she invests all at 8%, what does she earn?

    b. If she invests all at 5%, what does she earn?

c. If she invests half the money in each, what does she earn?

d. If she invests in a 1 to 2 ratio, what does she earn?

e. Comment on any patterns you observe.

38. a. If her total earnings are $1060, how much is invested in each?

    b. If her total earnings are $825, how much is invested in each?

    c. If her total earnings are $1005, how much is invested in each?

    d. If her total earnings are $1140, how much is invested in each?

    e. Comment on any patterns you observe.

## PROJECTS

39. **Flipping a Coin.** Flip a coin 15 times. Record the outcome—heads or tails—in a table as the input column.

    a. Make two output columns, headed "Cumulative Heads Percent" and "Cumulative Tails Percent," respectively. Calculate the percentages indicated. (The cumulative heads percent is calculated by dividing the total number of times the coins have come up heads by the total number of heads and tails together.) A sample table is shown below.

| Outcome | Cumulative Heads Percent | Cumulative Tails Percent |
|---|---|---|
| T | $\frac{0 \text{ heads}}{1 \text{ total}} = 0\%$ | $\frac{1 \text{ tail}}{1 \text{ total}} = 100\%$ |
| H | $\frac{1 \text{ head}}{2 \text{ total}} = 50\%$ | $\frac{1 \text{ tail}}{2 \text{ total}} = 50\%$ |
| H | $\frac{2 \text{ heads}}{3 \text{ total}} \approx 66.7\%$ | $\frac{1 \text{ tail}}{3 \text{ total}} \approx 33.3\%$ |
| H | $\frac{3 \text{ heads}}{4 \text{ total}} = 75\%$ | $\frac{1 \text{ tail}}{4 \text{ total}} = 25\%$ |

    b. Draw a set of axes, with numbers of flips on the $x$-axis and percents on the $y$-axis. Draw two lines, one for percent heads and one for percent tails.

    c. What do you observe about the two graphs? In what way should your graph be similar to the graphs in other people's projects? Why?

    d. Compare the effect of the fifteenth outcome on the percents to that of the second outcome.

40. **Spreadsheet.** Create a computer spreadsheet similar to the table at the top of the next column to model the quantity-value tables. Use it to do the related exercises in this section. To run a spreadsheet enter numbers in the positions

A2, A3, B2, and B3. If A2, A3, B2, or B3 information is missing in the problem, solve the problem by guessing and using the spreadsheet to check.

| Row | Column A | Column B | Column C |
|---|---|---|---|
| 1 | Quantity | Value | Item Total |
| 2 | A2 | B2 | +A2 * B2 |
| 3 | A3 | B3 | +A3 * B3 |
| 4 | +A2 + A3 | +(C2 + C3)/A4 | +C2 + C3 |
| 5 | "Calculated Average Value =" +B4 | | |
| 6 | "Product Across =" +A4 * B4 | | |
| 7 | "Item Sum Down =" +C2 + C3 | | |

41. **Concentration and Dilution.** Pouring water into a pitcher containing frozen orange juice concentrate dilutes the concentration of the orange juice. In contrast, removing water from orange juice at the processing plant increases the concentration of the orange juice. Adding pure antifreeze to a car radiator increases the concentration of the antifreeze in the radiator.

    The choice of word, *concentration* or *dilution*, depends on which ingredient is of importance. In a *dilution*, the relative amount of the ingredient of importance *gets smaller*. In a *concentration*, the relative amount of the ingredient of importance *gets larger*.

    a. Complete the table below. Calculate the percent orange juice concentrate by dividing the total orange juice concentrate by the total mixture in each row.

    b. Use the data to create a graph, with the percent orange juice concentrate as output and the total mixture as input.

    c. Calculate what percent the total water added is of the total mixture for each can of water.

    d. On the same axes as in part a, graph the percent water as output.

    e. Comment on the rate of change of the percents by comparing the effect of the first can of water on the percent with that of the last can of water.

| Total Orange Juice Concentrate | Total Water Added | Quantity (total mixture) | Value or Rate (percent OJ concentrate) |
|---|---|---|---|
| 12 ounces | 0 ounces | | |
| 12 ounces | 12 ounces | | |
| 12 ounces | 24 ounces | | |
| 12 ounces | 36 ounces | | |

## 4.5    AVERAGES, MIDPOINTS, AND CENTROIDS

OBJECTIVES    Find the mean, median, and mode of a set of data.  ♦  Determine the midpoint of a line segment, given its endpoints as coordinates.  ♦  Identify the centroid of a figure, given in coordinates.

WARM-UP    **1.** Solve for $x$: $\dfrac{0.78 + 0.70 + 0.90 + x}{4} = 0.80$.

**2.** Solve for $x$: $\dfrac{1.93 + 1.5x}{4} = 0.80$.

**3.** What is the area of a circle of radius 5 inches? Leave the answer in terms of $\pi$.

**4.** What is the area of a circle of radius 10 inches? Leave the answer in terms of $\pi$.
♦

---

### JUST FOR FUN

*What four consecutive letters of the alphabet form a word?* You may scramble the order of the letters. (*Note*: "Figh" is not a fruit, and "d'cab" is not what you call to get a ride to d'airport!) Determine a proportion that gives the word upon cross multiplication.

---

L oading an airplane, writing a test, and summarizing test results may seem to have little in common, but all are related to a branch of mathematics called *statistics*.

The airplane in the photograph on the chapter opener became so out of balance that it tipped. The situation was not only embarrassing, but also a graphic illustration of how important the location of the center of gravity is. The *center of gravity is the average position of weight*. The pilot determines the center of gravity from the weight of the luggage and cargo, the weight of the fuel, and an average weight assigned to each passenger. Additional fuel, as needed, is ordered by weight and loaded into the various fuel tanks so as to maintain an appropriate center of gravity. If the suitcases and cargo are loaded into the back of the plane before the plane is refueled, the loading may cause the plane to tip as shown.

When instructors write tests, they make a list of questions that sample the total knowledge of the students. A s*ample* is a small number of objects from a large set. A good test will provide a sample that accurately describes the students' knowledge. To summarize test results, instructors calculate an average test score (similar to the center of gravity on the plane). The average is used to determine how one individual performs compared with the class or to compare one class with another.

In this section we use the concept of an average in several ways, including numerically and graphically.

### BASIC STATISTICS: AVERAGES

Usually, people say "average" when they *add numbers together and divide by how many numbers were added*. This average is called the **mean**.

E X A M P L E   **1**   Baggage limits    What are the means of these sets of suitcase weights? Suppose each set represents one family's luggage, and the family is allowed to average the weights. Which sets would satisfy a 70-pound average domestic airline baggage limit?

**a.** 95, 58, 52, 88       **b.** 72, 47, 90       **c.** 98, 55, 58, 75, 63

SOLUTION   **a.** $(95 + 58 + 52 + 88) \div 4 = 73.25$, over the 70-pound average weight limit

**b.** $(72 + 47 + 90) \div 3 = 69.7$, under the 70-pound average

**c.** $(98 + 55 + 58 + 75 + 63) \div 5 = 69.8$, under the 70-pound average

♦ ♦ ♦ ♦ ♦ ♦ ♦ ♦ ♦ ♦ ♦ ♦ ♦ ♦ ♦ ♦ ♦ ♦ ♦ ♦ ♦ ♦ ♦ ♦ ♦ ♦ ♦ ♦ ♦ ♦ ♦ ♦

In Section 3.5 we calculated the score needed on a final exam to earn a specific grade. In Example 2 each test and the total homework are worth the same amount, so we can add the percent scores and divide by the number of scores to get the average. In general, however, we avoid averaging percents unless, as in Example 2, they represent percents of the same number.

E X A M P L E   **2**   Class average    Suppose an instructor gives equal weight to homework, each of two midterms, and the final exam. A student has 78% and 70% on the midterms and 90% on homework. What percent does this student need on the final to have an 80% mean average for the class?

SOLUTION

$$\frac{0.78 + 0.70 + 0.90 + x}{4} = 0.80$$

$$\frac{2.38 + x}{4} = 0.80$$

$$2.38 + x = 3.20$$

$$x = 0.82$$

The student needs 82% on the final exam.

♦ ♦ ♦ ♦ ♦ ♦ ♦ ♦ ♦ ♦ ♦ ♦ ♦ ♦ ♦ ♦ ♦ ♦ ♦ ♦ ♦ ♦ ♦ ♦ ♦ ♦ ♦ ♦ ♦ ♦ ♦ ♦

There are many circumstances in which the mean does not provide a good description of a set of numbers. In Example 1 we found the mean of 95, 58, 52, and 88 to be 73.25. To the airport baggage handlers, the mean is irrelevant because they have to work with the individual bags. In Example 3 the mean would not give a realistic description of the financial well-being of each family.

E X A M P L E   **3**   Mean income    Suppose the annual incomes of five families are $4000, $8000, $9000, $9000, and $100,000, respectively. Find the mean, and explain why the mean does not provide a good description of this set of families.

SOLUTION   The mean is

$$(\$4000 + \$8000 + \$9000 + \$9000 + \$100{,}000) \div 5 = \$26{,}000$$

The mean appears to indicate that all the families have an income well above the 1990 poverty level for families of four persons, $13,359. Yet in reality four of the five families are substantially below the poverty level.

♦ ♦ ♦ ♦ ♦ ♦ ♦ ♦ ♦ ♦ ♦ ♦ ♦ ♦ ♦ ♦ ♦ ♦ ♦ ♦ ♦ ♦ ♦ ♦ ♦ ♦ ♦ ♦ ♦ ♦ ♦ ♦

Because the mean average does not always provide a good description, statisticians invented other averages. The **median** *is an average found by selecting the middle number when the numbers are arranged in numerical order* (from smallest to largest or largest to smallest).

**EXAMPLE  4**    Median income    Find the median of each set of incomes.

**a.** $4000, $8000, $9000, $9000, $100,000

**b.** $50,000, $20,000, $8000, $16,000

SOLUTION    **a.** The median of $4000, $8000, $9000, $9000, and $100,000 is $9000, because the numbers are already arranged in order and $9000 is the middle number.

**b.** The second set of numbers must be rearranged into order: $8000, $16,000, $20,000, and $50,000. The set has no middle number. In this case we find the mean of the two middle numbers and use it as the median. The median is ($16,000 + $20,000) ÷ 2 = $18,000.

♦ ♦ ♦ ♦ ♦ ♦ ♦ ♦ ♦ ♦ ♦ ♦ ♦ ♦ ♦ ♦ ♦ ♦ ♦ ♦ ♦ ♦ ♦ ♦ ♦ ♦ ♦ ♦ ♦ ♦ ♦ ♦ ♦ ♦ ♦ ♦

The median is used to describe the center of a set of data that has many numbers close together and one number that is considerably different from the others.

The mode is another form of average. The **mode** of a set *is the number occurring most often*. The mode is applied to sets that contain repeated numbers.

Table 11 shows a tally by states of the age at which one is allowed unrestricted operation of private passenger cars. Many states with an age of 18 allow the applicant to drive at a lower age if he or she has taken a driver's education course.

| Driver's Age, Years | States (including the District of Columbia) |
|---|---|
| 15 | // |
| 16 | LHT LHT LHT LHT |
| 16.5 | / |
| 17 | /// |
| 18 | LHT LHT LHT LHT //// |
| 19 | / |

*Table 11  Age for Unrestricted Operation of Private Passenger Cars*

Data from *The World Almanac and Book of Facts 1992* (New York: Pharos Books, 1991), p. 678.

**EXAMPLE  5**    Driving age    What are the median and the mode for the data on driving age in Table 11?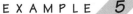

SOLUTION    The tally places the data in order by age, so we count to the middle tally mark. The middle of the 51 tally marks is the 26th tally mark, because it has 25 marks before it and 25 after it. The 26th tally mark is the last mark for age 17, so age 17 is the median. The mode is the number that is tallied most often: age 18.

♦ ♦ ♦ ♦ ♦ ♦ ♦ ♦ ♦ ♦ ♦ ♦ ♦ ♦ ♦ ♦ ♦ ♦ ♦ ♦ ♦ ♦ ♦ ♦ ♦ ♦ ♦ ♦ ♦ ♦ ♦ ♦ ♦ ♦ ♦ ♦

Summary

> ♦ The mean is the sum of the numbers divided by "how many" numbers.
> ♦ The median is the middle number when the numbers are arranged in order.
> ♦ The mode is the most frequently occurring number.

The averages thus far—mean, median, and mode—all relate to numbers. We now look at midpoints and centroids, which extend the concept of the average, or middle, number to lines, coordinates, and geometric shapes.

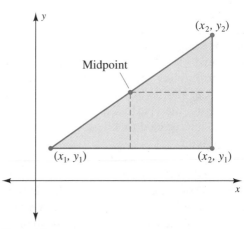

Figure 8

MIDPOINTS AND CENTROIDS

The **midpoint** *of a line is the center, or the point halfway between its end-points.* Figure 7 shows a line of length 10 units with a midpoint at the five-unit position. The point labeled 5 is halfway between a point labeled 0 and another labeled 10. The midpoint is the average of the coordinates of the endpoints; we can find the midpoint of any line segment by calculating this average.

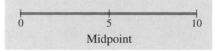
Midpoint

Figure 7

The **midpoint** *for coordinates is the mean in both the x and the y direc-tion.* Thus the midpoint of $(x_1, y_1)$ and $(x_2, y_2)$ is the average of $x_1$ and $x_2$ and the average of $y_1$ and $y_2$. Figure 8 shows the midpoint of the line connecting $(x_1, y_1)$ and $(x_2, y_2)$. The formula for finding midpoints is

$$\text{midpoint} = \left(\frac{x_1 + x_2}{2}, \frac{y_1 + y_2}{2}\right)$$

E X A M P L E   **6**

Find the midpoints for the segments formed by the following coordinates. First use the formula, then check the answer by sketching the line segment on a graph and plotting the midpoint.

**a.** $(0, 2)$ and $(5, 0)$    **b.** $(0, 0)$ and $(5, 2)$    **c.** $(7, 3)$ and $(9, -1)$

SOLUTION   **a.** For $(0, 2)$ and $(5, 0)$ the midpoint is

$$\left(\frac{x_1 + x_2}{2}, \frac{y_1 + y_2}{2}\right) = \left(\frac{0 + 5}{2}, \frac{2 + 0}{2}\right) = \left(\frac{5}{2}, 1\right) - (2.5, 1)$$

**b.** For $(0, 0)$ and $(5, 2)$ the midpoint is

$$\left(\frac{x_1 + x_2}{2}, \frac{y_1 + y_2}{2}\right) = \left(\frac{0 + 5}{2}, \frac{0 + 2}{2}\right) = \left(\frac{5}{2}, 1\right) = (2.5, 1)$$

**c.** For $(7, 3)$ and $(9, -1)$ the midpoint is

$$\left(\frac{x_1 + x_2}{2}, \frac{y_1 + y_2}{2}\right) = \left(\frac{7 + 9}{2}, \frac{3 + (-1)}{2}\right) = (8, 1)$$

The midpoint for the first two segments is the same, $(2.5, 1)$. It is labeled $M_a$ and $M_b$ in Figure 9. The midpoint for the third segment is $(8, 1)$ and is labeled $M_c$.

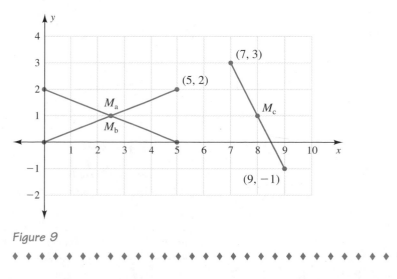

Figure 9

The midpoint locates the middle of a line, whereas the **centroid** *is the center of a flat or solid geometric shape.* In more complicated structures such as the human body, bicycles, and airplanes, the center is called the *center of mass* or *center of gravity.*

Centroids (and centers of gravity) are important because the behavior of an object in motion is dependent on where the centroid is located. Thus the flight of an airplane, the rotation of a tire, the spinning of a washing machine, and the wild motion of a carnival ride depend on a correct centering of weight. Many sports such as ice skating, platform diving, pole vaulting, gymnastics, and ski jumping require careful control of the body's center to achieve top performance. In more recent developments, the centroid of the area under a sensor data curve provides the single number needed to trigger an appropriate response from a computerized control circuit.

We think of the centroid as a balance point. Some centroids are easy to locate. A washing machine on spin cycle and an automobile wheel have circular motion. Clothes arranged evenly in a washing machine maintain the centroid of the load on the rotational axis of the machine; however, when the wet load piles up on one side, the centroid moves off the axis of rotation and the machine vibrates wildly. The little lead weights placed on automobile wheels act to keep the centroid of the wheel at the axle and prevent wobbling and uneven wear on the tires. Thus we can conclude the following.

---

The centroid of a circle is located at its center.

---

EXAMPLE  7

What is the centroid of the circle in Figure 10? The coordinates of the endpoints of one diameter are shown.

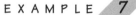

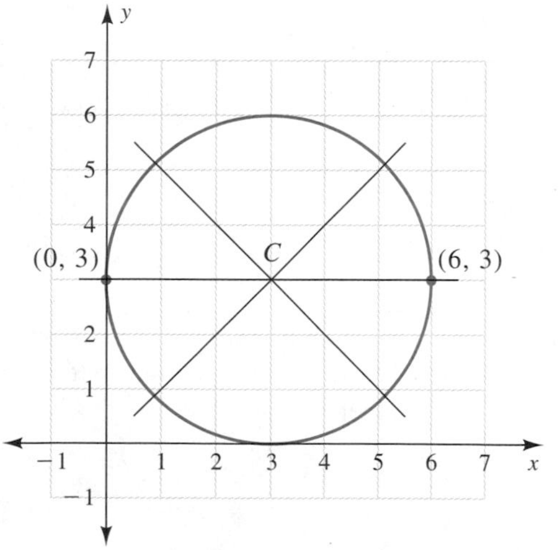

Figure 10

SOLUTION    The center of the circle is the midpoint of the diameter, so the centroid is

$$(x_c, y_c) = \left(\frac{x_1 + x_2}{2}, \frac{y_1 + y_2}{2}\right) = \left(\frac{0 + 6}{2}, \frac{3 + 3}{2}\right) = (3, 3)$$

♦ ♦ ♦ ♦ ♦ ♦ ♦ ♦ ♦ ♦ ♦ ♦ ♦ ♦ ♦ ♦ ♦ ♦ ♦ ♦ ♦ ♦ ♦ ♦ ♦ ♦ ♦ ♦ ♦ ♦ ♦ ♦ ♦ ♦ ♦ ♦ ♦ ♦ ♦ ♦ ♦

The relationship between averaging coordinates and the centroid works for many other flat shapes as well.

---

The centroid of a rectangle, square, or triangle is the average of the coordinates of its corners (vertices).

---

E X A M P L E **8**    Find the centroid of the rectangle in Figure 11.

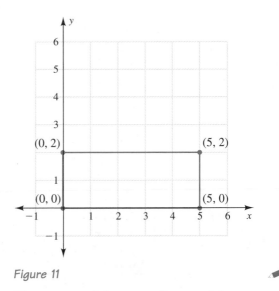

*Figure 11*

SOLUTION    The average of the *x*-coordinates of the corners is

$$x_c = \frac{0 + 5 + 5 + 0}{4} = \frac{10}{4} = 2.5$$

The average of the *y*-coordinates of the corners is

$$y_c = \frac{0 + 0 + 2 + 2}{4} = \frac{4}{4} = 1$$

The coordinates of the centroid are $(x_c, y_c) = (2.5, 1)$.

◆ ◆ ◆ ◆ ◆ ◆ ◆ ◆ ◆ ◆ ◆ ◆ ◆ ◆ ◆ ◆ ◆ ◆ ◆ ◆ ◆ ◆ ◆ ◆ ◆ ◆ ◆ ◆ ◆ ◆ ◆ ◆ ◆ ◆ ◆ ◆

E X A M P L E **9**    Find the position of the centroid for the triangle in Figure 12.

SOLUTION    The average of the *x*-coordinates of the vertices is

$$x_c = \frac{0 + 4 + 2}{3} = \frac{6}{3} = 2$$

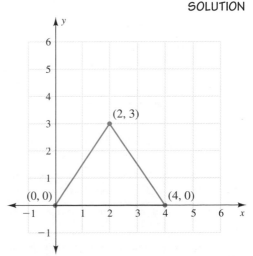

*Figure 12*

The average of the *y*-coordinates of the vertices is

$$y_c = \frac{0 + 0 + 3}{3} = 1$$

The coordinates of the centroid are $(x_c, y_c) = (2, 1)$.

◆ ◆ ◆ ◆ ◆ ◆ ◆ ◆ ◆ ◆ ◆ ◆ ◆ ◆ ◆ ◆ ◆ ◆ ◆ ◆ ◆ ◆ ◆ ◆ ◆ ◆ ◆ ◆ ◆ ◆ ◆ ◆ ◆

If the centroid is a center of balance, then it is reasonable that the centroid lies on a line of symmetry.

E X A M P L E   **10**    Draw in or identify lines of symmetry in the following figures to verify that they pass through the centroid.

     **a.** Figure 11      **b.** Figure 12

SOLUTION    **a.** The lines of symmetry intersect at the centroid.

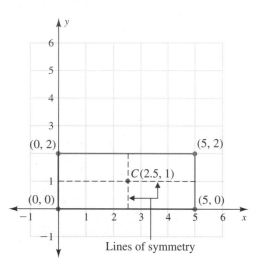

**b.** The line of symmetry is also the height of this triangle.

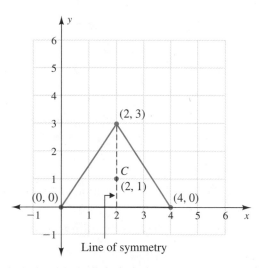

◆ ◆ ◆ ◆ ◆ ◆ ◆ ◆ ◆ ◆ ◆ ◆ ◆ ◆ ◆ ◆ ◆ ◆ ◆ ◆ ◆ ◆ ◆ ◆ ◆ ◆ ◆ ◆ ◆ ◆ ◆ ◆ ◆ ◆ ◆ ◆ ◆ ◆ ◆

### EXERCISES 4.5

In some elections the age of the candidate is an issue. In Exercises 1 to 4 calculate the mean, median, and mode age at inauguration of each group of Presidents of the United States.

1. 57, 61, 57, 57, 58 (Washington, Adams, Jefferson, Madison, Monroe)

2. 57, 61, 54, 68, 51 (Adams, Jackson, Van Buren, Harrison, Tyler)

3. 61, 52, 69, 64, 46 (Ford, Carter, Reagan, Bush, Clinton)

4. 65, 52, 56, 46, 54 (Buchanan, Lincoln, Johnson, Grant, Hayes)

5. If the mean is the same for two sets of data, does this imply that the numbers in the sets are the same? Explain.

6. How does one large piece of data affect the mean? the mode? the median?

7. Comment, using complete sentences, on the effect of one low grade or one high grade on the mean of the test scores.

8. Choose one: The mean is influenced by (every, most, few) measurements in the set. Explain.

9. Choose one: The median (is, is not) influenced by one large or small measurement. Explain.

10. If a fly ball or strikeout is worth 0 and a hit is worth 1, is a baseball batting average (hits divided by times at bat) a median, mode, or mean?

11. When might the median provide a better description of the average than the mean?

12. When might the mean provide a better description of the average than the median?

13. Is it possible to have $\frac{3}{4}$ of the students above the median test score? Explain.

14. What can you conclude about a set of data if the mean is larger than the median?

15. What can you conclude about a set of data if the median is larger than the mean?

16. List a set of numbers for which the mean is larger than the median. List a set of numbers for which the mean is smaller than the median. Explain how you determined your list.

**Find the midpoints of the line segments connecting the sets of points given in Exercises 17 and 18. Sketch the segments on coordinate axes to confirm that the midpoints are reasonable.**

17. a. $(0, 4)$, $(5, 2)$    b. $(-1, 3)$, $(3, -3)$

18. a. $(2, 3)$, $(8, 12)$    b. $(2, -5)$, $(-4, 3)$

**Find the midpoints of the line segments connecting the sets of points given in Exercises 19 and 20. Assume that the variables $a$ and $b$ are positive numbers.**

19. a. $(a, 0)$, $(0, b)$    b. $(a, a)$, $(0, 0)$

    c. $(0, b)$, $(0, 0)$

20. a. $(a, 0)$, $(0, a)$    b. $(0, 0)$, $(a, 0)$

    c. $(a, b)$, $(0, 0)$

**Find the midpoint of each side of the geometric figures in Exercises 21 and 22.**

21. a.

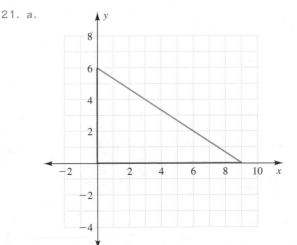

b.

22. a.

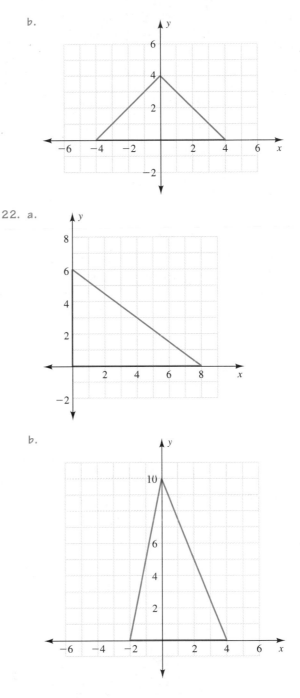

b.

**Sketch a set of coordinate axes for each of Exercises 23 to 26.**

23. Place a rectangle with base $b$ and height $h$ on the axes, with the lower left corner at the origin. Label the coordinates of all vertices and the midpoint of each side.

24. Place a square of side $n$ on the axes, with the lower left corner at the origin. Label the coordinates of all four vertices and the midpoint of each side.

25. Place a right triangle with base $b$ and height $h$ on the axes. The right angle should be at the origin. Label the coordinates of all three vertices and the midpoint of each side.

26. Place a circle with radius *r* on the axes, with the center at the point (*r*, *r*). Label the points where the circle touches the axes.

27. Lines bisect each other if they have the same midpoint. Use the midpoint formula to verify that the diagonals of a rectangle bisect each other. (*Hint*: See Exercise 23.)

28. Lines bisect each other if they have the same midpoint. Use the midpoint formula to verify that the diagonals of a square bisect each other. (*Hint*: See Exercise 24.)

The *median of a triangle is a line that connects a vertex to the middle of the side opposite that vertex*. There are three median lines in every triangle. For Exercises 29 and 30 do the following.

   a. Use the midpoint formula or reasoning to find the midpoint of each side.

   b. Use $y = mx + b$ to determine the equation of each of the median lines.

29.

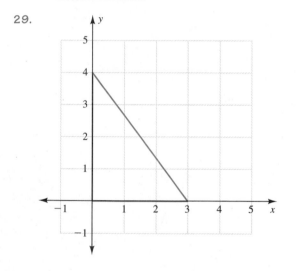

30.

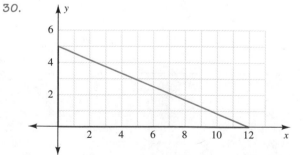

31. Why might the *median of a set of numbers* and the *median of a triangle* be given such similar names?

32. The guard rail or strip of ground between the traffic lanes of a freeway is called the *median*. Explain why this is an appropriate word.

33. *Mode* has the same root as *modern*, *model*, and *a la mode*, which are associated with current, fashionable, or most popular styles. Explain how *mode* as an average fits with these other words.

**For Exercises 34 to 36 find the centroids of the shapes. Show your calculations for rectangles and triangles. Give the answers as coordinates.**

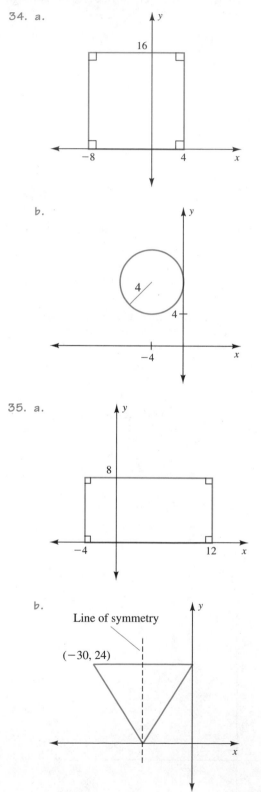

34. a.

b.

35. a.

b.

36. a.

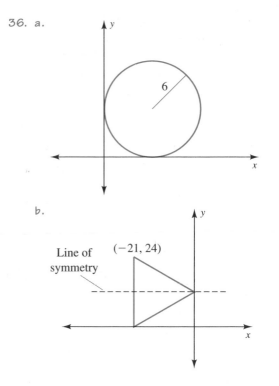

b.

| | Quantity (weight) | Value (score) | Item Total |
|---|---|---|---|
| Midterm | 0.25 | 0.78 | |
| Midterm | 0.25 | 0.70 | |
| Homework | 0.10 | 0.90 | |
| Final | 0.40 | $x$ | |
| Total | 1.00 | | |

38. **Center of Population.** The U.S. Center of Population is like a centroid of population. It would be the balance point for the United States if each person (assume equal weight) were standing at her or his home location on a rigid plate the size of the country.

a. Trace a small map of the United States, and guess where the population center started and how it moved during the period from 1790 to 1990.

b. The center is recalculated after each ten-year census. An almanac or other general census reference will give the center, decade by decade. Plot the 21 centers of population for 1790 to 1990, and connect them.

39. **Centroid Exploration**

a. Cut a large triangle from a stiff piece of paper or cardboard (notebook-size paper is good). Hold a pencil in a vertical position, and move the triangle around on the eraser until the triangle balances. Mark this point on the triangle. The balance point is the centroid of the triangle.

b. Measure the sides of the triangle, and mark the midpoint of each side. Connect each midpoint with the vertex opposite that side. The intersection of these three lines is the centroid. Does this centroid match that found in part a?

c. Use a paper punch to put a hole near each corner of the cardboard triangle. Place a pencil point through one of the holes, and hang a string with a weight tied at the bottom from the pencil point. The string and the triangle must be able to turn freely. Mark the point where the string crosses the opposite side of the triangle. Connect the hole and the mark with a line. Repeat for each of the other two holes. The intersection of the lines is the centroid. How close is this experimental centroid to the balance point? Where does the string cross the opposite side?

d. Make an arbitrary flat shape from stiff paper or cardboard. Find the centroid by repeating the process in part c. Use three arbitrary points near the edge.

40. **Probability, Sets, and Averages**

a. List the set containing all possible current denominations of U.S. coins.

b. Remove the coins from your pocket, purse, or wallet. This is your *set of coins, C.*

c. Count the *number of coins* in your set: $n(C) = ?$ Count the number of each denomination.

PROJECTS

37. **Weighted Grade Averages**

a. Suppose the instructor in Example 2 gives homework half the weight of a midterm and gives the final one and a half times the weight of a midterm. The weights are quantities, and the scores are values. Complete the following table. The weighted average is the sum of the item totals divided by the sum of the weights. Set up and solve an equation to find what score the student needs on the final exam to average 80%.

| | Quantity (weight) | Value (score) | Item Total |
|---|---|---|---|
| Midterm | 1 | 0.78 | |
| Midterm | 1 | 0.70 | |
| Homework | 0.5 | 0.90 | |
| Final | 1.5 | $x$ | |
| Total | 4.0 | | |

b. Suppose the instructor in Example 2 makes the homework worth 10% of the grade, each midterm worth 25%, and the final worth 40%. Complete the table at the top of the next column. Set up and solve an equation to find the score needed on the final exam to average 80%.

This weighted average method extends to finding centroids of geometric figures with any number of parts. For more information look up *centroids* or *center of gravity* in textbooks on airframe mechanics, welding, engineering mechanics, or applied physics. Pilots have computer or calculator programs based on this method to find the center of gravity for their airplanes.

*d.* Complete the table for *C*.

| Coin Denomination | Quantity | Value | Item Total |
|---|---|---|---|
| | | | |
| | | | |
| | | | |
| | | | |
| | | | |
| Total | | | |

*e.* What is the weighted average of the value of the coins in your set (sum of item totals divided by the total number of coins)? Does this number have any meaning?

*f.* Write a fraction that relates the quantity of each denomination to the total quantity of coins. These fractions describe the *probability* of drawing a coin of that denomination if you reached into your pocket, coin purse, or wallet and drew out one coin at random.

*g.* Explain in complete sentences the habits you have with coins that would make your set of coins different from other people's sets or unusual in any way.

## CHAPTER 4 SUMMARY

### Vocabulary

*For definitions and page references, see the Glossary/Index.*

| | |
|---|---|
| centroid | median of a triangle |
| continued ratio | midpoint |
| corresponding | mode |
| cross multiplication | percent |
| equilateral triangle | proportion |
| isosceles triangle | rate |
| mean | ratio |
| median | similar figures |

### Concepts

Division by zero is undefined.

Ratios are equivalent if they simplify to the same number or if they divide to the same decimal value.

Fractions containing variables or units are simplified in the same way as numbers:

$$\frac{a}{a} \text{ or } \frac{\text{inches}}{\text{inches}} \text{ equals } 1.$$

When a proportion contains units, the units must match either side to side or within numerator and denominator pairs.

Apply proportions to find the lengths of the sides in similar figures.

When we know the total length or quantity of something that is separated into two parts, then we use one variable to describe the two parts. If the whole is 10 and one part is $x$, then the other part is $10 - x$.

The sum of the angle measures in a triangle is 180°.

The centroid of a circle is located at the center.

The centroid of a square, rectangle, or triangle is the average of the coordinates of the vertices.

If a figure is symmetric, the centroid lies on the line of symmetry.

## CHAPTER 4 REVIEW EXERCISES

**For Exercises 1 and 2 simplify the ratios. Exclude the inputs for which the original expression will be undefined.**

1. $16x^2$ to $4x^4$

2. $\dfrac{x(x-3)}{(x-3)(x+3)}$

**Simplify the ratios in Exercises 3 to 5.**

3. 5 feet : 18 inches

4. 50 centimeters to 2 meters

5. *m* quarts to *n* gallons

**In Exercises 6 and 7 describe in complete sentences an appropriate situation for each rate.**

6. hours per revolution

7. dollars per day

8. Solve for *x*: $\dfrac{2}{x} = 2,000,000$.

9. Sales of beverages are in a 5:5:1 ratio for Pepsi® products, Coca Cola® products, and other brands. How many of each brand were sold if 121,000 cases were sold in total?

10. The angle measures in a triangle are in the ratio 1:2:3. What is the measure of each angle?

Nutrition experts recommend that no more than 30% of a person's daily calorie intake come from fat. The remaining calories should come from carbohydrates and protein. For Exercises 11 and 12 set up a proportion and solve it to answer the first question in each exercise.

11. In a 1500-calorie diet, how many calories should come from fat? At 9 calories per gram of fat, how many grams of fat would this be?

12. In a 2000-calorie diet, how many calories should come from carbohydrates and protein together? At 4 calories per gram of protein and carbohydrate, how many grams of protein and carbohydrate would this be?

**In Exercises 13 and 14 arrange the facts into a unit analysis that solves the problem.**

13. How many yards is 100 meters?
    3 feet = 1 yard
    12 inches = 1 foot
    39.37 inches = 1 meter

14. How many gallons are in the 1000-liter container shown in the figure?
    1.0567 quarts = 1 liter
    4 quarts = 1 gallon
    1000 liters is enough water for which of the following: a glass of lemonade, a hot tub, or a diving pool?

15. An access ramp needs to rise 30 inches. If the ramp is to have a slope ratio of 1:8, what horizontal distance is needed to build the ramp? Give the answer in feet.

16. A staircase needs to have a slope of 6.5 to 11. What horizontal distance is needed for an 8.5-foot vertical distance between floors?

17. The ratio of water to final mixture in orange juice is 3 to 4. How much water is needed to make 6 gallons of orange juice?

**For Exercises 18 to 21 solve the equations.**

18. $\dfrac{2}{3} = \dfrac{x}{17}$        19. $\dfrac{2}{6} = \dfrac{3}{x}$

20. $\dfrac{x+1}{2} = \dfrac{4x-1}{6}$     21. $\dfrac{2x+5}{2x-1} = \dfrac{7}{5}$

**For Exercises 22 and 23 solve the formulas.**

22. $\dfrac{a}{b} = \dfrac{c}{d}$ for $d$      23. $\dfrac{V_1}{C_2} = \dfrac{V_2}{C_1}$ for $V_2$

24. Determine a value or expression for $x$.

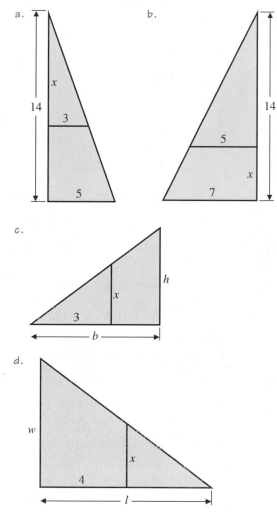

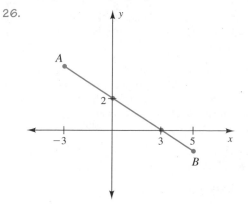

25. In which categories are all figures similar to each other: equilateral triangles, right triangles, circles, squares, rectangles, 3-4-5 right triangles, cubes, isosceles triangles? Explain your reasoning.

**For Exercises 26 and 27 determine the coordinates of $A$ and $B$ on the graph. Use slope, proportions, or counting, as necessary.**

26.

27.

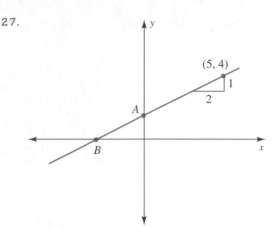

28. The model for an experimental airplane is $\frac{2}{3}$ full size. The wing span of the model is 20 feet. What is the wing span of the full-size plane?

29. Kei's research team catches and tags 250 bats. The next week they catch 300 bats from the same cave, and only 15 are tagged. Estimate the total bat population from this cave.

**Make a value-quantity table for each problem situation in Exercises 30 to 35. Determine appropriate totals and averages. Indicate units and whether any of the totals are meaningful. Suggest questions that could be asked.**

30. Debi invests $10,000 in a tax-free bond at 8% and $5000 in savings earning 3.5%.

31. A plane flies 500 miles at 130 mph and 1500 miles at 200 mph.

32. An aquarium is to be filled with 8 liters of hot water at 90° C and 50 liters of cold water at 5° C.

33. Max buys 6 gallons of 88-octane fuel and 10 gallons of 92-octane fuel.

34. Arlan earns an A on a 12-credit course and a B on a 4-credit course.

35. Karen has 6 dimes and 4 quarters.

36. Because of heat loss in uninsulated pipes, hot water in the bathtub is 130°. How many gallons of hot water need to be added to 9 gallons of cold water (35°) to raise the average temperature to 103°?

37. Rafael wants to earn an average of 6% on his investments. He needs to keep $10,000 in liquid investments (investments that are turned to cash easily without penalty). Liquid assets pay only 3%. How much money does he need to invest at 8% to reach his goal?

**Find the mean, median, and mode of each of the measurements in Exercises 38 to 41.**

38. 4.2 grams, 4.3 grams, 4.3 grams

39. 31.7 cm, 31.8 cm, 31.6 cm, 31.8 cm

40. 6.9 miles, 6.9 miles, 6.8 miles, 7.0 miles, 6.8 miles, 6.8 miles

41. 45.0 mL, 44.0 mL, 45.5 mL, 45.0 mL, 44.5 mL

42. Refer to the figure.

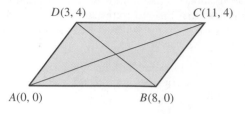

a. Find the midpoint of each line: *AD*, *BC*, *AC*, and *BD*.

b. Find the slope of each line: *AD*, *AB*, *CD*, and *BC*.

c. What is the name of the shape *ABCD*?

43. Refer to the figure.

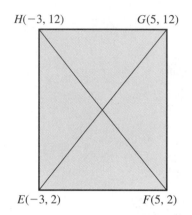

a. Find the midpoint of each line: *EG*, *EF*, *HE*, and *HF*.

b. Find the slope of each line: *HF* and *EG*.

c. What concept explains why the slope of *HE* is undefined? (*Hint*: Calculate the slope.)

d. What is the centroid of figure *EFGH*?

**For Exercises 44 and 45 find the centroid.**

44.

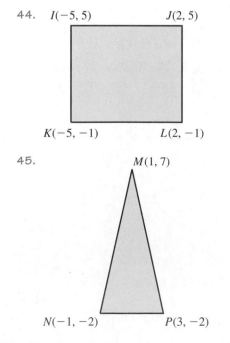

46. The table shows several examples of quantities and values. Several units in the item total column are completed. Fill in the missing units in the item total column.

| Quantity | Value or Rate | Item Total |
|---|---|---|
| Number of coins | Value of coin, cents per coin | |
| Time in hours | Miles per hour | Distance in miles |
| Dollars invested | Percent interest | |
| Credit hours | Point value of grade per credit hour | Points |
| Number of cans | Cost per can | |
| Number of pounds | Cost per pound | |
| Hours worked | Wages per hour | |
| Liters of acid | Molar value, moles per liter | Moles |
| Area, units$^2$ | Coordinate of centroid, units | Units$^3$ |

47. If an experimental aircraft model is $\frac{2}{3}$ full size, estimate the fraction of the full-size weight that the experimental model is. (*Hint*: Assume weight is proportional to volume.)

## CHAPTER 4 TEST

For Exercises 1 and 2 give three ratios that are equivalent to these ratios.

1. $\dfrac{3}{3 + 4}$     2. $\dfrac{3}{3 + 6}$

**Simplify the ratios in Exercises 3 to 6.**

3. $\dfrac{7ab^3}{28a^2b^2}$     4. $\dfrac{(a + b)}{(a + b)(a - b)}$

5. $\dfrac{10 \text{ ft}}{24 \text{ in.}}$     6. $\dfrac{h \text{ hr}}{m \text{ min}}$

7. Arrange these facts into a unit analysis that determines how many inches are in 5 meters:
   1 centimeter = 0.3937 inch
   100 centimeters = 1 meter

8. A $3\frac{1}{2}$-inch-by-5-inch photo is to be enlarged into a poster so that its longer side is 2 feet. How long will the other side be?

**In Exercises 9 and 10 solve the equations for $x$.**

9. $\dfrac{2}{5} = \dfrac{13}{x}$     10. $\dfrac{4}{x + 1} = \dfrac{3}{x - 3}$

11. Rosa wants a 90% average in math. She has 85% and 91% on her first two of three equal-value midterms. What does she need on the third test to obtain the 90?

**What are the median and the mean of each set of data in Exercises 12 and 13?**

12. Five ball point pens: $0.29, $0.29, $0.29, $0.29, $1.79

13. Five ball point pens: $0.69, $0.29, $0.19, $0.69, $1.09

14. Make some observations about the medians and means in Exercises 12 and 13 and what might have caused these results. Under what circumstances would each be a good description of the average cost of the pens?

**For Exercises 15 and 16, refer to the figure.**

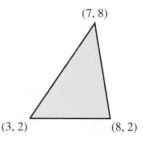

15. What is the midpoint of each side of the triangle?

16. What is the centroid of the triangle?

17. Eugene borrows $17,000 for a car at 5.8% interest and has a $2000 credit card balance at 14.9%. Make a quantity-value table. Explain whether the totals are meaningful, and if so, why. Is there a meaningful average value?

18. What is the length of the line labeled *x* in the triangle?

a.

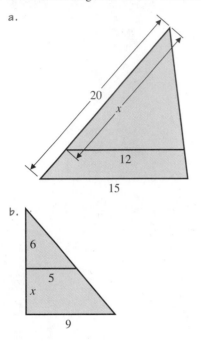

b.

19. Pavel wants to raise his grade point average (GPA). He currently has 60 credits, with a GPA of 3.40. How many credits of As, at 4.00 points per credit, does he need to raise his GPA to a 3.50?

20. The base unit of a portable telephone must be connected to an electrical outlet. The small box plugged into the outlet is a transformer that changes 120-volt ($V_1$) electricity into 12-volt ($V_2$) electricity. The output current ($I_2$) from the transformer is 100 mA. What is the input current ($I_1$) when the voltage and current are related by the formula $\dfrac{V_1}{V_2} = \dfrac{I_2}{I_1}$?

Are voltage ($V$) and current ($I$) proportional in this problem setting? Why or why not?

# Systems of Equations and Inequalities

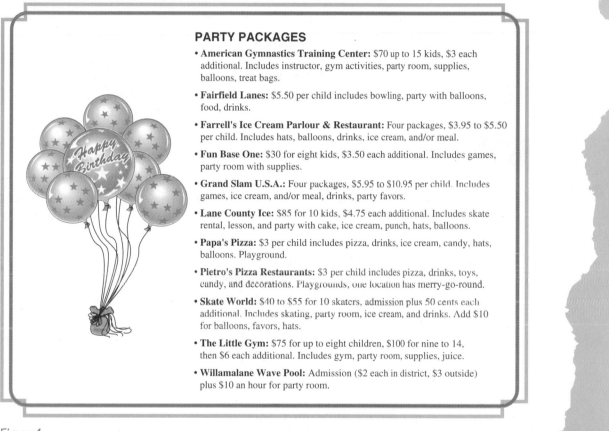

**PARTY PACKAGES**

- **American Gymnastics Training Center:** $70 up to 15 kids, $3 each additional. Includes instructor, gym activities, party room, supplies, balloons, treat bags.

- **Fairfield Lanes:** $5.50 per child includes bowling, party with balloons, food, drinks.

- **Farrell's Ice Cream Parlour & Restaurant:** Four packages, $3.95 to $5.50 per child. Includes hats, balloons, drinks, ice cream, and/or meal.

- **Fun Base One:** $30 for eight kids, $3.50 each additional. Includes games, party room with supplies.

- **Grand Slam U.S.A.:** Four packages, $5.95 to $10.95 per child. Includes games, ice cream, and/or meal, drinks, party favors.

- **Lane County Ice:** $85 for 10 kids, $4.75 each additional. Includes skate rental, lesson, and party with cake, ice cream, punch, hats, balloons.

- **Papa's Pizza:** $3 per child includes pizza, drinks, ice cream, candy, hats, balloons. Playground.

- **Pietro's Pizza Restaurants:** $3 per child includes pizza, drinks, toys, candy, and decorations. Playgrounds, one location has merry-go-round.

- **Skate World:** $40 to $55 for 10 skaters, admission plus 50 cents each additional. Includes skating, party room, ice cream, and drinks. Add $10 for balloons, favors, hats.

- **The Little Gym:** $75 for up to eight children, $100 for nine to 14, then $6 each additional. Includes gym, party room, supplies, juice.

- **Willamalane Wave Pool:** Admission ($2 each in district, $3 outside) plus $10 an hour for party room.

*Figure 1*

**WHETHER WE ARE PLANNING** a birthday party (with options as shown in Figure 1), a wedding reception, or a grand opening of a new business, we are faced with a number of questions. How much will the party cost? What is the cost per person? What choices do we have within a given budget? For how many guests will two different choices have the same cost?

(continued)

215

The techniques in this chapter provide a number of ways to answer these questions. In this chapter we review solving problems with guess-and-check tables. We use tables to identify relationships and to write equations and systems of equations. We extend our earlier skills in graphing to a system of two equations. We find solutions to systems of equations from a graph and then examine two symbolic techniques for solving these systems. Finally, we use inequalities in problem situations where the solutions may be intervals of inputs rather than a single number.

## 5.1    SOLVING SYSTEMS OF EQUATIONS WITH TABLES

**OBJECTIVES**    Solve word problems by using guess-and-check tables. ♦ Solve word problems by using tables to build equations. ♦ Translate word problems into equations using both one and two variables.

**WARM-UP**
1. What is the total amount of money in 20 one-dollar bills, 15 five-dollar bills, 5 ten-dollar bills, and 10 twenty-dollar bills?
2. Write eight-fifths as a decimal.
3. Write five-eighths as a decimal.
4. Write the formula for the perimeter of a rectangle.
5. Write six fractions that simplify to $\frac{8}{5}$.    ♦

In this section we use tables to organize a guess-and-check process and to build equations. Guess-and-check is a problem-solving technique. Guess-and-check frequently leads to a solution. When it fails to give a solution, the work provides an estimate of the solution and an understanding of the relationships within the problem.

### SOLUTIONS BY GUESS-AND-CHECK

A good guess-and-check process follows our four problem-solving steps:

We move toward *understanding* by reading carefully.

We *plan* by considering what might be reasonable inputs and preparing a table to record our guesses.

We *carry out the plan* by working through the problem with the chosen input.

We *check* by comparing the result with the conditions or requirements of the original problem.

We use a table to organize our steps and record our guesses so that we learn from each guess. *Each row in the table represents a guess.*

E X A M P L E  *1*   **Party packages**   Skate World costs $55 for up to 10 persons and $6 for each additional person. Lane County Ice costs $85 for up to 10 persons and $4.75 for each additional person. For what number of people will the cost be the same for a party at these two locations?

SOLUTION   We build Table 1 with columns for the number of people and the costs at the two locations. Observe that we show *how to find* the total cost in the columns, rather than just the total cost. This will help us build equations in later problems.

| Persons | Skate World Cost ($) | Lane County Ice Cost ($) |
|---|---|---|
| 10 | 55 | 85 |
| 20 | $55 + 6(20 - 10) = 115$ | $85 + 4.75(20 - 10) = 132.50$ |
| 30 | $55 + 6(30 - 10) = 175$ | $85 + 4.75(30 - 10) = 180$ |
| 40 | $55 + 6(40 - 10) = 235$ | $85 + 4.75(40 - 10) = 227.50$ |
| 35 | $55 + 6(35 - 10) = 205$ | $85 + 4.75(35 - 10) = 203.75$ |
| 34 | $55 + 6(34 - 10) = 199$ | $85 + 4.75(34 - 10) = 199$ |

*Table 1 Party Costs*

For 10 persons, the costs at the two locations are clearly not the same. Lane County Ice costs $30 more. We guess 20 persons and complete the row. The costs are still different but are less than $20 apart. At 30 persons the costs are within $5. At 40 persons Lane County Ice is less expensive, so we try a lower guess. For 35 persons, Lane County Ice is still cheaper but by less than $2. We guess 34. Now the two locations have the same cost: $199.

✦ ✦ ✦ ✦ ✦ ✦ ✦ ✦ ✦ ✦ ✦ ✦ ✦ ✦ ✦ ✦ ✦ ✦ ✦ ✦ ✦ ✦ ✦ ✦ ✦ ✦ ✦ ✦ ✦ ✦ ✦ ✦ ✦ ✦ ✦ ✦ ✦

Our next example is a geometry puzzle. We will solve it symbolically in Section 5.3.

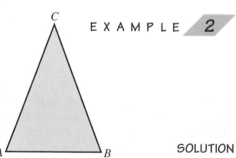

*Figure 2*

E X A M P L E  *2*   **Isosceles triangle**   If one angle of an isosceles triangle is 15° smaller than another angle, what are the three angle measures?

***Geometry notes***: An isosceles triangle has two equal sides with equal angles opposite these sides. Figure 2 shows an isosceles triangle with $AC = BC$ and angle $A$ = angle $B$. The sum of the measures of all three angles of the triangle is 180°.

SOLUTION   We build a table with a column for each angle measure and a column for the angle sum. The measures of the three angles should add to 180°. For a starting guess we suppose all three angles are equal. The angles would be 180° divided by 3, or 60° each. We assume angle $C$ is 15° smaller than angles $A$ and $B$ and fill in the first row of Table 2.

Because our first guess gave too small a total, we increase the second guess to 70°. Successive guesses should be slightly higher or slightly lower than the prior guess, depending on the total. In three guesses we conclude that angles $A$ and $B$ are each 65° and angle $C$ is 50°.

We assumed that angle $C$ is 15° smaller than the two equal angles, $A$ and $B$. We could have assumed that angles $A$ and $B$ are the smaller angles. This creates another solution and is left as an exercise.

| Angle $A$ | Angle $B$ | Angle $C$ | Total Degrees |
|---|---|---|---|
| 60° | 60° | 45° | 165° |
| 70° | 70° | 55° | 195° |
| 65° | 65° | 50° | 180° |

*Table 2 Isosceles Triangle*

✦ ✦ ✦ ✦ ✦ ✦ ✦ ✦ ✦ ✦ ✦ ✦ ✦ ✦ ✦ ✦ ✦ ✦ ✦ ✦ ✦ ✦ ✦ ✦ ✦ ✦ ✦ ✦ ✦ ✦ ✦ ✦ ✦ ✦ ✦ ✦ ✦

Example 3 describes a business decision.

**EXAMPLE 3**

**Onion bags**   A produce distributor supplies fresh fruit and vegetables to grocery stores. The distributor receives a shipment of 4050 pounds of onions and needs to bag them for retail sale. Past history shows that stores sell four times as many 5-pound bags of onions as 10-pound bags. Set up a table showing the number of each weight bagged and the pounds of onions packaged. Determine how many bags of each weight will use up the 4050 pounds.

**SOLUTION**   In Table 3 we start with 25 bags of the 10-pound size. To fit the third sentence of the problem statement, we enter four times 25 bags (100) as the number of 5-pound bags. The total weight is too low, so we double the number of bags. The weight is still too low, so we do this again. As we near 4000 total pounds, we reduce the change in our guesses.

| Quantity (10# bags) | Value (weight) | Item Total | Quantity (5# bags) | Value (weight) | Item Total | Total Pounds |
|---|---|---|---|---|---|---|
| 25 | 10 | 250 | 100 | 5 | 500 | 750 |
| 50 | 10 | 500 | 200 | 5 | 1000 | 1500 |
| 100 | 10 | 1000 | 400 | 5 | 2000 | 3000 |
| 150 | 10 | 1500 | 600 | 5 | 3000 | 4500 |
| 140 | 10 | 1400 | 560 | 5 | 2800 | 4200 |
| 135 | 10 | 1350 | 540 | 5 | 2700 | 4050 |

**Table 3**  *Onion Bags*

◆ ◆ ◆ ◆ ◆ ◆ ◆ ◆ ◆ ◆ ◆ ◆ ◆ ◆ ◆ ◆ ◆ ◆ ◆ ◆ ◆ ◆ ◆ ◆ ◆ ◆ ◆ ◆ ◆ ◆ ◆ ◆ ◆ ◆ ◆

An effective guess-and-check strategy finds the answer to many word-problem situations. In the next examples the solutions may or may not be whole numbers. We start with the guess-and-check table strategy, observe patterns, and then write equations. You will learn techniques for solving the equations in Sections 5.2, 5.3, and 5.4.

**FROM TABLES TO EQUATIONS**

**EXAMPLE 4**

**Party packages, continued**   For its party package, the American Gymnastics Training Center charges $70 for up to 15 children and $3 for each additional child. Farrell's Ice Cream Parlour and Restaurant charges $3.95 per child. Use a table to build equations to solve for the number of children for which parties at both locations will cost the same.

**SOLUTION**   We first guess 15 children, as shown in the first row of Table 4. This guess shows that Farrell's costs less for 15 (or fewer) children. The costs are quite different, so we increase our guess first to 20 and then 25. With 25 children the costs are almost equal. We could continue guessing, but a pattern has emerged in the table that allows us to write equations.

If we let $x$ = number of children and $y$ = cost of the party, the Gymnastics Center cost is

$$y = \$70 + \$3(x - 15)$$

and Farrell's cost is

$$y = \$3.95x$$

This gives a system of two equations in two variables, which we will solve in Section 5.3.

| Persons | American Gymnastics | Farrell's |
|---------|---------------------|-----------|
| 15 | 70 | $3.95(15) = 59.25$ |
| 20 | $70 + 3(20 - 15) = 85$ | $3.95(20) = 79.00$ |
| 25 | $70 + 3(25 - 15) = 100$ | $3.95(25) = 98.75$ |
| $x$ | $70 + 3(x - 15)$ | $3.95x$ |

*Table 4 Party Costs*

◆ ◆ ◆ ◆ ◆ ◆ ◆ ◆ ◆ ◆ ◆ ◆ ◆ ◆ ◆ ◆ ◆ ◆ ◆ ◆ ◆ ◆ ◆ ◆ ◆ ◆ ◆ ◆ ◆ ◆ ◆ ◆ ◆ ◆

A **system of equations** *is a set of two or more equations that are to be solved for the values of the variables making all the equations true.* It is possible for there to be no such number, in which case we say *the system has no real number solutions.* It is often easier to write a system of equations for a problem situation than to write a single equation. The next example is most easily described with two equations.

**EXAMPLE 5**

**Bank deposit**    Sesha sealed the bank deposit bag and then realized she had forgotten to record the coins on the deposit slip. She recalled that her 5-year-old daughter had counted 68 coins altogether. The total amount of money was $358.55. Because Sesha kept all the pennies and nickels for change and there were no half dollars, the bank bag contains only dimes and quarters. The unfinished deposit slip is shown in Figure 3.

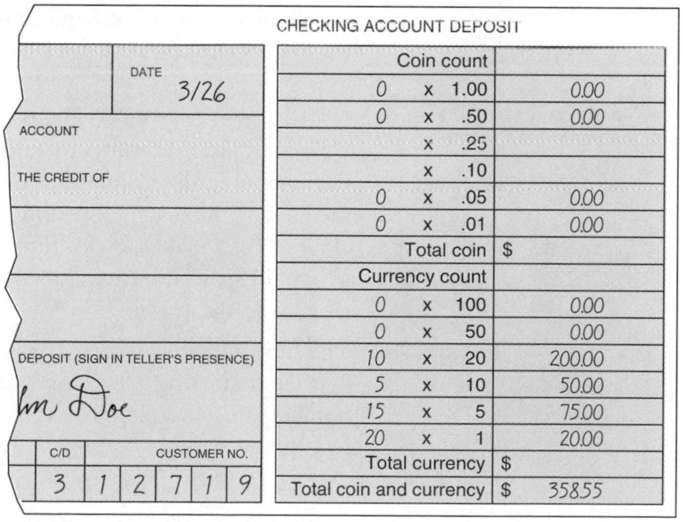

| CHECKING ACCOUNT DEPOSIT | | |
|---|---|---|
| DATE **3/26** | Coin count | |
| | 0   x   1.00 | 0.00 |
| | 0   x   .50 | 0.00 |
| ACCOUNT | x   .25 | |
| THE CREDIT OF | x   .10 | |
| | 0   x   .05 | 0.00 |
| | 0   x   .01 | 0.00 |
| | Total coin | $ |
| | Currency count | |
| | 0   x   100 | 0.00 |
| | 0   x   50 | 0.00 |
| DEPOSIT (SIGN IN TELLER'S PRESENCE) | 10   x   20 | 200.00 |
| *Jm Doe* | 5   x   10 | 50.00 |
| | 15   x   5 | 75.00 |
| | 20   x   1 | 20.00 |
| C/D   CUSTOMER NO. | Total currency | $ |
| 3  1  2  7  1  9 | Total coin and currency | $   358.55 |

*Figure 3*

**a.** What is the total value of the dimes and quarters?

**b.** Set up a table showing the possible numbers of dimes and quarters.

**c.** Find an equation that describes the quantities of dimes and quarters.

**d.** Find an equation that describes the values of the dimes and quarters in terms of the numbers of dimes and quarters.

**SOLUTION**    **a.** The dimes and quarters add to $13.55 because that is the difference between the sum of the paper currency, $345, and the total deposit, $358.55.

**b.** We start the guessing strategy in Table 5 by calculating the value for all dimes and then for all quarters. The third guess is half of each, because the $13.55 total is between $6.80 and $17.00.

| Quantity (dimes) | Value | Item Total | Quantity (quarters) | Value | Item Total | Total Money Value |
|---|---|---|---|---|---|---|
| 68 | 0.10 | 6.80 | 0 | 0.25 | 0 | 6.80 |
| 0 | 0.10 | 0 | 68 | 0.25 | 17.00 | 17.00 |
| 34 | 0.10 | 3.40 | 34 | 0.25 | 8.50 | 11.90 |
| 28 | 0.10 | 2.80 | 40 | 0.25 | 10.00 | 12.80 |
| $d$ | 0.10 | $0.10d$ | $q$ | 0.25 | $0.25q$ | $0.10d + 0.25q$ |

Table 5 *Bank Deposit*

In each row of Table 5 the total number of coins is 68. Each time the last column is filled, it is compared with the amount of money needed, $13.55. Why would we expect an odd number of quarters? Finding the solution, by guess-and-check, is left to the exercises.

**c.** In the last row we write variables $d$ and $q$ for dimes and quarters. The quantities of dimes and quarters are described by the equation $d + q = 68$.

**d.** The bottom right corner gives the total money value from dimes and quarters and is set equal to the total amount of money: $0.10d + 0.25q = 13.55$. We will solve this equation, along with $d + q = 68$, as a system of equations in Section 5.3.

◆ ◆ ◆ ◆ ◆ ◆ ◆ ◆ ◆ ◆ ◆ ◆ ◆ ◆ ◆ ◆ ◆ ◆ ◆ ◆ ◆ ◆ ◆ ◆ ◆ ◆ ◆ ◆ ◆

The art community recognizes that certain rectangles are more pleasing than others. These rectangles have ratios of sides that fit the so-called Golden Ratio of approximately 1.62 to 1.

**EXAMPLE 6**

*Golden ratio frames*    Pablo has some expensive frame material remaining from a project. He wants to calculate the dimensions needed to make a Golden Ratio rectangular frame that uses 130 centimeters of the material. He experiments with his calculator and finds that a ratio of 8 to 5 will give a decimal close to the Golden Ratio. Find the length and width of the frame. Use a table as an aid in writing equations to describe the ratios of the sides and the perimeter of the frame.

**SOLUTION**    The ratio numbers 8 and 5 are too small for the length and width, but they provide a starting place for guessing in Table 6. Using multiples of the ratio numbers in subsequent rows makes it easy to keep the proper ratio between the length and width.

| Length | Width | Ratio, Length : Width | Perimeter, $2l + 2w$ |
|---|---|---|---|
| 8 | 5 | 8:5 = 8:5 | 2(8) + 2(5) = 26 |
| 16 | 10 | 16:10 = 8:5 | 2(16) + 2(10) = 54 |
| 32 | 20 | 32:20 = 8:5 | 2(32) + 2(20) = 108 |
| 40 | 25 | 40:25 = 8:5 | 2(40) + 2(25) = 130 |
| $l$ | $w$ | $l:w = 8:5$ | $2l + 2w = 130$ |

Table 6 *Ratios and Perimeter*

A length of 40 centimeters and width of 25 centimeters give a perimeter of 130 centimeters. We obtain two equations:

**1.** The ratio of sides is $l:w = 8:5$, a proportion.

**2.** The formula for the perimeter of a rectangle gives $2l + 2w = 130$.

◆ ◆ ◆ ◆ ◆ ◆ ◆ ◆ ◆ ◆ ◆ ◆ ◆ ◆ ◆ ◆ ◆ ◆ ◆ ◆ ◆ ◆ ◆ ◆ ◆ ◆ ◆ ◆ ◆ ◆ ◆ ◆ ◆ ◆ ◆

## EXERCISES 5.1

1. Write an equation for the cost of a party at each location considered in Example 1. Let $x$ = number of persons above 10 and $y$ = total cost.

2. Compare the cost for a 2-hour rental of Willamalane Wave Pool party room, at $3 per person and $10 per hour, with a party at Papa's Pizza, at $3 per person. Will the costs ever be equal? Why?

3. Refer to Example 2. Suppose angle $C$ is 15° larger than the equal angles, $A$ and $B$. Use a guess-and-check method to find the three angle measures.

4. Suppose the produce distributor in Example 3 wants to make five times as many 5-pound bags of onions as 10-pound bags. Rework the problem to determine how many bags of each size are needed.

5. Refer to Table 5 in Example 5.

   a. Continue the guess-and-check method in Table 5 until you reach a total of $13.55.

   b. Place your numbers for $d$ (dimes) and $q$ (quarters) into the equations $d + q = 68$ and $0.10d + 0.25q = 13.55$ to show that they make both equations true.

6. Rework Example 6 with a 21 to 13 ratio of sides and 170 inches of frame material. Describe another way of solving the problem.

**In Exercises 7 to 20** *do not write equations.*

   a. **Use 4 to 5 guesses in a guess-and-check table to start a solution.**

   b. **Stop at 4 to 5 guesses, and try to write an expression for each column in the guess-and-check table. These expressions then become building blocks for the equation(s).**

**Building a good guessing strategy is more important than solving the problems at this time.**

7. Renee cuts a 10-meter hose into two lengths. One piece is 5 meters longer than the other. What is the length of each piece?

8. Jacques cuts a 12-decimeter submarine sandwich into two pieces. One piece is 6 decimeters longer than the other. What is the length of each piece?

9. Angle $A$ of a triangle is twice angle $B$. Angle $C$ is 20° more than angle $A$. What is the measure of each angle?

10. The two equal angles of an isosceles triangle are each twice the size of the third angle. What is the measure of each angle?

11. Celesta has $2.90 in quarters and nickels. She has 22 coins altogether. How many of each does she have?

12. Larry has 24 nickels and quarters. He has $2.60 altogether. How many nickels and how many quarters does he have?

13. Casey has 26 coins. He has $5.45 altogether. If he has only dimes and quarters, how many of each does he have?

14. Nancy has $3.90. She has only dimes and quarters. She has 27 coins altogether. How many of each does she have?

15. Marielena spent $620 on a watch, locket, and chain. She paid $20 more for the locket than for the chain. She paid twice as much for the watch as for the locket. How much did she pay for each?

16. Katreen earns $1375 one summer. She earns twice as much mowing lawns as shopping for the elderly. She earns $500 more scraping old paint on houses than mowing lawns. How much does she earn at each job?

17. The ratio of width to length of a fencing mat (the piste, see the figure) is 1 to 7. The perimeter of this rectangular surface is 32 meters. Find the width and length.

18. The ratio of length to width of a field hockey playing area is 5 to 3. The perimeter of the field is 320 yards. Find the length and width.

19. A party at Fun Base One costs $30 for 8 children and $3.50 for each additional child. A party at Papa's Pizza costs $3.00 per child. For how many children will the cost of a party at the two locations be equal?

20. Lane County Ice costs $85 for 10 children and $4.75 for each additional child. Grand Slam U.S.A. charges $5.95 per child for its basic party. For what number of children will the party costs be equal?

**Exercises 21 to 34 repeat Exercises 7 to 20. This time write two types of equations for each exercise. First use one variable and build one equation. Then use two (or three) variables and build a system of two (or three) equations. Some hints are provided.** *Do not solve the equations.*

21. Renee cuts a 10-meter hose into two lengths. One piece is 5 meters longer than the other. What is the length of each piece?

    a. Let $x$ = smaller and $10 - x$ = larger.

    b. Let $A$ = smaller and $B$ = larger.

22. Jacques cuts a 12-decimeter submarine sandwich into two pieces. One piece is 6 decimeters longer than the other. What is the length of each piece?

    a. Let $x$ = smaller and $12 - x$ = larger.

    b. Let $A$ = smaller and $B$ = larger.

23. Angle $A$ in a triangle is twice angle $B$. Angle $C$ is 20° more than angle $A$. What is the measure of each angle?

    a. Let $x$ = angle $B$; describe $A$ and $C$ using $x$.

    b. Let $A$, $B$, and $C$ be the variables.

24. The two equal angles of an isosceles triangle are each twice the size of the third angle. What is the measure of each angle?

    a. Let $x$ = one of the equal angles.

    b. Let $A$, $B$, and $C$ be the variables.

25. Celesta has $2.90 in quarters and nickels. She has 22 coins altogether. How many of each does she have?

    a. Let $x$ = nickels and $22 - x$ = quarters.

    b. Let $n$ = nickels and $q$ = quarters.

26. Larry has 24 nickels and quarters. He has $2.60 altogether. How many nickels and how many quarters does he have?

    a. Let $x$ = nickels and $24 - x$ = quarters.

    b. Let $n$ = nickels and $q$ = quarters.

27. Casey has 26 coins. He has $5.45 altogether. If he has only dimes and quarters, how many of each does he have?

    a. Let $x$ = dimes and $26 - x$ = quarters. Why does $26 - x$ describe the number of quarters?

    b. Let $d$ = dimes and $q$ = quarters.

28. Nancy has $3.90. She has only dimes and quarters. She has 27 coins altogether. How many of each does she have?

    a. Let $x$ = dimes and $27 - x$ = quarters. Why does $27 - x$ describe the number of quarters?

    b. Let $d$ = dimes and $q$ = quarters.

29. Marielena spent $620 on a watch, locket, and chain. She paid $20 more for the locket than for the chain. She paid twice as much for the watch as for the locket. How much did she pay for each?

    a. Let $x$ = locket.

    b. Let $L$, $W$, and $C$ be the variables.

30. Katreen earns $1375 one summer. She earns twice as much mowing lawns as shopping for the elderly. She earns $500 more scraping old paint on houses than mowing lawns. How much does she earn at each job?

    a. Let $x$ = mowing.

    b. Let $M$ (mowing), $S$ (shopping), and $P$ (scraping) be the variables.

31. The ratio of width to length of a fencing mat is 1 to 7. The perimeter of this rectangular surface is 32 meters. Find the width and length.

    a. Let $x$ = width and $7x$ = length.

    b. Let $w$ = width and $l$ = length.

32. The ratio of length to width of a field hockey playing area is 5 to 3. The perimeter of the field is 320 yards. Find the length and width.

    a. Let $x$ = width and $\frac{5}{3}x$ = length.

    b. Let $w$ = width and $l$ = length.

33. A party at Fun Base One costs $30 for 8 children and $3.50 for each additional child. A party at Papa's Pizza costs $3.00 per child. For how many children will the cost of a party at the two locations be equal?

    a. Let $x$ = number of children.

    b. Let $x$ = number of children and $y$ = total cost.

34. Lane County Ice costs $85 for ten children and $4.75 for each additional child. Grand Slam U.S.A. charges $5.95 per child for its basic party. For what number of children will the party costs be equal?

    a. Let $x$ = number of children.

    b. Let $x$ = number of children and $y$ = total cost.

35. The number of equations we write to help solve a problem is related to the number of variables we use. Look for a pattern in the examples and exercises, and complete these statements:

    a. When we use one variable, we need _____ equation(s).

    b. When we use two variables, we need _____ equation(s).

    c. When we use three variables, we need _____ equation(s).

*PROJECTS*

36. *Using Manipulatives to Solve Problems*

    **Part I.** One day Mr. McFadden decides to count his farm animals. He counts strangely and reports 10 heads and 28 feet. His animals consist of cows and ducks. How many of each does Mr. McFadden have? Use 10 coins, rubber bands, or buttons for heads. Use 28 toothpicks, paper clips, safety pins, or cotton swabs for feet. Build the animals.

    a. Is it possible to have 10 heads and 24 feet?

    b. Is it possible to have 10 heads and 30 feet?

    c. What is the largest number of feet possible with the 10 heads?

d. What is the smallest number of feet possible with the 10 heads?

e. What patterns do you observe if you organize your data into a table? Use the following table, adding more rows or columns as needed.

| Ducks | Cows | Heads | Feet |
|-------|------|-------|------|
|       |      |       |      |
|       |      |       |      |
|       |      |       |      |
|       |      |       |      |

**Part II.** Mr. McFadden's neighbor is Mr. Schaaf. Mr. Schaaf counts his animals the same way. He reports 12 heads and 28 feet. How many cows and ducks does Mr. Schaaf have?

f. What is the largest number of heads possible with 28 feet?

g. What is the smallest number of heads possible with 28 feet?

h. Organize your results in the table below to look for patterns. Add more rows or columns as needed.

| Ducks | Cows | Heads | Feet |
|-------|------|-------|------|
|       |      |       |      |
|       |      |       |      |
|       |      |       |      |
|       |      |       |      |

37. ***Building Shapes.*** Suppose we want to build rectangular shapes on a tabletop by laying out paper clips. The length of the rectangle is to be twice as many paper clips as the width.

a. What are the dimensions of the smallest rectangle possible? How many paper clips did it take? What is its area in square units?

b. What is the next smallest rectangle? How many paper clips did it take? What is its area?

c. Summarize the answers to the questions in the table, and guess about the next two rectangles in the pattern.

| Length | Width | Perimeter | Area |
|--------|-------|-----------|------|
|        |       |           |      |
|        |       |           |      |
|        |       |           |      |
|        |       |           |      |

d. How big a rectangle with length twice the width could we build with 100 paper clips?

38. ***Golden Ratio in Rectangles***

a. Measure 15 different rectangular pictures, frames, or posters. Make a list of your measurements, describing each rectangle (store-bought picture frame, poster of such-and-such musical group, etc.).

b. Enter the lengths and widths in a table.

c. Calculate the ratio of the sides of each rectangle by dividing the longer side by the shorter side.

d. Indicate which rectangles are close to a Golden Ratio.

---

## 5.2 SOLVING SYSTEMS OF EQUATIONS WITH GRAPHS

**OBJECTIVES**  Solve equations by graphing. ♦ Identify the solution as the point of intersection of two graphs. ♦ Identify a system of equations. ♦ Identify parallel lines by slope. ♦ Determine when a system has no solution. ♦ Identify coincident lines. ♦ Determine when a system has an infinite number of solutions. ♦ Solve nonlinear systems by graphing.

**WARM-UP**  Find the slope of the line connecting the two points given. Then find the equation of each line, using $y = mx + b$.

1. (0, 0) and (3, 4)

2. (6, 0) and (0, 4)

3. (0, 8) and (3, 0)

♦

G raphing has three particularly valuable uses. First, by plotting points and estimating lines or curves, we can find solutions to problem situations without using any equations. This is important because in many applications the equation is not known. Second, in mathematics courses such as trigonometry, graphing allows us to solve equations that cannot be solved using symbolic techniques. The third use is our focus in this section: Graphing gives a method of finding the numerical solution to equations and lends a visual meaning to the solution.

The **point of intersection** of two graphs identifies a solution to the two equations. *Because every point on the graph of an equation makes the equation true, the point of intersection makes both equations true.* In a graph drawn by hand the point of intersection can be located only with limited accuracy. Graphing calculators and computers permit us to locate coordinates of the point of intersection that are correct to many decimal places.

### SOLUTIONS BY GRAPHING

In Example 1 we write and solve a system of equations, and we use intervals and inequalities to describe the results.

E X A M P L E    *1*

Car rental    Two competing car rental agencies offer a similar car. Ava's Rent-A-Car charges $0.10 a mile plus $40 per day. Herr's Rent-A-Car charges $0.20 a mile plus $20 per day.

**a.** Write a system of equations to describe the rental costs for *one* day, where $x$ is the number of miles driven.

**b.** Match each of your equations to its graph in Figure 4. Use the graph to determine the solution to the system.

**c.** What is the meaning of the point of intersection in terms of the problem situation?

**d.** For what mileage will Ava's car be cheaper? Describe the $x$ inputs with an interval and with an inequality.

**e.** For what mileage will Herr's be cheaper? Describe the $x$ inputs with an interval and with an inequality.

SOLUTION    **a.** The two rental costs for one day are

$$\text{Ava:} \quad y = 0.10x + 40$$

$$\text{Herr:} \quad y = 0.20x + 20$$

**b.** The dashed line (2) is Ava's graph. The solid line (1) is Herr's graph. The graph has miles on its horizontal axis and total dollar cost on its vertical axis. The solution to the system is the point of intersection of the graphs, (200, 60).

**c.** The point of intersection (200, 60) gives the number of miles, $x = 200$, and the total dollar cost, $y = 60$, at which the two agencies' charges are the same on a one-day rental.

**d.** Ava's will be better for longer trips, with mileage in the interval $(200, +\infty)$, or $x > 200$. Of course, infinity $(+\infty)$ is not a reasonable mileage because there is a limit on the number of miles that can be driven in a day.

**e.** Herr's will be better for short trips, with mileage in the interval $[0, 200)$, or $0 \le x < 200$.

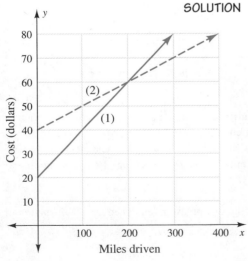

*Figure 4*

◆ ◆ ◆ ◆ ◆ ◆ ◆ ◆ ◆ ◆ ◆ ◆ ◆ ◆ ◆ ◆ ◆ ◆ ◆ ◆ ◆ ◆ ◆ ◆ ◆ ◆ ◆ ◆ ◆ ◆ ◆ ◆ ◆ ◆ ◆ ◆ ◆ ◆

**Graphing Calculator Technique**

Let X = miles traveled and Y = total dollar cost for Example 1. Enter $Y_1 = 0.20X + 20$ and $Y_2 = 0.10X + 40$. Set the window with X in the interval [0, 400], scale 100, and Y in the interval [0, 80], scale 10. Graph. Trace to the point of intersection, which shows $X \approx 200$ and $Y \approx 60$.

In Example 2 the solution is not whole numbers and the intersection is not as clear.

E X A M P L E  **2**

*Wage options*    The Appliance Works is reviewing employee wages. Management is considering two plans, both based on a fixed salary plus a percent commission on retail sales. Plan A is $300 salary plus 6% of sales, and Plan B is $150 salary plus 10% of sales.

**a.** Which category, total income or sales, should go on the horizontal axis? Why?

**b.** Write a system of equations for Plan A and Plan B.

**c.** Graph the system of equations.

**d.** For what level of sales will the plans be equal?

**e.** Under what conditions would an employee prefer Plan A?

**f.** What type of employee would prefer Plan B?

SOLUTION    **a.** Because total income depends on the amount of sales, sales should be $x$ on the horizontal axis.

**b.** Plan A:   $y = 300 + 0.06x$
Plan B:   $y = 150 + 0.10x$

**c.** The equations are graphed in Figure 5.

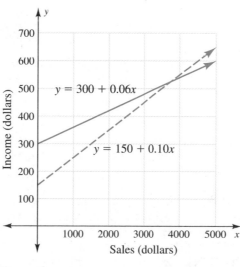

Figure 5

**d.** The lines intersect where the plans give the same total income, at or near (3750, 525).

**e.** Plan A yields more money with low sales. If business is slow or a salesperson is new, Plan A might be better.

**f.** A good salesperson would do well with Plan B.

◆ ◆ ◆ ◆ ◆ ◆ ◆ ◆ ◆ ◆ ◆ ◆ ◆ ◆ ◆ ◆ ◆ ◆ ◆ ◆ ◆ ◆ ◆ ◆ ◆ ◆ ◆ ◆ ◆ ◆ ◆ ◆ ◆ ◆ ◆ ◆ ◆ ◆ ◆ ◆

PARALLEL LINES, COINCIDENT LINES, AND NONLINEAR EQUATIONS

E X A M P L E    **3**

*More wage options*    Suppose Wage Plan A is given by $y = 300 + 0.06x$ and Wage Plan C is given by $y = 150 + 0.06x$.

**a.** Graph the wage plans.

**b.** Find the point of intersection.

**c.** Compare the two plans.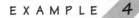

SOLUTION    **a.** The equations are graphed in Figure 6.

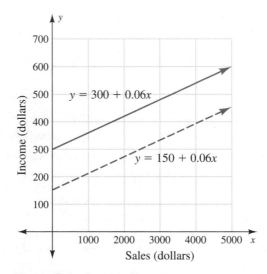

*Figure 6*

**b.** The two plans give parallel lines with no point of intersection; hence there is no real number solution to the system. The solution set is empty: { }.

**c.** The $150 salary plus 6% of sales would never be better than the $300 salary plus 6% of sales.

♦ ♦ ♦ ♦ ♦ ♦ ♦ ♦ ♦ ♦ ♦ ♦ ♦ ♦ ♦ ♦ ♦ ♦ ♦ ♦ ♦ ♦ ♦ ♦ ♦ ♦ ♦ ♦ ♦ ♦ ♦ ♦ ♦

In the next two examples the systems of equations contain a linear and a nonlinear equation.

E X A M P L E    **4**

Find the length and width of a rectangle with area 48 square centimeters and perimeter 38 centimeters. Write the problem in equations, and solve it by graphing.

SOLUTION    Let $x$ be the length and $y$ be the width. The area is $xy = 48$. The perimeter is $2x + 2y = 38$, or $x + y = 19$. The graphs are shown in Figure 7. The intersections are at (16, 3) and (3, 16). Thus the length and width are 16 centimeters and 3 centimeters. The mathematical results do not specify that length be the larger number, as is customary in everyday use.

*Check*: Area:                    $16(3) = 48 \text{ cm}^2$
        Perimeter:   $2(16) + 2(3) = 38 \text{ cm}$   ✔

The nonlinear graph, or curve, in Figure 7 is a *rectangular hyperbola*.

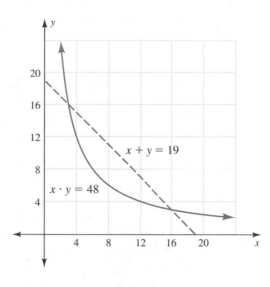

Figure 7

◆ ◆ ◆ ◆ ◆ ◆ ◆ ◆ ◆ ◆ ◆ ◆ ◆ ◆ ◆ ◆ ◆ ◆ ◆ ◆ ◆ ◆ ◆ ◆ ◆ ◆ ◆ ◆ ◆ ◆ ◆ ◆

E X A M P L E  **5**

**More party packages**   Skate World offers a Saturday afternoon party package: $45 for the first 10 people and $5 for each additional person. The Willamalane Wave Pool charges $30 for a three-hour rental of its party room and $3 per person. For how many people will the two locations cost the same? Make a table and graph to answer this question. What can we conclude about the costs?

SOLUTION   The first row of Table 7 shows the cost for 5 people to be $45 at each location. We would normally quit after this first guess, but since a five-person party might not be much of a party, we check some larger numbers to see what happens after the charge per person starts for Skate World. Surprisingly, there seems to be another intersection between 16 and 20 people.

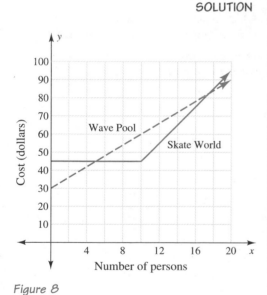

Figure 8

| Persons, $x$ | Skate World | Wave Pool |
|---|---|---|
| 5 | 45 | $3(5) + 30 = 45$ |
| 10 | 45 | $3(10) + 30 = 60$ |
| 15 | $45 + 5(15 - 10) = 65$ | $3(15) + 30 = 75$ |
| 20 | $45 + 5(20 - 10) = 95$ | $3(20) + 30 = 90$ |
| For $x \leq 10$ | 45 | $3x + 30$ |
| For $x > 10$ | $45 + 5(x - 10)$ | $3x + 30$ |

Table 7 *Party Costs*

The graph in Figure 8 shows two points of intersection: (5, 45) and about (17, 80). The wave pool is cheaper for fewer than 5 people and 18 or more people. Skate World is cheaper for $5 < x \leq 17$ people.

◆ ◆ ◆ ◆ ◆ ◆ ◆ ◆ ◆ ◆ ◆ ◆ ◆ ◆ ◆ ◆ ◆ ◆ ◆ ◆ ◆ ◆ ◆ ◆ ◆ ◆ ◆ ◆ ◆ ◆ ◆ ◆ ◆

The surprising results in Example 5 are due to the conditional nature of the Skate World prices. This realistic example illustrates why it is important that we graph conditional equations.

Two lines that have the same equation are **coincident lines**. Because coincident lines have all points in common, we say that as a system *they have an infinite number of solutions*. Coincident lines play an important role in applications. Suppose the management of the Appliance Works in Example 3 offers employees the package described by $2y = 600 + 0.12x$, instead of $y = 300 + 0.06x$. How should they react? If we graphed the two equations separately, we would end up with the same line. This means the equations are the same! The first offer is the same as the second with every number multiplied by 2.

**Summary**

In graphing a system of two linear equations we have three possible outcomes:

1. The equations describe lines that intersect in exactly one point. The system of equations has exactly one solution: the point of intersection.
2. The equations describe lines that are parallel. The graphs will have no point of intersection; hence the system will have no real number solution.
3. The equations describe the same line. The graphs are coincident. The system will have an infinite number of solutions, because every point on the graph of the first equation will make the second equation true.

In a graph of any system of two or more equations, the coordinates of the points of intersection are the numbers that make all of the equations true.

## EXERCISES 5.2

1. Refer to Example 1. Write a sentence to explain the meaning of the *y*-intercepts of the graphs.

2. Refer to Example 1. Write a sentence to explain the meaning of the slopes of the lines.

3. In Example 2, for what sales would Plan A produce more income? Describe the inputs, *x*, with an interval and with an inequality.

4. In Example 2, for what sales would Plan B produce more income? Describe the inputs, *x*, with an interval and with an inequality.

5. Refer to Example 2. Write a sentence to explain the meaning of the slopes of the lines.

6. Refer to Example 2. If the graphs were extended to the left, would the *x*-intercepts have any meaning? Explain in a sentence.

7. What do we know about the slopes of parallel lines? Use a full sentence.

8. In the equation $y = mx + b$, which letter represents slope?

9. Refer to Example 4. For what inputs, *x*, is the graph of $x + y = 19$ above the graph of $x \cdot y = 48$? Describe the inputs, using an interval and an inequality.

10. Refer to Example 4. For what inputs, *x*, is the graph of $x \cdot y = 48$ above the line $x + y = 19$? Describe the inputs, using intervals and inequalities.

11. In Example 4, does the curve $xy = 48$ have an *x*-intercept?

12. In Example 4, does the curve $xy = 48$ have a *y*-intercept?

13. In Example 5, why do we subtract 10 from the number of persons before multiplying by $5?

14. In Example 5, how would the graph for the wave pool change if the party room were rented for 2 hours instead of 3?

**Which of the pairs of equations in Exercises 15 to 20 form parallel lines? (Graphing may not be necessary.)**

15. $2x + 2y = 100$, $y = 20 - x$

16. $D = t + 55$, $D = t + 25$

17. $D = 55t$, $D = 25t$

18. $y = 10 - x$, $y = 5 + x$

19. $P = 4x$, $P = \frac{1}{4}(12 + 16x)$

20. $y = 300 - 60x$, $y = 60 - 300x$

**In Exercises 21 to 24, which of the pairs of equations form coincident lines?**

21. $y + 60x = 300$, $y = 60 - 300x$

22. $y + x = 20$, $2x = 40 - 2y$

23. $C = x + 0.15x$, $C = 1.15x$

24. $y = 150 + 0.10x$, $y = 150x + 0.10$

**In economics, the *equilibrium point E* is where a supply curve intersects a demand curve. Supply describes how much of a product is available; demand describes how much people want of the product. The following questions provide practice in reading and interpreting supply and demand graphs. No knowledge of economics is needed to answer the questions.**

Exercises 25 to 30 refer to the figure below. The supply line indicates how much gold the mine owners are willing to produce. The demand line indicates how much gold investors are willing to buy.

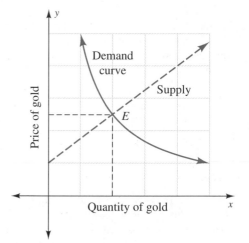

25. If the price of gold falls, what happens to the quantity of gold supplied by the mine owners?

26. If the price of gold falls, what happens to the quantity demanded by investors?

27. If the price of gold rises, what happens to the quantity of gold demanded by investors?

28. If the price of gold rises, what happens to the quantity supplied by the mine owners?

29. Is the slope of the supply graph positive, zero, or negative?

30. If the demand curve were approximated by a straight line, would its slope be positive, zero, or negative?

Exercises 31 to 38 refer to the figure below. The $D_0$ curve is demand on the day *before* Picasso died. The $D_1$ curve is demand on the day *after* Picasso died.

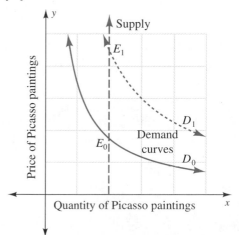

31. Does the quantity supplied change in this graph?

32. If there is no increase in the quantity available, what happens to price?

33. If demand moves upward from graph position $D_0$ to graph position $D_1$, what happens to $E$, the equilibrium point?

34. A few days later, a huge quantity of Picasso paintings are discovered in a warehouse. If the starting demand graph is $D_1$ and the ending demand graph is $D_0$, what happens to $E$, the equilibrium point?

35. As we go from $E_1$ to $E_0$, what happens to price?

36. As we go from $E_1$ to $E_0$, what happens to quantity?

37. As we go from $E_0$ to $E_1$, what happens to quantity?

38. As we go from $E_0$ to $E_1$, what happens to price?

Exercises 39 to 50 refer to the figure below, which shows two lines: revenue and cost. The revenue is earned from registration fees at a nonprofit student career conference in 1985. The costs are from printing the promotional flyer and ordering lunches for $x$ participants. In business the point of intersection of a cost graph and a revenue (income) graph is the *breakeven point, B.E.* The profit is equal to revenue minus cost.

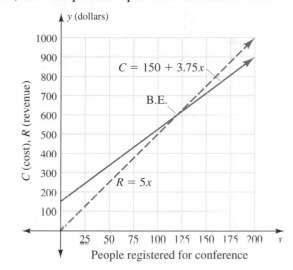

39. Suppose 100 students attend the conference.

 a. What is the total cost?    b. What is the total revenue?

40. What is the registration fee?

41. What is the cost of lunch?

42. What is the printing cost?

43. What is the profit at $x = 100$?

44. a. What is the total cost at $x = 150$?

 b. What is the revenue at $x = 150$?

 c. What is the profit at $x = 150$?

45. a. What is the total cost at $x = 200$?

 b. What is the revenue at $x = 200$?

 c. What is the profit at $x = 200$?

46. If total registration is below the breakeven point, which graph is on top?

47. If total registration is above the breakeven point, which graph is on top?

48. Estimate the breakeven point coordinates.

49. Let $5x = 150 + 3.75x$, and find the $x$-coordinate of the breakeven point.

50. What is the total cost at the breakeven point?

51. Refer to the figure below.

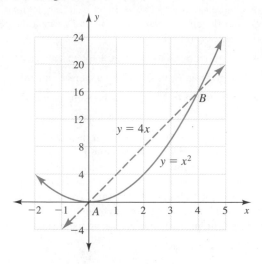

a. What are the coordinates of A and B?

b. What single equation is solved by the points A and B?

c. What system of equations is solved by the points A and B?

52. Refer to the figure below.

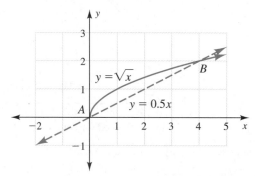

a. What are the coordinates of A and B?

b. What single equation is solved by the points A and B?

c. What system of equations is solved by the points A and B?

a. What are the coordinates of A and B?

b. What single equation is solved by the points A and B?

c. What system of equations is solved by the points A and B?

53. Graph the equations $x + y = 68$ and $0.10x + 0.25y = 13.55$ for the bank deposit problem in Example 5 of Section 5.1, with dimes as $x$ and quarters as $y$. Estimate the point of intersection.

54. In the bank deposit problem in Example 5 of Section 5.1 let dimes be $x$ and quarters be $68 - x$. What equation describes the total value of the money? Solve the equation for $x$ (dimes).

PROJECT

55. **Racing Times**

a. In a 500-mile Indy-car street race, Lyn St. James averaged 150 miles per hour for the entire race. Her main competitor, Bobby Rahal, averaged 160 miles per hour for the first 250 miles but, because of a slight brush with the safety wall, averaged 140 miles per hour for the second 250 miles of the race. Guess which driver won the race.

b. Draw a graph for Lyn with $D = 150t$. Suppose $t_1 = $ time for Bobby to drive the first 250 miles at 160 miles per hour. Calculate $t_1$. Draw a conditional graph for Bobby with $D = 160t$ for $t < t_1$ and $D = 140(t - t_1) + 250$ for $t > t_1$.

c. Assume each driver completed the 500 miles. How long did it take Lyn to finish the race? How long did it take Bobby to finish the race?

d. What is the difference in elapsed time, if any, between the drivers? Use unit analysis to change the time to seconds.

e. At what time after the race started did the winner take the lead?

---

## 5.3    SOLVING SYSTEMS OF EQUATIONS BY SUBSTITUTION

**OBJECTIVES**    Solve an equation for one variable in terms of a second variable.   ♦  Solve a system by isolating one variable and substituting.   ♦  Solve a system when two equations are in $y = mx + b$ form.   ♦  Solve a system of three equations in three variables.

**WARM-UP**    Solve for the indicated variables.

**1.** $2w + 2l = 130$ for $w$ in terms of $l$

**2.** $0.10x + 0.25y = 13.55$ for $y$ in terms of $x$

**3.** $l : w = 8 : 5$ for $l$ in terms of $w$                                    ♦

Mathematics is flexible. There are many ways of doing problems, and some work better at times than others. We have used both tables and graphs to solve some equations. Other equations have been solved just one way. For still other equations the solution cannot be easily found with either tables or graphs; often such a solution would be too time consuming.

A system of equations such as $x + y = 9$, $x + z = 1$, and $y + z = 2$ would require a complex table. It is not possible to solve the system by graphing on the coordinate plane because of the third variable, $z$. Yet it may be solved quite simply by the symbolic methods discussed in this and the next section. We now consider the substitution method of solving equations. Later we will use this method to solve the above system of three equations.

### SUBSTITUTION METHOD FOR SOLVING TWO EQUATIONS

The name **substitution** comes from the replacement process: *We replace variables with expressions or numbers.* Mathematical substitution resembles the substitution of players in a sports event or the substitution of ingredients in a recipe. The substitutions are alike in that there is a "taking out" and a "putting in" process in sports, cooking, and mathematics. The substitutions are different in that mathematics uses an equal replacement, whereas sports and cooking use a replacement that only approximates the player or ingredient removed. When we check answers to equations, for instance, we substitute the numbers for the variables in the equations.

The first step in substitution solutions is to solve an equation for one of its variables. Example 1 reviews this skill.

EXAMPLE 1

**a.** Solve $d + q = 68$ for $d$ in terms of $q$.

**b.** Solve $x + y = 9$ for $y$ in terms of $x$.

SOLUTION **a.** Subtract $q$ from both sides of $d + q = 68$: $d = 68 - q$.

**b.** Subtract $x$ from both sides of $x + y = 9$: $y = 9 - x$.

♦ ♦ ♦ ♦ ♦ ♦ ♦ ♦ ♦ ♦ ♦ ♦ ♦ ♦ ♦ ♦ ♦ ♦ ♦ ♦ ♦ ♦ ♦ ♦ ♦ ♦ ♦ ♦ ♦ ♦ ♦ ♦ ♦ ♦ ♦ ♦ ♦

In the second example we return to the bank deposit problem of Example 5 in Section 5.1 and solve the equations by substitution.

EXAMPLE 2

**Bank deposit** In Table 5 of Section 5.1 we observe two patterns. First, the total quantity of dimes and quarters is 68, which means that $d + q = 68$. Second, the total worth of the coins is \$13.55, or $0.10d + 0.25q = 13.55$.

**a.** Estimate the solution to the system from the graph in Figure 9.

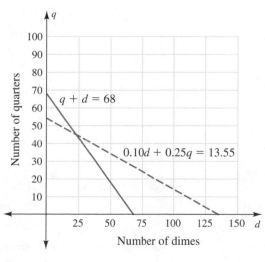

*Figure 9*

**b.** Solve the system of equations by replacing $d$ in the second equation with an equivalent expression from the first equation.

SOLUTION

**a.** The two lines intersect at approximately $d = 25$ and $q = 45$.

**b.** The two equations are

$$d + q = 68$$
$$0.10d + 0.25q = 13.55$$

Suppose we solve $d + q = 68$ for $d$:

$$d = 68 - q$$

Because $d$ and $(68 - q)$ are equal, we replace $d$ with $(68 - q)$ in the second equation and solve for $q$:

$$
\begin{aligned}
0.10(68 - q) + 0.25q &= 13.55 \\
6.80 - 0.10q + 0.25q &= 13.55 \quad \text{Distributive property} \\
6.80 + 0.15q &= 13.55 \quad \text{Add like terms.} \\
0.15q &= 6.75 \\
q &= 6.75 \div 0.15 \\
q &= 45
\end{aligned}
$$

We substitute $q = 45$ into $d = 68 - q$ to obtain $d = 23$.

***Check***: We substitute $d = 23$ and $q = 45$ into both equations:

$$23 + 45 \overset{?}{=} 68$$
$$0.10(23) + 0.25(45) \overset{?}{=} 13.55 \quad \checkmark$$

♦ ♦ ♦ ♦ ♦ ♦ ♦ ♦ ♦ ♦ ♦ ♦ ♦ ♦ ♦ ♦ ♦ ♦ ♦ ♦ ♦ ♦ ♦ ♦ ♦ ♦ ♦ ♦ ♦ ♦ ♦ ♦ ♦ ♦ ♦ ♦ ♦ ♦ ♦

One purpose of solving systems by graphing is to remind us that the solution is a coordinate point representing two numbers. A common error is to solve a system for only one of the two (or more) variables.

Examples 3 and 4 illustrate how easy it is to solve a system when one or both equations are already solved for a variable, in this case $y$. We repeat the wage option problem of Example 2 in Section 5.2 because the point of intersection was difficult to determine from the graph.

EXAMPLE **3**

SOLUTION

*Wage options*    The Appliance Works wage options are Plan A, $y = 300 + 0.06x$, and Plan B, $y = 150 + 0.10x$. Solve the system for the dollar sales amount, $x$, and the income level, $y$, that equate the two options.

Both equations have been solved for the variable $y$. We substitute $y = 300 + 0.06x$ into the second equation:

$$
\begin{aligned}
y &= 150 + 0.10x \\
300 + 0.06x &= 150 + 0.10x \\
150 + 0.06x &= 0.10x \quad \text{Subtract 150 from both sides.} \\
150 &= 0.04x \quad \text{Subtract 0.06x from both sides.} \\
x &= 3750
\end{aligned}
$$

To find $y$ we substitute $x = 3750$ into the first equation:

$$
\begin{aligned}
y &= 300 + 0.06(3750) \\
y &= 525
\end{aligned}
$$

***Check***: We substitute $x = 3750$ and $y = 525$ into both equations:

$$525 \overset{?}{=} 300 + 0.06(3750)$$
$$525 \overset{?}{=} 150 + 0.10(3750)$$

On sales of \$3750 each method gives the same income: \$525.  ✔

♦ ♦ ♦ ♦ ♦ ♦ ♦ ♦ ♦ ♦ ♦ ♦ ♦ ♦ ♦ ♦ ♦ ♦ ♦ ♦ ♦ ♦ ♦ ♦ ♦ ♦ ♦ ♦ ♦ ♦ ♦ ♦ ♦ ♦ ♦ ♦ ♦ ♦ ♦

We return to parties located at the American Gymnastics Training Center and Farrell's Ice Cream Parlour (Example 4, Section 5.1).

**EXAMPLE 4**

*Party packages* In Table 4 we found that the cost for 15 or more children at American Gymnastics is $y = 70 + 3(x - 15)$ dollars, and at Farrell's it is $y = 3.95x$ dollars. Solve this system of equations. Find the number of children that makes the costs equal, and determine the cost.

**SOLUTION** Because both equations are solved for $y$, we substitute $3.95x$ for $y$ in the Gymnastics equation:

$$
\begin{aligned}
y &= 70 + 3(x - 15) &&\text{Condition: } x \geq 15 \\
3.95x &= 70 + 3(x - 15) &&\text{Substitute.} \\
3.95x &= 70 + 3x - 45 &&\text{Distributive property} \\
3.95x - 3x &= 70 + 3x - 45 - 3x &&\text{Subtract 3x from both sides.} \\
0.95x &= 25 \\
0.95x \div 0.95 &= 25 \div 0.95 \\
x &\approx 26.3
\end{aligned}
$$

Our solution satisfies the condition $x \geq 15$.

A decimal portion of a child does not make sense, so we round down to a whole number—26 children makes the costs approximately equal. At Farrell's a party for 26 children costs \$102.70; at the Gymnastics center it costs \$103.

♦ ♦ ♦ ♦ ♦ ♦ ♦ ♦ ♦ ♦ ♦ ♦ ♦ ♦ ♦ ♦ ♦ ♦ ♦ ♦ ♦ ♦ ♦ ♦ ♦ ♦ ♦ ♦ ♦ ♦ ♦ ♦ ♦ ♦ ♦ ♦ ♦ ♦ ♦

### SUBSTITUTION METHOD FOR SOLVING THREE EQUATIONS

Calculator technology now provides easy solutions to complicated systems of three or more linear equations. The next two examples illustrate short solutions to systems of three equations.

In Example 5 each of the three equations contains a common letter. We solve two of the equations in terms of that common letter and substitute them into the third equation.

**EXAMPLE 5**

*Isosceles triangle* We return to the triangle problem in Example 2 of Section 5.1: If one angle of an isosceles triangle is 15° smaller than another, what are the three angle measures? We might have written three equations to describe the problem:

Two angles are equal: $A = B$.

One angle is 15° less than either $A$ or $B$: $C = A - 15$.

The three angles add to 180°: $A + B + C = 180$.

Solve this system of three equations by substitution.

**SOLUTION** We scan the three equations to see whether one letter appears in all three. Each equation contains an $A$, so we solve the system *in terms of A*.

We start with the equation containing all three letters. We replace variables $B$ and $C$ with their expressions *in terms of A* from the other two equations:

$$
\begin{aligned}
A + B + C &= 180 \\
A + A + (A - 15) &= 180 \\
3A - 15 &= 180 \\
3A &= 195 \\
A &= 195 \div 3 \\
A &= 65
\end{aligned}
$$

Now we use $A = 65$ in the other two equations to find $B$ and $C$:

$$B = A = 65$$
$$C = A - 15 = 65 - 15 = 50$$

*Check*:
$$65 + 65 + 50 \overset{?}{=} 180$$
$$65 \overset{?}{=} 65$$
$$50 \overset{?}{=} 65 - 15 \quad \checkmark$$

♦ ♦ ♦ ♦ ♦ ♦ ♦ ♦ ♦ ♦ ♦ ♦ ♦ ♦ ♦ ♦ ♦ ♦ ♦ ♦ ♦ ♦ ♦ ♦ ♦ ♦ ♦ ♦ ♦ ♦ ♦ ♦ ♦ ♦ ♦ ♦ ♦

At the beginning of this section we posed a problem with three equations in the variables $x$, $y$, and $z$. We now solve it.

**EXAMPLE 6**

**The introductory problem**    Solve the system

$$x + y = 9, \quad x + z = 1, \quad y + z = 2$$

**SOLUTION**    This time there is not one variable that appears in all three equations, nor is there one equation containing all three variables. Yet we again focus on one variable—say, $x$.

We solve the first two equations for $y$ and $z$ in terms of $x$:

$$y = 9 - x$$
$$z = 1 - x$$

Then we substitute both equations into the third equation:

$$y + z = 2$$
$$9 - x + 1 - x = 2$$
$$10 - 2x = 2$$
$$-2x = -8$$
$$x = 4$$

Now we substitute $x = 4$ into each of the first two equations:

$$y = 9 - x \qquad \text{and} \qquad z = 1 - x$$
$$y = 9 - 4 \qquad\qquad\qquad z = 1 - 4$$
$$y = 5 \qquad\qquad\qquad\quad z = -3$$

*Check*: We substitute $x = 4$, $y = 5$, and $z = -3$ into all three equations:

$$4 + 5 \overset{?}{=} 9$$
$$4 + (-3) \overset{?}{=} 1$$
$$5 + (-3) \overset{?}{=} 2 \quad \checkmark$$

♦ ♦ ♦ ♦ ♦ ♦ ♦ ♦ ♦ ♦ ♦ ♦ ♦ ♦ ♦ ♦ ♦ ♦ ♦ ♦ ♦ ♦ ♦ ♦ ♦ ♦ ♦ ♦ ♦ ♦ ♦ ♦ ♦ ♦ ♦ ♦ ♦

In Example 6 we might have solved two of the equations for $x$ and $y$ in terms of $z$ and substituted them into $x + y = 9$. Or, we might have solved two of the equations for $x$ and $z$ in terms of $y$ and substituted them into $z + x = 1$. These two approaches are included as exercises.

None of the examples thus far illustrate the symbolic solution of the non-linear systems encountered earlier. After additional symbolic work we will solve nonlinear systems with substitution. You will learn some of these techniques in Chapters 7 and 8, and others in subsequent courses.

## EXERCISES 5.3

Solve each of the equations in Exercises 1 to 6 for the indicated variable.

1. $L = 2W$ for $W$
2. $L = 3W$ for $W$
3. $C = 2\pi r$ for $r$
4. $D = 2r$ for $r$
5. $C = \pi d$ for $d$
6. $P = 4x$ for $x$

In Exercises 7 to 10 solve for $x$.

7. $2x + 3(2) = 12$
8. $9x - 2(6) = -3$
9. $-5x - 6(-3) = 3$
10. $7 = -2x - 13(-1)$

In Exercises 11 to 18 identify the variable with a coefficient of 1 or −1. Solve for that variable in terms of the other variable.

11. $3x + y = 4$
12. $x + 3y = 26$
13. $x - 4y = 5$
14. $2y - x = 7$
15. $5y - x = 9$
16. $x - 3y = 3$
17. $3x - y = -2$
18. $2x - y = 13$

For Exercises 19 and 20 write the equation, $y = mx + b$, for line $a$ and line $b$. Identify the intersection and substitute it into both equations to show that it is the solution to the system.

19.

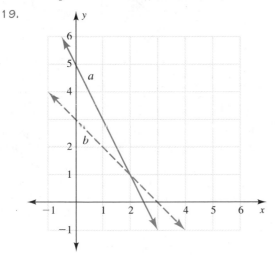

20.

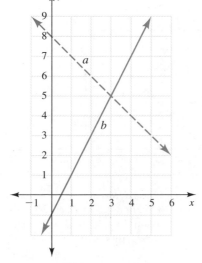

Use substitution to solve the systems of equations in Exercises 21 to 34 for all variables.

21. $y = x - 8$
    $3x + y = 4$

22. $x = 26 - 3y$
    $x - 4y = 5$

23. $5x + 5 = y$
    $y - 3x = 9$

24. $4x - 2y = 20$
    $x = 2 - y$

25. $4x + 5y = 11$
    $x = 9 + 5y$

26. $-x + 2y = 7$
    $x = 3 + 3y$

27. $2x - y = 1$
    $2y = 3x + 3$

28. $3y + x = -1$
    $2x + 6 = -5y$

29. $2x + 3y = 0$
    $3x + y = 7$

30. $y = 3x - 2$
    $y = -2x + 13$

31. $A + B + C = 180$
    $A = 2B$
    $B = 3C$

32. $A + B + C = 180$
    $A = 3B$
    $B = 2C$

33. $y = \frac{4}{3}x$
    $y = -\frac{8}{3}x + 8$

34. $y = -\frac{8}{3}x + 8$
    $y = -\frac{2}{3}x + 4$

Use substitution and any other techniques that seem appropriate to solve the "pictorial" equations in Exercises 35 to 38.

35. $\square + \square + ☺ = ♥ + ♥ + ♥ + ♥ + ♥ + ♥ + ♥$
    $\square + ♥ = ☺$
    What is $\square$ in terms of ♥?
    What is ☺ in terms of ♥?

36. $✿ + ✿ + § = ♦ + ♦ + ♦ + ♦ + ♦ + ♦$
    $✿ + ♠ + ♠ + ♠ = §$
    What is ✿ in terms of ♦?
    What is § in terms of ♦?

37. $♠ + ♠ = ♣ + ♣ + ♣ + ♥ + ♥ + ♥ + ♥$
    $♣ + ♣ + ♥ = ♠$
    What is ♣ in terms of ♥?
    What is ♠ in terms of ♥?

38. $\cap + \cap + ☺ = ♥ + ♥ + ♥ + ♥ + ☺$
    $☺ + ☺ + ♥ = \cap + \cap + ☺$
    What is ☺ in terms of ♥?
    What is $\cap$ in terms of ♥?

For Exercises 39 to 52 choose variables and identify what they represent, then build equations and solve the resulting system.

39. Stephanie cuts a 20-yard ribbon. One piece is 3 yards longer than the other. What is the length of each piece?

40. Delores cuts a 16-inch salami. One piece is 4 inches longer than the other. What is the length of each piece?

41. Yoko has $4.60 in quarters and nickels. She has 24 coins altogether. How many of each does she have?

42. Bart has 30 nickels and quarters. He has $3.70 altogether. How many of each coin does he have?

43. Chen Chen has 28 coins. She has $4.45 altogether. If she has only dimes and quarters, how many of each does she have?

44. Janice has $5.05. She has only dimes and quarters. She has 25 coins altogether. How many of each does she have?

45. Hassan plans to spend no more than $100 on his birthday party. He decides to go to Skate World in the afternoon, when the cost is $40 for 10 people and $4.50 for each additional person. How many people may attend the party?

46. Joaquin has budgeted $80 for his daughter's party. Fun Base One costs $30 for 8 children and $3.50 for each additional child. How many children may attend if the party is at Fun Base One?

47. The perimeter of the front of a 15-ounce Cheerios® box is 40 inches. The height is 4 inches more than the width. What are the width and height of the front of the box?

48. The perimeter of the front of a 3-ounce Jello® box is 32 centimeters. The height is 2 centimeters less than the width. What are the width and height of the front of the box?

49. The perimeter of the front of a videotape box is 58 centimeters. The height of the front is 1 less than twice the width. What are the height and width of the box front?

50. The perimeter of the front of a 7-ounce Jiffy Muffin Mix® box is 44 centimeters. The height of the front is 2 less than twice the width. What are the height and width of the box front?

51. The two equal angles of an isosceles triangle are each half the size of the third angle. What is the measure of each angle?

52. The two equal angles of an isosceles triangle are each twice the size of the third angle. What is the measure of each angle?

53. In Example 6 we solved the system

$$x + y = 9, \quad x + z = 1, \quad y + z = 2$$

Solve the second and third equations for $x$ and $y$ in terms of $z$, and substitute them into $x + y = 9$. Find $x$, $y$, and $z$, and compare the solution with that of Example 6.

54. In Example 6 we solved the system

$$x + y = 9, \quad x + z = 1, \quad y + z = 2$$

Solve the first and third equations for $x$ and $z$ in terms of $y$, and substitute them into the second equation, $z + x = 1$. Find $x$, $y$, and $z$, and compare the solution with that of Example 6.

### PROBLEM SOLVING

*Reminder*: When we solve an equation or formula for a certain variable, we use the phrase "in terms of" to describe the relationship. $A = l \cdot w$ is area *in terms of* length and width. Use substitution to make the requested formula from the two given formulas.

55. Area in terms of base, using $A = \dfrac{bh}{2}$ and $h = \dfrac{b\sqrt{3}}{2}$

56. Area in terms of height, using $A = \dfrac{bh}{2}$ and $b = h$

57. Area in terms of circumference and $\pi$, using $A = \pi r^2$ and $C = 2\pi r$ (*Hint*: Solve $C = 2\pi r$ for $r$.)

58. Area in terms of diameter and $\pi$, using $A = \pi r^2$ and $d = 2r$

59. Diameter in terms of circumference and $\pi$, using $d = 2r$ and $C = 2\pi r$

60. Area of a square in terms of perimeter, using $A = x^2$ and $P = 4x$

### PROJECTS

61. **Intersection of Diagonals**

   a. Guess the coordinates of the point of intersection of the diagonals of the square in the figure below.

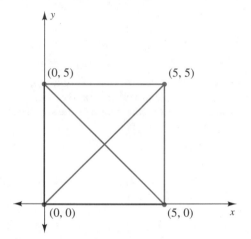

   b. Find the intersection of the diagonals of this square by writing a linear equation for each diagonal and then solving for $x$ and $y$.

   c. Find the intersection of the diagonals of any square by writing an equation for each diagonal in the figure below. Use the $y = mx + b$ form, and then solve for $x$ and $y$.

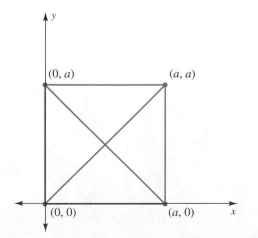

   d. You should recognize the coordinates of the intersection of the diagonals from Section 4.5 of Chapter 4. Use the name of the intersection to complete this sentence: The diagonals of a square cross at the _____ of each diagonal.

62. **U.S. Currency.** Who is pictured on each denomination of U.S. currency? Currency is printed in denominations of $1, $2, $5, $10, $20, $50, $100, $500, $1000, $5000, and $10,000. Select an appropriate variable for each person pictured on the bills, as named in the following clues. Where two people's names start with the same letter, hints for variables are suggested. Write an equation for each relevant clue. Solve by substitution.

a. 10 Clevelands ($C_l$) equal 1 Chase ($C_h$).

b. Cleveland, Franklin, and Chase total $11,100.

c. 10 Hamiltons make a Franklin.

d. 3 Washingtons plus a Jefferson ($J_e$) equal a Lincoln.

e. 4 Lincolns make a Jackson ($J_a$).

f. Hamilton, Franklin, and Chase were never president.

g. 2 Washingtons make a Jefferson.

h. 10 Grants make a McKinley ($M_c$).

i. 2 Grants make a Franklin.

j. Cleveland was born on March 18.

k. 2 Hamiltons make a Jackson.

l. Madison's picture is on the $5000 bill.

m. Washington is on the $1 bill.

---

## MID-CHAPTER 5 TEST

**In Exercises 1 and 2 solve for the indicated variable.**

1. $2x - y = 5000$ for $y$

2. $V = \dfrac{\pi r^2 h}{3}$ for $h$

**Solve the systems in Exercises 3 and 4 for the $x$ and $y$ that make both equations true.**

3. $x + y = 5000$
   $3x - 2y = -2500$

4. $x + y = 5000$
   $3x - 2y = 2500$

5. a. The point $(3, 4)$ is the solution to which two equations in the figure below?

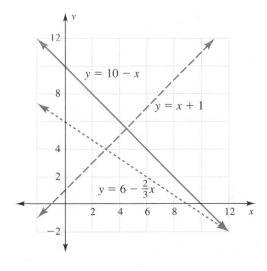

b. Estimate the intersection of $y = x + 1$ and $y = 10 - x$.

c. Use substitution to find the coordinates of the intersection of $y = 10 - x$ and $y = x + 1$.

d. Use substitution to find the coordinates of the intersection of $y = 10 - x$ and $y = 6 - \frac{2}{3}x$.

**Use guess-and-check to start a solution in Exercises 6 to 9. Define your variables, and write a system of equations. Solve the system if your guess-and-check did not reach a solution.**

6. A green turtle lays 12 times more eggs than an ostrich. If the sum of the eggs laid is 195, how many did each lay?

7. A produce distributor receives 10,000 pounds of potatoes and needs to bag them for retail sale. Local grocery stores sell three times as many 5-pound bags as 10-pound bags. How many bags of each weight should be prepared?

8. The ratio of length to width of a tennis court is 13 to 6. The perimeter of the court is 114 feet. Find the length and width.

9. The record number of children born to one mother is 69. There were 27 sets of births, including quadruplets, triplets, and twins. There were no single births. The woman bore four times as many sets of twins as sets of quadruplets. There were nine fewer sets of triplets than sets of twins. How many sets of twins, triplets, and quadruplets were delivered? (This problem may take three equations.)

10. Why does the intersection of two graphs solve the system of equations describing the graphs?

---

**5.4**

## SOLVING SYSTEMS OF EQUATIONS BY ELIMINATION

**OBJECTIVES**   Solve a system of two equations by addition or subtraction. ◆ Solve a system of two equations by multiplying to obtain opposite coefficients on variables. ◆ Convert wind and current problems into quantity and value form. ◆ Solve wind and current problems with a system of two equations.

**WARM-UP**
1. Write the equation formed by adding 5 to both sides of $x - 5 = 55$.
2. Write the equation formed by subtracting 4 from both sides of $x + 4 = 24$.
3. Suppose $x = 5$. Write the equation formed by adding $x$ to the left side and 5 to the right side of $y - x = 15$. Have we added the same expression to both sides of $y - x = 15$? Have we added equal quantities to both sides?
4. Suppose $y = -3$. Write the equation formed by adding $y$ to the left side and $-3$ to the right side of $x - y = 10$. Have we added the same expression to both sides of $x - y = 10$? Have we added equal quantities to both sides?
5. Suppose $x + y = 5$. Write the equation formed by adding $x + y$ to the left side and 5 to the right side of $x - y = 11$. Have we added equal quantities to both sides? What are $x$ and $y$?
6. Suppose $x - y = 8$. Write the equation formed by adding $x - y$ to the left side and 8 to the right side of $x + y = -2$. Have we added equal quantities to both sides? What are $x$ and $y$?     ♦

S olving a system of equations by the substitution method is convenient when it is easy to solve for one variable and there are relatively few equations. This is usually not the case. Application problems often contain fractions or decimals as coefficients. Systems of a hundred or more equations are common in business and economics applications.

### ELIMINATION IN A SYSTEM OF TWO EQUATIONS

We now introduce *elimination*, another symbolic method of solving systems of equations. Elimination is the foundation for techniques programmed into calculators and computers for solving large systems in applications. These techniques are topics in business, economics, mathematics, and sociology courses. The techniques include matrices, linear programming, and the simplex method.

**Elimination** *is a process in which one variable is removed from a system of equations by adding (or subtracting) the respective sides of two equations.* Elimination depends on the equality of the left and right sides of an equation. We begin with two puzzle problems in order to focus on the process of variable elimination. We then move on to a more realistic application in Example 3.

**EXAMPLE 1**

*A puzzle problem*    Two numbers add to 500. Their difference is 900. What are the two numbers?

**SOLUTION**    Each sentence in the problem suggests an equation:

$$x + y = 500 \quad \text{and} \quad x - y = 900$$

The left sides of the two equations contain $y$, but with opposite signs. If those left sides are added, the $y$ variables are eliminated. Because each equation is balanced over the equal sign, we are able to add the left sides and add the right sides.

$$x + y = 500$$
$$x - y = 900$$
$$(x + y) + (x - y) = 500 + 900 \quad \text{Add the two equations.}$$
$$2x = 1400 \quad \text{Combine like terms.}$$
$$x = 700$$

To find $y$ we substitute $x = 700$ into one of the original equations:

$$x + y = 500$$
$$700 + y = 500$$
$$y = -200$$

***Check***: We substitute the values found for both $x$ and $y$ into the original equations:

$$x + y = 500$$
$$700 + (-200) \overset{?}{=} 500$$
$$x - y = 900$$
$$700 - (-200) \overset{?}{=} 900 \quad \checkmark$$

◆ ◆ ◆ ◆ ◆ ◆ ◆ ◆ ◆ ◆ ◆ ◆ ◆ ◆ ◆ ◆ ◆ ◆ ◆ ◆ ◆ ◆ ◆ ◆ ◆ ◆ ◆ ◆ ◆ ◆ ◆ ◆ ◆ ◆

In Example 1 the coefficients of $y$ on the left sides of the equations were already opposites. Equations seldom have variables with opposite coefficients. This is where multiplication helps. In Example 2 we multiply the first equation by the opposite of the numerical coefficient of $c$ in the second equation and thereby obtain equations with opposite coefficients on the $c$ terms.

**E X A M P L E  2**

**McFadden's farm**   Mr. McFadden has a farm where he raised cows and ducks. There are 20 total heads among the animals and 50 total feet. How many cows and how many ducks are there?

**SOLUTION**   We start with guess-and-check in Table 8, but we change to equations when a pattern shows up that leads to an equation.

| Cow Heads | Cow Feet | Duck Heads | Duck Feet | Total Feet |
|-----------|----------|------------|-----------|------------|
| 0 | 0 | 20 | 40 | 40 |
| 20 | 80 | 0 | 0 | 80 |
| 10 | 40 | 10 | 20 | 60 |
| $c$ | $4c$ | $d$ | $2d$ | $4c + 2d$ |

*Table 8* Cows and Ducks

If $c =$ number of cow heads (and hence number of cows) and $d =$ number of duck heads (and hence number of ducks), then the system of equations is

$$c + d = 20$$
$$4c + 2d = 50$$

In order to solve by elimination we want variables with opposite coefficients.

| | |
|---|---|
| $-4c - 4d = -80$ | Multiply the first equation by $-4$. |
| $4c + 2d = 50$ | Write the second equation. |
| $(-4c - 4d) + 4c + 2d = -80 + 50$ | Add the two equations. |
| $-2d = -30$ | Add like terms. |
| $d = 15$ | |

We substitute $d = 15$ into the first equation to find $c$:

$$c + 15 = 20$$
$$c = 5$$

***Check***: We substitute $c = 5$ and $d = 15$ into both equations:

$$5 + 15 \overset{?}{=} 20 \quad \text{Total heads}$$
$$4(5) + 2(15) \overset{?}{=} 50 \quad \text{Total feet} \quad \checkmark$$

There is flexibility in solution steps. Here is another approach.

$$-2c - 2d = -40 \quad \text{Multiply the first equation by } -2.$$
$$4c + 2d = 50 \quad \text{Write the second equation.}$$
$$(-2c - 2d) + 4c + 2d = -40 + 50 \quad \text{Add the two equations.}$$
$$2c = 10 \quad \text{Add like terms.}$$
$$c = 5$$

We substitute $c = 5$ into the first equation to find $d$:

$$5 + d = 20$$
$$d = 15$$

These are the same results as before.

◆ ◆ ◆ ◆ ◆ ◆ ◆ ◆ ◆ ◆ ◆ ◆ ◆ ◆ ◆ ◆ ◆ ◆ ◆ ◆ ◆ ◆ ◆ ◆ ◆ ◆ ◆ ◆ ◆ ◆ ◆ ◆ ◆

**EXAMPLE  3**

*Ration balancing*    Suppose the local animal science agent at the Extension Service advises that your cattle need a 20% protein supplement. Ingredients for this supplement are soybean meal (44% protein) and corn (9% protein). Use Table 9 to determine how much of each ingredient is needed to produce 2000 pounds of supplement.

|  | Quantity (pounds) | Value (percent protein) | Protein Total |
|---|---|---|---|
| Soybean Meal | $x$ | 0.44 | $0.44x$ |
| Corn | $y$ | 0.09 | $0.09y$ |
| Total | 2000 pounds | 0.20 (average) | Two expressions |

*Table 9  Protein Supplement*

**a.** What equation describes the quantity of each ingredient?

**b.** What two expressions describe the lower right corner of the table? Set these expressions equal in order to form an equation.

**c.** Solve the equations from a and b by the elimination method.

**d.** Summarize your results in terms of the problem situation.

**SOLUTION**    **a.** The two ingredients add to 2000 pounds: $x + y = 2000$.

**b.** The product across the total row is $2000(0.20) = 400$. The sum of the protein total column is $0.44x + 0.09y$. Because the two expressions are equal, $0.44x + 0.09y = 400$.

**c.** To solve $x + y = 2000$ and $0.44x + 0.09y = 400$ by elimination of the $x$ term, we multiply the first equation by $-0.44$:

$$-0.44(x + y) = -0.44(2000)$$
$$-0.44x - 0.44y = -880$$

We now add the two equations—left side to left side and right side to right side—and solve for $y$:

$$(-0.44x - 0.44y) + 0.44x + 0.09y = -880 + 400$$
$$-0.35y = -480$$
$$y \approx 1371 \text{ lb of corn}$$

We substitute $y$ into $x + y = 2000$ and find that $x \approx 629$ pounds of soybean meal.

**Check**: $0.44(629) + 0.09(1371) \overset{?}{=} 400$

$$629 + 1371 \overset{?}{=} 2000 \quad \text{✓}$$

**d.** A ration composed of 629 pounds of soybean meal at 44% protein and 1371 pounds of corn at 9% protein will provide a total of 400 pounds of protein.

◆ ◆ ◆ ◆ ◆ ◆ ◆ ◆ ◆ ◆ ◆ ◆ ◆ ◆ ◆ ◆ ◆ ◆ ◆ ◆ ◆ ◆ ◆ ◆ ◆ ◆ ◆ ◆ ◆ ◆ ◆ ◆ ◆ ◆ ◆ ◆ ◆ ◆

In commercial livestock operations other equations would be included to consider alternative ingredients as well as relative costs of ingredients.

In Examples 2 and 3 we multiplied the first equation by the opposite of one coefficient in the second equation so that the two equations contained like terms with opposite signs. When the equations were added, the two like terms dropped out, leaving only one variable. In the next example both equations must be multiplied by a number to find a common term with opposite signs.

**E X A M P L E  4**

*Calories*  Dietitians know that 3 grams of protein and 4 grams of fat make 48 calories. Also, 4 grams of protein and 3 grams of fat make 43 calories. Write each fact as an equation. Solve for the calories in a gram of protein and a gram of fat by the elimination method.

**SOLUTION**  The equations are $3p + 4f = 48$ and $4p + 3f = 43$. If we multiply the first equation by 3 and the second by $-4$, the terms containing $f$ will be opposites.

$$9p + 12f = 144$$
$$-16p - 12f = 172$$
$$(9p + 12f) + (-16p - 12f) = 144 + (-172)$$
$$-7p = -28$$
$$p = 4$$

To find $f$ we substitute $p = 4$ into one of the original equations:

$$3p + 4f = 48$$
$$3(4) + 4f = 48$$
$$12 + 4f = 48$$
$$4f = 48 - 12$$
$$4f = 36$$
$$f = 9$$

**Check**: We substitute into both original equations:

$$3(4) + 4(9) \overset{?}{=} 48$$
$$4(4) + 3(9) \overset{?}{=} 43 \quad \text{✓}$$

◆ ◆ ◆ ◆ ◆ ◆ ◆ ◆ ◆ ◆ ◆ ◆ ◆ ◆ ◆ ◆ ◆ ◆ ◆ ◆ ◆ ◆ ◆ ◆ ◆ ◆ ◆ ◆ ◆ ◆ ◆ ◆ ◆ ◆ ◆ ◆ ◆ ◆

In Example 4 we chose to multiply by 3 and $-4$ out of convenience, because they made the terms containing $p$ opposites. There are other pairs of numbers that would work as well. We could multiply the first equation by $-4$ and the second equation by 3, or we could multiply the first equation by 4 and the second equation by $-3$. Confirming these results is left to the exercises.

WIND AND CURRENT PROBLEMS

A strong wind called the *jet stream* flows eastbound across the United States (see Figure 10). The jet stream is generally at an altitude of 12 kilometers (40,000 feet). In winter the jet stream flows at about 130 kilometers per hour across the southern part of the country; in summer it flows at only about 65 kilometers per hour across the northern part. Airplanes traveling eastbound can often take advantage of the jet stream, whereas those flying westbound must allow extra time and fuel for the flight.

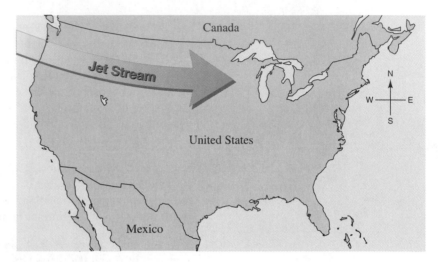

**Figure 10** North America

EXAMPLE  5    Aircraft speeds    What is the speed of each aircraft?

**a.** A jet flying 500 km/hr eastbound in winter

**b.** A jet flying 500 km/hr westbound in winter

**c.** A jet flying 500 km/hr eastbound in summer

**d.** A jet flying 500 km/hr westbound in summer

SOLUTION    We assume the jet is flying directly into or along with the jet stream.

**a.** Total speed is 500 + 130 km/hr = 630 km/hr.

**b.** Total speed is 500 − 130 km/hr = 370 km/hr.

**c.** Total speed is 500 + 65 km/hr = 565 km/hr.

**d.** Total speed is 500 − 65 km/hr = 435 km/hr.

♦ ♦ ♦ ♦ ♦ ♦ ♦ ♦ ♦ ♦ ♦ ♦ ♦ ♦ ♦ ♦ ♦ ♦ ♦ ♦ ♦ ♦ ♦ ♦ ♦ ♦ ♦ ♦ ♦ ♦ ♦ ♦ ♦

When a plane flies directly into the wind—a *headwind*—the plane is slowed down by the wind speed, and the net speed is the rate minus the wind. When a plane flies with the wind—a *tailwind*—the speed is increased by the wind speed, and the net speed is the rate plus the wind.

In Example 6 we organize the information into a table. We then build equations using the same quantity (time) and value (rate of travel) relationships presented in Section 3.4. The wind in Example 6 is due to the cold weather systems that head south from the Arctic Ocean and Canada during the winter.

EXAMPLE 6     Airplane travel    A plane leaves Fargo, North Dakota, and flies south to Dallas, Texas (see Figure 11). Another plane flies the opposite route. The planes fly the same airspeed, $r$, but there is a wind, $w$, from the north that gives the southbound plane a net rate of $r + w$ and slows the northbound plane to a net rate of $r - w$. The Fargo-to-Dallas plane takes $4\frac{2}{5}$ hours for the 1100-mile trip. The Dallas-to-Fargo plane takes $7\frac{1}{3}$ hours. What is the airspeed of the two planes, and what is the wind speed?

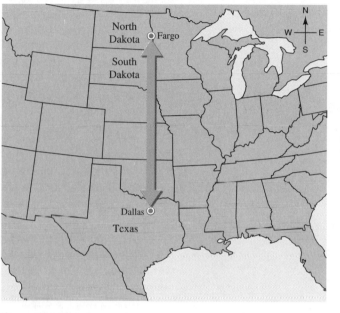

Figure 11 *Midwest*

SOLUTION     First we change the time to improper fractions:

$$4\tfrac{2}{5} = \tfrac{22}{5} \quad \text{and} \quad 7\tfrac{1}{3} = \tfrac{22}{3}$$

Then we enter the time, rate, and distance information into Table 10.

| Quantity (time) | Value (net rate relative to ground) | Distance (Time · Net rate) |
|---|---|---|
| $\frac{22}{5}$ hr | $r + w$ | 1100 mi |
| $\frac{22}{3}$ hr | $r - w$ | 1100 mi |

Table 10 *Travel Rates*

The distance is the product of time and rate, so each row of Table 10 yields one equation:

$$\tfrac{22}{5}(r + w) = 1100$$
$$\tfrac{22}{3}(r - w) = 1100$$

Rather than multiply the flying times using the distributive property, we can multiply both sides of each equation by the reciprocal of the improper fraction (or divide both sides by the improper fraction):

$$r + w = 1100 \cdot \left(\tfrac{5}{22}\right)$$
$$r - w = 1100 \cdot \left(\tfrac{3}{22}\right)$$

The equations simplify to

$$r + w = 250$$
$$r - w = 150$$

Adding the equations will eliminate the $w$:

$$2r = 400$$
$$r = 200$$

Substituting back into either equation gives $w = 50$.

Both planes have an airspeed of 200 miles per hour, but the 50-mile-per-hour wind dramatically changes their speeds relative to the ground. The southbound plane, flying *with* the wind, travels at a net rate of 250 miles per hour. The northbound plane travels at a net rate of only 150 miles per hour *against* the same wind.

◆ ◆ ◆ ◆ ◆ ◆ ◆ ◆ ◆ ◆ ◆ ◆ ◆ ◆ ◆ ◆ ◆ ◆ ◆ ◆ ◆ ◆ ◆ ◆ ◆ ◆ ◆ ◆ ◆ ◆ ◆ ◆ ◆ ◆ ◆ ◆

---

The current in a river affects boats in the same way as wind affects airplanes. Going downstream—*with* the current—adds to the speed at which a boat travels in still water. Going upstream—*against* the current—subtracts from the speed at which a boat travels in still water.

---

If the wind exceeds the airspeed, the plane will move backwards; similarly, if the current exceeds the speed in still water, the boat will move backwards. This may sound like fantasy, but just imagine swimming 3 miles per hour against a flood traveling 20 miles per hour.

In this section the direction of motion is always assumed to be in line with the direction of the wind or current. In Chapter 6 we consider motion that is perpendicular to the direction of the wind or current. For any other combination of wind (or current) and direction of motion we need trigonometry.

**Summary**

> To solve equations by elimination:
>
> 1. Arrange the equations so that like terms line up.
> 2. Multiply one or both equations by a number that makes the coefficients in one pair of like terms opposites.
> 3. Add the equations to eliminate one variable, and solve for the first variable.
> 4. Use substitution to find the second variable.

## EXERCISES 5.4

**Solve the systems of equations in Exercises 1 to 16 by the elimination method.**

1. $x + y = -2$
   $x - y = 8$

2. $x + y = -15$
   $x - y = 3$

3. $m + n = 3$
   $-m + n = -11$

4. $p - q = 9$
   $p + q = -5$

5. $2x + y = -1$
   $x + 2y = 4$

6. $3x + y = -6$
   $x + 2y = -7$

7. $2a + b = -5$
   $a + 3b = 35$

8. $3c + d = 28$
   $c + 3d = -12$

9. $2x + 3y = 3$
   $3x - 4y = -21$

10. $3m - 2n = 22$
    $2m + 3n = -7$

11. $5p - 2q = -6$
    $2p + 3q = 9$

12. $4x - 3y = -8$
    $3x + 5y = -6$

13. $7 = m \cdot 5 + b$
    $3 = m \cdot 3 + b$

14. $3 = m \cdot (-1) + b$
    $-1 = m \cdot (6) + b$

15. $0.5x + 0.2y = 1.8$
    $0.2x - 0.3y = -0.8$

16. $0.4x - 0.3y = 3.0$
    $0.5x + 0.2y = 2.6$

**In Exercises 17 to 20 identify two appropriate variables, and use the variables to write equations. Use either the elimination or the substitution method to solve the equations.**

17. The sum of two numbers is 25. Their difference is 8. Find the numbers.

18. The sum of two numbers is 35. Their difference is 10. Find the numbers.

19. The sum of two numbers is 20. Twice the second less twice the first is 21. Find the numbers.

20. The sum of two numbers is 12. Twice the larger less four times the smaller is 2. Find the numbers.

**In Exercises 21 and 22 solve Example 4 using the alternative methods mentioned at the end of the example. The equations are $3p + 4f = 48$ and $4p + 3f = 43$.**

21. Multiply the first equation by $-4$ and the second equation by 3.

22. Multiply the first equation by 4 and the second equation by $-3$.

**Recall that substituting a point on a graph into the equation producing that graph makes a true statement. In Exercises 23 to 28 substitute the coordinate points $(x, y)$ into $y = mx + b$ to obtain equations. Solve the equations for $m$ and $b$, and use $y = mx + b$ to state the equation of the line connecting the two given points.**

23. $(2, 5)$ and $(-2, 1)$

24. $(-2, 2)$ and $(3, 1)$

25. $(4, 0)$ and $(0, -2)$

26. $(-3, 0)$ and $(0, -1)$

27. $(-3, 2)$ and $(2, 2)$

28. $(5, -2)$ and $(0, -2)$

**For Exercises 29 to 38, identify two appropriate variables, and use the variables to write equations. A quantity-value table may be helpful. Use either the elimination or the substitution method to solve the equations.**

29. A snack of 4 sugar cookies and 2 ginger snaps makes 296 calories. Another of 3 sugar cookies and 10 ginger snaps makes 329 calories. How many calories in one of each kind of cookie?

30. A candy selection of 5 large gumdrops and 8 caramels makes 511 calories. Another selection of 3 large gumdrops and 10 caramels makes 525 calories. How many calories in a piece of each kind of candy?

31. A fruit plate of 15 large cherries and 22 red grapes has 126 calories. Another assortment with 20 large cherries and 11 red grapes has 113 calories. How many calories in one of each kind of fruit?

32. A bag of dried fruit, 10 dates and 3 figs, has 415 calories. Another bag of 6 dates and 5 figs has 425 calories. How many calories in one of each fruit?

33. An appetizer of 8 green olives and 5 ripe olives makes 285 calories. Another of 4 green olives and 10 ripe olives makes 330 calories. How many calories in one of each type of olive?

34. A vegetable snack has 6 small carrots and 4 stalks of celery for a total of 132 calories. Another snack of 3 small carrots and 10 stalks of celery makes 90 calories. How many calories in a small carrot and in a stalk of celery?

35. A set of 6 adult tickets and 3 student tickets to a basketball game costs $58.50. Another set, 5 adult tickets and 4 student tickets, costs $54. What is the cost of one of each type of ticket?

36. A group of 3 adult tickets and 8 student tickets to a football game costs $67.50. For 4 adult tickets and 5 student tickets to the same game, the cost is $64.50. What is the cost of one of each type of ticket?

37. Jordan buys 3 identical shirts and 2 identical ties for $109.95. Gabe buys 4 of the same shirts and a tie for $119.95. What is the cost of one of each?

38. At the bookstore, 2 notebooks and 5 pens cost $14.91. At the same time, 6 notebooks and 3 pens cost $32.85. What is the cost of one of each?

39. A flight from Cleveland, Ohio, to Washington, D.C., takes 4.5 hours against a wind, whereas the return flight on the same small airplane takes only 3 hours with the same wind. The flight covers 360 miles. What is the airspeed (speed in still air) of the airplane? What is the speed of the wind?

40. A flight from Lincoln, Nebraska, to Dallas, Texas, takes 2.5 hours with the wind. The return flight takes 3.4 hours against the same wind. The distance between the cities is 612 miles. Find the airspeed of the airplane (speed in still air) and the speed of the wind.

41. A fishing boat goes 20 miles upstream, against a current, in 5 hours. The same boat goes 20 miles downstream, with the current, in 2 hours. What is the speed of the boat (in still water)? What is the speed of the current?

42. A jet boat goes 24 miles upstream, against a current, in 3 hours. It travels downstream the same distance in 2 hours. What is the speed of the boat (in still water)? What is the speed of the current?

43. Why is it possible to make forward progress when walking 3 miles per hour against a 40-mile-per-hour wind but not when swimming 3 miles per hour against a 6-mile-per-hour current?

44. Why is there a limit to acceptable wind levels in a short race (100 meters) but not in a long race that takes several laps of an oval track to complete?

45. What factors would help you decide whether to use elimination, substitution, or graphing to solve a system of equations?

*PROJECT*

**46. *Changing Repeating Decimals to Fractions.*** We change repeating decimals into fractions with a process similar to the elimination method. Instead of using multiplication and subtraction to eliminate a variable, we use them to eliminate decimal portions of a number.

***Example a***: To change $0.33333$ to a fraction, we let $f$ represent the fraction equivalent to the given repeating decimal:

$$f = 0.33333$$

Because only one digit is being repeated, we multiply both sides of the equation by 10. This shifts the decimal point one place to the right. Then we subtract the original equation.

$$
\begin{array}{r}
10f = \phantom{-}3.33333 \\
-f = -0.33333 \\
\hline
9f = 3 \\
f = \tfrac{3}{9}, \text{ or } \tfrac{1}{3}
\end{array}
$$

If the decimal repeated two digits, we would multiply both sides by 100 and repeat the process in Example a.

***Example b***: To change $0.45454\overline{545}$ to a fraction, we let $f = 0.45454\overline{545}$. Because two digits are repeated, we multiply both sides by 100 to move the decimal point two places to the right.

$$
\begin{array}{r}
100f = \phantom{-}45.454545 \\
-f = -0.454545 \\
\hline
99f = 45 \\
f = \dfrac{45}{99} = \dfrac{9 \cdot 5}{9 \cdot 11} = \dfrac{5}{11}
\end{array}
$$

Change these repeating decimals to fractions. Simplify the fractions.

a. $0.4444\ldots$    b. $0.7777\ldots$    c. $0.151515\ldots$

d. $0.161616\ldots$    e. $0.243243243\ldots$

f. $0.270270270\ldots$

---

## 5.5   INEQUALITIES: GRAPHS AND SOLUTIONS

**OBJECTIVES**    Plot points that satisfy inequalities in two variables. ♦ Determine boundaries of inequality regions. ♦ Graph an equation, and determine by test point the side of the line satisfying a given inequality. ♦ Determine the solution to inequalities in one variable by graphing left and right sides. ♦ Explore inequalities numerically. ♦ Solve inequalities using addition and subtraction. ♦ Solve inequalities using multiplication and division by positive numbers. ♦ Solve inequalities using properties based on multiplication and division with negative numbers.

**WARM-UP**    Review Section 2.5 before doing the following.

1. Sketch a number-line graph for each of the following inequalities:

   **a.** $x \geq 2$    **b.** $x < 0$    **c.** $x \leq 5$    **d.** $x \leq 0$

2. **a.** Solve for $y$: $40x + 20y = 160$.

   **b.** Solve for $x$: $-x = x - 2$.

   **c.** Solve for $x$: $2x + 3 = -3x - 2$.    ♦

In this section we return to inequalities, introduced in Section 2.5, and relate them to coordinate graphs. Graphing calculators have shading features that permit creation of the graphs shown in Examples 3, 5, 6, and 7. Because of the availability of technology, we want to understand the concepts but not spend excessive time drawing graphs. We close the section by solving inequalities graphically and symbolically.

### GRAPHING TWO-VARIABLE INEQUALITIES ON THE COORDINATE PLANE

**EXAMPLE 1**    *Counting calories*    Sens-a-diet allows a maximum of 160 calories in snacks. Caramels have 40 calories each, and ginger snaps have 20 calories each. Assume a dieter eats only whole candies or whole cookies.

**a.** Use a table to identify some of the possible combinations of each snack.

**b.** Express the numbers of caramels and ginger snaps as $(x, y)$ pairs, and plot your combinations on a graph.

**c.** What expression describes the total number of calories from the two snacks?

SOLUTION   **a.** The 160 calories is an upper limit to the number of calories. Table 11 shows five combinations of snacks that fit the problem situation. The first three rows in the table use up all 160 calories; the last two leave some calories, possibly for a third snack option.

| Caramels, $x$ | Calories | Ginger Snaps, $y$ | Calories | Total Calories |
|:---:|:---:|:---:|:---:|:---:|
| 4 | 160 | 0 | 0 | 160 |
| 0 | 0 | 8 | 160 | 160 |
| 3 | 120 | 2 | 40 | 160 |
| 3 | 120 | 1 | 20 | 140 |
| 2 | 80 | 2 | 40 | 120 |
| $x$ | $40x$ | $y$ | $20y$ | $40x + 20y$ |

**Table 11** *Calories in Snacks*

**b.** Figure 12 shows the graph of the $(x, y)$ pairs.

**c.** The expression for the total calories is $40x + 20y$.

◆ ◆ ◆ ◆ ◆ ◆ ◆ ◆ ◆ ◆ ◆ ◆ ◆ ◆ ◆ ◆ ◆ ◆ ◆ ◆ ◆ ◆ ◆ ◆ ◆ ◆ ◆ ◆ ◆ ◆ ◆ ◆ ◆ ◆ ◆ ◆ ◆ ◆

EXAMPLE **2**

*Counting calories, continued*   The total calories for several snack combinations are shown in Figure 13. Determine where on the graph we find combinations of snacks that total 160 or fewer calories. Describe the location of solutions to the 160-calorie limit.

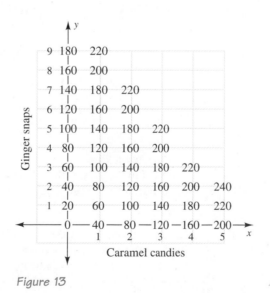

**Figure 13**

SOLUTION   It is not possible to eat a negative number of snacks, so the $x$-axis and the $y$-axis form two boundaries to the set of possible snacks. All of the 160-calorie combinations lie in a straight line. This line forms a third boundary and lies between the snack points $(0, 8)$ and $(4, 0)$.

◆ ◆ ◆ ◆ ◆ ◆ ◆ ◆ ◆ ◆ ◆ ◆ ◆ ◆ ◆ ◆ ◆ ◆ ◆ ◆ ◆ ◆ ◆ ◆ ◆ ◆ ◆ ◆ ◆ ◆ ◆ ◆ ◆ ◆ ◆ ◆ ◆ ◆ ◆ ◆

EXAMPLE **3**

Use inequalities to describe the solution set for Example 1. The graph in Figure 14 might help. Use test points to check your answers.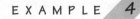

SOLUTION

From Table 11, the expression $40x + 20y$ describes the number of calories, so the equation describing the combinations of snacks yielding the maximum calorie count is $40x + 20y = 160$. The points $(0, 8)$ and $(4, 0)$ make the calorie boundary true:

$$40(0) + 20(8) = 160 \quad \text{and} \quad 40(4) + 20(0) = 160$$

Any combination of snacks with fewer calories is also a solution, so we have the inequality $40x + 20y \leq 160$. We check the inequality with a test point—say, $(3, 1)$:

$$40(3) + 20(1) \overset{?}{\leq} 160$$
$$140 \leq 160$$

As indicated in Example 2, the axes form boundaries that may also be described by inequalities. Points on or to the right of the $y$-axis have zero or positive $x$-coordinates, or $x \geq 0$. Points on or above the $x$-axis have zero or positive $y$-coordinates, or $y \geq 0$. The test point $(3, 1)$ makes both $x \geq 0$ and $y \geq 0$ true.

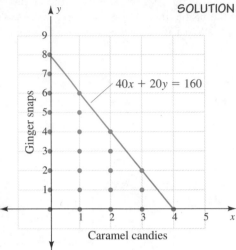

*Figure 14*

◆ ◆ ◆ ◆ ◆ ◆ ◆ ◆ ◆ ◆ ◆ ◆ ◆ ◆ ◆ ◆ ◆ ◆ ◆ ◆ ◆ ◆ ◆ ◆ ◆ ◆ ◆ ◆ ◆ ◆ ◆ ◆ ◆ ◆ ◆ ◆ ◆ ◆ ◆ ◆ ◆

**Graphing Calculator Technique**

### Shading and Inequalities

Graphing calculators contain shading features. Check your instruction manual for specifics. (The subtle differences between $>$ and $\geq$ and between $<$ and $\leq$ are not distinguished by these shading features.)

To approximate the graph in Figure 14, first solve $40x + 20y = 160$ for $y$. It is not necessary to simplify. Set $Y_1 = (160 - 40X)/20$, and set the window with X in the interval $[-1, 5]$, scale 1, and Y in the interval $[-1, 9]$, scale 1. Then, for the TI 82, use the commands

| 2nd | DRAW | Shade( | 0, | 2nd | Y-VARS | ENTER | ENTER |, 1, 0, 5 | ENTER |

The first zero means $y \geq 0$, the lower boundary. The next expression, $Y_1$, enters the equation, the upper boundary. The 1 gives the resolution—how darkly the graph is to be shaded. The second zero means $x \geq 0$, the left boundary. The 5 means $x \leq 5$, an arbitrary right boundary.

If necessary, use | 2nd | DRAW | 1 to clear a prior drawing.

The inequalities in Example 3 form boundaries for a set of points in a region. This region, rather than just a single point or set of points on a line, is the solution to the problem.

---

Inequalities may be used to describe regions. Equations are used to determine the boundaries of these regions. If coordinate pairs are not limited to integers, then we use shading to indicate regions.

---

EXAMPLE **4**

Determine which pairs of points make $y \geq x + 2$ true. Plot the points, then compare the coordinates to $y = x + 2$.

**a.** $(0, 0)$    **b.** $(2, 1)$    **c.** $(-2, 0)$    **d.** $(0, 2)$    **e.** $(-3, 2)$

**f.** $(0, 4)$    **g.** $(1, 3)$    **h.** $(-3, -1)$

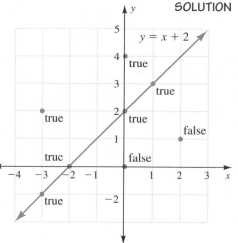

Figure 15

SOLUTION   **a.** The point (0, 0) makes a false inequality, $0 \geq 0 + 2$.

**b.** The point (2, 1) makes a false inequality, $1 \geq 2 + 2$.

**c.** The point (−2, 0) makes a true inequality, $0 \geq -2 + 2$.

**d.** The point (0, 2) makes a true inequality, $2 \geq 0 + 2$.

**e.** The point (−3, 2) makes a true inequality, $2 \geq -3 + 2$.

**f.** The point (0, 4) makes a true inequality, $4 \geq 0 + 2$.

**g.** The point (1, 3) makes a true inequality, $3 \geq 1 + 2$.

**h.** The point (−3, −1) makes a true inequality, $-1 \geq -3 + 2$.

The points (−2, 0), (0, 2), (1, 3), and (−3, −1) lie on the line $y = x + 2$ (see Figure 15). The points (−3, 2) and (0, 4) are true and lie to the left of the line. The false points (0, 0) and (2, 1) are both to the right of the line.

◆ ◆ ◆ ◆ ◆ ◆ ◆ ◆ ◆ ◆ ◆ ◆ ◆ ◆ ◆ ◆ ◆ ◆ ◆ ◆ ◆ ◆ ◆ ◆ ◆ ◆ ◆ ◆ ◆ ◆ ◆ ◆

E X A M P L E   **5**   Graph $y \geq x + 2$ on coordinate axes. Confirm the region described by the inequality with a test point.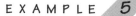

SOLUTION   The graph of $y \geq x + 2$ combines the graph of $y = x + 2$ with the region determined by $y > x + 2$. The phrase *greater than or equal to* implies that we want numbers satisfying either the inequality or the equation. We first graph the equation $y = x + 2$ as shown in Figure 16. Because of the *equal to* in the inequality sign, the line is part of the solution set.

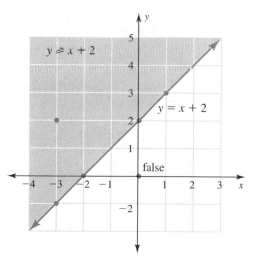

Figure 16

To graph the inequality we shade one side of the line, but we need a test point to determine which side. Suppose we test the origin, (0, 0). As we determined in Example 4, the inequality $0 \geq 0 + 2$ is false, so the origin is *not* in the solution set. This implies that the solution set is all points on the other side of the line.

◆ ◆ ◆ ◆ ◆ ◆ ◆ ◆ ◆ ◆ ◆ ◆ ◆ ◆ ◆ ◆ ◆ ◆ ◆ ◆ ◆ ◆ ◆ ◆ ◆ ◆ ◆ ◆ ◆ ◆ ◆ ◆

*The region shaded on one side of a line* is called a **half-plane**. The graph of a line determines two half-planes. Only one of these half-planes will make an inequality true. Examples 5 and 6 illustrate situations where the shaded region is limited to the first quadrant.

EXAMPLE 6    Party budget    Audrey has $65 to spend on a birthday party for *x* people.

**a.** Write Audrey's budget as an inequality using *y*, and graph it.

**b.** What other boundaries besides the $65 must be considered in a party situation? Include these boundaries on the graph.

SOLUTION    **a.** Audrey's budget is $y \leq \$65$. The boundary line at $65 is solid because Audrey can spend up to or exactly $65.

**b.** Both dollars, *y*, and people invited, *x*, are positive. This is a first-quadrant graph with $x > 0$ and $y > 0$, as shown in Figure 17.

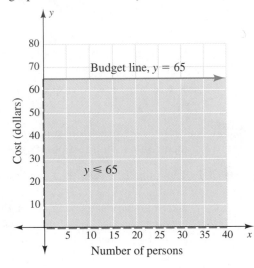

Figure 17

◆ ◆ ◆ ◆ ◆ ◆ ◆ ◆ ◆ ◆ ◆ ◆ ◆ ◆ ◆ ◆ ◆ ◆ ◆ ◆ ◆ ◆ ◆ ◆ ◆ ◆ ◆ ◆ ◆ ◆ ◆ ◆ ◆

EXAMPLE 7    Competition    The manager of Fun Corporation wants to set the amusement center's rental fees below those of Willamalane Wave Pool. Willamalane has a $3-per-person admission fee and a $10-per-hour party room rent. Make a graph showing the guest and cost options for a 2-hour party at Willamalane. Shade the region that represents Fun Corporation's rental options if their total cost is to be less than that of a 2-hour party at Willamalane for any number of persons. Assume that the number of guests and the cost of the party are positive.

SOLUTION    The cost of renting Willamalane for 2 hours is $20. The cost boundary, $y = 20 + 3x$, is graphed in Figure 18. Because Fun Corporation's total cost is

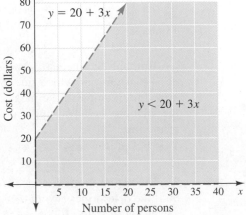

Figure 18

to be less than Willamalane's, $y = 20 + 3x$ is graphed as a dashed line. Select $(15, 40)$ as a test point. The test point makes $y < 20 + 3x$ true, so we shade below the cost line.

The positive number of guests means $x > 0$, and the positive cost means $y > 0$. Thus we have a first-quadrant graph shaded below the cost line, above the $x$-axis, and to the right of the $y$-axis.

♦ ♦ ♦ ♦ ♦ ♦ ♦ ♦ ♦ ♦ ♦ ♦ ♦ ♦ ♦ ♦ ♦ ♦ ♦ ♦ ♦ ♦ ♦ ♦ ♦ ♦ ♦ ♦ ♦ ♦ ♦ ♦ ♦ ♦ ♦ ♦ ♦ ♦

**Summary**

> ♦ On coordinate axes we include points by using a solid line or a shaded region. We exclude points by using a dashed (or dotted) line or an unshaded region.
>
> ♦ We check the direction of the graph or shaded half-plane by finding a true test point.

### SOLVING ONE-VARIABLE INEQUALITIES FROM A GRAPH

The inequalities in Examples 2 through 7 produced regions as solutions. We now consider graphical solutions to inequalities with one variable. This time the solutions will be sets described by inequalities in one variable only.

EXAMPLE  **8**

**More party plans**    Suppose Audrey decides to have a two-hour party at Willamalane Wave Pool, where the cost is given by $y = 3x + 20$. Her budget is $y = \$65$.

**a.** Graph each equation. Identify the point of intersection.

**b.** Write an inequality that describes the number of people Audrey can take to the wave pool.

**c.** Use the graph to solve the inequality.

SOLUTION    **a.** The graphs intersect at $(15, 65)$ in Figure 19.

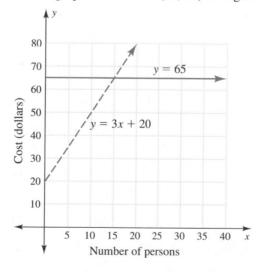

READING NOTE: In Figures 19 and 20 one of the lines is dashed to make it look different from the other line. The dashed line does not indicate an inequality, as it did in the graphs with shaded regions.

*Figure 19*

**b.** The number of people in the party must lie where the cost is less than or equal to the budget, $3x + 20 \leq 65$.

**c.** The line $y = 3x + 20$ is below the line $y = 65$ to the left of the intersection. "Left of the intersection" means for all $x < 15$. Thus we say $3x + 20 \leq 65$ for all $x \leq 15$.

♦ ♦ ♦ ♦ ♦ ♦ ♦ ♦ ♦ ♦ ♦ ♦ ♦ ♦ ♦ ♦ ♦ ♦ ♦ ♦ ♦ ♦ ♦ ♦ ♦ ♦ ♦ ♦ ♦ ♦ ♦ ♦ ♦ ♦ ♦ ♦ ♦ ♦

EXAMPLE  9    **Another party package**    Suppose Julian wishes to find the number of persons for which a 2-hour rental at Willamalane will be cheaper than Farrell's at $3.95 per person.

**a.** State and graph the cost equations.

**b.** Write an inequality describing the cost comparison.

**c.** Solve the inequality from the graph.

SOLUTION    **a.** The cost equations are $y = 3x + 20$ and $y = 3.95x$. They are graphed in Figure 20.

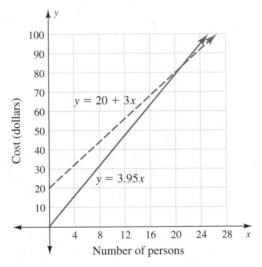

Figure 20

**b.** Willamalane will be cheaper when $3x + 20 < 3.95x$.

**c.** Willamalane will be cheaper when the graph $y = 3x + 20$ is below the graph $y = 3.95x$. The point of intersection is approximately (21, 83). The graph $y = 3x + 20$ is below $y = 3.95x$ to the right of the point of intersection. Thus $3x + 20 < 3.95x$ for $x > 21$.

♦ ♦ ♦ ♦ ♦ ♦ ♦ ♦ ♦ ♦ ♦ ♦ ♦ ♦ ♦ ♦ ♦ ♦ ♦ ♦ ♦ ♦ ♦ ♦ ♦ ♦ ♦ ♦ ♦ ♦ ♦ ♦ ♦ ♦ ♦ ♦ ♦ ♦ ♦ ♦ ♦ ♦

If we are given an inequality, we can solve it by graphing each side as a separate equation.

EXAMPLE  10    Solve $2x + 3 \le -3x - 2$ by graphing each side separately.

SOLUTION    The graphs of $y = 2x + 3$ and $y = -3x - 2$ are in Figure 21. The lines intersect at $(-1, 1)$. The line $y = 2x + 3$ is below the line $y = -3x - 2$ to the left of the intersection. Thus all inputs less than $-1$ make the equation true. Because of the equality implied by the $\le$ sign, we also have the point of intersection itself making the inequality true. Hence

$$2x + 3 \le -3x - 2 \quad \text{for } x \le -1$$

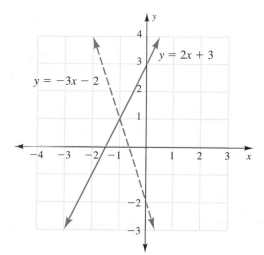

Figure 21

♦ ♦ ♦ ♦ ♦ ♦ ♦ ♦ ♦ ♦ ♦ ♦ ♦ ♦ ♦ ♦ ♦ ♦ ♦ ♦ ♦ ♦ ♦ ♦ ♦ ♦ ♦ ♦ ♦ ♦ ♦ ♦ ♦ ♦ ♦ ♦ ♦ ♦

SOLVING INEQUALITIES USING SYMBOLS

Example 11 indicates a surprising property of inequality statements. This property will lead to the only difference between solving equations and solving inequalities.

EXAMPLE **11**

An exploration     Perform the indicated operation on each true statement. Is the resulting statement true or false?

a.   $6 < 7$       Multiply both sides by 3.

b.  $-5 < 4$       Add $-6$ to both sides.

c.  $-3 < 5$       Subtract 10 from both sides.

d.  $-2 < 1$       Multiply both sides by $-3$.

e.  $-4 < -2$      Divide both sides by $-2$.

SOLUTION   a.  $-18 < 21$       True

b.  $-11 < -2$       True

c.  $-13 < -5$       True

d. $6 < -3$         False

e. $2 < 1$           False

When we multiplied by a negative number in part d and divided by a negative number in part e, we ended up with false statements. These statements would be true if we reversed the inequality signs:

$$6 > -3 \qquad \text{True}$$
$$2 > 1 \qquad \text{True}$$

♦ ♦ ♦ ♦ ♦ ♦ ♦ ♦ ♦ ♦ ♦ ♦ ♦ ♦ ♦ ♦ ♦ ♦ ♦ ♦ ♦ ♦ ♦ ♦ ♦ ♦ ♦ ♦ ♦ ♦ ♦ ♦ ♦ ♦ ♦ ♦ ♦ ♦

Thus the one difference between solving an equation and solving an inequality can be summarized as follows.

When we multiply or divide an inequality by a negative number, we reverse the inequality sign. All other additions, subtractions, multiplications, and divisions leave the inequality sign unchanged.

In many cases we can avoid multiplication and division by a negative if we keep the coefficient on the variable positive. In Example 12 it is natural to keep the variable positive.

**E X A M P L E   12**

Solve for $x$: $20 + 3x < 3.95x$. Compare the solution with that in Example 9.

SOLUTION

$$20 + 3x < 3.95x$$
$$20 + 3x - 3x < 3.95 - 3x \qquad \text{Subtract } 3x.$$
$$20 < 0.95x$$
$$20 \div 0.95 < 0.95x \div 0.95 \qquad \text{Divide by } 0.95.$$
$$21.05 < x$$

Solving the inequality gives the same result we obtained by graphing. The inequality may be read either as 21.05 is less than $x$ ($21.05 < x$) or as $x$ is greater than 21.05 ($x > 21.05$). In this situation we round to the nearest whole number and read it as $x > 21$ people, which is more natural than saying "21 people is less than $x$." Willamalane is cheaper for more than 21 people.

♦ ♦ ♦ ♦ ♦ ♦ ♦ ♦ ♦ ♦ ♦ ♦ ♦ ♦ ♦ ♦ ♦ ♦ ♦ ♦ ♦ ♦ ♦ ♦ ♦ ♦ ♦ ♦ ♦ ♦ ♦ ♦ ♦ ♦ ♦

We will solve the inequality in Example 13 two ways, to illustrate division by a negative.

**E X A M P L E   13**

Solve for $x$ in two different ways: $2x + 3 \leq -3x - 2$. Compare the solutions with that found in Example 10.

SOLUTION

First we will use division by a negative.

$$2x + 3 \leq -3x - 2$$
$$3 \leq -5x - 2 \qquad \text{Subtract } 2x.$$
$$5 \leq -5x \qquad \text{Add } 2.$$
$$5 \div -5 \geq -5x \div -5 \qquad \text{Divide by } -5, \text{ and reverse the inequality sign.}$$
$$-1 \geq x$$

Now we will solve the problem again, keeping the variable positive.

$$2x + 3 \leq -3x - 2$$
$$5x + 3 \leq -2 \qquad \text{Add } 3x.$$
$$5x \leq -5 \qquad \text{Subtract } 3.$$
$$5x \div 5 \leq -5 \div 5 \qquad \text{Divide by } +5.$$
$$x \leq -1 \qquad \text{The inequality sign is not changed.}$$

The pointed end of the inequality faces $x$ in both solutions. Thus $-1 \geq x$ and $x \leq -1$ represent the same set of numbers. It is useful to be able to recognize an inequality in either direction.

♦ ♦ ♦ ♦ ♦ ♦ ♦ ♦ ♦ ♦ ♦ ♦ ♦ ♦ ♦ ♦ ♦ ♦ ♦ ♦ ♦ ♦ ♦ ♦ ♦ ♦ ♦ ♦ ♦ ♦ ♦ ♦ ♦ ♦ ♦

Keeping the variable's coefficient positive has the distinct advantage of eliminating the need to change the inequality sign. It is possible to keep the variable positive in solving most inequalities.

**EXERCISES 5.5**

1. A dieter is allowed 140 calories for a snack.

   a. What are the possible combinations of apricots at 20 calories each and tangerines at 35 calories each?

   b. Plot the boundary lines, and describe the region that shows sensible solutions.

   c. Describe the solution set using three inequalities, and identify the appropriate type of number for inputs.

2. A dieter limits a snack to 60 calories.

   a. What are the possible combinations of small carrots at 20 calories each and medium celery stalks at 3 calories each?

   b. Plot the boundary lines, and describe the region that shows sensible solutions.

   c. Describe the solution set using three inequalities, and identify the appropriate type of number for inputs.

**Graph the inequalities in Exercises 3 to 6 on coordinate axes. Verify the direction of the half-plane shading with a test point. Check with a graphing calculator.**

3. $y < -3x + 2$         4. $y < 2x - 11$

5. $y \leq 2 - 2x$         6. $y \leq 3 - 3x$

7. Describe the region shown in the figure.

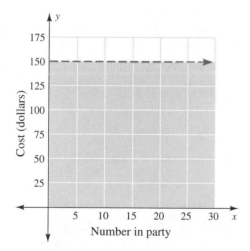

8. What party scenario might the figure describe?

**In Example 3, $x \geq 0$ and $y \geq 0$ describe points in the first quadrant, or on the positive portions of the axes. For Exercises 9 to 11 identify the regions described by the inequalities. Try shading the resulting quadrants on the graphing calculator.**

9. $x > 0$ and $y \leq 0$         10. $x < 0$ and $y \geq 0$

11. $x < 0$ and $y > 0$

**What inequalities describe the regions in Exercises 12 to 14?**

12. The second quadrant without axes

13. The fourth quadrant without axes

14. The third quadrant together with the negative $x$-axis and the negative $y$-axis

**Use the figure to solve the inequalities in Exercises 15 to 18.**

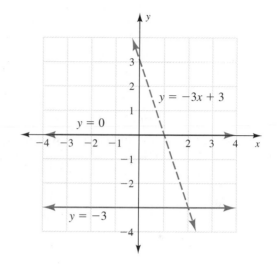

15. $3 > -3x + 3$         16. $-3 < -3x + 3$

17. $0 < -3x + 3$         18. $0 > -3x + 3$

**Use the figure to solve the inequalities in Exercises 19 to 22.**

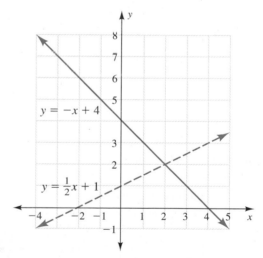

19. $-x + 4 < \frac{1}{2}x + 1$         20. $-x + 4 > \frac{1}{2}x + 1$

21. $-x + 4 > 0$         22. $\frac{1}{2}x + 1 > 0$

**For Exercises 23 to 26 solve the inequalities by using the graphs of the left and right sides. Use a graphing calculator if one is available.**

23. $-1 < -3x + 2$         24. $3 < 2x - 1$

25. $-2 \leq 2 - 2x$         26. $0 \leq 3 - 3x$

27. Describe in complete sentences how to determine which side of a line to shade when graphing an inequality.

28. Describe the effect on $-3 < 4$ of multiplying by $-2$.

29. Describe the effect on $-8 < -6$ of dividing by $-2$.

30. What is the advantage of keeping the variable's coefficient positive when solving an inequality?

**In Exercises 31 to 42 solve each inequality symbolically. Use either method shown in Example 13.**

31. $-3 > -3x + 3$   32. $-3 < -3x + 3$

33. $0 < -3x + 3$    34. $0 > -3x + 3$

35. $-x + 4 < \frac{1}{2}x + 1$   36. $-x + 4 > \frac{1}{2}x + 1$

37. $-x + 4 > 0$   38. $\frac{1}{2}x + 1 > 0$

39. $-1 < -3x + 2$   40. $3 < 2x - 1$

41. $-2 \leq 2 - 2x$   42. $0 \leq 3 - 3x$

**Solve Exercises 43 to 46 symbolically. Check your work on a graphing calculator, if available.**

43. $-x < -3x + 2$   44. $x > 2x - 1$

45. $2 - 2x < 3 - 3x$   46. $2 - 2x > 3 - 3x$

PROJECT

47. ***Pricing to undersell the competition.*** The Fun Corporation is building a bowling alley to compete with Fairfield Lanes. Fairfield charges $5.50 per person. Fun Corporation wants to attract parties with budgets of $100 or less and to keep the per-person cost below that charged by Fairfield.

 a. Let the number of people be the input, and let total cost be the output. Draw a line representing the $100 budget limit. Add another line and shading to represent the pricing options for Fun Corporation.

 b. If a party were held at Fairfield Lanes for $100, how many people could attend?

 c. Suppose Fun Corporation charged $5 per person. Mark and identify the points that show how many people could attend for $50 and for $100.

 d. What is the Fun Corporation charge per person if (10, 35) is a point on their cost line?

---

# CHAPTER 5 SUMMARY

### Vocabulary

*For definitions and page references, see the Glossary/Index.*

coincident lines    point of intersection

elimination    substitution

half-plane    system of equations

### Concepts

Use *guessing* to help understand a word problem. Organize guesses into a table. Look for patterns to determine expressions for building an equation.

 1. Use one variable, and build one equation.
 2. Use two (or three) variables, and build two (or three) equations.

When two expressions describe the same idea, the principle of **substitution** allows us to set the expressions equal to each other.

**Coincident lines** have all points in common, so we say their equations have an infinite number of solutions. The slopes and *y*-intercepts of coincident lines are the same.

When a system of equations contains *parallel lines*, we say the system has no point of intersection and hence no solution.

### Solving Inequalities

In solving an inequality from a graph we find the inputs, *x*, that make the inequality true.

 1. Graph $y_1$ = left side of inequality and graph $y_2$ = right side of inequality.

 2. Find the point of intersection of the two graphs, $x = a$. The point $x = a$ will be where $y_1 = y_2$.

 3. If $y_1 > y_2$, the graph of $y_1$ is above the graph of $y_2$. If $y_1 < y_2$, the graph of $y_1$ is below the graph of $y_2$.

  Describe the inputs, *x*, using an inequality that makes the original inequality true. Answers will be in the form $x > a$, $x < a$, $x \leq a$, or $x \geq a$.

In solving an inequality symbolically, if we keep the variable terms positive, the inequality may be solved like an equation. However, if we multiply or divide by a negative when solving the inequality, we must *reverse* the inequality sign.

### Graphing Inequalities in Two Variables

The inequality $x \geq 0$ represents all points on the *y*-axis or in Quadrants I and IV. The inequality $x < 0$ represents all points to the left of the *y*-axis, in Quadrants II and III.

The inequality $y \geq 0$ represents all points on the *x*-axis or in Quadrants I and II. The inequality $y < 0$ represents all points below the *x*-axis, in Quadrants III and IV.

To graph an inequality $y \leq mx + b$ or $y \geq mx + b$, first graph the line $y = mx + b$ with a *solid* line. Then select a test point on one side of the line. Shade that half-plane if the test point makes the inequality true. If the test point makes the inequality false, shade the half-plane on the other side of the line.

To graph an inequality $y < mx + b$ or $y > mx + b$, first graph the line $y = mx + b$ with a *dashed* line (thereby indicating the lack of an equality sign). Select a test point on one side of the line, and shade as directed.

# CHAPTER 5 REVIEW EXERCISES

Throughout this review decide for yourself whether it is appropriate to solve a system of equations by elimination or by substitution. If a graphing calculator is available, a graphical solution is permissible. For a graphical solution provide documentation of equations in $y = mx + b$ form and the viewing window dimensions containing the solution to the nearest hundredth.

**In Exercises 1 to 4 solve for the indicated variable.**

1. $y - 2x = 5000$ for $y$        2. $A = bh$ for $h$

3. $C = 2\pi r$ for $r$        4. $M = \dfrac{mgl}{\pi r^2 s}$ for $g$

Identify each pair of lines in Exercises 5 to 8 as parallel, coincident (the same line), or neither. If neither, solve for the point of intersection.

5. $2x - 3y = 7$        6. $2x - 3y = 7$
   $4x + 3y = -1$           $-2x + 3y = 4$

7. $3y - 4x = 2$        8. $3y - 4x = 2$
   $8x - 6y + 4$           $8x = 6y - 4$

9. Solve for the $m$ and $b$ that make both equations true:

$$3 = m \cdot (-2) + b$$
$$2 = m \cdot 3 + b$$

**Solve the systems in Exercises 10 and 11 for the $x$ and $y$ that make both equations true.**

10. $x + y = 5000$
    $3x - 2y = 12500$

11. $x + y = 5000$
    $3x - 2y = -7500$

12. Give an equation of a line that has the same graph as $x + y = 5$. Solve the system formed by the two equations, and indicate at which point in the algebra we know the number of solutions is infinite.

13. The total amount budgeted by Nikki and Boris for books and movies is $30 per month. They spend twice as much on books as on movies. The graph in the figure shows three lines.

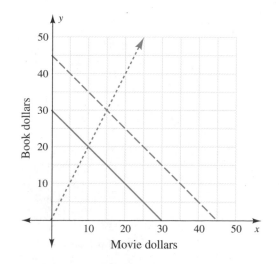

a. Find the line that represents the possible ways to spend the $30 budgeted. Identify four coordinate points on the line, and explain what the coordinates mean. Include the $x$- and $y$-intercepts as two of the points.

b. Find the line that represents spending twice as much on books as on movies. Identify three points on the line, and tell what total spending is implied by each point.

c. Estimate the point of intersection of the budget and spending lines. Explain what the intersection means.

d. Describe what sort of line would represent a $20 book and movie budget.

e. Estimate the point of intersection of the spending line and the line passing through $(45, 0)$ and $(0, 45)$ as shown in the graph. Explain what this intersection means.

**For Exercises 14 to 17 define your variables, and write a system of equations. Solve the system.**

14. The record centipede has 356 fewer legs than the record millipede. Together the two have 1064 legs. How many legs does each have?

15. A Boeing 747 holds 279 more people than a Boeing 707. If the two planes together carry 721 people, how many does each plane carry? (The total people includes crew.)

16. A snack of 7 dried figs and 3 dried dates contains 460 calories. Another snack of 2 dried figs and 7 dried dates contains 285 calories. How many calories are there in each?

17. A ton (2000 pounds) of dairy cow ration needs to be made from corn that is 9% protein and soybean meal that is 44% protein. The final ration must be 20% protein. How many pounds each of corn and soybean meal need to be mixed to prepare the ration?

18. A trout travels upstream 8 miles in 0.8 hour. Another trout travels downstream 16 miles in the same time. Assume both trout travel the same speed in still water. What is their speed in still water? What is the speed of the current?

19. The perimeter of an isosceles triangle is 32 inches. The two equal sides are each 2.5 inches longer than the third side. What is the length of each side?

**For Exercises 20 to 22 start with guess-and-check. Use your guesses to define variables and write a system of three equations to describe each problem situation. You do not need to solve the equations if you found the answers by guessing.**

20. One angle of a triangle is 16° less than a second angle. A third angle equals the sum of the first two angles. The total angle measure of all three angles is 180°. What is the measure of each angle?

21. One serving of 7 medium shrimp, fried in vegetable shortening, contains 198 calories. Protein and carbohydrates make 4 calories per gram; fat makes 9 calories per gram. The total weight of protein, fat, and carbohydrates in the shrimp is 37 grams. There are 5 more grams of protein than of carbohydrates. Find the number of grams of each (protein, fat, and carbohydrates) in this serving of shrimp.

22. A cup of peanuts roasted in oil contains 903 calories. The total weight in protein, fat, and carbohydrates is 137 grams. Fat contains 9 calories per gram; protein and carbohydrates each contain 4 calories per gram. There are 44 more grams of fat than of carbohydrates. How many grams of each (protein, fat, and carbohydrates) are there in the peanuts?

23. Why does the intersection of two graphs solve the system of equations describing the graphs?

24. A diet has a 400-calorie snack limit. Vincente chooses to eat olives for his snack. Green olives contain 20 calories each, and ripe (black) olives contain 25 calories each. Define the variables. What are four different combinations of olives with fewer than 400 calories? What are four different combinations with exactly 400 calories? Plot your solutions on a graph. Shade the portion of the graph that shows all combinations that satisfy the diet.

25. Which pairs of points make $y \le x - 3$ true?

   a. $(4, 2)$

   b. $(-1, -6)$

   c. $(3, 8)$

   d. $(0, 0)$

26. For what inputs, $x$, will the graph of $y = 2$ be below the graph of $y = x - 3$? Sketch the graph.

27. Solve $2 < x - 3$ with symbols.

28. Solve $x - 4 > 2 - x$ with symbols.

29. Which line in the figure is $y = 15 - 2x$, and which line is $y = x - 6$? Label the equations near the lines. For what inputs, $x$, is $15 - 2x < x - 6$?

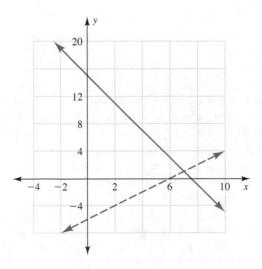

30. Solve $2x - 1 \le 2 - x$ with a coordinate graph and with algebra.

31. Solve $x + 1 \ge 9 - 3x$ with a coordinate graph and with algebra.

32. Describe how the intersection of the graphs of $y = 15 - 2x$ and $y = x - 6$ is used to determine the solution to the inequality $15 - 2x > x - 6$.

33. The graph of $y = 2x - 1$ is above the graph of $y = 2 - x$ for $x > 1$. What inequality does $x > 1$ solve?

34. A can of paint covers 300 square feet. Mark four coordinates on the graph in the figure that show possible lengths and widths of a rectangular floor to be painted. Connect your points. Where on the graph will points be located that represent floor sizes requiring less than a full can of paint?

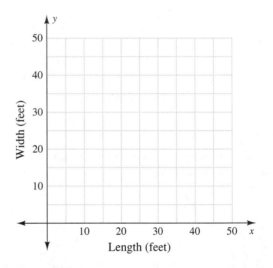

35. Emerson has a $150 party budget. At $85 for 10 people and $4.75 for each additional, how many people can attend his party at Lane County Ice? Assume that at least 10 people attend. Write an inequality, and solve it from the graph in the figure.

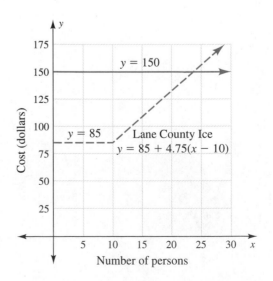

36. A bowling party at Fairfield Lanes costs $5.50 per hour. For how many people will a Fairfield bowling party be cheaper than a 3-hour rental of Willamalane Wave Pool at a cost of $y = \$30 + \$3x$? Write an inequality, and solve it from the graph in the figure.

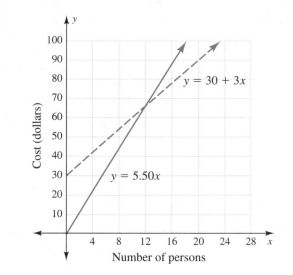

---

## CHAPTER 5 TEST

Decide for yourself whether it is appropriate to solve a system of equations by elimination or by substitution. For a graphical solution provide documentation of equations in $y = mx + b$ form and the viewing window dimensions containing the solution to the nearest hundredth.

In Exercises 1 to 3 solve the equation for the indicated variable.

1. $3x - y = 400$ for $y$

2. $A = \dfrac{bh}{2}$ for $b$

3. $I = \dfrac{AH}{T}$ for $T$

In Exercises 4 to 6 solve for the point of intersection, if any.

4. $3x - 4y = 3$
   $5x - 2y = 47$

5. $3x + 2y = 5700$
   $x - y = 900$

6. $x + 2y = -9$
   $4x + 3y = -76$

7. Give an example of equations of parallel lines. Solve them, and indicate at which point in the algebra we know there is no solution to the system of parallel equations.

8. The point (6, 2) in the figure is the point of intersection of which two lines? What will be the result when you substitute the intersection coordinates into the two equations?

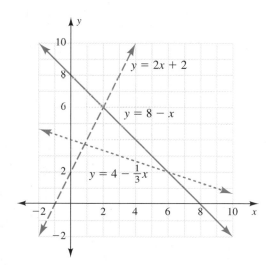

9. Solve for the $m$ and $b$ that make both equations true:

$$2 = m \cdot 5 + b$$
$$-1 = m \cdot 1 + b$$

In Exercises 10 to 12 identify variables and write equations to describe the problem situation. Solve the problem either by guess-and-check or by using your equations.

10. A typical caterpillar has ten more legs than a butterfly. Eight butterflies and six caterpillars have a gross (144) of legs. How many does each have?

11. A snack of 16 peanuts and 5 cashews contains 135 calories. Another snack of 20 peanuts and 25 cashews contains 405 calories. How many calories in each peanut and each cashew?

12. A dolphin travels 135 miles in 3 hours with an ocean current. Another dolphin travels 100 miles in 4 hours against the same current. Assume the dolphins travel the same speed in still water. What is the speed of the current, and what is the speed of the dolphins in still water?

13. a. Identify the lines $y = x - 1$ and $y = 5 - x$ on the graph in the figure.

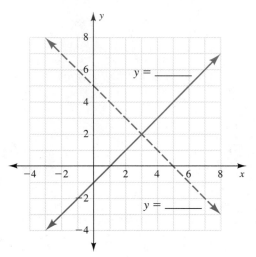

b. What is the point of intersection of the two graphs?

c. Use the graphs to solve the inequality $5 - x < x - 1$.

14. Solve $3x - 5 \leq 3 - x$ with a coordinate graph and with algebra.

15. Describe how you decide whether to use elimination or substitution to solve a system of equations.

## Polynomial

## Expressions

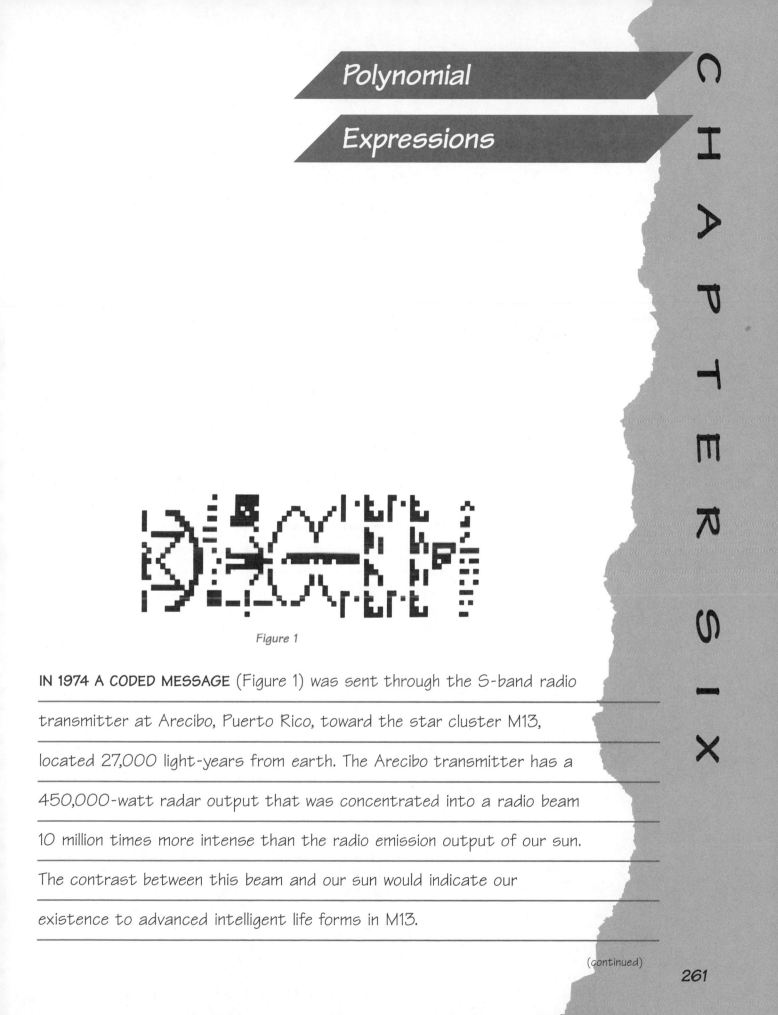

Figure 1

**IN 1974 A CODED MESSAGE** (Figure 1) was sent through the S-band radio

transmitter at Arecibo, Puerto Rico, toward the star cluster M13,

located 27,000 light-years from earth. The Arecibo transmitter has a

450,000-watt radar output that was concentrated into a radio beam

10 million times more intense than the radio emission output of our sun.

The contrast between this beam and our sun would indicate our

existence to advanced intelligent life forms in M13.

(continued)

The coded message is a set of 1679 ones and zeros. This particular number is factorable into only two primes, 23 and 73. When the ones and zeros are arranged as black and white squares in a rectangle of dimensions 23 by 73, the message appears. The decoded message was designed by Professor Frank Drake, then Director of the Arecibo Observatory.

## 6.1 POLYNOMIALS: TERMS AND PERIMETER, FACTORS AND AREA

**OBJECTIVES**  Identify monomials, binomials, and trinomials. ◆ Arrange expressions in ascending and descending order. ◆ Find the perimeter of geometric figures by adding like terms. ◆ Determine missing sides of geometric figures. ◆ Identify the largest common monomial to factor expressions. ◆ Factor numbers into primes. ◆ Determine the length and width of a rectangle, given an area in symbols. ◆ Solve problems in which the area under a graph represents work, total cost, or distance traveled. ◆ Distinguish between terms and factors.

**WARM-UP**  The concepts covered in this Warm-up are from Section 3.3.
Use the distributive property to multiply.

1. $7(x + 5)$
2. $-2(x - 3)$
3. $-6a(a - b - 2c)$
4.

| Multiply | $2a$ | $-2b$ | $+c$ |
|----------|------|-------|------|
| $5a$     |      |       |      |

Reverse the distributive property and factor.

5. $3a + 3b$
6.

| Factor |         |         |        |
|--------|---------|---------|--------|
|        | $2x^3$  | $+4x^2$ | $+6xy$ |

7. $6 \text{ cm} + 5 \text{ cm} - 2 \text{ cm}$
8. $5 \text{ kg} + 11 \text{ kg} - 2 \text{ kg}$

Add like terms. Recall that a term usually contains a sign, a number, and a variable or a variable expression.

9. $\frac{1}{2}x^2 + 3x^2 + 0.5x^2$
10. $-2ab + 3ab + \frac{1}{2}ab$
11. $5 \text{ mL} + 12 \text{ mL} - 3 \text{ mL}$
12. $4\spadesuit + 5\spadesuit - 2\spadesuit$
13. $4(x + 2) + 5(x + 2) - 2(x + 2)$

◆

In this chapter we turn our attention to computation with symbols. The relationship between computations and realistic applications is more obscure here than it was in earlier chapters. When realistic applications do occur, the complexity of the expressions and equations requires that the symbolic work be done by computers rather than by hand.

As we work through this chapter, we will focus on two items. First, we will examine how basic symbolic computations are done so that results from computers or calculators may be interpreted appropriately. Second, we will look for patterns. Many of these patterns have historical and cultural significance.

We start with addition, subtraction, multiplication, and factoring of expressions. Although the skills involved in these operations are already familiar (from Section 3.3), we will use new vocabulary and applications.

## POLYNOMIALS: LIKE TERMS AND PERIMETER

**Polynomials** *are expressions containing one or more terms being added or subtracted, where the terms are limited to numbers, constants, and/or variables with positive integer exponents.* The prefix *poly* means "many."

Historically, polynomials were used to approximate mathematical relationships such as the trigonometric functions $\sin x$, $\cos x$, and $\tan x$ or the key element in continuous compound interest, $e^x$:

$$e^x \approx 1 + x + \frac{x^2}{2} + \frac{x^3}{6} + \frac{x^4}{24} + \frac{x^5}{120}$$

These functions are now evaluated by calculators or computers, using secret algorithms known only to the manufacturers. Polynomials still have diverse and important roles in mathematics and applications, though. For example,

Party costs: $\qquad\qquad\qquad\qquad\qquad\qquad y = 70 + 3(x - 15),\ x \geq 15$

Resistance to bending by a structural beam: $\quad M_x = \dfrac{wx}{2}(l - x) - \dfrac{lwx}{2} - \dfrac{wx^2}{2}$

Deflection of a beam: $\qquad\qquad\qquad\qquad d = \dfrac{wx}{24EI}(l^3 - 2lx^2 + x^3)$

Surface area of a cylinder: $\qquad\qquad\quad \mathrm{SA} = 2\pi r^2 + 2\pi rh$

Height attained by a projectile: $\qquad\qquad h = -\frac{1}{2}gt^2 + v_0 t + h_0$

Polynomial expressions containing one, two, or three terms are called, respectively, **monomials**, **binomials**, or **trinomials**. No special names are given to expressions containing four, five, or more terms—all are polynomials. *Monomial, binomial, trinomial,* and *polynomial* are words often used to describe patterns and to give directions for exercises.

Generally, the terms in polynomials are arranged in a particular order. Listing terms in increasing order of the exponents on $x$—as in $M_x = \dfrac{lwx}{2} - \dfrac{wx^2}{2}$ (the resistance to bending by a structural beam)—is called arranging terms in **ascending order**. When we list the terms with highest exponents first, we have **descending order**. The polynomials describing the surface area of a cylinder and the height of a projectile are in descending order of $r$ and $t$, respectively. Writing polynomials in either ascending or descending order makes it easier to compare several results. Either order provides a systematic way to organize our work and to prevent careless errors.

In Section 3.4 we added like terms as a step in simplifying expressions within equations. We review that skill now in the context of polynomials and perimeter.

EXAMPLE 1    Identify the like terms in these monomial expressions.

**a.** $3x^2, -\frac{1}{2}x^2, 2x, 3x^3, 0.5x^2, 5$    **b.** $-2ab, -a^2b, +3ab, -5ab^2, \frac{1}{2}ab, \frac{1}{2}$

SOLUTION    **a.** The three like terms contain $x^2$: $3x^2, -\frac{1}{2}x^2$, and $0.5x^2$. The other terms, $2x$, $3x^3$, and 5, are not alike.

**b.** The three like terms contain $ab$: $-2ab, +3ab$, and $\frac{1}{2}ab$. The other terms, $-a^2b, -5ab^2$, and $\frac{1}{2}$, are not alike. As with the two-variable expressions discussed earlier, the variables in polynomials are arranged in alphabetical order.

♦ ♦ ♦ ♦ ♦ ♦ ♦ ♦ ♦ ♦ ♦ ♦ ♦ ♦ ♦ ♦ ♦ ♦ ♦ ♦ ♦ ♦ ♦ ♦ ♦ ♦ ♦ ♦ ♦ ♦ ♦ ♦ ♦ ♦ ♦ ♦ ♦

Addition and subtraction of like terms follow the addition and subtraction rules for positive and negative numbers. When all addition and subtraction of like terms within a polynomial is complete, the expression is said to be *simplified*.

EXAMPLE 2    Simplify by addition and/or subtraction. Identify the simplified expression as a monomial, a binomial, or a trinomial.

**a.** $3a + 4b - 2a - 8b + 4a$

**b.** $(x^2 - 2x + 1) - (x^2 + 2x + 1)$

**c.** $(2ab - 8a^2b + 5ab^2) - (6ab^2 - 9a^2b - 5ab)$

SOLUTION    **a.** $3a + 4b - 2a - 8b + 4a = 3a + (-2a) + 4a + 4b + (-8b) = 5a - 4b$
The final expression is a binomial.

**b.** $(x^2 - 2x + 1) - (x^2 + 2x + 1) = x^2 - 2x + 1 - x^2 - 2x - 1 = -4x$
The final expression is a monomial.

**c.** $(2ab - 8a^2b + 5ab^2) - (6ab^2 - 9a^2b - 5ab)$
        $= 2ab - 8a^2b + 5ab^2 - 6ab^2 + 9a^2b + 5ab = 7ab + 1a^2b - 1ab^2$
The final expression is a trinomial.

♦ ♦ ♦ ♦ ♦ ♦ ♦ ♦ ♦ ♦ ♦ ♦ ♦ ♦ ♦ ♦ ♦ ♦ ♦ ♦ ♦ ♦ ♦ ♦ ♦ ♦ ♦ ♦ ♦ ♦ ♦ ♦ ♦ ♦ ♦ ♦ ♦

We return to the concept of perimeter as an application of addition of monomials. Recall that the perimeter is the sum of the lengths of the sides of the figure. The lengths of the sides may be numbers, expressions, or numbers with units of measurement. As with terms, only numbers with like measurement units may be added.

EXAMPLE 3    Find the perimeters of these figures.

**a.** Rectangle:    **b.** Equilateral triangle:    **c.**

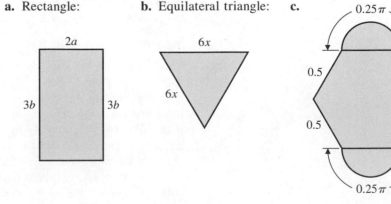

**SOLUTION**  **a.** Perimeter $= 2a + 3b + 2a + 3b = 4a + 6b$ units

**b.** Perimeter $= 6x + 6x + 6x = 18x$ units

**c.** Perimeter $= 0.5 + 0.5 + 0.25\pi + 0.5 + 0.5 + 0.25\pi = 2 + 0.5\pi$ units
(Pi may be approximated with 3.14. The perimeter is approximately 3.57 units.)

◆ ◆ ◆ ◆ ◆ ◆ ◆ ◆ ◆ ◆ ◆ ◆ ◆ ◆ ◆ ◆ ◆ ◆ ◆ ◆ ◆ ◆ ◆ ◆ ◆ ◆ ◆ ◆ ◆ ◆ ◆ ◆ ◆

POLYNOMIALS: COMMON FACTORS AND AREA

When we first added like terms, we reversed the distributive property as follows: $3a - 2a + 4a = a(3 - 2 + 4) = 5a$. The variable $a$ is a common factor in $3a - 2a + 4a$. When you are asked to *factor* an expression, you must factor out the largest possible monomial. It is customary to place the common factor in front of the parentheses so that it does not get confused with an exponent.

**EXAMPLE  4**

Factor the following expressions.

**a.** $x^2 - x$

**b.** $2ab - 8a^2b + 6ab^2$

**c.** $x^2y^2 + 2xy^3 - y^3$

**d.** $4a^3 + 8a^2 - 16a$

**e.**

| Factor | | |
|---|---|---|
| | $x^2$ | $+3x$ |

**SOLUTION**  **a.** $x^2 - x = x(x - 1)$

**b.** $2ab - 8a^2b + 6ab^2 = 2ab(1 - 4a + 3b)$

**c.** $x^2y^2 + 2xy^3 - y^3 = y^2(x^2 + 2xy - y)$

**d.** $4a^3 + 8a^2 - 16a = 4a(a^2 + 2a - 4)$

**e.**

| Factor | $x$ | $+3$ |
|---|---|---|
| $x$ | $x^2$ | $+3x$ |

◆ ◆ ◆ ◆ ◆ ◆ ◆ ◆ ◆ ◆ ◆ ◆ ◆ ◆ ◆ ◆ ◆ ◆ ◆ ◆ ◆ ◆ ◆ ◆ ◆ ◆ ◆ ◆ ◆ ◆ ◆ ◆ ◆

The table form of factoring in part e of Example 4 may be related to area if the factors are seen as representing the length and width of a rectangle. Area is modeled with algebra tiles in Figure 2 and with a rectangle in Figure 3. The width is $x$, and the length is $x + 3$. If we know the expression for the area, we can factor to find the length and width.

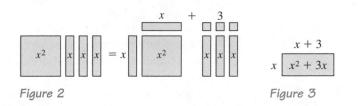

Figure 2                                    Figure 3

E X A M P L E   **5**

Find the length and width of each of the following rectangles. Because we do not know which expression is the larger, it is not necessary to specify which factor is length and which is width.

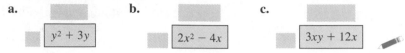

**a.**   $y^2 + 3y$     **b.**   $2x^2 - 4x$     **c.**   $3xy + 12x$

SOLUTION

The factors of each rectangle are shown, along with the factorization in symbolic form.

**a.** $y^2 + 3y = y(y + 3)$        **b.** $2x^2 - 4x = 2x(x - 2)$

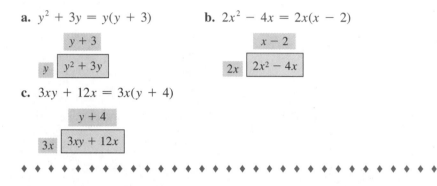

**c.** $3xy + 12x = 3x(y + 4)$

◆ ◆ ◆ ◆ ◆ ◆ ◆ ◆ ◆ ◆ ◆ ◆ ◆ ◆ ◆ ◆ ◆ ◆ ◆ ◆ ◆ ◆ ◆ ◆ ◆ ◆ ◆ ◆ ◆ ◆ ◆ ◆ ◆ ◆ ◆

In Example 6 we are reminded of the number of different rectangles that may be formed with a given area. This is an important geometric concept, and it is the basis for several investigations in later mathematics courses. The method of listing factors shown in Example 6 is used in the next section.

E X A M P L E   **6**

Find all the whole number factor pairs for these rectangular areas:

**a.** 45 square inches        **b.** 60 square meters
**c.** 28 square centimeters   **d.** 36 square feet

SOLUTION

Arranging the factor pairs in numerical order will help us make a complete list.

| **a.** Inches: | **b.** Meters: | **c.** Centimeters: | **d.** Feet: |
|---|---|---|---|
| $1 \cdot 45$ | $1 \cdot 60$ | $1 \cdot 28$ | $1 \cdot 36$ |
| $3 \cdot 15$ | $2 \cdot 30$ | $2 \cdot 14$ | $2 \cdot 18$ |
| $5 \cdot 9$ | $3 \cdot 20$ | $4 \cdot 7$ | $3 \cdot 12$ |
| | $4 \cdot 15$ | | $4 \cdot 9$ |
| | $5 \cdot 12$ | | $6 \cdot 6$ |
| | $6 \cdot 10$ | | |

◆ ◆ ◆ ◆ ◆ ◆ ◆ ◆ ◆ ◆ ◆ ◆ ◆ ◆ ◆ ◆ ◆ ◆ ◆ ◆ ◆ ◆ ◆ ◆ ◆ ◆ ◆ ◆ ◆ ◆ ◆ ◆ ◆ ◆ ◆

**Area Applications.**   The next three examples show applications of areas bounded by expressions on coordinate axes. The expressions are either numbers or variables, but the concepts apply to any polynomial used as a boundary. In the first two examples the areas are rectangles determined by connecting the described coordinate point to the two axes.

E X A M P L E   **7**

**Work done in lifting**   If one axis is labeled with the weight of an object and the other with the vertical distance an object is lifted, then each point $(x, y)$ on the graph is (weight, height lifted). The area of the rectangle formed by connecting $(x, y)$ with the axes is the work done in lifting the object.

**a.** Find the work done in lifting a 25-pound child to a height of 4 feet.

**b.** Find the work done in lifting an *x*-pound child to a height of 4 feet.

**c.** Find the work done in lifting a 25-pound child to a height of *y* feet.

SOLUTION   The work done in lifting the child is the area.

**a.** 25 lb · 4 ft = 100 foot-pounds
See Figure 4.

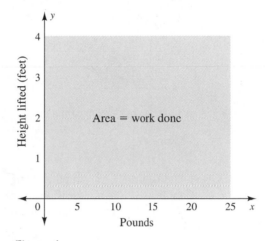

Figure 4

**b.** *x* lb · 4 ft = 4*x* foot-pounds

**c.** 25 lb · *y* ft = 25*y* foot-pounds

NOTE   The units for work are foot-pounds, not feet-pounds.

◆ ◆ ◆ ◆ ◆ ◆ ◆ ◆ ◆ ◆ ◆ ◆ ◆ ◆ ◆ ◆ ◆ ◆ ◆ ◆ ◆ ◆ ◆ ◆ ◆ ◆ ◆ ◆ ◆ ◆ ◆ ◆ ◆ ◆ ◆

EXAMPLE   **8**

Total cost   In economics, if one axis is labeled with items purchased and the other with cost per item, then each coordinate (*x*, *y*) is (item, cost). The rectangular area formed by drawing lines from the point (*x*, *y*) to the axes has an area equal to the total cost.

**a.** Find the total cost of 100 items purchased at a cost of $50 per item.

**b.** Suppose we need a minimum of 100 items. Find the total cost of 100 + *x* items purchased at a cost of $50 per item.

SOLUTION   The total cost is the area.

**a.** $50 · 100 = $5000
See Figure 5.

**b.** $50(100 + *x*) = $5000 + $50*x*

◆ ◆ ◆ ◆ ◆ ◆ ◆ ◆ ◆ ◆ ◆ ◆ ◆ ◆ ◆ ◆ ◆ ◆ ◆ ◆ ◆ ◆ ◆ ◆ ◆ ◆ ◆ ◆ ◆ ◆ ◆ ◆ ◆ ◆ ◆

NOTE   The cost in an economics model need not be a horizontal line. It could be a line with a negative slope, implying that the cost per item decreases as we increase the number of items purchased. The cost graph may have many other shapes as well.

In the next example the area model provides a way to solve a problem that is not as easily solved by other methods. If the axes are labeled with time and rate, then the area represents the total distance traveled. Because the rate is changing, the usual *D* = *rt* formula for distance will not work.

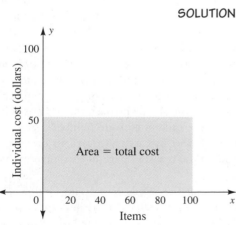

Figure 5

EXAMPLE 9

**Distance traveled**　　Figure 6 shows a constant acceleration from 0 to 55 miles per hour in one minute. Find the total distance traveled, in feet.

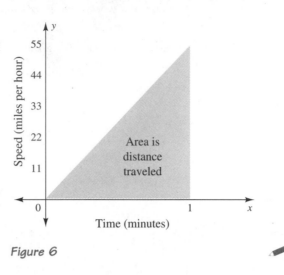

Figure 6

SOLUTION　　The figure on the graph is a triangle rather than a rectangle. The area of a triangle is $\frac{1}{2} \cdot b \cdot h$, or $\frac{1}{2}$ times 1 minute times 55 miles per hour.

We need to use unit analysis because the rate is in miles per hour and the time is in minutes.

$$\text{Distance traveled} = \frac{1}{2} \cdot (1 \text{ min}) \left( 55 \frac{\text{mi}}{\text{hr}} \right) \left( \frac{1 \text{ hr}}{60 \text{ min}} \right) \left( \frac{5280 \text{ ft}}{1 \text{ mi}} \right) = 2420 \text{ ft}$$

♦ ♦ ♦ ♦ ♦ ♦ ♦ ♦ ♦ ♦ ♦ ♦ ♦ ♦ ♦ ♦ ♦ ♦ ♦ ♦ ♦ ♦ ♦ ♦ ♦ ♦ ♦ ♦ ♦ ♦ ♦ ♦ ♦ ♦

Areas under graphs also have application in probability, electronics, and physics.

### VOCABULARY

In Section 6.2 we will factor polynomial expressions into primes. **Prime numbers** *are numbers that are divisible by only two factors, 1 and the number itself. An expression is prime if the largest common factor is 1.* When we factor an expression completely, we are finding *prime factors.* The expression $abc$ has three prime factors: $a$, $b$, and $c$. It may surprise you that we can think of expressions as prime. As mentioned earlier, $(a - b)(a + b)$ is a two-factor expression. Each factor, $(a - b)$ and $(a + b)$, is a prime expression having 1 as its largest common factor.

Numbers such as 91, 111, 129, 1001, and 1679 are composite numbers because they can be factored. **Composite numbers** *are formed by multiplying prime numbers. Composite* has a similar meaning in police work, where artists (and now computer programs) help witnesses identify suspects by building composite faces from a variety of facial features. The prime numbers might be thought of as the "facial features" building each composite number.

The factors of 1679 are 23 and 73. These factors were used in designing the radio telescope message described on the opening page of this chapter and in the Back to the Future box.

As mentioned in Chapter 3, confusing the words *terms* and *factors* is a common error. *Factors* are the numbers or variables being multiplied in a product. *Terms* are the expressions being added or subtracted. The product $(a - b)(a + b)$ has two factors, and each factor is a two-term expression.

---

**JUST FOR FUN**

Scientists assume that intelligent extra-terrestrials will factor 1679 into $23 \cdot 73$ and find the message in Figure 1 on the chapter-opening page.

Using zeros and ones to transmit data is common on earth. Dot matrix printers, an early form of printer for personal computers, relied on a rectangular array of dots to form each letter or character. Suppose a dot matrix printer head has a rectangular array, 8 dots wide by 9 dots tall. Decode the following message.

0011110001000010010000100011110001000010010000100011110000000000000000000

*Solution*: Set up a rectangular array, 8 numbers wide and 9 numbers tall. Rearrange the numbers into the new shape:

00111100
01000010
01000010
00111100
01000010
01000010
00111100
00000000
00000000

The digit 8 appears.

---

EXAMPLE **10**  How many terms are there in these expressions?

**a.** $x^2 + 4x + 2$        **b.** $-b + \sqrt{d}$    **c.** $3 - \sqrt{2}$
**d.** $1 + 3y + 3y^2 + y^3$    **e.** $x^2 - y^2$

SOLUTION  **a.** 3 terms    **b.** 2 terms    **c.** 2 terms    **d.** 4 terms    **e.** 2 terms

◆ ◆ ◆ ◆ ◆ ◆ ◆ ◆ ◆ ◆ ◆ ◆ ◆ ◆ ◆ ◆ ◆ ◆ ◆ ◆ ◆ ◆ ◆ ◆ ◆ ◆ ◆ ◆ ◆ ◆ ◆

EXAMPLE **11**  How many factors are there in each of these products?

**a.** $wxyz$        **b.** $(x + y)(x - y)$            **c.** $x(x + 1)(x + 2)$
**d.** $n(n - 1)$    **e.** $(x - 1)(x^2 + x + 1)$

SOLUTION  **a.** 4 factors    **b.** 2 factors    **c.** 3 factors    **d.** 2 factors    **e.** 2 factors

◆ ◆ ◆ ◆ ◆ ◆ ◆ ◆ ◆ ◆ ◆ ◆ ◆ ◆ ◆ ◆ ◆ ◆ ◆ ◆ ◆ ◆ ◆ ◆ ◆ ◆ ◆ ◆ ◆ ◆ ◆

---

**EXERCISES 6.1**

---

**Exercises 1 and 2 review integer addition and subtraction.**

1. a. $-9 + 4$      b. $-8 + 3$      c. $-5 - 8$
   d. $-7 - 3$      e. $-9 - (-5)$   f. $3 - (-4)$

2. a. $-4 + 6$      b. $-5 + 1$      c. $-4 - 3$
   d. $-3 - 9$      e. $9 - (-8)$    f. $-5 - (-1)$

**In Exercises 3 and 4 add like terms.**

3. a. $5a + 3b - 2c - 4a - 6c + 9b$

   b. $6m + 2n - 6p - 3m - 3n + 6p$

   c. 8 cm + 5 cm + 5 cm − 5 cm + 8 cm

   d. $(x^2 + 3x) - (4x + 12)$

   e. $(x^2 - 2x + 3) - (2x^2 - 4x + 6)$

4. a. $a - 8c - d + a + 4d - 6c$

   b. $5m - 4p + 8n - 9p + 8m - 7n$

   c. 9 ft + 7 ft + 7 ft + 8 ft − 2 ft

   d. $(2x^2 + 4x) - (3x - 4)$

   e. $(y^2 + 3y - 2) - (3y^2 + 9y - 6)$

**Borrowing may be needed in Exercises 5 to 8 so that no answer has a negative term.**

5. (4 yards + 2 feet + 5 inches) − (2 yards + 2 feet + 3 inches)

6. (7 yards + 2 feet + 11 inches) − (3 yards + 1 foot + 8 inches)

7. (3 days + 3 hours + 10 minutes) −
   (1 day + 8 hours + 5 minutes)

8. (5 days + 2 hours + 25 minutes) −
   (4 days + 1 hour + 35 minutes)

**In Exercises 9 to 12 add like terms.**

9. a. $x^2 + 2x + 3x + 6$    b. $3x^2 + 6x + x + 2$

   c. $5 - 4(x - 3)$    d. $6x^2 + 2x + 3x + 1$

   e. $a^2 - ab + ab - b^2$

10. a. $x^2 + x + 2x + 2$    b. $2x^2 + 3x + 4x + 6$

    c. $3 + 4(x - 5)$    d. $2x^2 - 2x + 3x - 3$

    e. $a^2 + b^2 + ab + ab$

11. a. $2y(-x + 2y) + x(x - 2y)$

    b. $x^2 - 4y^2 - 2xy + 2xy$

    c. $x^3 + 2x^2 + x + x^2 + 2x + 1$

    d. $x(4 + 4x + x^2) - 2(4 + 4x + x^2)$

    e. $b^3 - ab^2 + a^2b + a^3 - a^2b + ab^2$

12. a. $3y(3y + x) - x(x + 3y)$

    b. $-9y^2 + 3xy + x^2 - 3xy$

    c. $-x^2 + 2x - 1 + x^3 - 2x^2 + x$

    d. $2(4 - 2x + x^2) + x(4 - 2x + x^2)$

    e. $a^3 - a^2b + ab^2 - b^3 - ab^2 + a^2b$

13. Identify each answer to Exercises 9 and 11 as a monomial, a binomial, a trinomial, or a polynomial with more than three terms.

14. Identify each answer to Exercises 10 and 12 as a monomial, a binomial, a trinomial, or a polynomial with more than three terms.

15. Describe in complete sentences how to determine whether two terms are "like" terms.

16. Describe in complete sentences how to determine a common monomial factor.

**Factor the expressions in Exercises 17 to 24.**

17. $x^3 + 4x^2 + 4x$    18. $x^2y + xy^2 + y^3$

19. $a^2b + ab^2 + b^3$    20. $a^3 - a^2b + ab^2$

21. $6x^2 + 2x$    22. $12y^2 + 6y$

23. $15y^2 - 3y$    24. $8x^2 - 4x$

**Factor the expressions in Exercises 25 to 28 in two ways. First, use a positive monomial factor. Second, use a negative monomial factor.**

25. $-4x - 12$    26. $-2xy + 4y^2$

27. $-2x^2 - 8x - 8$    28. $-x^2y - xy^2 - y^3$

**In Exercises 29 to 32 factor each table.**

29.

| Factor | | | |
|---|---|---|---|
| | $y^3$ | $+2xy^2$ | $+x^2y$ |

30.

| Factor | | | |
|---|---|---|---|
| | $x^2y^3$ | $-x^3y^2$ | $+x^2y^2$ |

31.

| Factor | | | |
|---|---|---|---|
| | $2a^2b$ | $-4ab^2$ | $+6ab^3$ |

32.

| Factor | | | |
|---|---|---|---|
| | $4ab^3$ | $-8a^2b^3$ | $-16a^2b^2$ |

**Guess-and-check on a calculator to factor the numbers in Exercises 33 to 36 into primes.**

33. 111    34. 91    35. 1001    36. 129

**Factor and simplify the expressions in Exercises 37 and 38.**

37. a. $\dfrac{36}{99}$   b. $\dfrac{66}{990}$   c. $\dfrac{185}{999}$   d. $\dfrac{mn}{mp}$   e. $\dfrac{4np}{24mn}$

38. a. $\dfrac{407}{999}$   b. $\dfrac{108}{999}$   c. $\dfrac{39}{1001}$   d. $\dfrac{xz}{xy}$   e. $\dfrac{3xy}{9yz}$

**Make an organized list of all possible whole number length and width pairs for the rectangles whose areas are given in Exercises 39 to 44.**

39. 24 square feet    40. 35 square inches

41. 48 square meters    42. 56 square centimeters

43. 54 square yards    44. 64 square miles

**Find the length and width for the areas in Exercises 45 and 46.**

45. a.    b.

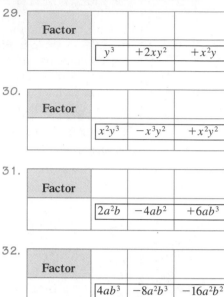

$2b$

$2a$ $(2a - 4ab)$     $x^2y + xy^2$

46. a.    b.

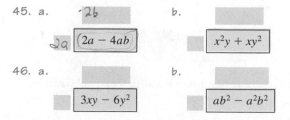

$3xy - 6y^2$     $ab^2 - a^2b^2$

47. Find the perimeter of each shape.

a.

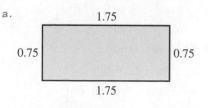

1.75

0.75        0.75

1.75

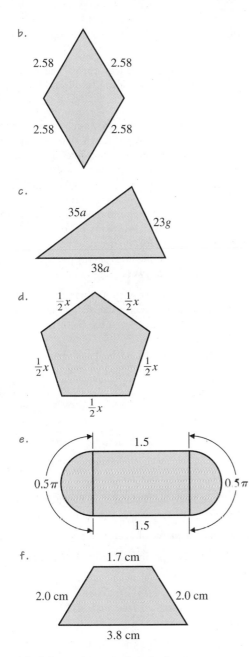

b.

2.58    2.58

2.58    2.58

c.

35a    23g

38a

d.

$\frac{1}{2}x$    $\frac{1}{2}x$

$\frac{1}{2}x$    $\frac{1}{2}x$

$\frac{1}{2}x$

e.

1.5

$0.5\pi$    $0.5\pi$

1.5

f.

1.7 cm

2.0 cm    2.0 cm

3.8 cm

48. Find the perimeter of each shape.

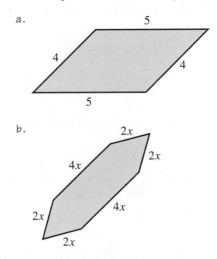

a.

5

4    4

5

b.

2x

4x    2x

2x    4x

2x

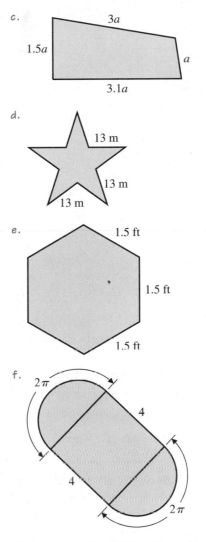

c.

3a

1.5a    a

3.1a

d.

13 m

13 m

13 m

e.

1.5 ft

1.5 ft

1.5 ft

f.

$2\pi$

4

4

$2\pi$

49. Find an expression for the side or sides labeled with a question mark in each shape below.

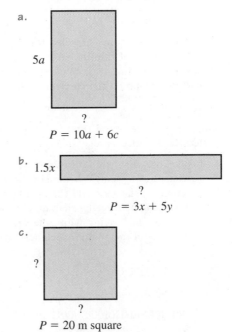

a.

5a

?

$P = 10a + 6c$

b.    1.5x

?

$P = 3x + 5y$

c.

?

?

?

$P = 20$ m square

50. Find an expression for the side or sides labeled with a question mark in each shape below.

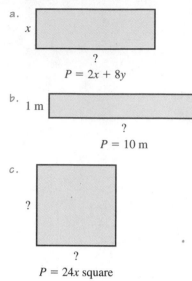

a.

$x$

?

$P = 2x + 8y$

b.

1 m

?

$P = 10$ m

c.

?

?

$P = 24x$ square

**For Exercises 51 to 55, use the reasoning in Example 7 to determine how much work is done.**

51. Lifting a 10-pound baby to a height of 5 feet

52. Lifting a 20-pound baby to a height of 3 feet

53. Lifting a 5-pound baby to a height of 4 feet

54. Lifting a 30-pound child to a height of $y$ feet

55. Lifting an $x$-pound child to a height of $y$ feet

56. Lifting a 30-pound child to a height of 4 feet is the same amount of work as lifting a 20-pound child to a height of how many feet?

**For Exercises 57 and 58 refer back to Example 8 and suppose the price per item increases from \$50 to (50 + $y$) dollars.**

57. How much will 100 items cost?

58. How much will 100 + $x$ items cost?

**In Example 9 we found the total distance traveled in accelerating from 0 to 55 miles per hour in 1 minute. In Exercises 59 to 61 use the given times for total time, and calculate the distance traveled. Assume 0 to 55 mph on the vertical axis.**

59. 8 seconds    60. 10 seconds    61. $x$ seconds

62. Suppose in Example 9 we accelerated from 0 to $y$ mph in 10 seconds. Calculate the distance traveled in terms of $y$.

63. Suppose a 15-digit coded message, similar to the one on the opening page of this chapter, was sent. What are the factors of 15? Into what rectangular shapes could the data be arranged? Decode the following message by shading a square in the rectangle for each 1 and leaving white the squares representing zeros. Only one of the possible rectangles will make sense.

101011110110101

64. Repeat Exercise 63 for a 50-digit message.

11000101000100010101110001011101000100011101010001

65. Why was the coded message on the chapter opener sent in symbols rather than in words?

66. There is a saying "You cannot add apples and oranges." How does this saying apply to adding like terms?

PROJECTS

67. **Dessert Distribution.** At a party an unknown number of people are seated at a table of unknown size. A platter of small desserts is passed around the table. Each person takes one dessert and passes the platter on to the next person. The platter keeps getting passed around the table until all the desserts are gone. The platter originally contained 62 desserts. Gloria took the first and the next to last dessert (and possibly others as well). How many people were seated at the table? There are several answers. Find as many answers as possible. Determine how many desserts Gloria received for each of your answers.

68. **Factors.** The smallest number with just one factor is 1. The smallest number with exactly two factors is 2 (1 · 2). The smallest number with exactly three factors is 4 (1 · 4, 2 · 2). The smallest number with exactly four factors is 6 (1 · 6, 2 · 3). Find the smallest number with exactly

a. five factors    b. six factors    c. seven factors

d. eight factors    e. nine factors    f. ten factors

What do all numbers with an odd number of factors have in common?

69. **Rectangles: Areas and Perimeters**

a. On graph paper, draw the rectangles in Example 6, parts b and d, to scale.

b. Determine the perimeter of each rectangle.

c. What is the smallest perimeter possible for each given area?

d. Is there any smaller perimeter for each area? (*Hint*: Suppose we used decimal lengths and widths instead of whole numbers.)

e. Describe a rule for finding the smallest perimeter for a given rectangular area.

f. Test your rule on a rectangle with an area of 72 square centimeters.

70. **CD Earnings.*** Eli and Shana are calculating the value of a certificate of deposit (CD) at the end of each year. The rate of interest, $r$, is 4%. The starting amount in the CD, $P$, is \$500. Eli's formulas are shown in the first table. Shana finds another set of formulas, shown in the second table.

a. Calculate the amount of money after each year, using the two formulas.

b. What do you observe?

c. Look for a pattern in each table in order to calculate the value of the CD for 4, 5, and 10 years.

d. Which table would you rather use?

e. Could either table be used to make a formula for $t$ years?

*This project was suggested by Charlotte Hutt, Southwestern Oregon Community College.

| Years, $n$ | Rule | Amount, $ |
|:---:|:---:|:---:|
| 1 | $P + Pr$ | |
| 2 | $P + 2Pr + Pr^2$ | |
| 3 | $P + 3Pr + 3Pr^2 + Pr^3$ | |
| 4 | | |
| 5 | | |
| 10 | | |
| $t$ | | |

*Eli's Formulas*

| Years, $n$ | Rule | Amount, $ |
|:---:|:---:|:---:|
| 1 | $P(1 + r)$ | |
| 2 | $P(1 + r)^2$ | |
| 3 | $P(1 + r)^3$ | |
| 4 | | |
| 5 | | |
| 10 | | |
| $t$ | | |

*Shana's Formulas*

## 6.2 MULTIPLYING POLYNOMIALS AND FACTORING

**OBJECTIVES**    Use the distributive property in a table twice to multiply a binomial times either a binomial or a trinomial. ♦ Identify the product of binomials as a binomial pair. ♦ Observe patterns in the products of binomial pairs. ♦ Observe the equivalent product of diagonal terms in the table. ♦ Use the equivalent diagonal property to factor trinomials in a table.

**WARM-UP**    Complete the table.

| $m$ | $n$ | $m + n$ | $m \cdot n$ |
|:---:|:---:|:---:|:---:|
| 3 | 4 | | |
| | | 8 | 12 |
| | | 8 | 15 |
| 1 | 15 | | |
| 4 | 6 | | |
| | | 11 | 24 |
| 2 | 12 | | |
| $-4$ | 6 | | |
| $-4$ | $-6$ | | |
| 2 | $-12$ | | |
| $-2$ | $-12$ | | |
| | | 5 | $-24$ |
| | | $-11$ | 24 |
| | | $-4$ | $-12$ |
| | | $-8$ | 15 |

♦

We now focus on multiplication of polynomial expressions and on the opposite operation—factoring. Polynomials may be multiplied and factored by any one of several methods. We will emphasize an extension of the table method used in earlier sections. If you are familiar with another way to multiply and factor polynomial expressions, try the table method in this section and then decide which you like better.

TABLE METHOD FOR MULTIPLICATION

The table method for multiplying polynomial expressions is based on finding the area of rectangles that have polynomial expressions for length and width.

In Example 8 of Section 6.1 the total cost of a purchase of 100 items at $50 each was the area inside a rectangle that was placed on coordinate axes labeled with number of items and individual cost.

E X A M P L E    *1*    **Total cost**    If the number of items purchased increases by $x$ units and the cost increases by $y$ dollars, as shown in Figure 7, what is the new cost?

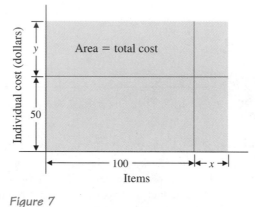

Figure 7

SOLUTION    The new cost is shown by the areas in Figure 8. The areas are the product $(100 + x)(50 + y)$, or $5000 + 50x + 100y + xy$.

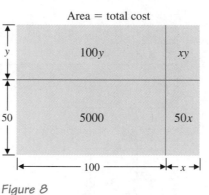

Figure 8

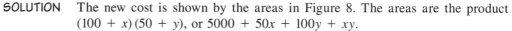

If we use the area of a rectangle as a model, we may then write any polynomial multiplication as the product of a length and a width. The polynomial expressions are placed in a table to keep everything lined up. The expressions representing length and width are placed across the top and down the left side, respectively. The area is the inner part of the table, surrounded by the black box.

E X A M P L E    *2*    Multiply $(x + 2)(x + 3)$.

SOLUTION

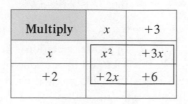

| Multiply | $x$ | $+3$ |
|----------|-----|------|
| $x$ | $x^2$ | $+3x$ |
| $+2$ | $+2x$ | $+6$ |

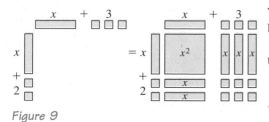

Figure 9

The product obtained with multiplication by table is also obtained with algebra tiles in Figure 9.

After using the distributive property twice to obtain the areas, we add like terms. Thus

$$(x + 2)(x + 3) = x^2 + 2x + 3x + 6 = x^2 + 5x + 6$$

◆ ◆ ◆ ◆ ◆ ◆ ◆ ◆ ◆ ◆ ◆ ◆ ◆ ◆ ◆ ◆ ◆ ◆ ◆ ◆ ◆ ◆ ◆ ◆ ◆ ◆ ◆ ◆ ◆ ◆ ◆ ◆ ◆ ◆ ◆ ◆

Example 2 illustrated the product of binomials. We use the same idea to multiply any two polynomials. In Example 3 we multiply to obtain the product of a binomial and a trinomial, and in Example 4, the product of two trinomials.

**E X A M P L E   3**

Set up a table, and multiply $(x - 2)(x^2 + 2x + 3)$.

**SOLUTION**

| Multiply | $x^2$ | $+2x$ | $+3$ |
|----------|-------|-------|------|
| $x$      | $x^3$ | $+2x^2$ | $+3x$ |
| $-2$     | $-2x^2$ | $-4x$ | $-6$ |

We add like terms to obtain the product. Thus

$$(x - 2)(x^2 + 2x + 3) = x^3 + 2x^2 - 2x^2 + 3x - 4x - 6 = x^3 - x - 6$$

◆ ◆ ◆ ◆ ◆ ◆ ◆ ◆ ◆ ◆ ◆ ◆ ◆ ◆ ◆ ◆ ◆ ◆ ◆ ◆ ◆ ◆ ◆ ◆ ◆ ◆ ◆ ◆ ◆ ◆ ◆ ◆ ◆ ◆ ◆ ◆

**E X A M P L E   4**

Set up a table, and multiply $(x^2 + 2x + 1)(x^2 - x - 2)$.

**SOLUTION**

| Multiply | $x^2$ | $-x$ | $-2$ |
|----------|-------|------|------|
| $x^2$    | $x^4$ | $-x^3$ | $-2x^2$ |
| $+2x$    | $+2x^3$ | $-2x^2$ | $-4x$ |
| $+1$     | $+x^2$ | $-x$ | $-2$ |

All like terms are located along diagonal lines from lower left to upper right. Adding the small areas gives

$$x^4 + x^3 - 3x^2 - 5x - 2$$

a five-term polynomial.

◆ ◆ ◆ ◆ ◆ ◆ ◆ ◆ ◆ ◆ ◆ ◆ ◆ ◆ ◆ ◆ ◆ ◆ ◆ ◆ ◆ ◆ ◆ ◆ ◆ ◆ ◆ ◆ ◆ ◆ ◆ ◆ ◆ ◆ ◆ ◆

### PATTERNS IN BINOMIAL PRODUCTS

We have noted that the like terms are located along diagonal lines. This is true if the polynomial terms are listed in descending order. We now look at several more patterns in these tables. To find our next pattern we look at several products of the form $(x + a)(x + b)$. *The product $(x + a)(x + b)$ is called a* **binomial pair**.

EXAMPLE 5    Use the tables to calculate the following binomial products. Look for a pattern between the numbers in the original factors and the numbers in the answers. Comment on the results.

**a.** $(x + 1)(x + 12)$          **b.** $(x + 2)(x + 6)$

| Multiply | $x$ | $+12$ |
|---|---|---|
| $x$ | | |
| $+1$ | | |

| Multiply | $x$ | $+6$ |
|---|---|---|
| $x$ | | |
| $+2$ | | |

**c.** $(x + 3)(x + 4)$          **d.** $(x - 3)(x - 4)$

| Multiply | $x$ | $+4$ |
|---|---|---|
| $x$ | | |
| $+3$ | | |

| Multiply | $x$ | $-4$ |
|---|---|---|
| $x$ | | |
| $-3$ | | |

**e.** $(x - 2)(x - 6)$          **f.** $(x - 1)(x - 12)$

| Multiply | $x$ | $-6$ |
|---|---|---|
| $x$ | | |
| $-2$ | | |

| Multiply | $x$ | $-12$ |
|---|---|---|
| $x$ | | |
| $-1$ | | |

SOLUTION

| *Binomial pair* | | *Trinomial product* |
|---|---|---|
| **a.** $(x + 1)(x + 12)$ | $=$ | $x^2 + 13x + 12$ |
| **b.** $(x + 2)(x + 6)$ | $=$ | $x^2 + 8x + 12$ |
| **c.** $(x + 3)(x + 4)$ | $=$ | $x^2 + 7x + 12$ |
| **d.** $(x - 3)(x - 4)$ | $=$ | $x^2 - 7x + 12$ |
| **e.** $(x - 2)(x - 6)$ | $=$ | $x^2 - 8x + 12$ |
| **f.** $(x - 1)(x - 12)$ | $=$ | $x^2 - 13x + 12$ |

Each product is of the form $x^2 \pm \Box x + 12$, with the middle term formed by the sum of the like terms on a diagonal. The symbol $\pm$ means "plus or minus." The coefficient on the middle term is the sum of the numbers in the binomial pair. The last term in the product is obtained by multiplying the numbers in the binomial pair.

♦ ♦ ♦ ♦ ♦ ♦ ♦ ♦ ♦ ♦ ♦ ♦ ♦ ♦ ♦ ♦ ♦ ♦ ♦ ♦ ♦ ♦ ♦ ♦ ♦ ♦ ♦ ♦ ♦ ♦ ♦ ♦ ♦ ♦ ♦ ♦

In Example 6 we explore a trinomial similar to those in Example 5.

EXAMPLE 6    List all binomial pairs whose trinomial product is $x^2 \pm \Box x - 12$. (The symbol $\pm$ indicates either addition or subtraction.) Comment on any number patterns in the products.

SOLUTION    *Binomial pair*          *Trinomial product*

$$(x + 1)(x - 12) \ = \ x^2 - 11x - 12$$
$$(x + 2)(x - 6) \ = \ x^2 - 4x - 12$$
$$(x + 3)(x - 4) \ = \ x^2 - 1x - 12$$
$$(x - 3)(x + 4) \ = \ x^2 + 1x - 12$$
$$(x - 2)(x + 6) \ = \ x^2 + 4x - 12$$
$$(x - 1)(x + 12) \ = \ x^2 + 11x - 12$$

The coefficient of the middle term in the product is the sum of the numbers in the binomial pair.

♦ ♦ ♦ ♦ ♦ ♦ ♦ ♦ ♦ ♦ ♦ ♦ ♦ ♦ ♦ ♦ ♦ ♦ ♦ ♦ ♦ ♦ ♦ ♦ ♦ ♦ ♦ ♦ ♦ ♦ ♦ ♦ ♦ ♦ ♦ ♦ ♦

The terms in the inner diagonal have another pattern, as indicated in Example 7.

EXAMPLE  **7**    Multiply $(3x + 1)$ and $(2x - 3)$. Calculate the product of the two expressions forming the diagonals of the answer section of the table. Compare the diagonal products.

SOLUTION

| **Multiply** | $2x$ | $-3$ |
|---|---|---|
| $3x$ | $6x^2$ | $-9x$ |
| $+1$ | $+2x$ | $-3$ |

The product is

$$(3x + 1)(2x - 3) = 6x^2 - 9x + 2x - 3 = 6x^2 - 7x - 3$$

Within the answer section of the table, the diagonal containing $6x^2$ and $-3$ has the product $-18x^2$. The other diagonal, containing $2x$ and $-9x$, also has the product $-18x^2$. The diagonal products are equal.

♦ ♦ ♦ ♦ ♦ ♦ ♦ ♦ ♦ ♦ ♦ ♦ ♦ ♦ ♦ ♦ ♦ ♦ ♦ ♦ ♦ ♦ ♦ ♦ ♦ ♦ ♦ ♦ ♦ ♦ ♦ ♦ ♦ ♦ ♦ ♦ ♦

Does the same pattern of equal diagonal products hold true for the binomial product tables in Examples 1 and 2? Is there a similar pattern in the answers in Examples 3 and 4?

---

Within the table for multiplication of binomials, the diagonals have the same product.

---

### FACTORING

The examples thus far in this section have focused on multiplication of polynomials. We now examine the products, or answers to the multiplications, and work backward to the original factors.

The table form of factoring is called the **factor by table method**, and it provides a means to determine the factors in a systematic way. A key step in our method is to *factor the tables*, or *remove common factors or terms from each row and column in the answer section*.

EXAMPLE  **8**     Factor $6x^2 - 15x + 4x - 10$, using a table.

| Factor | | |
|---|---|---|
| | $6x^2$ | $-15x$ |
| | $4x$ | $-10$ |

**a.** Factor the common monomial from each row and column.

**b.** Write the original binomial pair.

**c.** Verify that diagonal terms multiply to the same product.

**d.** Summarize the table with an equivalent expression and its factored form.

SOLUTION  **a.**

| Factor | $2x$ | $-5$ |
|---|---|---|
| $3x$ | $6x^2$ | $-15x$ |
| $+2$ | $+4x$ | $-10$ |

**b.** The binomial pair is $(3x + 2)(2x - 5)$.

**c.** The diagonals both multiply to $-60x^2$.

**d.** The expression $6x^2 - 11x - 10$ has factors $(3x + 2)$ and $(2x - 5)$.

*Check*: The factors multiply to give the inner "area" of the table.  ✔

♦ ♦ ♦ ♦ ♦ ♦ ♦ ♦ ♦ ♦ ♦ ♦ ♦ ♦ ♦ ♦ ♦ ♦ ♦ ♦ ♦ ♦ ♦ ♦ ♦ ♦ ♦ ♦ ♦ ♦ ♦ ♦ ♦ ♦ ♦ ♦

In Example 8 all four parts of the answer section were given. In the next examples we fill in the answer section by finding what terms were added to get the middle term.

EXAMPLE  **9**     Factor $x^2 + 11x + 24$ into a binomial pair, $(x + \Box)(x + \Box)$, using the table method.

SOLUTION     The terms $x^2$ and 24 are placed directly into the table, but $11x$ is the sum of two diagonal terms and must be split up.

| Factor | | |
|---|---|---|
| | $x^2$ | |
| | | $+24$ |

Although we might guess the terms that add to $11x$, it is better to focus on the product of the other diagonal terms, $24x^2$. We list the factors of $24x^2$ and look for factors that add to $11x$:

$$1x \cdot 24x$$
$$2x \cdot 12x$$
$$3x \cdot 8x$$
$$4x \cdot 6x$$

Because $3x$ and $8x$ add to $11x$, we enter $3x$ and $8x$ on the remaining diagonal and factor as before. The order of the terms $3x$ and $8x$ does not matter.

| Factor | | |
|--------|--------|--------|
| | $x^2$ | $+3x$ |
| | $+8x$ | $+24$ |

| Factor | $x$ | $+3$ |
|--------|--------|--------|
| $x$ | $x^2$ | $+3x$ |
| $+8$ | $+8x$ | $+24$ |

The factors of $x^2 + 11x + 24$ are $(x + 8)$ and $(x + 3)$.

***Check***: The factors multiply to give the inner "area" of the table.   ✔

♦ ♦ ♦ ♦ ♦ ♦ ♦ ♦ ♦ ♦ ♦ ♦ ♦ ♦ ♦ ♦ ♦ ♦ ♦ ♦ ♦ ♦ ♦ ♦ ♦ ♦ ♦ ♦ ♦ ♦ ♦ ♦ ♦ ♦ ♦ ♦ ♦

**E X A M P L E    10**    Factor $3x^2 + 13x + 12$ into a binomial pair, $(\square x + \square)(\square x + \square)$, using the table method.

**SOLUTION**    Enter $3x^2$ and 12. Determine the two terms that add to $13x$ by first finding the diagonal product, $(3x^2)(12) = 36x^2$.

| Factor | | |
|--------|--------|--------|
| | $3x^2$ | |
| | | $+12$ |

List factors of the diagonal product, $36x^2$, and look for those that add to $13x$:

$$1x \cdot 36x$$
$$2x \cdot 18x$$
$$3x \cdot 12x$$
$$4x \cdot 9x$$
$$6x \cdot 6x$$

The factors adding to $13x$ are $4x$ and $9x$. Enter $4x$ and $9x$ in the other diagonal, and factor to complete the table.

| Factor | | |
|--------|--------|--------|
| | $3x^2$ | $+9x$ |
| | $+4x$ | $+12$ |

| Factor | $x$ | $+3$ |
|--------|--------|--------|
| $3x$ | $3x^2$ | $+9x$ |
| $+4$ | $+4x$ | $+12$ |

Thus

$$3x^2 + 13x + 12 = (3x + 4)(x + 3)$$

***Check***: The factors multiply to give the inner "area" of the table.   ✔

♦ ♦ ♦ ♦ ♦ ♦ ♦ ♦ ♦ ♦ ♦ ♦ ♦ ♦ ♦ ♦ ♦ ♦ ♦ ♦ ♦ ♦ ♦ ♦ ♦ ♦ ♦ ♦ ♦ ♦ ♦ ♦ ♦ ♦ ♦ ♦ ♦

EXAMPLE  **11**     Factor $4x^2 - 9$ into a binomial pair, $(\Box x + \Box)(\Box x + \Box)$.

SOLUTION

| Factor | | |
|---|---|---|
| | $4x^2$ | |
| | | $-9$ |

Place $4x^2$ and $-9$ in the diagonal. There is no middle term, so the other diagonal terms add to zero. The diagonal product is $-36x^2$, so start listing factors of $-36x^2$. Continue until the factors add to zero.

$$1x \cdot -36x$$
$$2x \cdot -18x$$
$$3x \cdot -12x$$
$$4x \cdot -9x$$
$$6x \cdot -6x$$

The factors adding to zero are $6x$ and $-6x$. Enter $+6x$ and $-6x$ in the other diagonal, and factor to complete the table.

| Factor | | |
|---|---|---|
| | $4x^2$ | $-6x$ |
| | $+6x$ | $-9$ |

| Factor | $2x$ | $-3$ |
|---|---|---|
| $2x$ | $4x^2$ | $-6x$ |
| $+3$ | $+6x$ | $-9$ |

Thus

$$4x^2 - 9 = (2x - 3)(2x + 3)$$

***Check***: The factors multiply to give the inner "area" of the table.  ✔

♦ ♦ ♦ ♦ ♦ ♦ ♦ ♦ ♦ ♦ ♦ ♦ ♦ ♦ ♦ ♦ ♦ ♦ ♦ ♦ ♦ ♦ ♦ ♦ ♦ ♦ ♦ ♦ ♦ ♦ ♦ ♦ ♦ ♦ ♦

---

Always remember to remove the common monomial factor.

---

The expression $4x^2 - 10x - 6$ has 2 as a common monomial factor.

EXAMPLE  **12**     Factor $4x^2 - 10x - 6$, using the table method. First remove the common monomial factor.

SOLUTION     We begin with $4x^2 - 10x - 6 = 2(2x^2 - 5x - 3)$. This time enter only the terms from $2x^2 - 5x - 3$.

| Factor | | |
|---|---|---|
| | $2x^2$ | |
| | | $-3$ |

Place $2x^2$ and $-3$ in the diagonal. The diagonal product is $-6x^2$. Look for the factors of $-6x^2$ that add to $-5x$. The factors of $-6x^2$ are

$$-6x \cdot 1x$$
$$-3x \cdot 2x$$
$$-2x \cdot 3x$$
$$-1x \cdot 6x$$

The factors $-6x$ and $1x$ add to $-5x$. Enter them in the other diagonal, and factor.

| Factor | | |
|--------|--------|--------|
| | $2x^2$ | $-6x$ |
| | $+1x$ | $-3$ |

| Factor | $x$ | $-3$ |
|--------|--------|--------|
| $2x$ | $2x^2$ | $-6x$ |
| $+1$ | $+1x$ | $-3$ |

Thus

$$4x^2 - 10x - 6 = 2(2x + 1)(x - 3)$$

***Check***: The two binomial factors, $(2x + 1)$ and $(x - 3)$, multiply to the inner "area" of the table as required. This area times the original monomial factor, 2, gives the trinomial $4x^2 - 10x - 6$. ✔

◆ ◆ ◆ ◆ ◆ ◆ ◆ ◆ ◆ ◆ ◆ ◆ ◆ ◆ ◆ ◆ ◆ ◆ ◆ ◆ ◆ ◆ ◆ ◆ ◆ ◆ ◆ ◆ ◆ ◆ ◆ ◆ ◆ ◆ ◆

**Factoring $adx^2 + bdx + cd$ by Table**

1. First remove the common monomial factor, $d$, so that $ax^2 + bx + c$ remains: $d(ax^2 + bx + c)$.
2. Enter $ax^2$ and $c$ in one diagonal.
3. Multiply $ax^2$ and $c$ to obtain the diagonal product.
4. Find factors of the diagonal product that add to $bx$.
5. Enter the factors in the other diagonal.
6. Factor out the monomials, and write the binomial pair.
7. Multiply the factors to check that the inner "area" is $ax^2 + bx + c$.

**EXERCISES 6.2**

**Multiply the expressions in Exercises 1 to 12, using the table method.**

1. $(x - 2)(x - 2)$
2. $(x - 1)(x - 1)$
3. $(x + 2)(x - 2)$
4. $(x - 1)(x + 1)$
5. $(a + 5)(a + 5)$
6. $(b - 4)(b + 4)$
7. $(b + 5)(b - 5)$
8. $(a + 4)(a + 4)$
9. $(a + b)(a - b)$
10. $(a - b)(a + b)$
11. $(a - b)(a - b)$
12. $(a + b)(a + b)$

**Visualizing the table, multiply the expressions in Exercises 13 to 24 in your head.**

13. $(x + 1)(x + 7)$
14. $(x + 1)(x - 7)$
15. $(x - 1)(x + 7)$
16. $(x - 1)(x - 7)$
17. $(b + 1)(b + 1)$
18. $(a + 7)(a - 7)$
19. $(a - 7)(a + 7)$
20. $(b - 7)(b - 7)$
21. $(x + y)(x + y)$
22. $(x - y)(x - y)$
23. $(x + y)(x - y)$
24. $(x - y)(x + y)$

25. What binomial pair product is shown in the figure?

26. What binomial pair product is shown in the figure?

The product $(a + b)(a + b) = a^2 + 2ab + b^2$ is called a perfect square trinomial.

27. Multiply $(a - b)(a - b)$; multiply $(a + b)(a + b)$. What is the same about the products? What is different? Explain the difference.

28. How does the figure for Exercise 25 suggest the name *perfect square trinomial*? Which ten of Exercises 1 to 24 contain *perfect square trinomials*?

The product $(a - b)(a + b) = a^2 - b^2$ is called a **difference of perfect squares.**

29. Multiply $(a - b)(a + b)$; multiply $(x - y)(x + y)$. What is the same about the products? Why might their product be called a *difference of perfect squares*?

30. Which ten of Exercises 1 to 24 give products that are a *difference of perfect squares*?

**For Exercises 31 to 36 finish the multiplications. Do the products mentally or with a table, as needed.**

| *Binomial pair* | | *Trinomial product* |
|---|---|---|
| 31. $(x - 1)(x + 12)$ | = | $x^2 + \underline{\hspace{1cm}}x - 12$ |
| 32. $(x - 2)(x + 6)$ | = | $x^2 + \underline{\hspace{1cm}}x - 12$ |
| 33. $(x - 3)(x + 4)$ | = | $x^2 + \underline{\hspace{1cm}}x - 12$ |
| 34. $(x + 3)(x - 4)$ | = | $x^2 - \underline{\hspace{1cm}}x - 12$ |
| 35. $(x + 2)(x - 6)$ | = | $x^2 - \underline{\hspace{1cm}}x - 12$ |
| 36. $(x + 1)(x - 12)$ | = | $x^2 - \underline{\hspace{1cm}}x - 12$ |

**In Exercises 37 to 42 multiply, using a table.**

37. $(x - 1)(x^2 - 2x + 1)$    38. $(x - 2)(x^2 + 2x + 4)$

39. $(a - b)(a^2 + ab + b^2)$    40. $(x + y)(x^2 - xy + y^2)$

41. $(x + y)(x^2 + 2xy + y^2)$    42. $(a + b)(a^2 + 2ab + b^2)$

**Multiply the binomial pairs in Exercises 43 to 50, either by table or mentally.**

43. $(2x + 3)(2x + 3)$    44. $(2x - 3)(2x - 3)$

45. $(2x - 3)(2x + 3)$    46. $(2x + 3)(2x - 3)$

47. $(3x - 2)(3x - 2)$    48. $(3x + 2)(3x + 2)$

49. $(3x + 2)(3x - 2)$    50. $(3x - 2)(3x + 2)$

51. What do you observe in the answers to Exercises 43 to 50 when the signs in the binomials are different?

52. What do you observe in the answers to Exercises 43 to 50 when the signs in the binomials are alike?

**For Exercises 53 to 56 complete the tables to factor the expressions into a binomial pair. Summarize the results by writing the factorizations.**

53.
| Factor | | |
|---|---|---|
| | $x^2$ | $+5x$ |
| | $+4x$ | $+20$ |

54.
| Factor | | |
|---|---|---|
| | $x^2$ | $+5x$ |
| | $-4x$ | $-20$ |

55.
| Factor | | |
|---|---|---|
| | $x^2$ | $+2x$ |
| | $-10x$ | $-20$ |

56.
| Factor | | |
|---|---|---|
| | $x^2$ | $-2x$ |
| | $+10x$ | $-20$ |

57. Describe why we obtain the coefficient $-2$ on the middle term of the product in $(x + 3)(x - 5) = x^2 - 2x - 15$.

58. Describe why we obtain the coefficient $+1$ on the middle term of the product in $(x - 4)(x + 5) = x^2 + x - 20$.

**Factor the binomials and trinomials in Exercises 59 to 96. Algebra tiles may be helpful. The table method may be used anytime, but it will be most useful beginning with Exercise 73.**

59. $x^2 + 6x + 9$    60. $x^2 + 10x + 25$

61. $x^2 + 11x + 30$    62. $x^2 + 17x + 30$

63. $x^2 + 13x - 30$    64. $x^2 + x - 30$

65. $x^2 - 6x - 16$    66. $x^2 + 6x - 16$

67. $x^2 + 15x - 16$    68. $x^2 - 15x - 16$

69. $x^2 - 25$    70. $x^2 - 36$

71. $x^2 - 81$. What pattern do you observe in the factors here and in Exercise 69?

72. $x^2 - 100$. What pattern do you observe in the factors here and in Exercise 70?

73. $2x^2 + 11x + 12$    74. $3x^2 + 13x + 12$

75. $2x^2 - 3x - 9$    76. $2n^2 + 3n - 9$

77. $2n^2 + n - 3$    78. $2x^2 - 5x - 3$

79. $3x^2 + 5x - 2$    80. $3a^2 + a - 2$

81. $3a^2 - 11a - 4$    82. $3x^2 - 4x - 4$

83. $9x^2 - 49$

84. $9x^2 - 25$

85. $16x^2 - 9$

86. $16x^2 - 81$

87. $6x^2 + x - 2$

88. $6x^2 - 7x - 5$

89. $6x^2 + 5x - 6$

90. $6x^2 - 13x + 5$

91. $2n^2 + 9n - 5$

92. $2n^2 + 11n - 6$

93. $25x^2 - 36$

94. $100x^2 - 49$

95. $3x^2 + 12x + 9$

96. $2x^2 + 6x + 4$

97. "Prove" the equal diagonal products property by multiplying $(ax + b)(cx + d)$ in a table. Compare the diagonal products.

PROJECTS

98. **Factoring Cubic Polynomials**

a. Complete the tables to factor the expressions into a binomial times a trinomial. Summarize the results by writing the factorizations.

| Factor | | | |
|--------|--------|--------|--------|
| | $x^3$ | $-2x^2$ | $+x$ |
| | $-x^2$ | $+2x$ | $-1$ |

| Factor | | | |
|--------|--------|--------|--------|
| | $a^3$ | $-2a^2b$ | $+ab^2$ |
| | $-a^2b$ | $+2ab^2$ | $-b^3$ |

| Factor | | | |
|--------|--------|--------|--------|
| | $a^3$ | $-a^2b$ | $+ab^2$ |
| | $a^2b$ | $-ab^2$ | $+b^3$ |

| Factor | | | |
|--------|--------|--------|--------|
| | $x^3$ | $-2x^2$ | $+4x$ |
| | $+2x^2$ | $-4x$ | $+8$ |

b. Use a table to multiply:

$$(x + 1)(x^2 + 2x + 1)$$
$$(x + 3)(x^2 + 6x + 9)$$
$$(x - 2)(x^2 + 2x + 4)$$

c. What patterns do you observe in trinomial multiplication that might help in factoring expressions such as $x^3 - 1$, $x^3 - 3x^2 + 3x - 1$, and $x^3 - 27$?

99. **Common Monomial Factors.** If a trinomial contains no common monomial factor, is it possible for its binomial factors to contain a common monomial factor? That is, if $2x^2 - 9x + 10$ factors, is it possible for one of the factors to be $(2x - 2)$ or $(2x + 2)$? Give several examples, and explain in complete sentences how you came to your conclusion.

6.3 APPLICATIONS: POLYNOMIAL MULTIPLICATION AND BINOMIAL THEOREM

OBJECTIVES    Determine the number of ways to get from one point to another on a restricted street grid. ◆ Determine the possible birth orders of a given number of children. ◆ Determine the powers of binomial expressions. ◆ Calculate Pascal's Triangle to the row starting 1 10 .... ◆ Use the Binomial Theorem (informal) to write selected powers of binomials. ◆ Identify the line of symmetry in geometric figures. ◆ Identify the line of symmetry in Pascal's Triangle. ◆ Identify a symmetry of opposites in the variables in powers of binomials.

WARM-UP    Multiply the following.

1. $(x - 2)(x - 2)$

2. $(x - 3)(x + 2)$

3. $(x - 3)(x + 3)$

4. $(x + 1)(x - 1)$

Factor these expressions.

5. $x^2 + 5x + 4$

6. $x^2 - 6x + 9$

7. $x^2 + 2x - 3$

8. $x^2 - 16$

◆

This section introduces an arrangement of numbers that appears in many applications of mathematics including genetics, probability, and transportation. This symmetrical arrangement of numbers will give us a shortcut for finding powers of binomials.

### THE BINOMIAL THEOREM

We begin this section with three questions, followed by three examples that suggest how we might answer the questions.

**QUESTION 1**

The lines in Figure 10 represent city streets. If one can travel *only south or east* on the streets, how many different ways are there to get from point $A$ to point $B$? (This question is similar to one in a homework set in a transportation engineering course taken by the author.)

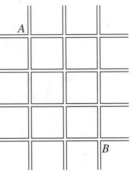

Figure 10

**QUESTION 2**

A family with 5 children has 2 boys and 3 girls. Assume there are no twins. What are all the possible birth orders? (This question is related to probability and statistics as well as to genetics.)

**QUESTION 3**

What is $(a + b)^5$? (This question refers to the fifth power of the binomial $a + b$. The Hardy-Weinberg principle, $p^2 + 2pq + q^2 = 1$, is a well-known application of binomial powers in the genetics field. It predicts the frequencies of certain genetic occurrences over several generations.)

To solve each question we might *consider an easier problem*. This returns us to our four-step strategy: understand, plan, carry out the plan, and check. Solving a simpler but related problem is a good strategy for building understanding of a problem and designing a plan to solve it. The bonus in this set of three questions is that all have related solutions.

**EXAMPLE 1**

Street grid    For Question 1 consider the number of ways to get to each street intersection, one at a time. Use Figure 11.

**a.** Find the number of ways from $A$ to $B_1$.

**b.** Find the number of ways from $A$ to $B_2$.

**c.** Find the number of ways from $A$ to $B_3$.

**d.** Using the results from parts a, b, and c, try Question 1 again. Carry out a plan to record a number at each intersection leading to point $B$.

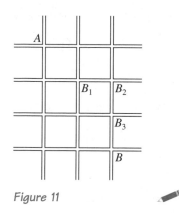

*Figure 11*

SOLUTION    **a.** As shown in Figure 12, there are 2 ways to leave position $A$, 1 to the right and 1 down. We record a 1 at each adjacent intersection. Because we are limited to moving to the right or down, each of the ways gives 1 route to $B_1$. Thus from $A$ to $B_1$ there are 2 routes. We record a 2 at $B_1$ to indicate the 2 routes.

**b.** If we count routes to $B_2$ and leave our earlier results recorded, we find there is 1 additional route to $B_2$. Thus there are 3 ways to get from $A$ to $B_2$. We record a 3 at $B_2$.

**c.** Continuing to record numbers on the intersections confirms that there are 6 routes to $B_3$ from $A$.

**d.** There are 10 ways to get from $A$ to $B$, as shown in Figure 13.

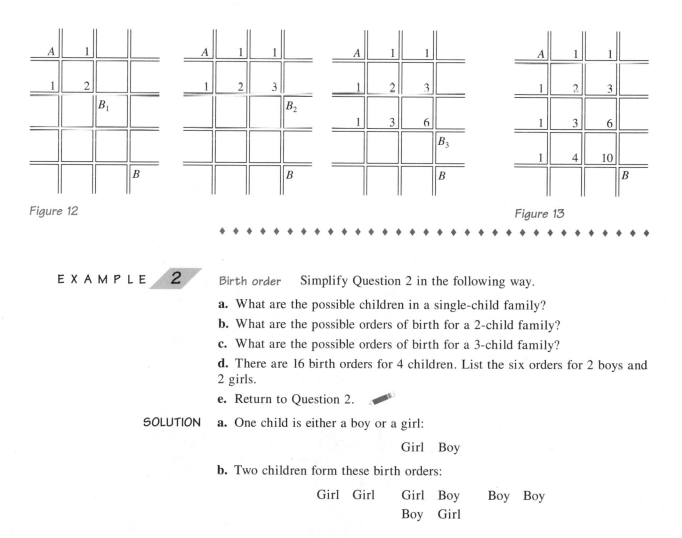

*Figure 12*                                                                      *Figure 13*

◆ ◆ ◆ ◆ ◆ ◆ ◆ ◆ ◆ ◆ ◆ ◆ ◆ ◆ ◆ ◆ ◆ ◆ ◆ ◆ ◆ ◆ ◆ ◆ ◆ ◆ ◆ ◆ ◆ ◆ ◆ ◆ ◆ ◆ ◆

E X A M P L E    **2**    Birth order    Simplify Question 2 in the following way.

**a.** What are the possible children in a single-child family?

**b.** What are the possible orders of birth for a 2-child family?

**c.** What are the possible orders of birth for a 3-child family?

**d.** There are 16 birth orders for 4 children. List the six orders for 2 boys and 2 girls.

**e.** Return to Question 2.

SOLUTION    **a.** One child is either a boy or a girl:

Girl    Boy

**b.** Two children form these birth orders:

Girl   Girl        Girl   Boy        Boy   Boy
                   Boy    Girl

**c.** Three children form these birth orders:

|        |        |        |        |
|--------|--------|--------|--------|
| GGG    | GGB    | GBB    | BBB    |
|        | GBG    | BGB    |        |
|        | BGG    | BBG    |        |

**d.** Two girls and 2 boys may be born in these six orders:

|        |        |
|--------|--------|
| GGBB   | BBGG   |
| GBGB   | BGBG   |
| GBBG   | BGGB   |

**e.** There are 10 birth orders for 3 girls and 2 boys:

|        |        |
|--------|--------|
| GGGBB  | GBBGG  |
| GGBGB  | BGGGB  |
| GGBBG  | BGGBG  |
| GBGGB  | BGBGG  |
| GBGBG  | BBGGG  |

◆ ◆ ◆ ◆ ◆ ◆ ◆ ◆ ◆ ◆ ◆ ◆ ◆ ◆ ◆ ◆ ◆ ◆ ◆ ◆ ◆ ◆ ◆ ◆ ◆ ◆ ◆ ◆ ◆ ◆ ◆ ◆ ◆ ◆

**E X A M P L E   3**

*Binomial powers*    In Question 3 we take the products, one power at a time.

**a.** Find $(a + b)^2$, using a table or mentally.

**b.** Find $(a + b)^3$ by multiplying the answer in part a by $(a + b)$.

**c.** Find $(a + b)^4$ by multiplying the answer in part b by $(a + b)$.

**d.** Find $(a + b)^5$ by multiplying the answer in part c by $(a + b)$.

**SOLUTION**    **a.** $(a + b)^2$ means $(a + b)(a + b)$. In Section 6.2 we multiplied expressions with a table.

| **Multiply** | $a$    | $+b$    |
|--------------|--------|---------|
| $a$          | $a^2$  | $+ab$   |
| $+b$         | $+ab$  | $+b^2$  |

Thus

$$(a + b)^2 = a^2 + 2ab + b^2$$

**b.** $(a + b)^3 = (a + b)(a + b)(a + b)$

$\phantom{(a + b)^3} = (a + b)(a + b)^2$

$\phantom{(a + b)^3} = (a + b)(a^2 + 2ab + b^2)$

We take the answer from the table in part a and multiply it by $(a + b)$ in a new table.

| **Multiply** | $a^2$    | $+2ab$    | $+b^2$    |
|--------------|----------|-----------|-----------|
| $a$          | $a^3$    | $+2a^2b$  | $+ab^2$   |
| $+b$         | $+a^2b$  | $+2ab^2$  | $+b^3$    |

Thus

$$(a + b)^3 = a^3 + 3a^2b + 3ab^2 + b^3$$

**c.** Similarly,

$$(a + b)^4 = (a + b)(a + b)^3$$
$$= (a + b)(a^3 + 3a^2b + 3ab^2 + b^3)$$

Multiply this product in a table. You should have five terms in your answer.

**d.** Use part c to obtain $(a + b)^5$.

$$(a + b)^5 = 1a^5 + 5a^4b + 10a^3b^2 + 10a^2b^3 + 5ab^4 + 1b^5$$

What do you observe about the exponents on $a$? What do you observe about the exponents on $b$?

♦ ♦ ♦ ♦ ♦ ♦ ♦ ♦ ♦ ♦ ♦ ♦ ♦ ♦ ♦ ♦ ♦ ♦ ♦ ♦ ♦ ♦ ♦ ♦ ♦ ♦ ♦ ♦ ♦ ♦ ♦ ♦ ♦ ♦

We now examine the three questions together.

EXAMPLE **4**    What do the answers to Questions 1, 2, and 3 have in common?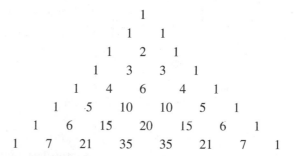

SOLUTION    The number 10 appears in each answer. This is not coincidental. Look at the pattern of numbers in this summary of each step in Example 3.

$$(a + b)^0 = \mathbf{1}$$
$$(a + b)^1 = \mathbf{1}a + \mathbf{1}b$$
$$(a + b)^2 = \mathbf{1}a^2 + \mathbf{2}ab + \mathbf{1}b^2$$
$$(a + b)^3 = \mathbf{1}a^3 + \mathbf{3}a^2b + \mathbf{3}ab^2 + \mathbf{1}b^3$$
$$(a + b)^4 = \mathbf{1}a^4 + \mathbf{4}a^3b + \mathbf{6}a^2b^2 + \mathbf{4}ab^3 + \mathbf{1}b^4$$
$$(a + b)^5 = \mathbf{1}a^5 + \mathbf{5}a^4b + \mathbf{10}a^3b^2 + \mathbf{10}a^2b^3 + \mathbf{5}ab^4 + \mathbf{1}b^5$$

If we compare the numerical coefficients in the expressions with the numbers in Figure 14, an expanded street grid from Question 1, we see the same set of numbers. The coefficients of $(a + b)^5$ are the same as the diagonal numbers at the intersections five blocks from $A$. These numbers are part of an array called **Pascal's Triangle**.

Here are the first few rows of Pascal's Triangle. This array of numbers is filled with patterns, and it could be a course of study all to itself.

```
                    1
                 1     1
              1     2     1
           1     3     3     1
        1     4     6     4     1
     1    ·5   10    10     5     1
   1     6    15    20    15     6     1
 1     7    21    35    35    21     7     1
```

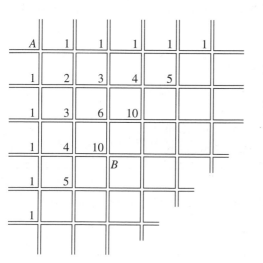

*Figure 14*

Compare the numbers of birth orders listed in parts a, b, and c of Example 2 with Pascal's Triangle. Each set of birth orders matches a row of the triangle. This pattern implies these birth orders for a family of five: 1 way for all girls, 5 ways for 4 girls and 1 boy, 10 ways for 3 girls and 2 boys, 10 ways for 2 girls and 3 boys, 5 ways for 1 girl and 4 boys, and 1 way for all boys.

♦ ♦ ♦ ♦ ♦ ♦ ♦ ♦ ♦ ♦ ♦ ♦ ♦ ♦ ♦ ♦ ♦ ♦ ♦ ♦ ♦ ♦ ♦ ♦ ♦ ♦ ♦ ♦ ♦ ♦ ♦ ♦ ♦ ♦

EXAMPLE 5      What does the 21 in the row starting 1 7 ... mean in each question setting?

SOLUTION    **a.** In the setting of Question 1, there are 21 ways to get from $A$ to $B_4$ in the street grid. $B_4$ is 7 blocks from $A$. $B_4$ is 5 streets below and 2 streets to the right of $A$, as shown in Figure 15.

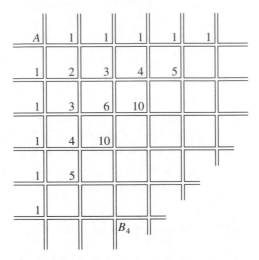

*Figure 15*

**b.** In the setting of Question 2, there are 21 birth orders for 5 girls and 2 boys.

**c.** In the setting of Question 3, 21 is the coefficient of the $a^5b^2$ term in the seventh power of $(a + b)$.

♦ ♦ ♦ ♦ ♦ ♦ ♦ ♦ ♦ ♦ ♦ ♦ ♦ ♦ ♦ ♦ ♦ ♦ ♦ ♦ ♦ ♦ ♦ ♦ ♦ ♦ ♦ ♦ ♦ ♦ ♦ ♦ ♦ ♦ ♦ ♦

Through Pascal's Triangle each question is related to a concept called the **binomial theorem**. This theorem (derived fact) dealing with the powers of binomials permits us to write out $(a + b)$ to any power we want. The binomial theorem works for powers of all binomials, but at this time we will consider only binomials like $(y - 1)$, $(x + y)$, or $(1 - a)$.

Binomial Theorem

> The binomial theorem for $(a + b)^n$ includes these properties:
>
> ♦ The numerical coefficients are the row starting 1 $n$ ... in Pascal's Triangle.
>
> ♦ The exponents on $a$ decrease, term by term.
>
> ♦ The exponents on $b$ increase, term by term.

The last two properties answer the two questions at the end of the solution to Example 3.

EXAMPLE 6      Write out the indicated power of the given binomial. Obtain the numbers from the appropriate row of Pascal's Triangle. Use descending exponents on the first term of the binomial and ascending exponents on the second. Note how subtraction creates alternating signs within the answers to parts c and d.

**a.** $(x + y)^3$      **b.** $(x + 1)^4$      **c.** $(y - 1)^3$      **d.** $(1 - a)^4$

Multiplying out these expressions using the table method as a check is left to the exercises.

SOLUTION   **a.** $(x + y)^3 = 1x^3 + 3x^2y + 3xy^2 + 1y^3$

The number coefficients (1, 3, 3, and 1) come from Pascal's Triangle. The exponents on $x$ *descend*—the first term contains $x^3$, and the last term contains no $x$. The exponents on $y$ *ascend*—there is no $y$ in the first term, and $y^3$ appears in the last term.

**b.** $(x + 1)^4 = 1x^4 + 4x^3 \cdot 1^1 + 6x^2 \cdot 1^2 + 4x \cdot 1^3 + 1 \cdot 1^4$
$$= 1x^4 + 4x^3 + 6x^2 + 4x + 1$$

The number coefficients (1, 4, 6, 4, and 1) come from Pascal's Triangle. The exponents on $x$ descend—the first term contains $x^4$, and the last term contains no $x$. The powers of 1 all equal 1, so they are not written out.

**c.** $(y - 1)^3 = 1y^3 + y^2 \cdot (-1)^1 + y^1 \cdot (-1)^2 + (-1)^3$
$$= 1y^3 - 3y^2 + 3y - 1$$

The number coefficients (1, 3, 3, and 1) come from Pascal's Triangle. The $y$ exponents descend. The expression $(y - 1)$ is the same as $[y + (-1)]$. The powers of $(-1)$ alternate between positive and negative, so the signs in the answer alternate.

**d.** $(1 - a)^4 = 1 + 4(-a)^1 + 6(-a)^2 + 4(-a)^3 + 1(-a)^4$
$$= 1 - 4a + 6a^2 - 4a^3 + 1a^4$$

The number coefficients (1, 4, 6, 4, and 1) come from Pascal's Triangle. The exponents on $a$ are in ascending order, and $(1 - a) = [1 + (-a)]$. Observe that if the exponent on $(-a)$ is even, the term is positive. If the exponent on $(-a)$ is odd, the term is negative.

♦ ♦ ♦ ♦ ♦ ♦ ♦ ♦ ♦ ♦ ♦ ♦ ♦ ♦ ♦ ♦ ♦ ♦ ♦ ♦ ♦ ♦ ♦ ♦ ♦ ♦ ♦ ♦ ♦ ♦ ♦ ♦

We close this section with a note on symmetry.

### SYMMETRY

Observe that Pascal's Triangle is symmetric. A vertical line passing between the 1s in the top row cuts the triangle into two identical but reflected parts. That line is a *line of symmetry* (or axis of symmetry).

E X A M P L E   **7**     Draw in the line of symmetry.

SOLUTION

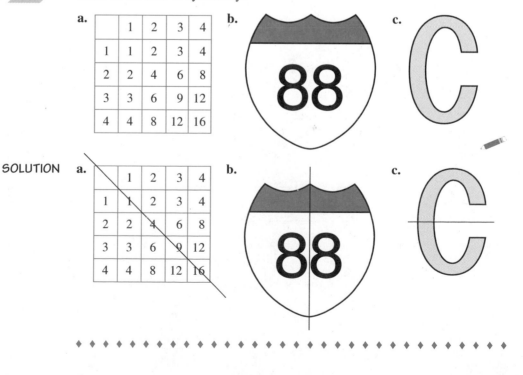

♦ ♦ ♦ ♦ ♦ ♦ ♦ ♦ ♦ ♦ ♦ ♦ ♦ ♦ ♦ ♦ ♦ ♦ ♦ ♦ ♦ ♦ ♦ ♦ ♦ ♦ ♦ ♦ ♦ ♦ ♦ ♦ ♦ ♦

There is also a form of symmetry in the variables of the powers of binomials.

$$(a + b)^0 = \qquad\qquad\qquad\qquad 1$$
$$(a + b)^1 = \qquad\qquad\qquad\qquad 1a + 1b$$
$$(a + b)^2 = \qquad\qquad\qquad 1a^2 + 2ab + 1b^2$$
$$(a + b)^3 = \qquad\qquad 1a^3 + 3a^2b + 3ab^2 + 1b^3$$
$$(a + b)^4 = \qquad 1a^4 + 4a^3b + 6a^2b^2 + 4ab^3 + 1b^4$$
$$(a + b)^5 = 1a^5 + 5a^4b + 10a^3b^2 + 10a^2b^3 + 5ab^4 + 1b^5$$

A vertical line through the entire set of expressions is not an axis of symmetry because expressions on each side—$a$ and $b$, $a^2$ and $b^2$, $a^3$ and $b^3$, and so forth—are not identical. However, there is a symmetry of opposites in that the variables $a$ and $b$ are interchanged as we move across the vertical line.

E X A M P L E   **8**     Identify the symmetry of opposite variables in $(x + y)^5$.

SOLUTION     The lines connect the opposite variables.

$$x^5 + 5x^4y^1 + 10x^3y^2 + 10x^2y^3 + 5x^1y^4 + y^5$$

♦ ♦ ♦ ♦ ♦ ♦ ♦ ♦ ♦ ♦ ♦ ♦ ♦ ♦ ♦ ♦ ♦ ♦ ♦ ♦ ♦ ♦ ♦ ♦ ♦ ♦ ♦ ♦ ♦ ♦ ♦ ♦ ♦ ♦ ♦ ♦

E X A M P L E   **9**     Use Pascal's Triangle and the symmetry of opposites to write out $(a + b)^6$. (*Hint*: Start with $1a^6$.)

SOLUTION     From Pascal's Triangle we obtain the numbers 1, 6, 15, 20, 15, 6, and 1 as coefficients of the terms. The exponents on variable $a$ are descending. The exponents on variable $b$ are ascending. We start with $a^6$ and end with $b^6$.

$$(a + b)^6 = 1a^6 + 6a^5b + 15a^4b^2 + 20a^3b^3 + 15a^2b^4 + 6ab^5 + 1b^6$$

Because the middle term, $20a^3b^3$, has symmetric exponents, it needs no term with opposing symmetry.

♦ ♦ ♦ ♦ ♦ ♦ ♦ ♦ ♦ ♦ ♦ ♦ ♦ ♦ ♦ ♦ ♦ ♦ ♦ ♦ ♦ ♦ ♦ ♦ ♦ ♦ ♦ ♦ ♦ ♦ ♦ ♦ ♦ ♦ ♦ ♦

E X A M P L E   **10**     Shade each design in the figures to show a symmetry of opposites.

**a.**                           **b.**

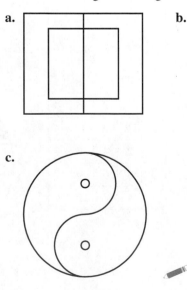

**c.**

SOLUTION

a.

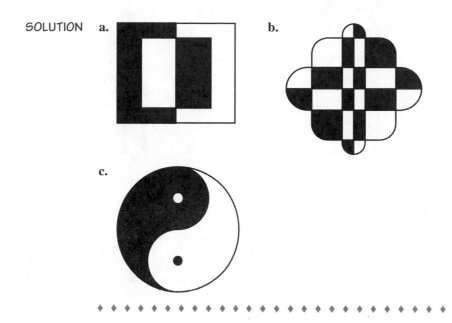

b.

c.

◆ ◆ ◆ ◆ ◆ ◆ ◆ ◆ ◆ ◆ ◆ ◆ ◆ ◆ ◆ ◆ ◆ ◆ ◆ ◆ ◆ ◆ ◆ ◆ ◆ ◆ ◆ ◆ ◆ ◆ ◆ ◆ ◆ ◆

## EXERCISES 6.3

Use the patterns started in Example 1 to determine the number of ways from *A* to *B* on the street grids in Exercises 1 to 4.

1.

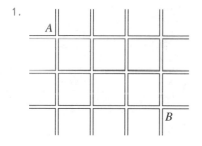

2.

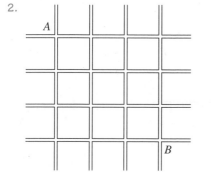

3.

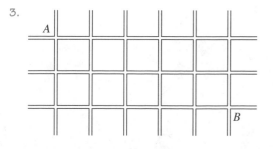

4.

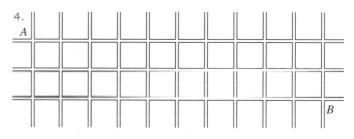

5. Complete Pascal's Triangle to the row starting 1 10 . . . . Note that each row begins and ends with a 1 and that the other terms in the row are related to the row above by addition.

6. One number pattern in Pascal's Triangle is 1, 3, 6, 10, 15, . . . .

   a. Give the next five numbers.

   b. Complete the table. Compare the number pattern above with the outputs obtained.

   | Input: $x$ | 1 | 2 | 3 | 4 | 5 | 6 | 7 | 8 | 9 |
   |---|---|---|---|---|---|---|---|---|---|
   | Output: $x(x + 1)$ | | | | | | | | | |

7. Evaluate the first ten powers of 2: $2^1 = 2$, $2^2 = 4$, $2^3 = 8$, etc. Add the numbers in each row of Pascal's Triangle. What do you observe?

8. There are five birth orders for 4 girls and 1 boy. List them.

9. There are sixteen birth orders for 4 children. They fall into five different groups, starting with GGGG and ending with BBBB. There are four arrangements with 3 girls and 1 boy and four with 3 boys and 1 girl. List the possible birth orders for 4 children.

In Example 6 the binomial expressions were expanded using the binomial theorem. Check the expansions in Exercises 10 to 13 by doing the last step with the table method of multiplication.

10. $(x + y)^3 = (x + y)(x + y)(x + y)$
$= (x + y)(x^2 + 2xy + y^2)$
$=$

11. $(1 + r)^4 = (1 + r)(1 + r)(1 + r)(1 + r)$
$= (1 + r)(1 + r)(1 + 2r + r^2)$
$= (1 + r)(1 + 3r + 3r^2 + r^3)$
$=$

12. $(y - 1)^3 = (y - 1)(y - 1)(y - 1)$
$= (y - 1)(y^2 - 2y + 1)$
$=$

13. $(1 - a)^4 = (1 - a)(1 - a)(1 - a)(1 - a)$
$= (1 - a)(1 - a)(1 - 2a + a^2)$
$= (1 - a)(1 - 3a + 3a^2 - a^3)$
$=$

Use Pascal's Triangle and the patterns formed in Example 6 to write out the binomial powers in Exercises 14 to 19.

14. $(1 + z)^4$

15. $(x + y)^6$

16. $(x + 1)^6$

17. $(b - 1)^4$

18. $(b - 1)^5$

19. $(1 + z)^5$

20. Describe how to get from one row to the next in Pascal's Triangle. How is a row of Pascal's Triangle related to the table method of multiplying polynomials?

21. Why do the signs alternate in Exercises 17 and 18?

22. It is a common error to say $(a + b)^2 = a^2 + b^2$. Explain what is wrong with that statement. Which row of Pascal's Triangle shows the coefficients of $(a + b)^2$? How does a table method of multiplying $(a + b)^2$ show what is wrong with the statement?

In Exercises 23 to 26 use symmetry of opposites and Pascal's Triangle to finish the expressions.

23. $(x + y)^8 = 1x^8 + 8x^7y^1 + 28x^6y^2 + 56x^5y^3 +$

24. $(a + b)^9 = 1a^9 + 9a^8b^1 + 36a^7b^2 + 84a^6b^3 +$

25. $(x + y)^9 = 1x^9 + 9x^8y^1 + 36x^7y^2 + 84x^6y^3 +$

26. $(a + b)^8 = 1a^8 + 8a^7b^1 + 28a^6b^2 + 56a^5b^3 +$

27. What exponent on $(p + q)$ is contained in the Hardy-Weinberg principle, $p^2 + 2pq + q^2 = 1$?

28. Let $p =$ the probability of having long eyelashes, $1/10$ (one in ten). Let $q =$ the probability of having short eyelashes, $9/10$. Verify that these numbers satisfy the Hardy-Weinberg principle.

In Exercises 29 and 30 draw in all lines of symmetry. Some letters (numbers) may have no lines of symmetry.

29.
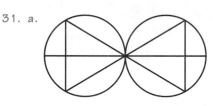

30.
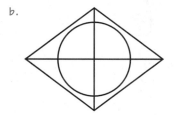

Shade each figure in Exercises 31 and 32 to have "opposite" symmetry.

31. a.
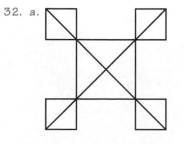

b.

32. a.

b.
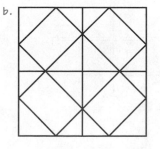

PROJECTS

33. **Other Binomial Powers.** There is additional complexity in finding powers of binomials that contain numbers other than 1. Multiply the binomials below, using tables as needed. Compare the results with those for $(x + 1)^4$ and $(x + y)^4$. Indicate where the answers are the same and where they are different.

a. $(x + 2)^4$ [*Hint*: Start with $(x + 2)(x + 2)$.]

b. $(x + 3)^4$ [*Hint*: Start with $(x + 3)(x + 3)$.]

34. Research the Hardy-Weinberg principle in genetics or biology books. Related information may also be found under the *Punnett-square method*.

35. **Polynomials and Volume.** The area of a square with side $x$ is $x^2$. The area of a square with side $(x + y)$ is $(x + y)^2$. The area as shown in the figure is

$$x^2 + xy + xy + y^2 = x^2 + 2xy + y^2$$

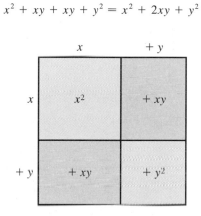

The volume of a cube with side $x$ is $x^3$. The volume of a cube with side $(x + y)$ is $(x + y)^3$. The parts making up the volume may also be shown geometrically with a three-dimensional table or stacks.

a. The figure below illustrates $(x + y)^3$. It is split into a front stack of boxes and a back stack of boxes, shown separately. Write out the volume for $(x + y)^3$.

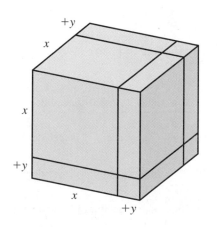

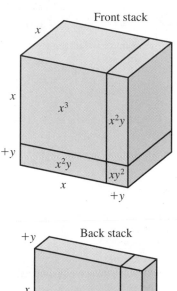

Front stack

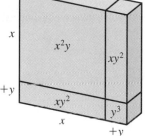

Back stack

b. Sketch the two stacks (front and back) of boxes that would show $(x + 2)^3$.

c. Sketch the two stacks (front and back) of boxes that would show $(x + 3)^3$.

d. Sketch the three stacks of boxes that would show $(x + y + z)^3$.

36. **Coin Toss Probabilities.** If 5 coins are tossed, what is the probability that there will be 3 heads and 2 tails, in any order? The **probability of an outcome** *is the number of ways in which that outcome can occur, divided by the total number of possible outcomes.*

a. Is there a simpler problem you might answer first?

b. Look for a pattern in the outcomes from tossing 1 coin, 2 coins, and then 3 coins. (*Hint*: Compare with the strategy used in Question 2.)

c. What is the relationship of this pattern to Pascal's Triangle?

d. The total number of outcomes is the sum of each row of Pascal's Triangle. What pattern do the sums form?

## MID-CHAPTER 6 TEST

1. Find the perimeter and area of the figure.

$$2x + 1$$

$3x$ ▭

2. Find the missing side of the figure.

$$x + 4$$

▭ $3x^2 + 12x$

3. Multiply $-2(x^2 - 2x + 1)$.

4. Identify the common monomial factor, and factor the expressions.

   a. $6x^2 - 2x + 8$      b. $2abc - 3ac + 4ab$

5. Complete these tables. Write the binomial pair and trinomial product for each.

a.

| Multiply | $3x$ | $-2$ |
|----------|------|------|
| $5x$     |      |      |
| $+3$     |      |      |

b.

| Factor |       |       |
|--------|-------|-------|
|        | $6x^2$ | $+8x$ |
|        | $-15x$ | $-20$ |

6. Multiply.

   a. $(x - 7)(x + 2)$      b. $(2x + 1)(2x + 3)$

7. Factor.

   a. $x^2 + 4x + 4$      b. $x^2 - 8x + 16$

   c. $2x^2 - 3x - 2$      d. $2x^2 - 8$

8. Multiply by any method, or use the binomial theorem.

   a. $(x + y)^3$      b. $(x - 1)^4$

---

## 6.4    EXPONENTS AND SCIENTIFIC NOTATION

**OBJECTIVES**    Simplify exponent expressions containing zero, negatives, and one-half as exponents. ♦ Simplify expressions containing multiplication, division, and powers of exponent expressions. ♦ Convert numbers to scientific notation. ♦ Convert scientific notation to regular decimal form. ♦ Explain why given expressions are not in correct scientific notation form. ♦ Use a calculator to do operations in scientific notation. ♦ Translate calculator output to scientific notation form.

**WARM-UP**    Do *not* use a calculator for this Warm-up.

1. What is the number $10^3$ in words?

2. What is the number $10^2$ in words?

3. What is the number $10^4$ in words?

4. What is the number $10^5$ in words?

5. In multiplying $2.78(1000)$, how many decimal places does the decimal point move?

6. In multiplying $1.6(100,000)$, how many decimal places does the decimal point move?

7. In multiplying $1.6(0.00001)$, how many decimal places does the decimal point move?

8. In multiplying $2.78(0.0001)$, how many decimal places does the decimal point move?

9. Evaluate $(1.2)(10^3)(3.0)(10^2)$.

10. Evaluate $(1.5) \div (3.0)$.

♦

In the last section we multiplied binomials, obtaining expressions with positive integers as exponents. We now examine the meaning of zero, negative integers, and one-half as exponents. The second part of the section introduces scientific notation, a common application of expressions containing negative integers as exponents.

## SPECIAL EXPONENTS: ZERO, NEGATIVE, ONE-HALF

We examined the following definition of exponents in Section 2.3:

> In $x^n$ the positive integer exponent, $n$, indicates the number of times the base, $x$, is used as a factor. Thus $x^n = x \cdot x \cdot x \cdot x \cdot \ldots \cdot x$ has $n$ factors of $x$.

However, this exponent definition makes no sense at all if the exponent is zero, negative, or a fraction, because we cannot write zero factors of the base, a negative number of factors, or even a fraction of one factor. Use the following Calculator Exploration to discover the meaning of $0$, $-1$, and $\frac{1}{2}$ as exponents.

**CALCULATOR EXPLORATION**

Complete Table 1. The exponent keys are $\boxed{\quad\wedge\quad}$, $\boxed{\quad y^x \quad}$, and $\boxed{\quad x^y \quad}$.

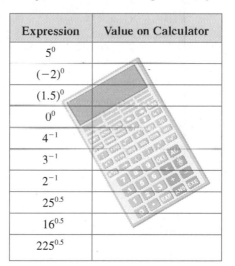

| Expression | Value on Calculator |
|------------|---------------------|
| $5^0$ | |
| $(-2)^0$ | |
| $(1.5)^0$ | |
| $0^0$ | |
| $4^{-1}$ | |
| $3^{-1}$ | |
| $2^{-1}$ | |
| $25^{0.5}$ | |
| $16^{0.5}$ | |
| $225^{0.5}$ | |

Table 1

In spite of the senselessness of these expressions in terms of the original definition of exponents, the calculator readily gives an answer for all table entries except $0^0$.

Zero to the zero power is undefined.

**EXAMPLE 1**

Describe patterns in the answers in Table 1.

**SOLUTION**

The *zero* exponents (except $0^0$) all gave 1 as the output. The *negative one* exponents gave decimal outputs that may not be meaningful until changed to fractions:

$$4^{-1} = 0.25 = \tfrac{1}{4}, \qquad 3^{-1} = 0.3\overline{3} = \tfrac{1}{3}, \qquad 2^{-1} = 0.5 = \tfrac{1}{2}$$

These may be recognized as reciprocals of the numbers 4, 3, and 2. The 0.5 (one-half) exponents seem to be giving the square roots.

◆ ◆ ◆ ◆ ◆ ◆ ◆ ◆ ◆ ◆ ◆ ◆ ◆ ◆ ◆ ◆ ◆ ◆ ◆ ◆ ◆ ◆ ◆ ◆ ◆ ◆ ◆ ◆ ◆ ◆ ◆ ◆

The patterns described in Example 1 suggest these additional definitions of exponential expressions.

> ♦ For zero exponents, $x^0 = 1$. The expression $0^0$ is not defined.
>
> ♦ For $-1$ exponents, $x^{-1} = \dfrac{1}{x}$, the reciprocal of $x$.
>
> ♦ For 0.5 or $\frac{1}{2}$ exponents, $x^{0.5} = x^{1/2} = \sqrt{x}$, the square root of $x$.

Other exponential forms will be defined as needed. The operations with exponents, from Section 2.3, are still valid for these new types of exponents and are repeated here.

> To multiply numbers with like bases, add the exponents:
> $$x^a \cdot x^b = x^{a+b}$$
> To divide numbers with like bases, subtract the exponents:
> $$\frac{x^a}{x^b} = x^{a-b} \quad \text{for } x \neq 0$$
> To apply an exponent to a number already in exponential form, multiply the exponents:
> $$(x^a)^b = x^{a \cdot b}$$
> An exponent outside the parentheses applies to all parts of a product or quotient inside the parentheses.
> $$(x \cdot y)^a = x^a \cdot y^a \qquad \left(\frac{x}{y}\right)^a = \frac{x^a}{y^a} \quad \text{for } y \neq 0$$

We expand the meaning of *simplify* to indicate that the definitions and properties of exponential expressions are to be used to remove zero, negative, or fractional exponents, as well as to remove parentheses, combine expressions with like bases, and, where possible, change expressions into numbers or expressions without exponents.

**EXAMPLE 2**   Simplify these exponential expressions by the means specified.

**a.** $5^1 \cdot 5^{-1}$, by adding exponents   **b.** $5^1 \cdot 5^{-1}$, by changing to fractions

SOLUTION   **a.** $5^1 \cdot 5^{-1} = 5^{1+(-1)} = 5^0 = 1$   **b.** $5^1 \cdot 5^{-1} = 5 \cdot \frac{1}{5} = 1$

♦ ♦ ♦ ♦ ♦ ♦ ♦ ♦ ♦ ♦ ♦ ♦ ♦ ♦ ♦ ♦ ♦ ♦ ♦ ♦ ♦ ♦ ♦ ♦ ♦ ♦ ♦ ♦ ♦ ♦ ♦ ♦ ♦ ♦ ♦ ♦ ♦ ♦

Example 2 reminds us that the product of a number and its reciprocal is 1 and illustrates why $5^0$ equals 1.

**EXAMPLE 3**   Simplify these exponential expressions.

**a.** $(x + y)^0$   **b.** $(3x^2y^2 + 5x^{20})^0$

SOLUTION   Recall that $0^0$ is undefined, so we must avoid that expression.

**a.** If $x$ and $y$ do not add to zero, then $(x + y)^0 = 1$.

**b.** If the expression inside the parentheses does not equal zero, then $(3x^2y^2 + 5x^{20})^0 = 1$.

♦ ♦ ♦ ♦ ♦ ♦ ♦ ♦ ♦ ♦ ♦ ♦ ♦ ♦ ♦ ♦ ♦ ♦ ♦ ♦ ♦ ♦ ♦ ♦ ♦ ♦ ♦ ♦ ♦ ♦ ♦ ♦ ♦ ♦ ♦ ♦ ♦ ♦

Example 4 illustrates why the $\frac{1}{2}$ power is the square root.

EXAMPLE 4

Simplify these exponential and radical expressions.

a. $36^{\frac{1}{2}} \cdot 36^{\frac{1}{2}}$, by adding exponents   b. $\sqrt{36} \cdot \sqrt{36}$

c. $25^{\frac{1}{2}} \cdot 25^{\frac{1}{2}}$, by adding exponents   d. $\sqrt{25} \cdot \sqrt{25}$

SOLUTION   a. $36^{\frac{1}{2}} \cdot 36^{\frac{1}{2}} = 36^{\left(\frac{1}{2} + \frac{1}{2}\right)} = 36^1 = 36$   b. $\sqrt{36} \cdot \sqrt{36} = 6 \cdot 6 = 36$

c. $25^{\frac{1}{2}} \cdot 25^{\frac{1}{2}} = 25^{\left(\frac{1}{2} + \frac{1}{2}\right)} = 25^1 = 25$   d. $\sqrt{25} \cdot \sqrt{25} = 5 \cdot 5 = 25$

♦ ♦ ♦ ♦ ♦ ♦ ♦ ♦ ♦ ♦ ♦ ♦ ♦ ♦ ♦ ♦ ♦ ♦ ♦ ♦ ♦ ♦ ♦ ♦ ♦ ♦ ♦ ♦ ♦ ♦ ♦ ♦

EXAMPLE 5

Simplify these expressions without a calculator. Write the answers as fractions or decimals, where appropriate.

a. $25^{-1}$   b. $9^{-1}$   c. $\left(\frac{2}{3}\right)^{-1}$   d. $(0.25)^{-1}$   e. $6.25^{-1}$

f. $10^{-1}$   g. $x^{-1}$   h. $\dfrac{1}{x^{-1}}$   i. $\left(\dfrac{a}{b}\right)^{-1}$   j. $(x + y)^{-1}$

SOLUTION   a. $25^{-1} = \frac{1}{25} = 0.04$   b. $9^{-1} = \frac{1}{9} = 0.1111\ldots$

c. $\left(\frac{2}{3}\right)^{-1} = \left(\frac{3}{2}\right) = 1.5$   d. $(0.25)^{-1} = \left(\frac{1}{4}\right)^{-1} = \frac{4}{1} = 4$

e. $6.25^{-1} = \left(\frac{25}{4}\right)^{-1} = \frac{4}{25} = 0.16$   f. $10^{-1} = \frac{1}{10} = 0.1$

g. $x^{-1} = \dfrac{1}{x}$   h. $\dfrac{1}{x^{-1}} = \dfrac{1}{\frac{1}{x}} = 1 \div \frac{1}{x} = 1 \cdot \dfrac{x}{1} = x$

i. $\left(\dfrac{a}{b}\right)^{-1} = \dfrac{b}{a}$   j. $(x + y)^{-1} = \dfrac{1}{x + y}$

♦ ♦ ♦ ♦ ♦ ♦ ♦ ♦ ♦ ♦ ♦ ♦ ♦ ♦ ♦ ♦ ♦ ♦ ♦ ♦ ♦ ♦ ♦ ♦ ♦ ♦ ♦ ♦ ♦ ♦ ♦ ♦

The expression $x^{-1}$, or $1/x$, is the formal mathematical expression for the reciprocal of $x$. Scientific calculators have a key for reciprocals, $\boxed{x^{-1}}$ or $\boxed{1/x}$. To practice using the reciprocal key, repeat Example 5, parts a to f, with a calculator.

EXAMPLE 6

a. With a calculator, verify the equality of the two expressions:

$$\left(\tfrac{25}{4}\right)^{-1} \quad \text{and} \quad \left(\tfrac{4}{25}\right)$$

b. Write the numbers as exponential expressions containing 2 as an exponent.

c. What do the numbers suggest about a $-2$ exponent?

SOLUTION   a. Both expressions are equivalent to 0.16.

b. $\left(\dfrac{25}{4}\right)^{-1} = \left(\dfrac{5^2}{2^2}\right)^{-1} = \dfrac{5^{-2}}{2^{-2}} = \left(\dfrac{5}{2}\right)^{-2}$

$\left(\dfrac{4}{25}\right) = \dfrac{2^2}{5^2} = \left(\dfrac{2}{5}\right)^2$

c. The equality of the original expressions indicates that $\left(\frac{5}{2}\right)^{-2} = \left(\frac{2}{5}\right)^2$. The $-2$ exponent on $\frac{5}{2}$ gives the reciprocal of $\frac{5}{2}$ squared.

♦ ♦ ♦ ♦ ♦ ♦ ♦ ♦ ♦ ♦ ♦ ♦ ♦ ♦ ♦ ♦ ♦ ♦ ♦ ♦ ♦ ♦ ♦ ♦ ♦ ♦ ♦ ♦ ♦ ♦ ♦ ♦

Both $-1$ and $-2$ as exponents create reciprocals. This leads us to a more general definition of exponents involving negative integer exponents.

> The exponential expression $x^{-n}$ is equivalent to $n$ factors of the reciprocal of $x$:
>
> $$x^{-n} = (x^{-1})^n = \underbrace{\left(\frac{1}{x}\right) \cdot \left(\frac{1}{x}\right) \cdot \ldots \cdot \left(\frac{1}{x}\right)}_{n \text{ factors}}$$

HINT   $x$ is a variable and can represent a fraction as well as a whole number. You may find $\left(\dfrac{a}{b}\right)^{-n} = \left(\dfrac{b}{a}\right)^{n}$ to be a more useful way of thinking about the definition of negative exponents.

EXAMPLE 7   Simplify these expressions by replacing negative exponent expressions with equivalent positive exponent forms.

**a.** $\left(\dfrac{2x}{3}\right)^{-2}$    **b.** $\left(\dfrac{4x^2}{y^3}\right)^{-2}$    **c.** $\dfrac{1}{a^{-2}}$    **d.** $\dfrac{b}{a^{-3}}$

SOLUTION  **a.** $\left(\dfrac{2x}{3}\right)^{-2} = \left(\dfrac{3}{2x}\right)\left(\dfrac{3}{2x}\right) = \dfrac{9}{4x^2}$    **b.** $\left(\dfrac{4x^2}{y^3}\right)^{-2} = \left(\dfrac{y^3}{4x^2}\right)\left(\dfrac{y^3}{4x^2}\right) = \dfrac{y^6}{16x^4}$

**c.** $\dfrac{1}{a^{-2}} = \left(\dfrac{1}{a}\right)^{-2} = \left(\dfrac{a}{1}\right)\left(\dfrac{a}{1}\right) = a^2$

**d.** $\dfrac{b}{a^{-3}} = b\left(\dfrac{1}{a^{-3}}\right) = b\left(\dfrac{1}{a}\right)^{-3} = b\left(\dfrac{a}{1}\right)\left(\dfrac{a}{1}\right)\left(\dfrac{a}{1}\right) = ba^3$

◆ ◆ ◆ ◆ ◆ ◆ ◆ ◆ ◆ ◆ ◆ ◆ ◆ ◆ ◆ ◆ ◆ ◆ ◆ ◆ ◆ ◆ ◆ ◆ ◆ ◆ ◆ ◆ ◆ ◆ ◆ ◆ ◆ ◆ ◆ ◆ ◆ ◆ ◆

### SCIENTIFIC NOTATION

One of the most common uses of negative exponents is in scientific notation. **Scientific notation** is a short way of writing large and small numbers, such as the distance to a star or the size of a virus. Calculators automatically change into scientific notation when the number is too large or too small to fit the answer display.

EXAMPLE 8   Use a scientific calculator to evaluate these numbers.

**a.** $(3600)^4$    **b.** $(0.00063)^5$

SOLUTION  The calculator answers are in an unexpected form. They are written in scientific notation. There is some variation among calculators, but the scientific notation display will probably appear as one of these two expressions:

**a.** 1.6796    14    or    1.6796E14

**b.** 9.9244    −17    or    9.9244E−17

◆ ◆ ◆ ◆ ◆ ◆ ◆ ◆ ◆ ◆ ◆ ◆ ◆ ◆ ◆ ◆ ◆ ◆ ◆ ◆ ◆ ◆ ◆ ◆ ◆ ◆ ◆ ◆ ◆ ◆ ◆ ◆ ◆ ◆ ◆ ◆ ◆ ◆ ◆

Scientific notation is based on the powers of 10. Numbers written in scientific notation have two parts, as shown below.

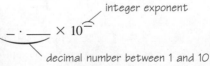

The first part is a decimal with one nonzero number to the left of the decimal point. The second part is a power of 10. The decimal number is multiplied by the power of 10. Calculator displays show the decimal part as well as the exponent but omit the symbol for multiplication and the base 10.

The answer to part a of Example 8 is 1.6796 times 10 with exponent 14, or

$$1.6796 \text{ times } 100{,}000{,}000{,}000{,}000$$

The 1.6796 is rounded, so the answer is approximately 167,960,000,000,000. The handwritten or textbook form for scientific notation is $1.6796 \times 10^{14}$. This large number has a positive exponent on the 10.

NOTE   We use the $\times$ sign for multiplication in scientific notation. The $\times$ is not a variable.

The answer to part b of Example 8 is 9.9244 times 10 with exponent $-17$, or

$$9.9244 \text{ times } 0.000\ 000\ 000\ 000\ 000\ 01$$

The 9.9244 is rounded, so the answer is approximately

$$0.000\ 000\ 000\ 000\ 000\ 099\ 244$$

The handwritten or textbook form is $9.9244 \times 10^{-17}$. This small number has a negative exponent on the 10.

Table 2 gives examples of scientific notation in a variety of settings and their relative sizes.

| Power of 10 | Value | Scientific Notation | Application |
|---|---|---|---|
| $10^9$ | 1,000,000,000 (billion) | $4.6 \times 10^9$ years ago | Estimated formation of earth |
| $10^8$ | 100,000,000 | $9.82 \times 10^8$ feet per second | Speed of light in a vacuum |
| $10^7$ | 10,000,000 | $2.46 \times 10^7$ years | Half-life of $^{236}$U, radioactive uranium |
| $10^6$ | 1,000,000 | $6.37 \times 10^6$ meters | Radius of earth |
| $10^5$ | 100,000 | $5.256 \times 10^5$ minutes | Minutes in a 365-day year |
| $10^4$ | 10,000 | $8.64 \times 10^4$ seconds | Seconds in one day |
| $10^3$ | 1,000 | $5.70 \times 10^3$ years | Half-life of $^{14}$C, radioactive carbon |
| $10^2$ | 100 | $1 \times 10^2$ meters | Length of 100-meter dash |
| $10^1$ | 10 | | |
| $10^0$ | 1 | $8.04 \times 10^0$ days | Half-life of $^{131}$I, radioactive iodine |
| $10^{-1}$ | $1/10 = 0.1$ | | |
| $10^{-2}$ | $1/100 = 0.01$ | | |
| $10^{-3}$ | $1/1000 = 0.001$ | $3 \times 10^{-3}$ meter | Size of a flea (3 mm) |
| $10^{-7}$ | 0.000 000 1 | $1 \times 10^{-7}$ meter | Length of HIV virus |
| $10^{-10}$ | | $1 \times 10^{-10}$ meter | 1 angstrom, measure of distance between atoms |
| $10^{-27}$ | | $1.66 \times 10^{-27}$ kilogram | The unit of atomic mass |

*Table 2  Applications of Scientific Notation*

To interpret answers from a calculator and to enter large and small numbers into a calculator, we need to be able to change numbers from scientific notation to regular decimal form and vice versa.

---

In changing from scientific notation to regular decimal notation, and the reverse, remember that large numbers have positive exponents on the 10 and small numbers between 0 and 1 have negative exponents on the 10.

---

EXAMPLE 9

Change each number to scientific notation.

**a.** The minimum distance from the sun to Mercury is 28,600,000 miles.

**b.** The maximum distance from the sun to Mercury is 43,400,000 miles.

**c.** The maximum distance from earth to Pluto is 4,644,000,000 miles.

**d.** The mass of an electron is

0.000 000 000 000 000 000 000 000 000 910 9 kilogram

**e.** The mass of a proton is

0.000 000 000 000 000 000 000 000 001 672 6 kilogram

SOLUTION

In parts a, b, and c the distances are large numbers, and the scientific notation will have a positive power of 10. To find the exponent on 10 we count the number of places the decimal point must move before only one place remains to the left of the decimal point. If there is no decimal point in the number, counting should start at the far right.

**a.** $2.86 \times 10^7$ miles    **b.** $4.34 \times 10^7$ miles    **c.** $4.644 \times 10^9$ miles

In parts d and e the masses are small numbers between 0 and 1. The scientific notation contains a negative power of 10. Count the number of places the decimal point must move before there is one place to the left of the decimal point.

**d.** $9.109 \times 10^{-31}$ kilogram    **e.** $1.6726 \times 10^{-27}$ kilogram

♦ ♦ ♦ ♦ ♦ ♦ ♦ ♦ ♦ ♦ ♦ ♦ ♦ ♦ ♦ ♦ ♦ ♦ ♦ ♦ ♦ ♦ ♦ ♦ ♦ ♦ ♦ ♦ ♦ ♦ ♦ ♦ ♦ ♦ ♦ ♦

EXAMPLE 10

Change each number from scientific notation to decimal notation.

**a.** The minimum distance from the sun to earth is $9.14 \times 10^7$ miles.

**b.** In chemistry Avogadro's number is a measure of number of molecules. It is $6.022 \times 10^{23}$ molecules per mole.

**c.** The estimated human population of the world in 1990 was $5.333 \times 10^9$.

**d.** The human population on earth in 1650 is estimated at $5.5 \times 10^8$.

**e.** The mass of a house spider is about $1 \times 10^{-4}$ kilogram.

SOLUTION

*Remember*: Positive exponents are associated with large numbers, and negative exponents are associated with small numbers between 0 and 1.

**a.** 91,400,000 miles

**b.** 602,200,000,000,000,000,000,000 molecules per mole

**c.** 5,333,000,000 people

**d.** 550,000,000 people

**e.** 0.0001 kilogram

♦ ♦ ♦ ♦ ♦ ♦ ♦ ♦ ♦ ♦ ♦ ♦ ♦ ♦ ♦ ♦ ♦ ♦ ♦ ♦ ♦ ♦ ♦ ♦ ♦ ♦ ♦ ♦ ♦ ♦ ♦ ♦ ♦ ♦ ♦ ♦ ♦ ♦ ♦

E X A M P L E **11**    Explain why each of the following is incorrect scientific notation. Guess what number is intended by each expression, and restate it in correct scientific notation.

**a.** $4.82 \ 10^{11}$    **b.** $5.23^{\ 5}$    **c.** $34.6 \times 10^{12}$    **d.** $0.465 \times 10^{-2}$

SOLUTION    **a.** The multiplication sign was left out. The correct form is $4.82 \times 10^{11}$.

**b.** The $\times 10$ was left out. Because many calculators display scientific notation answers in this form, this is a common error. The correct form is $5.23 \times 10^{5}$.

**c.** Correct scientific notation has only one nonzero digit to the left of the decimal point. The correct form is $3.46 \times 10^{13}$.

**d.** Correct scientific notation has one nonzero digit to the left of the decimal point. The correct form is $4.65 \times 10^{-3}$.

♦ ♦ ♦ ♦ ♦ ♦ ♦ ♦ ♦ ♦ ♦ ♦ ♦ ♦ ♦ ♦ ♦ ♦ ♦ ♦ ♦ ♦ ♦ ♦ ♦ ♦ ♦ ♦ ♦ ♦ ♦ ♦ ♦ ♦ ♦

Calculators are designed to do operations with scientific notation. We use the keys $\boxed{\text{EE}}$, $\boxed{\text{EXP}}$, and $\boxed{\text{EEX}}$—but not $\boxed{e^x}$—to enter a number in scientific notation. Negative exponents may need to have the negative sign, $\boxed{+/-}$, entered after the exponent. The TI 80 series permits a negative, $\boxed{(-)}$, before the exponent. Experiment with your calculator in Example 12 after calculating the results mentally.

E X A M P L E **12**    Use properties of exponents to do these operations. Then use a calculator to do the operations in scientific notation. *Remember*: The $\times$ is a multiplication sign.

**a.** $(1.2 \times 10^{3}) \cdot (3.0 \times 10^{2})$

**b.** $(1.5 \times 10^{-2}) \cdot (4.0 \times 10^{5})$

**c.** $(1.5 \times 10^{-4}) \div (5.0 \times 10^{2})$

**d.** $(2.4 \times 10^{8}) \div (3.0 \times 10^{-3})$

SOLUTION    *A possible mental approach:*

**a.** 1.2 times 3.0 is 3.6, and $10^3$ times $10^2$ is $10^5$. The answer is $3.6 \times 10^5$, or 360,000.

**b.** 1.5 times 4.0 is 6.0, and $10^{-2}$ times $10^5$ is $10^3$. The answer is $6.0 \times 10^3$, or 6000.

**c.** 1.5 divided by 5.0 is 0.3, and $10^{-4}$ divided by $10^2$ is $10^{-6}$. The answer is $0.3 \times 10^{-6} = 3.0 \times 10^{-1} \times 10^{-6} = 3 \times 10^{-7}$.

**d.** 2.4 divided by 3.0 is 0.8, and $10^8$ divided by $10^{-3}$ is $10^{11}$. The answer is $0.8 \times 10^{11} = 8.0 \times 10^{-1} \times 10^{11} = 8.0 \times 10^{10}$.

*A scientific calculator approach:*

**a.** $1.2 \boxed{\text{EE}} 3 \boxed{\times} 3.0 \boxed{\text{EE}} 2 \boxed{=} 3.6 \times 10^5$
Some calculators give 360,000.

**b.** $1.5 \boxed{\text{EE}} 2 \boxed{+/-} \boxed{\times} 4.0 \boxed{\text{EE}} 5 \boxed{=} 6.0 \times 10^3$
The calculator answer may be 6000.

**c.** $1.5 \boxed{\text{EE}} 4 \boxed{+/-} \boxed{\div} 5.0 \boxed{\text{EE}} 2 \boxed{=} 3.0 \times 10^{-7}$
The calculator answer may be 0.000 000 3.

**d.** $2.4 \boxed{\text{EE}} 8 \boxed{\div} 3.0 \boxed{\text{EE}} 3 \boxed{+/-} \boxed{=} 8.0 \times 10^{10}$
This is 80,000,000,000, but it is almost always shown in scientific notation.

## EXERCISES 6.4

**Evaluate the expressions in Exercises 1 to 6 without a calculator.**

1. a. $25^0$     b. $25^{-1}$     c. $25^{1/2}$     d. $25^{0.5}$

2. a. $4^{-1}$     b. $4^0$     c. $4^{0.5}$     d. $4^{1/2}$

3. a. $9^{1/2}$     b. $9^0$     c. $9^{0.5}$     d. $9^{-1}$

4. a. $36^{1/2}$     b. $36^{0.5}$     c. $36^0$     d. $36^{-1}$

5. a. $\left(\frac{1}{4}\right)^{-1}$     b. $\left(\frac{1}{4}\right)^0$     c. $\left(\frac{1}{4}\right)^{1/2}$     d. $\left(\frac{1}{4}\right)^{0.5}$

6. a. $\left(\frac{1}{9}\right)^{1/2}$     b. $\left(\frac{1}{9}\right)^{0.5}$     c. $\left(\frac{1}{9}\right)^0$     d. $\left(\frac{1}{9}\right)^{-1}$

**Use a calculator to evaluate the expressions in Exercises 7 and 8.**

7. a. $(0.25)^{-1}$     b. $(0.01)^{0.5}$     c. $(6.25)^{0.5}$
   d. $(0.25)^{0.5}$     e. $(0.02)^{-1}$     f. $(0.05)^{-1}$

8. a. $(2.25)^{0.5}$     b. $(0.36)^{0.5}$     c. $(0.01)^{-1}$
   d. $(0.1)^{-1}$     e. $(0.125)^{-1}$     f. $(0.0001)^{0.5}$

9. True or false: The reciprocal of a number is always smaller than the original number. Explain your reasoning.

10. True or false: The square root of a number is always smaller than the original number. Explain your reasoning.

**In Exercises 11 to 14 simplify the expressions to remove negative and zero exponents.**

11. a. $x^{-1}$     b. $\left(\frac{x}{y}\right)^{-1}$     c. $\left(\frac{y}{x}\right)^{-1}$
    d. $\left(\frac{a}{b}\right)^0$     e. $\left(\frac{a}{c}\right)^0$     f. $\left(\frac{a}{bc}\right)^{-1}$

12. a. $y^{-1}$     b. $\left(\frac{b}{c}\right)^{-1}$     c. $\left(\frac{c}{b}\right)^{-1}$
    d. $\left(\frac{c}{b}\right)^0$     e. $\left(\frac{x}{y}\right)^0$     f. $\left(\frac{c}{ab}\right)^{-1}$

13. a. $x^{-2}$     b. $y^{-3}$     c. $\left(\frac{y}{x}\right)^{-2}$
    d. $\left(\frac{a}{b}\right)^{-3}$     e. $\left(\frac{4a^2}{c}\right)^{-2}$     f. $\left(\frac{a}{b^2}\right)^{-3}$     g. $\frac{1}{b^{-3}}$

14. a. $y^{-2}$     b. $x^{-3}$     c. $\left(\frac{c}{b}\right)^{-2}$
    d. $\left(\frac{c}{b}\right)^{-3}$     e. $\left(\frac{2x^2}{y}\right)^{-2}$     f. $\left(\frac{2c}{a^2b}\right)^{-3}$     g. $\frac{1}{x^{-2}}$

15. Simplify these expressions. What do you observe?
    a. $(x^{-1})^{-1}$, by multiplying exponents
    b. $\dfrac{1}{x^{-1}}$
    c. $\dfrac{1}{\frac{1}{x}}$, by division of fractions

16. Multiply these expressions. Describe the pattern.
    a. $\frac{3}{4} \cdot \frac{4}{3}$     b. $\frac{8}{5} \cdot \frac{5}{8}$     c. $\frac{12}{7} \cdot \frac{7}{12}$
    d. $a^{-1} \cdot a^1$     e. $b^2 \cdot \left(\frac{1}{b^2}\right)$     f. $(ab)^{-2} \cdot (ab)^2$

17. Complete the table.

| $x$ | $10^x$ as fraction | $10^x$ as decimal |
|---|---|---|
| 0 | | |
| −1 | | |
| −2 | | |
| −3 | | |
| −4 | | |
| −5 | | |

**Complete the input–output tables in Exercises 18 to 21. Look for a pattern down the output column.**

18.

| Input: $x$ | Output: $3^x$ |
|---|---|
| 2 | |
| 1 | |
| 0 | |
| −1 | |
| −2 | |
| −3 | |

19.

| Input: $x$ | Output: $2^x$ |
|---|---|
| 2 | |
| 1 | |
| 0 | |
| −1 | |
| −2 | |
| −3 | |

20.

| Input: $x$ | Output: $4^x$ |
|---|---|
| 2 | |
| 1 | |
| 0 | |
| −1 | |
| −2 | |
| −3 | |

21.

| Input: $x$ | Output: $5^x$ |
|---|---|
| 2 | |
| 1 | |
| 0 | |
| −1 | |
| −2 | |
| −3 | |

When we apply the binomial theorem to complicated expressions, we obtain expressions that contain zero exponents. In Exercises 22 to 27 use the definitions of exponents to simplify the expressions to the right of the equal sign.

22. $(2x - 3)^2 = 1(2x)^2(-3)^0 + 2(2x)^1(-3)^1 + 1(2x)^0(-3)^2$

23. $(2x + 3)^3 = 1(2x)^3(3)^0 + 3(2x)^2(3)^1 + 3(2x)^1(3)^2 + 1(2x)^0(3)^3$

24. $(3x - 2)^3 = 1(3x)^3(-2)^0 + 3(3x)^2(-2)^1 + 3(3x)^1(-2)^2 + 1(3x)^0(-2)^3$

25. $(3x - 2)^4 = 1(3x)^4(-2)^0 + 4(3x)^3(-2)^1 + 6(3x)^2(-2)^2 +$
$\qquad 4(3x)^1(-2)^3 + 1(3x)^0(-2)^4$

26. $\left(x + \frac{1}{2}\right)^3 = 1(x)^3\left(\frac{1}{2}\right)^0 + 3(x)^2\left(\frac{1}{2}\right)^1 + 3(x)^1\left(\frac{1}{2}\right)^2 + 1(x)^0\left(\frac{1}{2}\right)^3$

27. $\left(x - \frac{1}{2}\right)^3 = 1(x)^3\left(-\frac{1}{2}\right)^0 + 3(x)^2\left(-\frac{1}{2}\right)^1 + 3(x)^1\left(-\frac{1}{2}\right)^2 + 1(x)^0\left(-\frac{1}{2}\right)^3$

**In Exercises 28 to 32 change each number to scientific notation (1 million = 1,000,000).**

28. The maximum distance from the sun to earth is 94.6 million miles.

29. The minimum distance from the sun to Pluto is 2756.4 million miles.

30. The mass of a bacteria is 0.000 000 000 000 1 kilogram.

31. The mass of a chicken is 1800 grams.

32. The mass of an average polar bear is 322 kilograms.

**In Exercises 33 to 36 change each number from scientific notation to decimal notation.**

33. Dinosaurs first appeared on earth about $2.0 \times 10^8$ years ago.

34. Dinosaurs were extinct by $6.5 \times 10^7$ years ago.

35. The mass of a neutron is $1.6750 \times 10^{-27}$ kilogram.

36. The projected human population of earth for 2025 is $8.17 \times 10^9$.

37. Refer to Example 9. Which has a smaller mass, an electron or a proton?

38. Refer to Examples 9 and 10. How many times farther than the minimum distance from Mercury to the sun is the minimum distance from earth to the sun?

**Change the "calculator outputs" in Exercises 39 to 46 from scientific notation to decimals.**

39. 2.34 −02       40. 3.14 03       41. 6.28 07

42. 9.01 −03       43. 4.56 08       44. 2.41 12

45. 6.34 −04       46. 1.02 −07

47. What is the largest number, in scientific notation, that your calculator will accept?

**In Exercises 48 to 53 calculate in scientific notation, either by calculator or mentally.**

48. $(3.6 \times 10^2) \cdot (2.0 \times 10^3)$

49. $(1.6 \times 10^{-3}) \cdot (3.0 \times 10^{-4})$

50. $(1.1 \times 10^{-4}) \cdot (5.0 \times 10^2)$

51. $(5.6 \times 10^4) \div (8.0 \times 10^2)$

52. $(7.2 \times 10^7) \div (9.0 \times 10^3)$

53. $(4.8 \times 10^{-5}) \div (1.2 \times 10^2)$

**Write the answers to Exercises 54 to 58 in scientific notation. The speed of light is 186,000 miles per second.**

54. How far, in miles, will light travel in a year?

55. The minimum distance from the sun to Mercury is 28,600,000 miles. How long does it take light to travel from the sun to Mercury?

56. The maximum distance from the sun to Mercury is 43,400,000 miles. How long does it take light to travel from the sun to Mercury?

57. The minimum distance from the sun to earth is $9.14 \times 10^7$ miles. How long does it take light to travel from the sun to earth?

58. The distance from the sun to Pluto is $2.7564 \times 10^9$ miles. How long does it take for light to travel from the sun to Pluto?

## PROJECTS

59. **Interest on CDs.** The interest on certificates of deposit (CDs) at many banks and credit unions is calculated with annual compounding. If you want $A$ dollars $n$ years in the future, you need to invest $P$ dollars now at interest rate $r$ (expressed as a decimal).

    a. Use the formula $P = A(1 + i)^{-n}$ in the table to calculate how much, $P$, you need to save now to have $50,000 in the future. Round your answers to the nearest dollar.

| Interest Rate | $P = 50000(1 + r)^{-10}$ ($n = 10$ years) | $P = 50000(1 + r)^{-20}$ ($n = 20$ years) |
|---|---|---|
| $r = 0.03$ | | |
| $r = 0.04$ | | |
| $r = 0.05$ | | |
| $r = 0.06$ | | |
| $r = 0.07$ | | |
| $r = 0.08$ | | |

    b. For what rate of interest over 10 years do we need to invest about half the $50,000?

    c. For what rate of interest over 20 years do we need to invest about half the $50,000?

60. **Exponent Puzzles.** The variables $A$, $B$, $C$, and $D$ in the equation $A^B C^D = ABCD$ represent numbers from the set {0, 1, 2, 3, 4, 5, 6, 7, 8, 9}. The expression $ABCD$ in the equation represents a number with four digits, not a product of the numbers $A$ and $B$ and $C$ and $D$. Use a calculator to guess the numbers. (*Hint*: $7^3 \cdot 8^3$ gives a six-digit answer, 175,616.) Digits may be used twice.

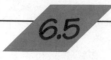

## 6.5  LINEAR POLYNOMIAL EQUATIONS, MORE APPLICATIONS

**OBJECTIVES**   Write expressions to show why puzzle problems yield the same output for all inputs. ♦ Solve consecutive integer problems. ♦ Solve word problems involving consecutive multiples. ♦ Solve problems involving complementary and supplementary angles.

**WARM-UP**   Multiply these expressions.

**1.** $(x + 4)(x - 3)$                          **2.** $(3x - 2)(2x + 3)$

**3.** $(x - 5)(x + 5)$                          **4.** $(x + 4)(x + 4)$

Factor these expressions.

**5.** $x^2 + 2x + 1$                            **6.** $x^2 - 3x - 4$

**7.** $x^2 - 4x + 4$                            **8.** $x^2 - 4$

**9.** $6x^2 + 13x + 6$                          **10.** $1020a + 102b$

I n this section we apply to various polynomials many of the skills that were developed earlier in the chapter. We start with "black hole" puzzles, move on to consecutive integer problems, and practice solving equations based on geometry facts.

### BLACK HOLES

Black holes are bodies in space that are so dense their gravitational attraction keeps most light from escaping. If an object passes too near a black hole, it will be drawn inside, no matter how fast it travels.

There are expressions in mathematics that result in numerical black holes.* No matter what number we use as an input, the expression always outputs the same result. Because the results are known in advance, these expressions are often the source of amusement at parties and mathematics conferences and for leisure magazine columnists.

**EXAMPLE 1**   Try the numerical black hole by completing Table 3. Determine why it works.

| Input Number | Multiply by 9 | Add 6 | Divide by 3 | Subtract 3 Times the Number |
|:---:|:---:|:---:|:---:|:---:|
| 4 | 36 | 42 | 14 | 2 |
| 6 |  |  |  |  |
| 9 |  |  |  |  |

*Table 3*

**SOLUTION**   The completed table is shown in Table 4.

---

*The term *black hole* was applied to the numerical form in the *NCTM Student Math Notes*, September 1991, edited by Carol Findell; National Council of Teachers of Mathematics, 1906 Association Drive, Reston, VA 22091.

| Input Number | Multiply by 9 | Add 6 | Divide by 3 | Subtract 3 Times the Number |
|:---:|:---:|:---:|:---:|:---:|
| 4 | 36 | 42 | 14 | 2 |
| 6 | 54 | 60 | 20 | 2 |
| 9 | 81 | 87 | 29 | 2 |

*Table 4*

Let $x$ be the chosen number. For any $x$ the steps are described by

$$(9 \cdot x + 6) \div 3 - 3x$$

This expression simplifies as follows:

$$(9 \cdot x + 6) \div 3 - 3x = \frac{9x + 6}{3} - 3x$$

$$= 3x + 2 - 3x$$

$$= 2$$

There is no input $x$ remaining in the simplified expression, so the answer does not depend on the input.

◆ ◆ ◆ ◆ ◆ ◆ ◆ ◆ ◆ ◆ ◆ ◆ ◆ ◆ ◆ ◆ ◆ ◆ ◆ ◆ ◆ ◆ ◆ ◆ ◆ ◆ ◆ ◆ ◆ ◆ ◆ ◆ ◆ ◆ ◆ ◆ ◆

**EXAMPLE 2**   Try this numerical black hole by completing Table 5. Determine why it works.

| Input Number | Add 8 | Multiply by 2 | Subtract 6 | Divide by 2 | Subtract the Original Number |
|:---:|:---:|:---:|:---:|:---:|:---:|
| 4 | 12 | 24 | 18 | 9 | 5 |
| 6 | | | | | |
| 9 | | | | | |

*Table 5*

**SOLUTION**   The completed table is shown in Table 6.

| Input Number | Add 8 | Multiply by 2 | Subtract 6 | Divide by 2 | Subtract the Original Number |
|:---:|:---:|:---:|:---:|:---:|:---:|
| 4 | 12 | 24 | 18 | 9 | 5 |
| 6 | 14 | 28 | 22 | 11 | 5 |
| 9 | 17 | 34 | 28 | 14 | 5 |

*Table 6*

Let $x$ be the chosen number. For any input $x$ the steps are

$$[(x + 8) \cdot 2 - 6] \div 2 - x$$

The parentheses are needed to group the sum of $x$ and 8 before the multiplication by 2. The brackets are needed to include the subtraction of 6 before the division is completed. This expression simplifies as follows:

$$[(x + 8) \cdot 2 - 6] \div 2 - x = \frac{[(x + 8) \cdot 2 - 6]}{2} - x$$

$$= \frac{[2x + 16 - 6]}{2} - x$$

$$= \frac{[2x + 10]}{2} - x$$

$$= x + 5 - x$$

$$= 5$$

Again, there is no input $x$ remaining in the simplified expression, so the answer does not depend on the input. This is why the black hole works.

♦ ♦ ♦ ♦ ♦ ♦ ♦ ♦ ♦ ♦ ♦ ♦ ♦ ♦ ♦ ♦ ♦ ♦ ♦ ♦ ♦ ♦ ♦ ♦ ♦ ♦ ♦ ♦ ♦ ♦ ♦ ♦ ♦ ♦ ♦ ♦

### CONSECUTIVE INTEGERS

If we go shopping on three consecutive days or make five consecutive shots in basketball, we are engaging in events that occur one after another without interruption. Appointments on four consecutive Mondays have the interruption of the intervening days, but the Mondays themselves are in a row. **Consecutive integers** *are integers that follow one after another without interruption.*

EXAMPLE **3**    Describe these years (in consecutive five-year intervals) using polynomials: 1990, 1995, 2000, 2005, 2010. Let $x =$ the first year, 1990.

SOLUTION    If $x = 1990$, then

$$1995 = x + 5$$
$$2000 = x + 10$$
$$2005 = x + 15$$
$$2010 = x + 20$$

♦ ♦ ♦ ♦ ♦ ♦ ♦ ♦ ♦ ♦ ♦ ♦ ♦ ♦ ♦ ♦ ♦ ♦ ♦ ♦ ♦ ♦ ♦ ♦ ♦ ♦ ♦ ♦ ♦ ♦ ♦ ♦ ♦ ♦ ♦ ♦

Polynomials permit us to describe any set of consecutive integers in terms of one number in the set. Any set of consecutive integers such as 4, 5, and 6 may be described in terms of the smallest number: $x$, $x + 1$, and $x + 2$. We could also describe the set in terms of either of the other two numbers.

EXAMPLE **4**    Write a polynomial description for 4, 5, and 6 with $x$ as the largest number. Repeat with $x$ as the middle number.

SOLUTION    If $x$ represents the largest number, 6, then $x - 1 = 5$ and $x - 2 = 4$. If $x$ is the middle number, 5, then $x - 1 = 4$ and $x + 1 = 6$.

♦ ♦ ♦ ♦ ♦ ♦ ♦ ♦ ♦ ♦ ♦ ♦ ♦ ♦ ♦ ♦ ♦ ♦ ♦ ♦ ♦ ♦ ♦ ♦ ♦ ♦ ♦ ♦ ♦ ♦ ♦ ♦ ♦ ♦ ♦ ♦

Recall that the numbers including 0, 2, 4, 6, . . . were described as the *even* integers and the numbers including 1, 3, 5, 7, . . . were the *odd* numbers. There is a surprising result in the polynomial description of these integers.

EXAMPLE **5**

Evaluate $x$, $x + 2$, and $x + 4$ for $x = 11$ and for $x = 12$. Identify the resulting numbers as consecutive even numbers, consecutive odd numbers, or neither.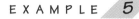

SOLUTION

If $x = 11$, then $x$, $x + 2$, and $x + 4$ give 11, 13, and 15, respectively, which are odd numbers. If $x = 12$, then $x$, $x + 2$, and $x + 4$ give 12, 14, and 16, respectively, which are even numbers.

♦ ♦ ♦ ♦ ♦ ♦ ♦ ♦ ♦ ♦ ♦ ♦ ♦ ♦ ♦ ♦ ♦ ♦ ♦ ♦ ♦ ♦ ♦ ♦ ♦ ♦ ♦ ♦ ♦ ♦ ♦ ♦ ♦ ♦

Example 5 demonstrates that the symbolic description—$x$, $x + 2$, $x + 4$—for consecutive even numbers is identical to the symbolic description for consecutive odd numbers. It is a common error to write consecutive odd numbers as $x$, $x + 1$, and $x + 3$.

Example 6 illustrates the use of consecutive integers in number puzzle problems. Such puzzles are important only because they provide practice in writing and solving equations.

EXAMPLE **6**

The sum of three consecutive odd integers is 447. Set up an equation that shows this relationship, and solve it.

SOLUTION

The equation is $x + (x + 2) + (x + 4) = 447$. Solving for $x$, we find

$$x + (x + 2) + (x + 4) = 447$$
$$3x + 6 = 447$$
$$3x = 441$$
$$x = 147$$

Thus the numbers are $x = 147$, $x + 2 = 149$, and $x + 4 = 151$.

♦ ♦ ♦ ♦ ♦ ♦ ♦ ♦ ♦ ♦ ♦ ♦ ♦ ♦ ♦ ♦ ♦ ♦ ♦ ♦ ♦ ♦ ♦ ♦ ♦ ♦ ♦ ♦ ♦ ♦ ♦ ♦ ♦ ♦

Many students solve consecutive-number problems mentally. How might this be possible?

ANGLES: GEOMETRY AND ALGEBRA

We now examine problems that illustrate geometric facts about angles. Some angle concepts are introduced, and then polynomial algebra is applied to the solution of related geometry problems.

A circle, or complete revolution, is traditionally described as a 360° angle. The number 360 is arbitrary and dates back to ancient civilizations, where it was believed that there were 360 days in a year. The use of 360° has continued in part because 360 has many factors and is easy to use but also because people are reluctant to abandon tradition.

All of our conclusions about the degree measures of certain angles and the sums of angles in a triangle are based on the arbitrary assumption of 360° in a circle.

---

**HISTORICAL NOTE**

The Fahrenheit temperature scale was devised by G. D. Fahrenheit (1686–1736). He based the scale on the temperature of a mixture of ice and salt (0°), the freezing point of water (32°), and what he believed to be normal human temperature (96°). The 180° interval between 32°F and 212°F, the freezing and boiling points of water, is a coincidence.

Our traditional protractor (see Figure 16) shows 180° for a half circle. The numbers on a protractor permit measuring angles from either the left side or the right side.

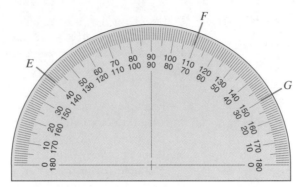

*Figure 16*

E X A M P L E   **7**

What pattern do the pairs of numbers labeled with the letters *E*, *F*, and *G* have in the protractor in Figure 16?

SOLUTION

At position *E*, 40 matches with 140. At position *F*, 110 matches with 70. At position *G*, 150 matches with 30. Each pair of numbers adds to 180.

♦ ♦ ♦ ♦ ♦ ♦ ♦ ♦ ♦ ♦ ♦ ♦ ♦ ♦ ♦ ♦ ♦ ♦ ♦ ♦ ♦ ♦ ♦ ♦ ♦ ♦ ♦ ♦ ♦ ♦ ♦ ♦ ♦ ♦ ♦ ♦

When we look at the two angles being described by each pair of numbers on a protractor, we find that *the angles share a side and their other sides form a straight line*. These angles are often called a **linear pair**. The first drawing in Figure 17 shows a linear pair.

The next two drawings in Figure 17, a parallelogram and parallel lines, show angles that may be rearranged to form a linear pair. *Two angles that add up to a linear pair* (180°) *are* **supplementary angles**. The justification for the fact that the angles in Figure 17 are supplementary is left to a geometry course.

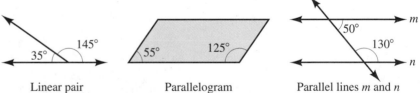

Linear pair        Parallelogram        Parallel lines *m* and *n*

*Figure 17*

E X A M P L E   **8**

Line of sight    The parallel lines in Figure 18 represent the flight path of an airplane and the ground below. The slanted line is a line of sight between the plane and a control tower. If angle *M* is *x* + 1 and angle *N* is 6*x* + 4, determine the measures of angles *M* and *N*.

SOLUTION

The angles *M* and *N* are supplementary angles, so they add to 180°.

$$M + N = 180°$$
$$x + 1 + 6x + 4 = 180°$$
$$7x + 5 = 180°$$
$$x = 25°$$
$$M = x + 1 = 26°$$
$$N = 6x + 4 = 154°$$

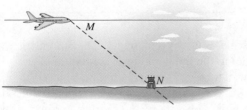

*Figure 18*

♦ ♦ ♦ ♦ ♦ ♦ ♦ ♦ ♦ ♦ ♦ ♦ ♦ ♦ ♦ ♦ ♦ ♦ ♦ ♦ ♦ ♦ ♦ ♦ ♦ ♦ ♦ ♦ ♦ ♦ ♦ ♦ ♦ ♦ ♦ ♦

Figure 19

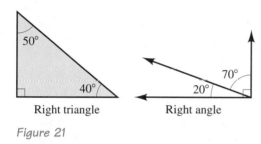

Figure 20

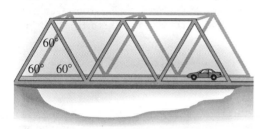

Right triangle        Right angle

Figure 21

Because any rectangle or square contains four right angles, we may conclude that the sum of its interior (inside) angles is 360°. One diagonal in a square or rectangle forms two triangles (see Figure 19), and thus we may also conclude that the sum of the interior angles of a triangle is 180°. Exercise 31 provides an activity to verify this angle sum using paper and scissors.

---

The sum of the angles in a triangle is 180°.

---

The sum of the angles in a triangle is of interest to civil engineers because a triangle, as a rigid figure, is an essential component of bridges and other structures. Many steel bridge trusses (see Figure 20) are composed of equilateral triangles because equal sides and equal angles simplify the designing and building processes. If the sum of the angles is 180°, then each angle of an equilateral triangle is 60°.

In engineering and physics a force such as the tension on a rope or cable or the compression in a member of a truss is separated into smaller forces in the x and y directions in order to do calculations. The separation process is dependent on trigonometry and on the fact that two acute (small) angles in a right angle or a right triangle add to 90°. *Any two angles that add to 90°* are **complementary angles**. Figure 21 shows complementary angles in both a right triangle and a right angle.

**EXAMPLE 9** The rope in Figure 22 is being held tight in such a way as to form an angle, D, that is twice the other angle, C. What are the measures of angles C and D?

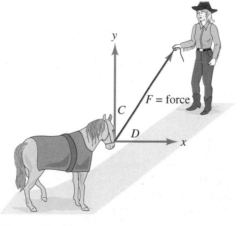

Figure 22

**SOLUTION** The x- and y-axes are perpendicular, so the two angles, C and D, form a right angle. They add to 90°. Let n be the number of degrees in angle C, and let 2n be the number of degrees in angle D.

$$C + D = 90°$$
$$n + 2n = 90°$$
$$3n = 90°$$
$$n = 30° = C$$
$$2n = 60° = D$$

◆ ◆ ◆ ◆ ◆ ◆ ◆ ◆ ◆ ◆ ◆ ◆ ◆ ◆ ◆ ◆ ◆ ◆ ◆ ◆ ◆ ◆ ◆ ◆ ◆ ◆ ◆ ◆ ◆ ◆ ◆ ◆ ◆ ◆ ◆

## EXERCISES 6.5

**The black hole problems in Exercises 1 and 2 are also known as mind-reading tricks.**

1. A number plus 6 is multiplied by 3. Then 6 is subtracted, and the result is divided by 3. The original number is then subtracted.

    a. Find the result using 4 then 6 as the initial number.

    b. Use algebra to show why the exercise always gives the same result.

2. A number is multiplied by 6, then 4 is subtracted. The result is divided by 2, then 5 is added. Three times the original number is subtracted.

    a. Find the result using 4 then 6 as the initial number.

    b. Use algebra to show why the exercise always gives the same result.

3. One way to write 4, 5, 6 with polynomials is $x$, $x + 1$, $x + 2$ if $x$ is 4. Write 4, 5, 6 another way, using polynomials, where $x = 6$.

4. One way to write $-7$, $-6$, $-5$ with polynomials is $x$, $x + 1$, $x + 2$ if $x$ is $-7$. Write $-7$, $-6$, $-5$ another way, using polynomials, where $x = -6$.

5. One way to write the consecutive odd integers 5, 7, 9 with polynomials is $x$, $x + 2$, $x + 4$ if $x$ is 5. Write them two other ways, with $x = 7$ and then with $x = 9$.

6. Write the consecutive even integers 4, 6, 8 in polynomials, first with $x = 6$ and then with $x = 8$.

7. Evaluate $x$, $x + 1$, $x + 3$ for $x = 1$ and $x = 2$. Why might someone incorrectly use these expressions to describe a set of odd numbers?

8. Write the consecutive odd integers 55, 57, 59 with polynomials in three different ways.

9. Compare the polynomial descriptions for the even consecutive integers and the odd consecutive integers.

10. Draw a number line with equally spaced numbers from 1 to 16. Circle the odd numbers, and connect each with the next odd number using a curved line above the number. Place a small square around the even numbers, and connect each to the next even number using a curved line below the number line. How does this show that the description of both even and odd numbers is $x$, $x + 2$, $x + 4$?

11. Three children in a family are born at 3-year intervals. If the first is born in 1939, what are the years of birth for the next two? If $1939 = x$, describe the other two birth years using polynomials.

12. If $2005 = x$, describe these years (consecutive 5-year intervals) using polynomials: 1990, 1995, 2000, 2005, 2010.

13. The U.S. presidential elections, the summer Olympics, and leap years all occur every 4 years. Furthermore, these years are evenly divisible by 4. Which year—2010 or 2012—is one of these special years? If $x$ represents the first year, give the polynomial description for the next three leap years.

14. The concept of events' taking place at regular intervals even appears in the U.S. Constitution: "The actual Enumeration shall be made within three Years after the first Meeting of the Congress of the United States, and within every subsequent term of ten Years, . . ." (Article One, Section 2.3). What important activity does this article describe?

15. The sum of three consecutive integers is 42. Write an equation and solve it to find the three integers. Describe any way you might have simply reasoned out this answer without an equation.

16. The sum of three consecutive integers is 288. Write an equation and solve it to find the three integers. Describe any way you might have simply reasoned out this answer without an equation.

17. A series of books is published at 7-year intervals. When the seventh book is issued, the sum of the publication years is 13,741. When was the first book published?

18. A series of books is published at 4-year intervals. When the fifth book is issued, the sum of the publication years is 10,020. When was the first book published?

19. The sum of three consecutive odd integers is 177. Write an equation and solve it to find the three integers. Describe any way you might have simply reasoned out this answer without an equation.

20. The sum of three consecutive odd integers is 429. Write an equation and solve it to find the three integers. Describe any way you might have simply reasoned out this answer without an equation.

21. Find the sum of the first six even numbers. How is this sum related to 6?

22. Find the sum of the first seven even numbers. How is this sum related to 7?

23. The angles of a triangle are consecutive multiples of 10. What are their measures?

24. The angles of a triangle are consecutive multiples of 15. What are their measures?

25. Two angles of a triangle are equal. The third angle is 30° smaller than the other two. What are the measures of the three angles?

26. Two angles of a triangle are equal. The third angle is 30° larger than the other two. What are the measures of the three angles?

27. One angle of a triangle is 2° larger than the smallest angle. The third angle is 39° larger than the smallest. What are the measures of the three angles?

28. One angle of a triangle is 5° smaller than the largest angle. Another angle is 24° smaller than the largest. What are the measures of the three angles?

29. Use complementary or supplementary angles to find the number of degrees in each angle labeled with a letter in the figure.

a.

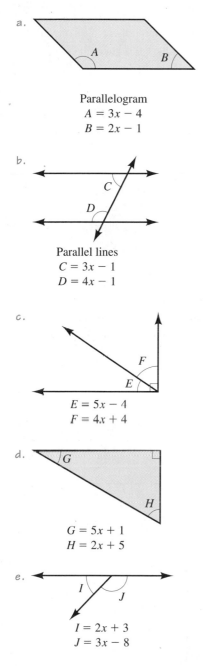

Parallelogram
$A = 3x - 4$
$B = 2x - 1$

b.

Parallel lines
$C = 3x - 1$
$D = 4x - 1$

c.

$E = 5x - 4$
$F = 4x + 4$

d.

$G = 5x + 1$
$H = 2x + 5$

e.

$I = 2x + 3$
$J = 3x - 8$

30. Use complementary or supplementary angles to find the number of degrees in each angle labeled with a letter in the figure.

a.

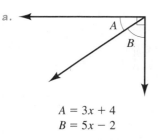

$A = 3x + 4$
$B = 5x - 2$

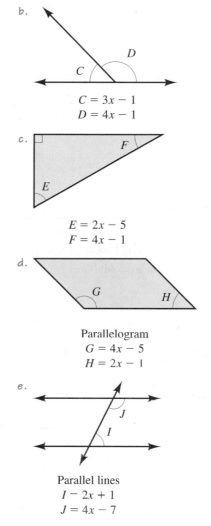

b.

$C = 3x - 1$
$D = 4x - 1$

c.

$E = 2x - 5$
$F = 4x - 1$

d.

Parallelogram
$G = 4x - 5$
$H = 2x - 1$

e.

Parallel lines
$I - 2x + 1$
$J = 4x - 7$

## PROJECTS

31. ***Interior Angles of a Triangle.*** Using a ruler, carefully draw three different triangles. Make one triangle with an angle larger than 90°. Make another triangle close to a right triangle. Cut out the triangles. Taking one triangle at a time, tear the triangle in three pieces to separate the corners. Arrange the points of the corners together to show the total angle measure of all three corners. What do you observe about each triangle? Tape your corners to your homework paper.

32. ***Piling Up Coins***

a. Arrange 25 coins into four piles that fit the following conditions: The second pile is 3 times the first pile. The third pile is 1 less than the second. The fourth pile is 2 more than the first. How many coins are in each pile? Write an equation that would solve the same problem.

b. Arrange 28 coins into four piles that fit the following conditions: The third pile is 3 more than the second pile. The first pile is twice the second pile. The fourth pile is 1 more than the first pile. How many coins are in each pile? Write an equation that would solve the same problem.

c. Describe a strategy to arrange the coins into the requested piles.

## CHAPTER 6 SUMMARY

### Vocabulary

*For definitions and page references, see the Glossary/Index.*

| | |
|---|---|
| ascending order | linear pair |
| binomial pair | monomials |
| binomials | Pascal's Triangle |
| binomial theorem | perfect square trinomial |
| complementary angles | polynomials |
| composite numbers | prime numbers |
| consecutive integers | probability of an outcome |
| descending order | scientific notation |
| difference of perfect squares | supplementary angles |
| factor by table method | trinomials |

### Concepts

Use the *largest common monomial* to factor a single term from an expression.

**To factor by the table method:**

1. First remove any common monomial factor, $d$, so that $ax^2 + bx + c$ remains.
2. Enter $ax^2$ and $c$ in one diagonal.
3. Multiply $ax^2$ and $c$ to obtain the diagonal product.
4. Find the factors of the diagonal product that add to $bx$.
5. Enter the factors in the other diagonal.
6. Factor out the monomials, and write the binomial pair.

In binomial multiplication by table, there are two patterns: The diagonal adds to the middle term of the trinomial answer, and the diagonals in the table form multiply to the same product.

The first eight rows of **Pascal's Triangle** are

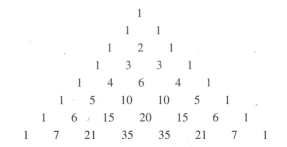

The **binomial theorem** for $(a + b)^n$ includes these properties:

The numerical coefficients come from the row starting $1$ $n$ ... in Pascal's Triangle.

The exponents on $a$ decrease term by term.

The exponents on $b$ increase term by term.

The key to changing from **scientific notation** to regular decimal notation and vice versa is to remember that large numbers have positive exponents on the 10 and small numbers between 0 and 1 have negative exponents on the 10.

## CHAPTER 6 REVIEW EXERCISES

1. Find the perimeter and area of each shape below.

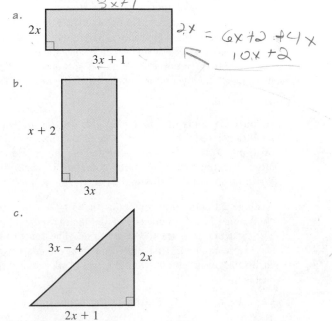

2. Determine the missing dimensions of each shape below.

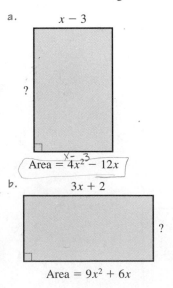

c.

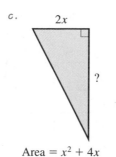

Area = $x^2 + 4x$

3. Multiply using the distributive property.

a. $4(x - 5)$

b. $-7(x^2 + 2x - 3)$

c. $-3x(3x^2 - 2x - 7)$

4. Identify the common monomial factor, and factor the polynomial.

a. $14x^2 + 7xh + 49h^2$

b. $32xy - 24xy^2 + 16x^2y^2$

c. $25abc - 15ac + 35bc$

5. Complete these tables. Write the multiplication problem each table describes.

a.

| Multiply | $2x$ | $+5$ |
|---|---|---|
| $3x$ | $6x^2$ | $15y$ |

b.

| Factor | $9$ $^2$ | $-3$ |
|---|---|---|
| $3c$ | $3a^2$ | $-9$ |

6. Complete these tables. Write the multiplication problem each table describes.

a.

| Factor | $-x^3$ | $49$ |
|---|---|---|
| $x$ | $x^2$ | $+7x$ |
| $-2x$ | $-2x$ | $-14$ |

b.

| Factor | | |
|---|---|---|
| | $6x^2$ | $+10x$ |
| | $+9x$ | $+15$ |

7. Multiply these polynomials.

a. $(x + 4)(x + 3)$

b. $(x + 4)(x - 3)$

c. $(2x - 5)(2x - 5)$

d. $(2x - 5)(2x + 5)$

8. Explain how the answers to a and b in Exercise 7 are the same. How are they different? Why?

9. Explain how the answers to c and d in Exercise 7 are the same. How are they different? Why? What caused the middle term to disappear in the answer to d?

10. a. What is the common product of the diagonal terms in Exercise 6a?

b. What are all the factor pairs for the diagonal product in Exercise 6a?

c. Which factor pairs would correspond to $(x + 14)(x - 1)$?

11. a. What is the common product of the diagonal terms in Exercise 6b?

b. What are all the factor pairs for the diagonal product in Exercise 6b?

c. Which factor pairs would correspond to $(2x + 15)(3x + 1)$?

12. Complete these tables. Write the original trinomial and its factors.

a. Factor $x^2 - 9x + 14$.

| Factor | | |
|---|---|---|
| | $x^2$ | |
| | | $+14$ |

b. Factor $2x^2 + 11x - 21$.

$(2x - 3)$ $(x + 7)$

| Factor | | |
|---|---|---|
| | $2x^2$ | |
| | | $-21$ |

c. Factor $12x^2 - x - 1$.

$(4x + 1)$ $(3x - 1)$

| Factor | | |
|---|---|---|
| | $12x^2$ | |
| | | $-1$ |

d. Factor $20x^2 + 24x - 9$.

| Factor | | |
|---|---|---|
| | | |
| | | |

e. Factor $25x^2 - 16$.

$(5x - 4)$ $(3x + 4)$

| Factor | | |
|---|---|---|
| | | |
| | | |

f. Factor $4x^2 - y^2$.

| Factor | | |
|--------|--------|--------|
| | $4x^2$ | |
| | | $-y^2$ |

13. Factor these trinomials and binomials. Use a table or another method. First remove common monomial factors, as needed.

    a. $x^2 + 5x - 6$       b. $2x^2 - 3x - 35$

    c. $4x^2 - 4x - 35$       d. $4x^2 - 8x + 4$

    e. $4x^2 - 4$       f. $x^3 + 4x^2 + 4x$

14. Multiply or use patterns to remove parentheses from these expressions.

    a. $(x + 1)^3$       b. $(y - 1)^4$

    c. $(x - y)^3$       d. $(x - 1)^5$

15. List the possible birth orders of a family of 6 children with 4 girls and 2 boys.

16. What is the equation of the line of symmetry in each figure? (There may be more than one.)

    a.

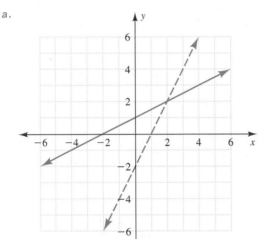

    b.

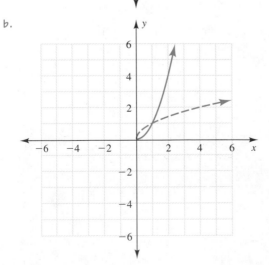

17. Use symmetry of opposites to shade the designs in the figures.

    a.                    b.

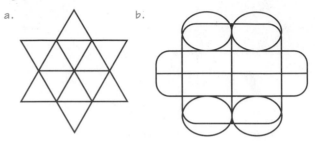

**Use the definitions of exponents to simplify each expression in Exercises 18 to 23. Try them without a calculator first.**

18. a. $49^{-1}$       b. $49^{1/2}$       c. $49^{0.5}$       d. $49^0$

19. a. $144^{1/2}$       b. $144^{-1}$       c. $144^0$       d. $144^{0.5}$

20. a. $\left(\frac{1}{25}\right)^{-1}$       b. $\left(\frac{1}{25}\right)^{1/2}$       c. $\left(\frac{1}{25}\right)^0$       d. $\left(\frac{1}{25}\right)^{0.5}$

21. a. $(0.36)^{1/2}$       b. $(0.36)^{-1}$       c. $(0.36)^{0.5}$       d. $(0.36)^0$

22. a. $(0.25)^{-1}$       b. $(0.25)^0$       c. $(0.25)^{1/2}$       d. $(0.25)^{0.5}$

23. a. $\left(\frac{a}{b}\right)^{-2}$, $a \neq 0, b \neq 0$       b. $\left(\frac{a}{b^2}\right)^0$, $a \neq 0, b \neq 0$

    c. $\left(\frac{a^2}{b^2}\right)^{1/2}$, $a \geq 0, b \geq 0$       d. $\left(\frac{2a}{b^2c}\right)^{-3}$, $a \neq 0, b \neq 0,$
    $c \neq 0$

24. Convert each number in the national debt and population columns of the table to scientific notation. Use a calculator's scientific notation key to calculate the last column (divide national debt by population). Write the last column in both decimal notation and scientific notation.

| Year | National Debt | Population | Debt per Person |
|------|--------------|-----------|-----------------|
| 1900 | 1,200,000,000 | 76,200,000 | |
| 1920 | 24,200,000,000 | 106,000,000 | |
| 1950 | 256,100,000,000 | 151,300,000 | |
| 1990 | 3,233,300,000,000 | 248,700,000 | |

25. Complete the table by changing each number into decimal form.

| Chemical | Symbol for Isotope | Half-life | Half-life as Decimal |
|----------|-------------------|-----------|----------------------|
| Potassium | $K^{40}$ | $1.4 \times 10^9$ years | |
| Calcium | $Ca^{41}$ | $1.2 \times 10^5$ years | |
| Radon | $Rn^{219}$ | $1.243 \times 10^{-7}$ year | |
| Polonium | $Po^{212}$ | $3.0 \times 10^{-7}$ second | |

26. Use unit analysis to change the radon half-life to seconds.

27. Use unit analysis to change the polonium half-life to years.

28. Use algebra to show why this mind-reading trick works: Take any number. Multiply it by 6. Add 9. Divide the result by 3. Subtract twice the original number. The result is 3.

29. The perimeter of a right triangle is 60 inches. The sides are three consecutive multiples of 5. What are the sides of the triangle? Write an equation to solve the problem.

30. Two consecutive angles of a parallelogram are supplementary. Because the three angles of a triangle add to 180°, the two smaller angles of a right triangle are complementary. Find the measure of each angle in the figures below.

a.

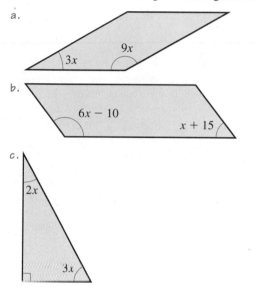

b.

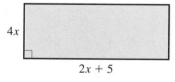

c.

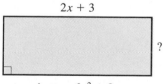

d.

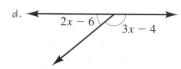

STRETCHERS

31. Use a calculator to make a list of the powers of 11 from $11^0$ to $11^6$. Compare your list with Pascal's Triangle. What do you observe? How is the list similar to Pascal's Triangle? How is it different?

32. Why does 10 $\boxed{EE}$ 3 give 10,000, whereas 10 $\boxed{y^x}$ 3 gives 1000? *Note*: Some calculators have $\boxed{EXP}$ or $\boxed{EEX}$ instead of $\boxed{EE}$ and some have $\boxed{x^y}$ or $\boxed{\wedge}$ instead of $\boxed{y^x}$.

33. Show that $9(x - 1)^2 - 25(x + 1)^2 = -4(4x + 1)(x + 4)$ in two different ways.

34. Factor $r^2 + 5r - 1400$.

---

# CHAPTER 6 TEST

1. Find the perimeter and area of the rectangle in the figure.

[Rectangle with height labeled $4x$ and base labeled $2x + 5$]

2. Determine the width of the rectangle in the figure.

[Rectangle with top labeled $2x + 3$, side labeled ?, Area = $6x^2 + 9x$]

Area = $6x^2 + 9x$

3. Multiply.

a. $(x - 4)(x + 7)$     b. $(x - 7)(x + 7)$

c. $(2x - 7)(2x - 7)$     d. $2(x - 4)(x - 4)$

e. $x(x + 7)(x + 7)$     f. $(x - 2)(x^2 + 2x + 4)$

4. Factor, using any method.

a. $14xy + 6x^2y - 18y^2$     b. $x^2 - 8x + 16$

c. $6x^2 - 54$     d. $9x^2 + 6x + 1$

e. $15x^2 - 14x + 3$     f. $10x^2 + 5x - 5$

5. Is it possible for a trinomial such as $6x^2 - x - 4$ to be factored into $(2x - 2)(\square \pm \square)$? Why or why not?

6. Simplify each expression.

a. $\left(\frac{1}{9}\right)^{-1}$     b. $(0.16)^{1/2}$

c. $\left(\frac{a}{2b}\right)^0$, $a \neq 0$, $b \neq 0$

d. $\left(\frac{9x^2}{25y^2}\right)^{-2}$, $x \neq 0$, $y \neq 0$

7. Divide. Write the answer in both decimal notation and scientific notation.

a. $\dfrac{2.25 \times 10^{18}}{6.25 \times 10^{-4}}$

b. $\dfrac{1.44 \times 10^{-15}}{1.8 \times 10^5}$

8. Multiply or use a pattern to remove parentheses from $(a + 1)^4$.

9. The perimeter of one right triangle is 156 inches. The sides are three consecutive multiples of 13. What are the sides of the triangle? Write an equation to solve the problem.

10. Two consecutive angles of the parallelogram in the figure are supplementary. Determine the angle measures in the parallelogram.

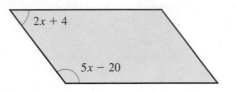

11. Draw in the line(s) of symmetry, if any, in each of the digits 0 to 9.

12. Why would $x$, $x + 1$, $x + 2$ be preferable notation to $a$, $b$, $c$ for describing three consecutive numbers?

13. The formula for evaluating a certificate of deposit of $P$ dollars for $t$ years at a yearly interest rate $r$ is $A = P(1 + r)^t$. What amount of money, $A$, will a $1000 certificate be worth in 10 years at 8% interest?

# Squares and Square Roots: Expressions and Equations

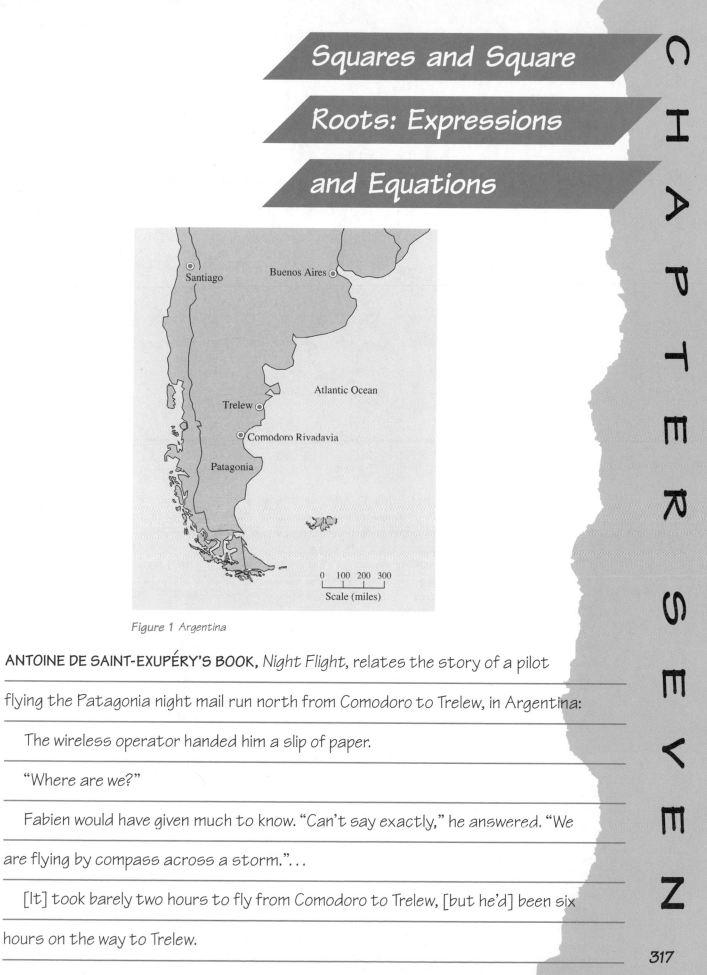

Figure 1 *Argentina*

**ANTOINE DE SAINT-EXUPÉRY'S BOOK,** *Night Flight*, relates the story of a pilot flying the Patagonia night mail run north from Comodoro to Trelew, in Argentina:

The wireless operator handed him a slip of paper.

"Where are we?"

Fabien would have given much to know. "Can't say exactly," he answered. "We are flying by compass across a storm."...

[It] took barely two hours to fly from Comodoro to Trelew, [but he'd] been six hours on the way to Trelew.

317

(continued)

Unbeknownst to Fabien, the wind was a gale from the west at a hundred feet per second, and they had been blown off course. The normal course is about 35° east of north. Estimate the distance between Comodoro and Trelew from the map in Figure 1. Estimate the normal travel rate. Where is the airplane after 2 hours?*

This chapter introduces the Pythagorean theorem, which we then use to suggest a better flight plan for Fabien's airplane. We review ratios, slope, factoring, and solving equations while we examine the algebra of radical and quadratic expressions and equations.

## 7.1    PYTHAGOREAN THEOREM AND RADICALS

**OBJECTIVES**    Confirm the Pythagorean theorem with measurement.  ♦  Use the converse of the Pythagorean theorem to determine whether triangles are right triangles.  ♦  Use the Pythagorean theorem to write equations and to solve for missing sides on a right triangle.  ♦  Identify radicands and irrational numbers.  ♦  Give the principal square root of a number.  ♦  Give the negative root or both ± roots of a number, when specified.  ♦  Simplify expressions using the radical property for products.  ♦  Simplify expressions using the radical property for quotients.

**WARM-UP**    List the values of these perfect squares:

| | | | |
|---|---|---|---|
| $1^2$ | $6^2$ | $11^2$ | $16^2$ |
| $2^2$ | $7^2$ | $12^2$ | $20^2$ |
| $3^2$ | $8^2$ | $13^2$ | $25^2$ |
| $4^2$ | $9^2$ | $14^2$ | $30^2$ |
| $5^2$ | $10^2$ | $15^2$ | $40^2$ |

Knowledge of these facts will be useful in our study of the Pythagorean theorem and work with radical expressions.  ♦

This section begins with the Pythagorean theorem, applies it to ladder safety and airplane navigation, and closes with multiplication and division of radical expressions.

---

*The excerpt on the previous page is from pages 110 and 131 of *Night Flight* (New York: The Century Company, c. 1932). Patagonia is a regional name applied to the southern part of South America, comprising portions of Argentina and Chile.

PYTHAGOREAN THEOREM

Right Triangles.   The **Pythagorean theorem** relates the lengths of the perpendicular sides (legs) of a right triangle to the length of the longest side (hypotenuse). The legs and hypotenuse are shown on the right triangle in Figure 2.

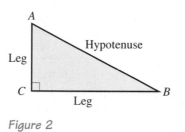

Figure 2

EXPLORATION   Measure the lengths of the sides of the triangles to the nearest 0.5 centimeter. Place your measures in Table 1, and complete the remaining columns.

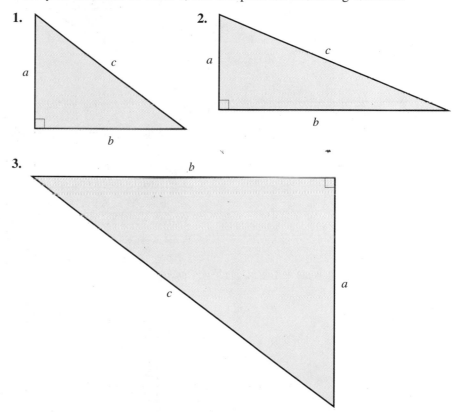

| Triangle | Leg $a$ | Leg $b$ | Hypotenuse $c$ | $a^2$ | $b^2$ | $a^2 + b^2$ | $c^2$ |
|---|---|---|---|---|---|---|---|
| 1 | | | | | | | |
| 2 | | | | | | | |
| 3 | | | | | | | |

Table 1

   The Pythagorean theorem appears in the last columns of Table 1, where $a^2 + b^2 = c^2$.

**The Pythagorean Theorem**

> If a triangle is a right triangle, then the sum of the squares of the two shortest sides (legs) is equal to the square of the longest side (hypotenuse).

*Changing the position of the* if *and* then *statements* in the Pythagorean theorem gives us the **converse of the Pythagorean theorem**. The converse is also true.

**Converse of the Pythagorean Theorem**

> If the sum of the squares of the two shortest sides (legs) is equal to the square of the longest side (hypotenuse), then the triangle is a right triangle.

E X A M P L E   **1**    *Right triangles*    Each triangle below is drawn to look like a right triangle. Which *is* a right triangle?

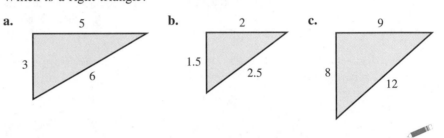

a.
5
3
6

b.
2
1.5
2.5

c.
9
8
12

SOLUTION    **a.** $3^2 + 5^2 = 9 + 25 = 34$, which does not equal 36, the square of the third side. This triangle is not a right triangle.

**b.** $1.5^2 + 2^2 = 2.25 + 4 = 6.25 = 2.5^2$. This is a right triangle, with the right angle where the shorter sides meet.

**c.** $8^2 + 9^2 = 64 + 81 = 145$, which does not equal 144, the square of the third side. This triangle is not a right triangle.

◆ ◆ ◆ ◆ ◆ ◆ ◆ ◆ ◆ ◆ ◆ ◆ ◆ ◆ ◆ ◆ ◆ ◆ ◆ ◆ ◆ ◆ ◆ ◆ ◆ ◆ ◆ ◆ ◆ ◆ ◆ ◆ ◆ ◆ ◆

E X A M P L E   **2**    *Missing sides*    What is the missing side, $n$, in each drawing below?

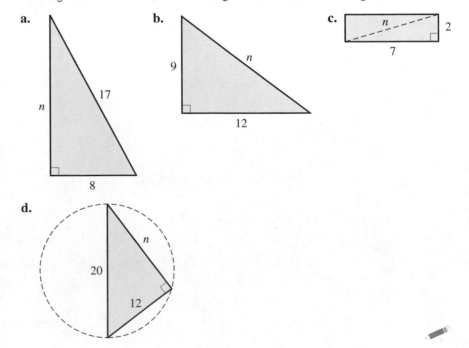

a.
17
$n$
8

b.
9
$n$
12

c.
$n$
2
7

d.
20
$n$
12

SOLUTION    **a.** The sides $n$ and 8 are perpendicular and are the legs, $a$ and $b$.

$$n^2 + 8^2 = 17^2$$

$$n^2 + 64 = 289$$

$$n^2 = 225$$

$$n = \sqrt{225}$$

$$n = 15$$

Although $-15$ also makes $n^2 = 225$ true, length must be positive, so we disregard $-15$. (More on this later.)

**b.** The sides 9 and 12 are perpendicular and are the legs.

$$9^2 + 12^2 = n^2$$

$$81 + 144 = n^2$$

$$225 = n^2$$

$$15 = n$$

**c.** The sides 2 and 7 are perpendicular.

$$2^2 + 7^2 = n^2$$

$$4 + 49 = n^2$$

$$53 = n^2$$

$$\sqrt{53} = n$$

When a square root does not give an exact value, we generally leave it in square root form or use the calculator to give a decimal approximation. In this case $n \approx 7.28$, to the nearest hundredth.

**d.** The sides $n$ and 12 are perpendicular.

$$n^2 + 12^2 = 20^2$$

$$n^2 + 144 = 400$$

$$n^2 = 256$$

$$n = 16$$

Observe that the diameter of the circle is the hypotenuse.

♦ ♦ ♦ ♦ ♦ ♦ ♦ ♦ ♦ ♦ ♦ ♦ ♦ ♦ ♦ ♦ ♦ ♦ ♦ ♦ ♦ ♦ ♦ ♦ ♦ ♦ ♦ ♦ ♦ ♦ ♦ ♦ ♦ ♦ ♦

In solving for the missing side we are using a new manipulation of equations: *taking the square root of both sides.* We will discuss this operation more formally in Section 7.2.

Ladder Safety.    In Section 4.2 we examined the 4 to 1 ladder safety ratio (height of the ladder on the wall to the distance of the base of the ladder from the wall). We now use the Pythagorean theorem to determine the lengths of the ladders needed.

EXAMPLE **3**     Ladder lengths    A ladder needs to reach 21 feet up a wall (Figure 3). The base of the ladder must be 5.25 feet from the wall for a safety ratio of 4 to 1. What length ladder is needed?

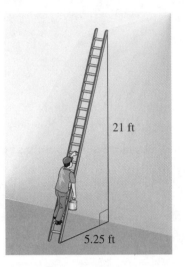

21 ft

5.25 ft

Figure 3

SOLUTION     The ground and the wall form the legs of a right triangle, and the ladder is the hypotenuse. The length of the ladder is $c$.

$$c^2 = (21 \text{ ft})^2 + (5.25 \text{ ft})^2$$
$$c^2 \approx 468.56 \text{ ft}^2$$
$$c \approx 21.6 \text{ ft}$$

Extension ladders are flexible in length, but usually in approximately 1-foot increments. Thus a 21.6-foot length may be difficult to obtain.

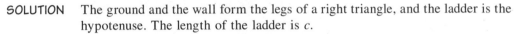

EXAMPLE **4**     Ladder positions    What is the safe ladder position for a 16-foot ladder (Figure 4)?

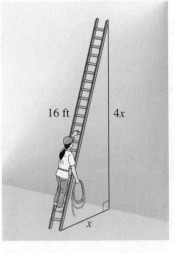

16 ft     $4x$

$x$

Figure 4

SOLUTION   The base-height ratio must be 1 to 4. Thus the legs of the right triangle are $x$ and $4x$. By the Pythagorean theorem,

$$(16 \text{ ft})^2 = x^2 + (4x)^2$$
$$256 \text{ ft}^2 = 17x^2$$
$$256 \div 17 = x^2$$
$$15.059 = x^2$$
$$x \approx 3.88 \text{ ft from the base of the wall}$$
$$4x \approx 15.52 \text{ ft up the wall}$$

◆ ◆ ◆ ◆ ◆ ◆ ◆ ◆ ◆ ◆ ◆ ◆ ◆ ◆ ◆ ◆ ◆ ◆ ◆ ◆ ◆ ◆ ◆ ◆ ◆ ◆ ◆ ◆ ◆ ◆ ◆ ◆ ◆

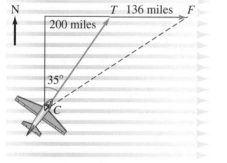

Figure 5

*Flight and Wind.*   In the chapter opener, Fabien set his plane to fly a course 35° east of north. Unfortunately, in 2 hours of flying time, the wind blew his plane to a point $F$ in Figure 5, 136 miles to the east of Trelew ($T$). The 100-foot-per-second (or 68-mph) influence of the wind amounted to 136 miles over 2 hours.

In 6 hours, without course correction, Fabien would be well beyond point $F$ on the line from Comodoro Rivadavia ($C$).

As we will see in Example 5, had Fabien known the wind speed, he would have flown due north because the wind's effect on that direction produced almost exactly the course needed to fly to Trelew. When we fly due north while the wind blows from the west, our actual direction of flight is to the northeast.

E X A M P L E   **5**

*Plane location*   If the wind is blowing at 68 miles per hour from the west and the plane is traveling 100 miles per hour due north, calculate the pilot's position after 1 hour (as shown in Figure 6).

Figure 6

SOLUTION   In 1 hour the plane will travel 100 miles north and 68 miles east. Because these directions are at right angles, the actual flight path will be the hypotenuse.

$$100^2 + 68^2 = x^2$$
$$14{,}624 = x^2$$
$$x = \sqrt{14{,}624} \approx 121 \text{ mi}$$

When we introduced slope, we also introduced a way to find the angle at $A$: `2nd` `tan` `(` `68` `÷` `100` `)` , or 34°. The wind will blow the plane 34° to the northeast. Because the required direction of the flight is 35°, flying due north would have been a better strategy.

◆ ◆ ◆ ◆ ◆ ◆ ◆ ◆ ◆ ◆ ◆ ◆ ◆ ◆ ◆ ◆ ◆ ◆ ◆ ◆ ◆ ◆ ◆ ◆ ◆ ◆ ◆ ◆ ◆ ◆ ◆ ◆ ◆ ◆ ◆

RADICALS

*Radical Vocabulary.*   The expressions $\sqrt{53}$ and $\sqrt{14{,}624}$ did not have exact decimal answers because 53 and 14,624 are not perfect squares. The radical forms $\sqrt{53}$ and $\sqrt{14{,}624}$ are **irrational numbers**. The box below shows several examples of rational numbers and irrational numbers. Together the rational and irrational numbers make up our *real number system*.

Real Numbers

> **Rational Numbers**
> Fractions: $\frac{1}{2}, \frac{1}{4}$
> Repeating decimals: $0.33\overline{3}, 0.12\overline{12}$
> Terminating decimals: 0.5, 0.125
> Integers: $-2, -1, 365$
> Whole numbers: 3, 18
> Radicals with exact decimal values: $\sqrt{16}, \sqrt{256}$
> **Irrational Numbers**
> Pi: 3.14159265...
> Radicals without exact decimal values: $\sqrt{2}, \sqrt{3}, \sqrt{5}$

If $n^2 = 53$, then $n = \sqrt{53}$ is called the principal square root. The **principal square root** *is the positive root of a number.* Unless otherwise directed, we always give the principal square root for a radical expression. If the negative form is required, a negative $(-)$ sign is placed in front of the radical. If both the positive and the negative form are required, a plus or minus $(\pm)$ sign appears in front of the radical.

EXAMPLE   **6**

Simplify these expressions.

    **a.** $\sqrt{25}$     **b.** $-\sqrt{49}$     **c.** $\pm\sqrt{121}$

SOLUTION   **a.** $\sqrt{25} = 5$         Give the positive, or principal, root.

    **b.** $-\sqrt{49} = -7$       The negative root is requested.

    **c.** $\pm\sqrt{121} = +11$ or $-11$    Both roots are requested.

♦ ♦ ♦ ♦ ♦ ♦ ♦ ♦ ♦ ♦ ♦ ♦ ♦ ♦ ♦ ♦ ♦ ♦ ♦ ♦ ♦ ♦ ♦ ♦ ♦ ♦ ♦ ♦ ♦ ♦ ♦ ♦ ♦

*The number or expression under the radical sign is called the* **radicand**. The square root of a negative radicand is neither rational nor irrational, so we say that the radical has no real number solution. Try $\sqrt{-16}$, using 16 $\boxed{+/-}$ $\boxed{\sqrt{\ }}$ or $\boxed{\sqrt{\ }}$ $\boxed{(-)}$ 16 on a scientific calculator. The calculator should respond with an error message, indicating that the square root of a negative is undefined. The square root of a negative does have another type of number solution, which is introduced later in algebra. This "imaginary" solution appears on some graphing calculators.

EXAMPLE   **7**

Simplify these radical expressions.

    **a.** $\sqrt{(-9)}$      **b.** $\sqrt{(-16)}$     **c.** $-\sqrt{25}$

SOLUTION   **a.** There is no real number solution because no real number times itself gives $-9$.

    **b.** There is no real number solution; the radicand is negative.

    **c.** The answer is the opposite of 5, or $-5$. The negative in front of the radical specifies the negative square root rather than the principal square root.

♦ ♦ ♦ ♦ ♦ ♦ ♦ ♦ ♦ ♦ ♦ ♦ ♦ ♦ ♦ ♦ ♦ ♦ ♦ ♦ ♦ ♦ ♦ ♦ ♦ ♦ ♦ ♦ ♦ ♦ ♦ ♦ ♦ ♦

Radical Operations. We now examine multiplication and division with radical expressions.

EXAMPLE  **8**

Use a calculator to investigate these expressions.

**a.** $\sqrt{4} \cdot \sqrt{9}$    **b.** $\sqrt{4 \cdot 9}$    **c.** $\sqrt{9 \cdot 25}$    **d.** $\sqrt{9} \cdot \sqrt{25}$

SOLUTION

**a.** $\sqrt{4} \cdot \sqrt{9} = 2 \cdot 3 = 6$    **b.** $\sqrt{4 \cdot 9} = \sqrt{36} = 6$
**c.** $\sqrt{9 \cdot 25} = \sqrt{225} = 15$    **d.** $\sqrt{9} \cdot \sqrt{25} = 3 \cdot 5 = 15$

♦ ♦ ♦ ♦ ♦ ♦ ♦ ♦ ♦ ♦ ♦ ♦ ♦ ♦ ♦ ♦ ♦ ♦ ♦ ♦ ♦ ♦ ♦ ♦ ♦ ♦ ♦ ♦ ♦ ♦ ♦

From these examples we see that the square root may be taken before or after multiplication. In Section 6.4 we found that $\frac{1}{2}$ or 0.5 as an exponent meant square root. Thus all the properties of exponents apply to square roots or radicals. Of interest at this time is the exponent property for products because it permits us to multiply radical expressions without first changing them to decimals.

**Product Property for Exponents**

$$(a \cdot b)^n = a^n \cdot b^n$$

When the exponent $n$ equals $\frac{1}{2}$ in the product property for exponents, we have

$$(a \cdot b)^{1/2} = a^{1/2} \cdot b^{1/2}$$

Because $\frac{1}{2}$ as an exponent means square root, the product property for exponents leads directly to the product property for radicals.

**Product Property for Radicals**

$$\sqrt{a \cdot b} = \sqrt{a} \cdot \sqrt{b}$$

if $a$ and $b$ are positive numbers or zero.

EXAMPLE  **9**

Simplify these expressions by using the properties of exponents and radicals. Recall that there is no real number solution to the square root of a negative, so assume that all variables represent positive numbers or zero.

**a.** $\sqrt{2} \cdot \sqrt{3}$    **b.** $\sqrt{3} \cdot \sqrt{3}$    **c.** $(\sqrt{2})^2$    **d.** $(2\sqrt{3})^2$
**e.** $(3\sqrt{2})^2$    **f.** $\sqrt{x}\sqrt{x}$    **g.** $\sqrt{25x^2}$    **h.** $\sqrt{49a^2}$

SOLUTION

**a.** $\sqrt{2} \cdot \sqrt{3} = \sqrt{(2 \cdot 3)} = \sqrt{6}$
**b.** $\sqrt{3} \cdot \sqrt{3} = \sqrt{(3 \cdot 3)} = \sqrt{9} = 3$
**c.** When we square a number, we can write it twice and multiply. Thus

$$(\sqrt{2})^2 = \sqrt{2} \cdot \sqrt{2} = \sqrt{(2 \cdot 2)} = \sqrt{4} = 2$$

**d.** We can write $(2\sqrt{3})$ twice or use $2^2 \cdot \sqrt{3^2}$:

$$(2\sqrt{3})^2 = 2\sqrt{3} \cdot 2\sqrt{3} = 2 \cdot 2 \cdot \sqrt{3} \cdot \sqrt{3} = 4 \cdot \sqrt{9} = 4 \cdot 3 = 12$$
$$(2\sqrt{3})^2 = 2^2 \cdot \sqrt{3^2} = 4 \cdot 3 = 12$$

**e.** We have two choices:

$$(3\sqrt{2})^2 = 3\sqrt{2} \cdot 3\sqrt{2} = 9 \cdot \sqrt{4} = 9 \cdot 2 = 18$$
$$(3\sqrt{2})^2 = 3^2 \cdot \sqrt{2^2} = 9 \cdot \sqrt{4} = 9 \cdot 2 = 18$$

**f.** $\sqrt{x} \cdot \sqrt{x} = \sqrt{(x^2)} = x$, when $x$ is positive.
**g.** $\sqrt{25x^2} = 5x$, when $x$ is positive.
**h.** $\sqrt{49a^2} = 7a$, when $a$ is positive.

♦ ♦ ♦ ♦ ♦ ♦ ♦ ♦ ♦ ♦ ♦ ♦ ♦ ♦ ♦ ♦ ♦ ♦ ♦ ♦ ♦ ♦ ♦ ♦ ♦ ♦ ♦ ♦ ♦ ♦ ♦ ♦ ♦

EXAMPLE **10**   Simplify.

a. $\dfrac{\sqrt{16}}{\sqrt{4}}$    b. $\sqrt{\dfrac{16}{4}}$    c. $\dfrac{\sqrt{144}}{\sqrt{16}}$    d. $\sqrt{\dfrac{144}{16}}$

SOLUTION   a. $\dfrac{\sqrt{16}}{\sqrt{4}} = \dfrac{4}{2} = 2$    b. $\sqrt{\dfrac{16}{4}} = \sqrt{4} = 2$

c. $\dfrac{\sqrt{144}}{\sqrt{16}} = \dfrac{12}{4} = 3$    d. $\sqrt{\dfrac{144}{16}} = \sqrt{9} = 3$

✦ ✦ ✦ ✦ ✦ ✦ ✦ ✦ ✦ ✦ ✦ ✦ ✦ ✦ ✦ ✦ ✦ ✦ ✦ ✦ ✦ ✦ ✦ ✦ ✦ ✦ ✦ ✦ ✦ ✦ ✦ ✦

Example 10 shows that, as with multiplication of square roots, we can simplify the square roots before or after division. Again, because the square root equals $\frac{1}{2}$ as an exponent, the quotient property of exponents leads to the quotient property of radicals.

**Quotient Property for Exponents**

$$\left(\dfrac{a}{b}\right)^n = \dfrac{a^n}{b^n}, \quad b \neq 0$$

**Quotient Property for Radicals**

$$\sqrt{\dfrac{a}{b}} = \dfrac{\sqrt{a}}{\sqrt{b}}$$

if $a$ and $b$ are positive.

EXAMPLE **11**   Simplify these expressions. You may find it useful to simplify fractions before taking the square roots. Assume the variables represent only positive numbers.

a. $\sqrt{\dfrac{36}{81}}$    b. $\sqrt{\dfrac{64x^2}{16}}$    c. $\sqrt{\dfrac{50}{18}}$    d. $\sqrt{\dfrac{4x^3}{36x}}$

SOLUTION   a. $\sqrt{\dfrac{36}{81}} = \dfrac{6}{9} = \dfrac{2}{3}$, or $\sqrt{\dfrac{36}{81}} = \sqrt{\dfrac{4 \cdot 9}{9 \cdot 9}} = \sqrt{\dfrac{4}{9}} = \dfrac{2}{3}$

b. $\sqrt{\dfrac{64x^2}{16}} = \dfrac{8x}{4} = 2x$, or $\sqrt{\dfrac{64x^2}{16}} = \sqrt{4x^2} = 2x$

c. $\sqrt{\dfrac{50}{18}} = \sqrt{\dfrac{25 \cdot 2}{9 \cdot 2}} = \sqrt{\dfrac{25}{9}} = \dfrac{5}{3}$

d. $\sqrt{\dfrac{4x^3}{36x}} = \sqrt{\dfrac{4 \cdot x^2 \cdot x}{4 \cdot 9 \cdot x}} = \sqrt{\dfrac{x^2}{9}} = \dfrac{x}{3}$

✦ ✦ ✦ ✦ ✦ ✦ ✦ ✦ ✦ ✦ ✦ ✦ ✦ ✦ ✦ ✦ ✦ ✦ ✦ ✦ ✦ ✦ ✦ ✦ ✦ ✦ ✦ ✦ ✦ ✦ ✦ ✦

## EXERCISES 7.1

Use the ⬚ $x^2$ key on a calculator to answer the questions in Exercises 1–6.

1. What is the largest three-digit number that gives a five-digit number when squared?

2. What is the largest four-digit number that gives a seven-digit number when squared?

3. What is the largest five-digit number that gives a nine-digit number when squared? Compare the digits with those in $\sqrt{10}$. Is this a coincidence?

4. What is the largest square possible with a five-digit number?

5. My grandmother was $x$ years old in the year $x^2$. She was still alive when I was born in 1945. Find the age $x$ and the year $x^2$. When was she born?

6. My great, great, great grandmother was $x$ years old in the year $x^2$. She was alive during the American Civil War. Find the age $x$ and the year $x^2$. When was she born?

**Which of the sets of three numbers in Exercises 7 to 16 could represent the sides of a right triangle?**

7. {7, 8, 9}       8. {12, 15, 18}       9. {12, 16, 20}

10. {8, 15, 17}       11. {7, 24, 25}       12. {9, 40, 41}

13. {3, $\sqrt{7}$, 4}       14. {8, $\sqrt{17}$, 9}       15. {6, $\sqrt{13}$, 7}

16. {10, $\sqrt{21}$, 11}

**In Exercises 17 and 18 solve for the side marked with an *x* in each figure.**

17. a.

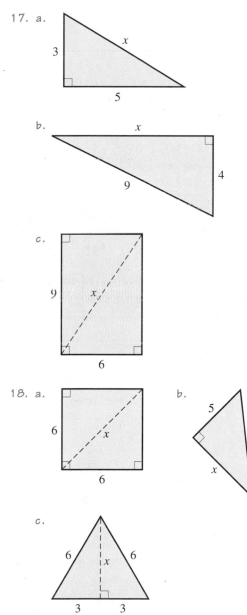

b.

c.

18. a.      b.

c.

19. A safe ladder position for reaching 12 feet up a wall is 3 feet from the base of the ladder to the wall. How long a ladder is needed?

20. A safe ladder position for reaching 8 feet up a wall is 2 feet from the base of the ladder to the wall. How long a ladder is needed?

21. A safe ladder position for reaching 9 feet up a wall is 2.25 feet from the base of the ladder to the wall. How long a ladder is needed?

22. A safe ladder position for reaching 10 feet up a wall is 2.5 feet from the base of the ladder to the wall. How long a ladder is needed?

23. What is the safe ladder position for reaching 14 feet up a wall? How long a ladder is needed?

24. What is the safe ladder position for reaching 19 feet up a wall? How long a ladder is needed?

**What are safe ladder positions for the ladder lengths in Exercises 25 to 28? Give your answers first with decimal portions of a foot and then estimate feet and inches to the nearest half inch.**

25. 12-foot ladder

26. 22-foot extension ladder

27. 18-foot extension ladder

28. 25-foot extension ladder

29. Estimate the results in Example 4 in feet and inches.

**In Exercises 30 to 32, draw triangles to scale. Measure to estimate answers.**

30. A plane flies for 3 hours on a heading due north from San Francisco. The plane is flying at 200 miles per hour while a 32 mile per hour wind is blowing from the west. What is the total distance flown? Use $\boxed{\text{2nd}}$ $\boxed{\text{tan}}$ to find the angle the plane drifted off course to the east, as in Example 5.

31. A plane flies south for 1 hour at 250 miles per hour from Montreal, Canada. A 30-mile-per-hour wind is blowing from the east. How far will the plane have flown? Use $\boxed{\text{2nd}}$ $\boxed{\text{tan}}$ to determine the angle it drifted to the west.

32. A plane flies east for 1 hour from Missoula, Montana, at 210 miles per hour. A 35-mile-per-hour wind from the north blows the plane off course. How far will the plane have traveled? Use $\boxed{\text{2nd}}$ $\boxed{\text{tan}}$ to determine the angle it drifted to the south.

33. What is the height of the house in the figure?

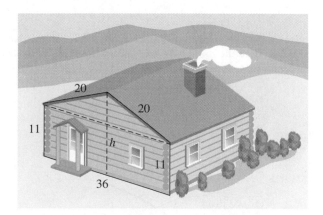

34. What are the lengths of the crossed pieces, *AB* and *CD*, needed to make the kite in the figure?

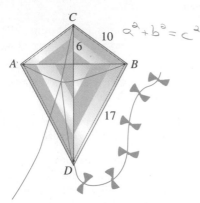

A rectangular house has a 25-foot by 40-foot floor plan (see the figure). For Exercises 35 to 38 assume (incorrectly) that the roof ends at the edge of the wall. Recall that pitch of a roof is height over span.

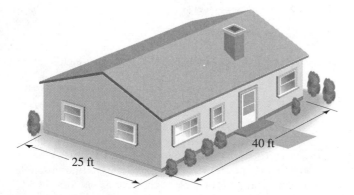

35. How many square feet of roofing material are needed for a pitch of 3 to 25?

36. How many square feet of roofing material are needed for a pitch of 5 to 25?

37. How many square feet of roofing material are needed for a pitch of 9 to 25?

38. How many square feet of roofing material are needed for a pitch of 10 to 25?

**Simplify the radical expressions in Exercises 39 and 40.**

39. a. $\sqrt{49}$   b. $-\sqrt{225}$   c. $\pm\sqrt{400}$

40. a. $\sqrt{121}$   b. $\pm\sqrt{64}$   c. $-\sqrt{169}$

**Simplify the exponent and radical expressions in Exercises 41 to 46. Assume the radicands are positive and any variables in the denominators are not zero.**

41. a. $\sqrt{5} \cdot \sqrt{3}$   b. $\sqrt{7} \cdot \sqrt{2}$   c. $(3\sqrt{5})^2$

    d. $(5\sqrt{3})^2$   e. $(2\sqrt{2})^2$

42. a. $\sqrt{3} \cdot \sqrt{7}$   b. $\sqrt{2} \cdot \sqrt{5}$   c. $(4\sqrt{2})^2$

    d. $(3\sqrt{7})^2$   e. $(7\sqrt{3})^2$

43. a. $\sqrt{a}\sqrt{a}$   b. $\sqrt{(b^2)}$   c. $\sqrt{121a^2}$

    d. $\sqrt{2x} \cdot \sqrt{18x}$

44. a. $\sqrt{b}\sqrt{b}$   b. $\sqrt{(c^2)}$   c. $\sqrt{400x^2}$

    d. $\sqrt{32a} \cdot \sqrt{2a}$

45. a. $\sqrt{\dfrac{x^2}{9}}$   b. $\sqrt{\dfrac{4}{25}}$   c. $\sqrt{\dfrac{45}{5}}$

    d. $\sqrt{\dfrac{28x}{7x^3}}$

46. a. $\sqrt{\dfrac{y^2}{4}}$   b. $\sqrt{\dfrac{64}{9}}$   c. $\sqrt{\dfrac{48}{3}}$

    d. $\sqrt{\dfrac{6y^3}{30y}}$

## PROJECTS

47. **Pythagorean Theorem**

    a. On graph paper draw a triangle with perpendicular sides of 3 units and 4 units. Leave about 6 units from any edge of the paper.

    b. On each leg of the triangle, form a square with area equal to the side squared.

    c. From a separate piece of graph paper cut a square that equals the square of the hypotenuse. Fasten it to the hypotenuse of your original triangle.

    d. Shade the area of the square on the hypotenuse that corresponds to the square of the smallest leg. Count the remaining squares to verify that they equal the square of the larger leg.

    e. How does the figure illustrate the Pythagorean theorem?

    f. Repeat for two other right triangles such as triangles with perpendicular sides of lengths 6 and 8, 5 and 12, or 8 and 15.

48. **Square Root Patterns**

    a. Use your knowledge of perfect squares and a calculator, as needed, to find a pattern in the values of these square roots. Describe your findings in words. (*Hint:* Your word description should relate to the number of zeros or decimal places in the radicand.)

| | | |
|---|---|---|
| $\sqrt{4}$ | $\sqrt{9}$ | $\sqrt{25}$ |
| $\sqrt{40}$ | $\sqrt{90}$ | $\sqrt{250}$ |
| $\sqrt{400}$ | $\sqrt{900}$ | $\sqrt{2500}$ |
| $\sqrt{4000}$ | $\sqrt{0.09}$ | $\sqrt{0.25}$ |
| $\sqrt{40,000}$ | $\sqrt{0.0009}$ | $\sqrt{0.025}$ |
| $\sqrt{4,000,000}$ | $\sqrt{0.000\,000\,9}$ | $\sqrt{0.000\,25}$ |

    b. Test your pattern statement on the following square roots. Check with a calculator.

| | | |
|---|---|---|
| $\sqrt{0.000\,000\,4}$ | $\sqrt{9,000,000}$ | $\sqrt{250,000,000}$ |
| $\sqrt{0.000\,000\,000\,04}$ | $\sqrt{0.000\,09}$ | $\sqrt{0.000\,000\,25}$ |

    c. Use your pattern on the following square roots. Suppose you know $\sqrt{8} = 2.8284$ and $\sqrt{80} = 8.9443$. Do not use a calculator.

| | | |
|---|---|---|
| $\sqrt{800,000}$ | $\sqrt{0.8}$ | $\sqrt{8000}$ |
| $\sqrt{8,000,000}$ | $\sqrt{0.0008}$ | $\sqrt{0.000\,000\,8}$ |

## 7.2 SOLVING QUADRATIC EQUATIONS BY SQUARE ROOT AND FACTORING

**OBJECTIVES**    Solve quadratic equations by graphing. ♦ Solve quadratic equations by taking the square root of both sides. ♦ Solve quadratic equations by applying the zero product rule to factors.

**WARM-UP**    **1.** Complete the table for $y = x^2$.

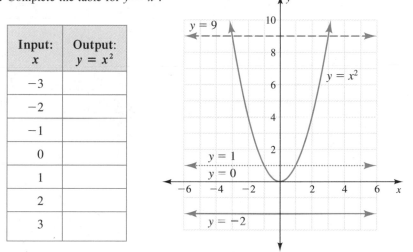

| Input: $x$ | Output: $y = x^2$ |
|---|---|
| $-3$ | |
| $-2$ | |
| $-1$ | |
| $0$ | |
| $1$ | |
| $2$ | |
| $3$ | |

Use the graph of $y = x^2$ in the figure to solve these equations.

**2.** $x^2 = 9$            **3.** $x^2 = 1$            **4.** $x^2 = 0$

**5.** $x^2 = -2$

Simplify these absolute value expressions.

**6.** $|-2|$            **7.** $|2|$            **8.** $|-3|$

Use guess-and-check to solve these absolute value equations.

**9.** $|x| = 2$            **10.** $|x| = 3$

Factor.

**11.** $x^2 - 3x - 4$            **12.** $x^2 - 4$            ♦

We have used several techniques to solve equations: tables, graphs, and symbols. With symbols we have added, subtracted, multiplied, or divided the same quantities on both sides of an equation. We now apply tables and graphs to solving quadratic equations. We then examine two new techniques for solving quadratic equations—taking the square root of both sides and applying the zero product rule to factors.

> Quadratic equations may be written in the form $y = ax^2 + bx + c$.

### SOLVING QUADRATICS FROM TABLES AND GRAPHS

In many applications quadratic equations contain variables other than $x$. The area of a circle is quadratic in terms of $r$: $A = \pi r^2$. The height of an object after $t$ seconds is quadratic in terms of $t$: $h = -gt^2 + vt + s$.

E X A M P L E    **1**    Use a table and a graph to solve $y = x^2 - 3x - 4$ for $x$ when $y = 0$. What name is given to the solutions of $x^2 - 3x - 4 = 0$ on the graph?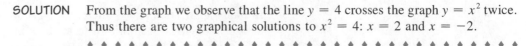

SOLUTION    When $y = 0$, the equation becomes $x^2 - 3x - 4 = 0$. In Table 2 $y$ is zero when $x = -1$ and $x = 4$. On the graph in Figure 7 $y = 0$ is the $x$-axis. The curve $y = x^2 - 3x - 4$ crosses the $x$-axis at $x = -1$ and $x = 4$.

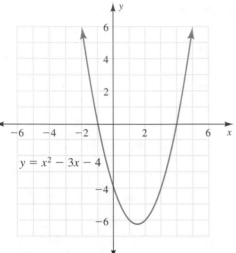

Figure 7

| Input: $x$ | Output: $y = x^2 - 3x - 4$ |
|---|---|
| $-2$ | 6 |
| $-1$ | 0 |
| 0 | $-4$ |
| 1 | $-6$ |
| 2 | $-6$ |
| 3 | $-4$ |
| 4 | 0 |

Table 2

The solutions to $x^2 - 3x - 4 = 0$ are the $x$-intercepts.

♦ ♦ ♦ ♦ ♦ ♦ ♦ ♦ ♦ ♦ ♦ ♦ ♦ ♦ ♦ ♦ ♦ ♦ ♦ ♦ ♦ ♦ ♦ ♦ ♦ ♦ ♦ ♦ ♦ ♦ ♦ ♦ ♦ ♦ ♦ ♦ ♦ ♦

Because many quadratic equations are solved in the form $ax^2 + bx + c = 0$, it is important to be able to find the $x$-intercepts. If graphing technology is available, trace the graph to find the $x$-intercepts.

In the Warm-up we solved the quadratic equation $y = x^2$ for several outputs. For example, the solution to $x^2 = 0$ was the $x$-intercept, $x = 0$. We now look more closely at solving $x^2 = 4$, the quadratic equation $y = x^2$ when $y = 4$.

E X A M P L E    **2**    Solve $x^2 = 4$, using the graph in Figure 8.

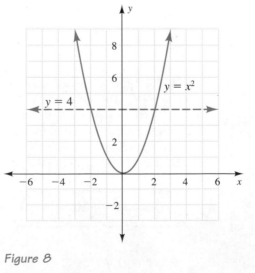

Figure 8

SOLUTION    From the graph we observe that the line $y = 4$ crosses the graph $y = x^2$ twice. Thus there are two graphical solutions to $x^2 = 4$: $x = 2$ and $x = -2$.

♦ ♦ ♦ ♦ ♦ ♦ ♦ ♦ ♦ ♦ ♦ ♦ ♦ ♦ ♦ ♦ ♦ ♦ ♦ ♦ ♦ ♦ ♦ ♦ ♦ ♦ ♦ ♦ ♦ ♦ ♦ ♦ ♦ ♦ ♦ ♦ ♦ ♦

### SOLVING QUADRATICS WITH SQUARE ROOTS

When we solved Pythagorean theorem problems, we often had to solve equations containing $x^2$. We review the solutions of these problems more formally now.

---

With some caution we can take the square root of both sides of an equation.

---

Caution is required because we must watch out for restrictions on square roots, specifically on $\sqrt{x^2}$. We defined the principal square root of $n$, $\sqrt{n}$, as the positive number that, when squared, gives $n$. For now, $n$ is $x^2$. Because $x$ is a variable, it is possible for $x$ to be negative. We need to maintain the definition of the square root as the principal square root, so we must force our output to $\sqrt{x^2}$ to be positive. Thus we have to use absolute value around the $x$ when we take the square root of $x^2$:

$$\sqrt{x^2} = |x|$$

**E X A M P L E   3**     Solve $x^2 = 4$ using symbols.

**SOLUTION**     We start by taking the square root of both sides of $x^2 = 4$:

$$\sqrt{x^2} = \sqrt{4}$$
$$\sqrt{x^2} = 2 \qquad \text{This time x may be either positive or negative. We}$$

still must have $\sqrt{x^2}$ be positive, and the only way to do this is with absolute value. Thus $\sqrt{x^2} = |x|$.

$$|x| = 2$$

There are two numbers that make $|x| = 2$ true: $x = 2$ and $x = -2$. Thus there are two solutions to $x^2 = 4$:

$$x - 2 \qquad \text{and} \qquad x = 2$$

♦ ♦ ♦ ♦ ♦ ♦ ♦ ♦ ♦ ♦ ♦ ♦ ♦ ♦ ♦ ♦ ♦ ♦ ♦ ♦ ♦ ♦ ♦ ♦ ♦ ♦ ♦ ♦ ♦ ♦ ♦ ♦ ♦ ♦ ♦ ♦

The absolute value gives a formally correct solution. If it is too formal, just remember this:

> The equation $x^2 = n$ has two solutions
> $$x = +\sqrt{n} \qquad \text{and} \qquad x = -\sqrt{n}$$

In Section 7.1 we ignored the second (negative) solution because we were working with physical objects, which do not have negative dimensions. In this section we want *both* solutions.

**E X A M P L E   4**     Solve $3x^2 = 75$.

**SOLUTION**
$$3x^2 = 75 \qquad \text{Divide both sides by 3.}$$
$$x^2 = 25$$
$$\sqrt{x^2} = \sqrt{25} \qquad \text{Take the square root of both sides.}$$
$$|x| = 5 \qquad \sqrt{x^2} = |x|, \text{ the formal step.}$$
$$x = 5 \qquad \text{and} \qquad x = -5 \quad \text{There are two solutions.}$$

♦ ♦ ♦ ♦ ♦ ♦ ♦ ♦ ♦ ♦ ♦ ♦ ♦ ♦ ♦ ♦ ♦ ♦ ♦ ♦ ♦ ♦ ♦ ♦ ♦ ♦ ♦ ♦ ♦ ♦ ♦ ♦ ♦ ♦ ♦ ♦ ♦ ♦

SOLVING QUADRATICS WITH FACTORING

Taking the square root of both sides of a quadratic equation such as $x^2 - 3x - 4 = 0$ will not give a solution. This brings us to our second symbolic method for quadratics—applying the zero product rule to factors.

Using a graph, we found that the solutions to $x^2 - 3x - 4 = 0$ were $x = 4$ and $x = -1$. The factors of $x^2 - 3x - 4$ are $(x - 4)$ and $(x + 1)$. The similarity between the factors and the solutions is not a coincidence. The key is the **zero product rule**:

Zero Product Rule

> If $A \cdot B = 0$, then either $A = 0$ or $B = 0$.

E X A M P L E    **5**

Describe the role of zero in these equations and expressions.

a. $A \cdot 5 = 0$

b. $-5 \cdot B = 0$

c. $(x - 4)(x + 1)$ if $x = 4$

d. $(x - 4)(x + 1)$ if $x = -1$

SOLUTION

a. $A = 0$, because zero multiplied by 5 gives a zero product.

b. $B = 0$, because zero multiplied by $-5$ gives a zero product.

c. $(4 - 4)(4 + 1) = 0 \cdot 5 = 0$. The product $(x - 4)(x + 1)$ is zero if $x = 4$.

d. $(-1 - 4)(-1 + 1) = (-5) \cdot 0 = 0$. The product $(x - 4)(x + 1)$ is zero if $x = -1$.

Parts a and b illustrate the zero product rule. Parts c and d illustrate that if a factor is zero, the product is zero. This is the converse of the zero product rule.

♦ ♦ ♦ ♦ ♦ ♦ ♦ ♦ ♦ ♦ ♦ ♦ ♦ ♦ ♦ ♦ ♦ ♦ ♦ ♦ ♦ ♦ ♦ ♦ ♦ ♦ ♦ ♦ ♦ ♦ ♦ ♦ ♦ ♦ ♦ ♦ ♦ ♦ ♦ ♦

The zero product rule indicates that if we factor an expression, we may solve for inputs that cause the expression to have a zero output.

E X A M P L E    **6**

Solve $x^2 - 3x - 4 = 0$ by factoring.

SOLUTION

$$x^2 - 3x - 4 = 0$$

$$(x - 4)(x + 1) = 0 \qquad \text{Factor.}$$

Either $(x - 4) = 0$ or $(x + 1) = 0$     Apply the zero product rule.

Either $\qquad x = 4$ or $\qquad x = -1$     Solve the factor equations.

For the check, see parts c and d of Example 5.

♦ ♦ ♦ ♦ ♦ ♦ ♦ ♦ ♦ ♦ ♦ ♦ ♦ ♦ ♦ ♦ ♦ ♦ ♦ ♦ ♦ ♦ ♦ ♦ ♦ ♦ ♦ ♦ ♦ ♦ ♦ ♦ ♦ ♦ ♦ ♦ ♦ ♦ ♦ ♦

In Example 6, $x$ represents only one input at a time. Thus we used *either* . . . *or* in our solution. This phrasing is important, and you are encouraged to use it whenever solving by factoring.

Factoring makes the reason for two solutions clearer than absolute value does.

EXAMPLE   **7**

Solve $x^2 = 4$ by factoring.

SOLUTION

$$x^2 = 4$$

$$x^2 - 4 = 0$$   Subtract 4 from both sides to obtain 0 on one side.

$$(x - 2)(x + 2) = 0$$   Factor.

Either   $x - 2 = 0$   or   $x + 2 = 0$   Apply the zero product rule.

Either   $x = 2$   or   $x = -2$   Solve the factor equations.

***Check***: $(2)^2 \overset{?}{=} 4$   and   $(-2)^2 \overset{?}{=} 4$   ✔

♦ ♦ ♦ ♦ ♦ ♦ ♦ ♦ ♦ ♦ ♦ ♦ ♦ ♦ ♦ ♦ ♦ ♦ ♦ ♦ ♦ ♦ ♦ ♦ ♦ ♦ ♦ ♦ ♦ ♦ ♦ ♦ ♦ ♦ ♦ ♦ ♦

EXAMPLE   **8**

*Spraying water*   Suppose a connector in a high-pressure water system has broken and a fine spray of water is shooting straight up into the air, according to the formula

$$y = -16t^2 + 48t$$

where $y$ is height and $t$ is seconds. In how many seconds after leaving the break will a droplet of water reach the 32-foot level? Solve from the graph in Figure 9 and by factoring.

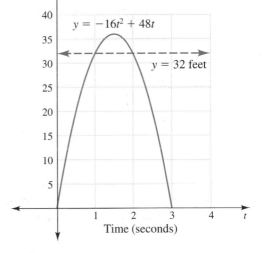

Figure 9

SOLUTION

According to the graph a water droplet reaches 32 feet at $t = 1$ second on the way up and again at $t = 2$ seconds on the way down.

The equation to solve by factoring is $32 = -16t^2 + 48t$.

$$32 = -16t^2 + 48t$$

$$16t^2 - 48t + 32 = 0$$   Change to equal zero.

$$16(t^2 - 3t + 2) = 0$$   Factor out the common monomial.

$$16(t - 2)(t - 1) = 0$$   Factor the remaining trinomial.

Either   $(t - 2) = 0$   or   $(t - 1) = 0$   Use the zero product rule.

Either   $t = 2$   or   $t = 1$   Solve the factor equations.

***Check***: $32 \overset{?}{=} -16(2)^2 + 48(2)$   and   $32 \overset{?}{=} -16(1)^2 + 48(1)$   ✔

♦ ♦ ♦ ♦ ♦ ♦ ♦ ♦ ♦ ♦ ♦ ♦ ♦ ♦ ♦ ♦ ♦ ♦ ♦ ♦ ♦ ♦ ♦ ♦ ♦ ♦ ♦ ♦ ♦ ♦ ♦ ♦ ♦ ♦ ♦ ♦ ♦

The graph in Figure 9 does not show the path of the water. For the graph to show such a path, the $x$-axis would have to be labeled in feet instead of time. The path of the water in this examples goes straight up and straight back down.

**EXERCISES 7.2**

Use the graphs to solve the equations in Exercises 1 and 2 for the $y$ values given.

1. $y = x^2 + x - 6$

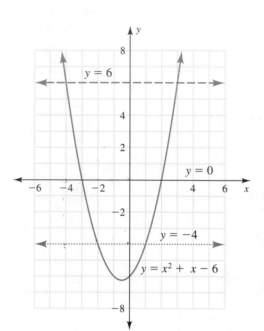

a. $y = 6$      b. $y = -4$      c. $y = -8$

d. $y = 0$

2. $y = x^2 + 2x - 3$

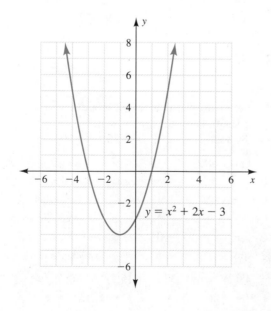

a. $y = 5$      b. $y = -3$      c. $y = -4$

d. $y = 0$

Solve the equations in Exercises 3 and 4 from the graphs.

3. $x^2 = 1$

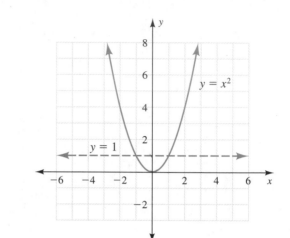

4. $x^2 - 3 = 1$

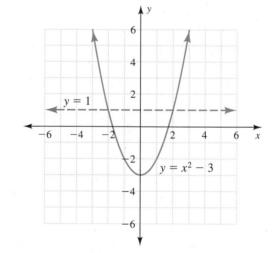

In Exercises 5 to 14 solve for all inputs, $x$, that make the statement true. The variable $x$ may represent any real number.

5. $x^2 = \frac{4}{25}$     6. $x^2 = \frac{36}{49}$     7. $x^2 = \frac{4}{100}$

8. $x^2 = \frac{25}{100}$     9. $x^2 = \frac{225}{49}$     10. $x^2 = \frac{169}{144}$

11. $x^2 = \frac{121}{36}$     12. $x^2 = \frac{64}{121}$     13. $x^2 = \frac{27}{75}$

14. $x^2 = \frac{12}{75}$

Solve the equations in Exercises 15 to 24 by factoring and applying the zero product rule.

15. $x^2 + x - 6 = 0$       16. $x^2 + 2x - 3 = 0$

17. $x^2 - 2x - 15 = 0$      18. $x^2 - 8x + 15 = 0$

19. $x^2 - 4x - 12 = 0$      20. $x^2 + x - 12 = 0$

21. $2x^2 - x - 3 = 0$       22. $2x^2 - 5x - 3 = 0$

23. $4x^2 - 25 = 0$        24. $9x^2 - 4 = 0$

25. In Example 8 the water is at ground level when the height, $y$, is zero. Use factoring to find the inputs, $t$, when $y = 0$. Do the results agree with the graph?

26. Estimate the maximum height reached by the water in Example 8. Estimate the input that gives maximum height. How could the input be used to find the maximum height?

27. An air-powered rocket is launched, and at time $t$ it is at height $y = -16t^2 + 64t$. Use the graph in the figure to find the times when the rocket is at 48 feet. Use factoring to find the time. Do the results agree?

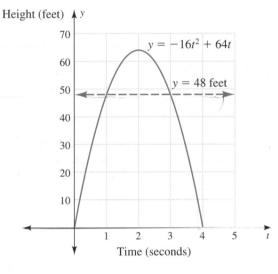

Height (feet)

$y = -16t^2 + 64t$

$y = 48$ feet

Time (seconds)

28. Estimate the maximum height reached by the rocket in Exercise 27 (see the figure). Estimate the input that gives maximum height. How could the input be used to find the maximum height?

29. Use factoring to determine when the rocket in Exercise 27 is at ground level, $y = 0$. Does your result agree with the graph?

30. Use the graph to estimate how high the rocket in Exercise 27 travels in the first second and how high it travels in the second second. If rate, or speed, is distance divided by time, what is the average speed of the rocket during the first second? What is the average speed of the rocket in the second second?

In $y = ax^2 + bx + c$ the letters $a$, $b$, and $c$ are parameters. *Parameters are constants or variables that identify a particular expression.* What are the parameters in the quadratic equations in Exercises 31 to 42? Identify the parameters as $a =$ ___, $b =$ ___, and $c =$ ___. Some parameters may be zero.

31. $y = 2x^2 + 3x + 1$   32. $y = 3x^2 + 2x - 1$

33. $y = r^2 - 4r + 4$   34. $y = r^2 - 6r + 9$

35. $y = x^2 - 4$   36. $y = x^2 - 25$

37. $y = 4t^2 - 8t$   38. $y = 3t^2 - 27t$

39. Height after time $t$: $h = -0.5gt^2 + vt + s$

40. Angle swept in time $t$ for circular motion: $A = 0.5at^2 + wt$

41. Area of circle: $A = \pi r^2$

42. Surface area of cylinder of height 1: $SA = 2\pi r^2 + 2\pi r$

43. a. Complete the table.

| $a$ | $b$ | $a^2 + b^2$ | $(a + b)^2$ | $\sqrt{(a + b)}$ | $\sqrt{a} + \sqrt{b}$ |
|---|---|---|---|---|---|
| 4 | 9 | | | | |
| 1 | 3 | | | | |
| 4 | 5 | | | | |
| 3 | 6 | | | | |

b. Explain why a student might think that $a^2 + b^2$ and $(a + b)^2$ were equal. How does the table show that they are not equal?

c. Explain why a student might think that $\sqrt{(a + b)}$ and $\sqrt{a} + \sqrt{b}$ were equal. How does the table show that they are not equal?

44. a. Complete the table.

| $a$ | $b$ | $a^2 \cdot b^2$ | $(a \cdot b)^2$ | $\sqrt{(a \cdot b)}$ | $\sqrt{a} \cdot \sqrt{b}$ |
|---|---|---|---|---|---|
| 4 | 9 | | | | |
| 1 | 3 | | | | |
| 4 | 5 | | | | |
| 3 | 6 | | | | |

b. What does the table indicate about $a^2 \cdot b^2$ and $(a \cdot b)^2$?

c. What does the table indicate about $\sqrt{(a \cdot b)}$ and $\sqrt{a} \cdot \sqrt{b}$?

PROBLEM SOLVING

Suppose we find the solutions given in Exercises 45 to 48 from a graph. Find an equation of the form $y = x^2 + bx + c$ that has these solutions. (*Hint:* Work backwards. If $x = n$, then $x - n$ is a factor of the equation.)

45. $x = -3$ and $x = -2$   46. $x = -1$ and $x = 2$

47. $x = 4$ and $x = -3$   48. $x = 5$ and $x = 2$

PROJECT

49. *Quadratic Formula.* Not all quadratic equations factor easily. Some do not factor at all. When a quadratic equation is written in the form $ax^2 + bx + c = 0$, it may be solved for $x$ with the quadratic formula. The quadratic formula is shown as two equations:

$$x = \frac{-b + \sqrt{b^2 - 4ac}}{2a} \quad \text{and} \quad x = \frac{-b - \sqrt{b^2 - 4ac}}{2a}$$

a. Use the quadratic formula on these equations to check that you are using it correctly. They are equations from earlier exercises.

| | |
|---|---|
| $x^2 - 4x - 12 = 0$ | Answer: $x = -2$, $x = 6$ |
| $2x^2 - x - 3 = 0$ | Answer: $x = 1.5$, $x = -1$ |
| $4x^2 - 25 = 0$ | Answer: $x = 2.5$, $x = -2.5$ |

b. Use the quadratic formula to solve these equations.

$2x^2 + 5x - 3 = 0$         $3x^2 + 5x - 4 = 0$

$3x^2 - 2x - 4 = 0$         $3x^2 - 2x - 3 = 0$

### PYTHAGOREAN THEOREM AND QUADRATIC FORMULA: PROOFS AND PATTERNS

**OBJECTIVES**    Review polynomial multiplication using the table method or mental multiplication. ♦ Identify perfect square trinomials. ♦ Identify the difference of perfect squares. ♦ Factor perfect square trinomials and the difference of perfect squares. ♦ Prove the Pythagorean theorem. ♦ Prove the quadratic formula. ♦ Use the quadratic formula to solve quadratic equations. ♦ Prove that the {3, 4, 5} right triangle is the only right triangle with consecutive integers as its sides.

**WARM-UP**    Multiply these binomials. State any patterns you observe.

**1.** $(x - 1)(x - 1)$    **2.** $(x + 3)(x + 3)$    **3.** $(2x + y)(2x + y)$

**4.** $(2ax + b)(2ax + b)$    **5.** $(x - 1)(x + 1)$    **6.** $(x + 3)(x - 3)$

**7.** $(y - 6)(y + 6)$    **8.** $(2x + 3y)(2x - 3y)$    ♦

This section provides review in multiplying and factoring polynomials. After an introduction to formal algebraic proofs, we apply the quadratic formula to solving quadratic equations and then close the section with patterns and a proof that there is only one set of consecutive positive integers that forms a right triangle.

#### PERFECT SQUARE TRINOMIALS AND THE DIFFERENCE OF SQUARES

In Chapter 6 we multiplied expressions of the form $(a + b)(a + b)$ and $(a + b)(a - b)$ using a table. We review these multiplications now and identify some patterns.

**EXAMPLE    1**    Multiply these expressions. Describe any patterns you observe.

**a.** $(x - 1)(x - 1)$    **b.** $(x + 3)(x + 3)$

**c.** $(2x + y)(2x + y)$    **d.** $(2ax + b)(2ax + b)$

**SOLUTION**    **a.** $(x - 1)(x - 1) = x^2 - 2x + 1$    **b.** $(x + 3)(x + 3) = x^2 + 6x + 9$

**c.** $(2x + y)(2x + y) = 4x^2 + 4xy + y^2$

**d.** $(2ax + b)(2ax + b) = 4a^2x^2 + 4abx + b^2$

In each problem we multiplied a binomial by itself. The answer always starts and ends with a perfect square. The middle term is twice the product of the terms in one factor.

♦ ♦ ♦ ♦ ♦ ♦ ♦ ♦ ♦ ♦ ♦ ♦ ♦ ♦ ♦ ♦ ♦ ♦ ♦ ♦ ♦ ♦ ♦ ♦ ♦ ♦ ♦ ♦ ♦ ♦ ♦ ♦ ♦ ♦ ♦

If we work this problem in general form, $(a + b)(a + b)$, we see a description of our answer to Example 1.

**EXAMPLE    2**    Multiply $(a + b)(a + b)$ with a table.

**SOLUTION**

| Multiply | $a$ | $+b$ |
|----------|-----|------|
| $a$ | $a^2$ | $+ab$ |
| $+b$ | $+ab$ | $+b^2$ |

Adding terms in the table gives $a^2 + ab + ab + b^2 = a^2 + 2ab + b^2$.

♦ ♦ ♦ ♦ ♦ ♦ ♦ ♦ ♦ ♦ ♦ ♦ ♦ ♦ ♦ ♦ ♦ ♦ ♦ ♦ ♦ ♦ ♦ ♦ ♦ ♦ ♦ ♦ ♦ ♦ ♦ ♦ ♦ ♦ ♦ ♦

The variables $a$ and $b$ may represent any number or expression. The first and last terms are perfect squares, and the middle term is twice the product of $a$ and $b$. Because the two factors are identical, the product $(a + b)(a + b)$ is a perfect square, and the resulting trinomial, $a^2 + 2ab + b^2$, is called a **perfect square trinomial**. The product $(a + b)(a + b)$ is written in *factored* form.

**Perfect Square Trinomial**

$$(a + b)(a + b) = (a + b)^2 = a^2 + 2ab + b^2$$

The expression $a^2 + 2ab + b^2$ has three terms: the square of $a$, twice the product of $a$ and $b$, and the square of $b$.

**EXAMPLE 3**

Multiply these expressions. Describe any patterns you observe.

**a.** $(x - 1)(x + 1)$     **b.** $(x + 3)(x - 3)$

**c.** $(y - 6)(y + 6)$     **d.** $(2x + 3y)(2x - 3y)$

**SOLUTION**    **a.** $(x - 1)(x + 1) = x^2 - 1$     **b.** $(x + 3)(x - 3) = x^2 - 9$

**c.** $(y - 6)(y + 6) = y^2 - 36$     **d.** $(2x + 3y)(2x - 3y) = 4x^2 - 9y^2$

In each problem we multiplied two binomials. The binomials contained the same numbers and letters, but one was an addition and the other was a subtraction. The answers always start and end with a perfect square and have a subtraction between them. There is no middle term.

♦ ♦ ♦ ♦ ♦ ♦ ♦ ♦ ♦ ♦ ♦ ♦ ♦ ♦ ♦ ♦ ♦ ♦ ♦ ♦ ♦ ♦ ♦ ♦ ♦ ♦ ♦ ♦ ♦ ♦ ♦ ♦ ♦ ♦ ♦ ♦

If we work this problem in general form, $(a + b)(a - b)$, we see a description of our answer to Example 3.

**EXAMPLE 4**

Multiply $(a + b)(a - b)$ with a table.

**SOLUTION**

| Multiply | $a$ | $-b$ |
|----------|------|------|
| $a$ | $a^2$ | $-ab$ |
| $+b$ | $+ab$ | $-b^2$ |

Adding terms in the table gives $a^2 - ab + ab - b^2 = a^2 - b^2$. The first and last terms are perfect squares, and the like terms, $-ab$ and $+ab$, add to zero.

♦ ♦ ♦ ♦ ♦ ♦ ♦ ♦ ♦ ♦ ♦ ♦ ♦ ♦ ♦ ♦ ♦ ♦ ♦ ♦ ♦ ♦ ♦ ♦ ♦ ♦ ♦ ♦ ♦ ♦ ♦ ♦ ♦ ♦ ♦ ♦

The product $(a + b)(a - b) = a^2 - b^2$ is called the **difference of squares**. The product $(a + b)(a - b)$ is written in *factored* form.

**Difference of Squares**

$$(a + b)(a - b) = a^2 - b^2$$

The expression $a^2 - b^2$ has two terms: the square of $a$ and the square of $b$.

EXAMPLE 5

Identify each expression as a perfect square trinomial or a difference of squares or neither. Change each perfect square trinomial or difference of squares to factored form. Explain why the remaining expressions are neither.

**a.** $x^2 + 2x + 1$    **b.** $y^2 - 4y + 4$    **c.** $x^2 - 5x + 25$

**d.** $4x^2 - 25$    **e.** $x^2 + 4x + 8$    **f.** $9x^2 + 25$

**g.** $y^2 + 2yz + z^2$

SOLUTION

**a.** Perfect square trinomial: $x^2 + 2x + 1 = (x + 1)(x + 1)$.

**b.** Perfect square trinomial: $y^2 - 4y + 4 = (y - 2)(y - 2)$.

**c.** Neither; a perfect square trinomial needs $-10x$, not $-5x$.

**d.** Difference of squares: $4x^2 - 25 = (2x - 5)(2x + 5)$.

**e.** Neither; a perfect square trinomial needs 4, not 8.

**f.** Neither; a difference of squares needs subtraction, not addition.

**g.** Perfect square trinomial: $y^2 + 2yz + z^2 = (y + z)(y + z)$.

♦ ♦ ♦ ♦ ♦ ♦ ♦ ♦ ♦ ♦ ♦ ♦ ♦ ♦ ♦ ♦ ♦ ♦ ♦ ♦ ♦ ♦ ♦ ♦ ♦ ♦ ♦ ♦ ♦ ♦ ♦ ♦ ♦ ♦ ♦

## PROOF OF THE PYTHAGOREAN THEOREM

A **proof** *is a logical argument that demonstrates the truth of a statement.* It is based on previously known facts or agreed-upon statements. The Greek mathematician Pythagoras is credited with being the first to record a proof of the relationship for right triangles given his name. However, documents from Chinese history indicate that the relationship was known long before Pythagoras's time.

Example 6 proves the Pythagorean theorem by showing that it is true for any *a*, *b*, and *c* that satisfy the conditions of the theorem: *a* and *b* are legs of a right triangle, and *c* is the hypotenuse.

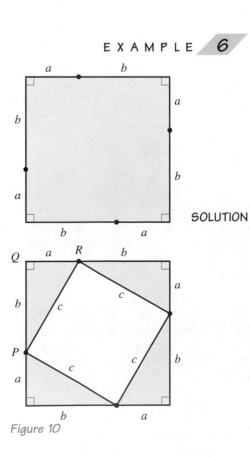

Figure 10

EXAMPLE 6

Use the following steps to prove the Pythagorean theorem where $\triangle PQR$ is a right triangle.

**a.** What is the area of the first square in Figure 10 that has $a + b$ on each side?

**b.** Write the area of each of the five parts of the second square in Figure 10, and find the sum of the areas.

**c.** Because both figures are squares with sides $a + b$, their areas are equal. Set the expressions for their areas equal, and show that $a^2 + b^2 = c^2$.

SOLUTION

**a.** The square with side $a + b$ has area $(a + b)^2$.

**b.** The second square has four right triangles and an inner square of side $c$. The area of each right triangle is $\frac{1}{2}ab$. The area of the inner square is $c^2$. Thus the total area is $4\left(\frac{1}{2}ab\right) + c^2$.

**c.** Set the areas equal:

$$(a + b)^2 = 4\left(\tfrac{1}{2}ab\right) + c^2$$

$$a^2 + 2ab + b^2 = 2ab + c^2 \qquad \text{Multiply to remove the parentheses.}$$

$$a^2 + b^2 = c^2 \qquad \text{Subtract } 2ab \text{ from both sides.}$$

Therefore, for right triangle *PQR*, $a^2 + b^2 = c^2$.

♦ ♦ ♦ ♦ ♦ ♦ ♦ ♦ ♦ ♦ ♦ ♦ ♦ ♦ ♦ ♦ ♦ ♦ ♦ ♦ ♦ ♦ ♦ ♦ ♦ ♦ ♦ ♦ ♦ ♦ ♦ ♦ ♦ ♦ ♦

PROOF OF THE QUADRATIC FORMULA

In Section 7.2 we solved quadratic equations by graphing, by taking the square root, and by factoring. These three methods are limited in application. The quadratic formula is more general. We may use the **quadratic formula** to solve for the variable $x$ in any equation of the form $ax^2 + bx + c = 0$.

The quadratic formula may be shown as two equations:

**Quadratic Formula**

For $ax^2 + bx + c = 0$,

$$x = \frac{-b + \sqrt{b^2 - 4ac}}{2a} \quad \text{and} \quad x = \frac{-b - \sqrt{b^2 - 4ac}}{2a}$$

In the proof of the Pythagorean theorem, we squared a binomial, $(a + b)$, to obtain the perfect square trinomial, $a^2 + 2ab + b^2$. In proving the quadratic formula, we will build a perfect square trinomial and factor it back to the square of a binomial:

$$4a^2x^2 + 4abx + b^2 = (2ax + b)(2ax + b) = (2ax + b)^2$$

**EXAMPLE 7**

The steps below show how we obtain the quadratic formula from a quadratic equation in the form $ax^2 + bx + c = 0$. Explain what was done to obtain each step.

$$ax^2 + bx + c = 0$$

**a.** $4a^2x^2 + 4abx + 4ac = 0$

**b.** $\qquad 4a^2x^2 + 4abx = -4ac$

**c.** $\quad 4a^2x^2 + 4abx + b^2 = b^2 - 4ac$

**d.** $\qquad\qquad (2ax + b)^2 = b^2 - 4ac$

**e.** Either $2ax + b = \sqrt{b^2 - 4ac}$ $\qquad$ or $2ax + b = -\sqrt{b^2 - 4ac}$

**f.** Either $\qquad 2ax = -b + \sqrt{b^2 - 4ac}$ or $\qquad 2ax = -b - \sqrt{b^2 - 4ac}$

**g.** Either $\qquad x = \dfrac{-b + \sqrt{b^2 - 4ac}}{2a}$ or $\qquad x = \dfrac{-b - \sqrt{b^2 - 4ac}}{2a}$

**SOLUTION**
**a.** Multiply both sides of $ax^2 + bx + c = 0$ by $4a$.

**b.** Subtract $4ac$ from both sides of the equation.

**c.** Add $b^2$ to both sides. This creates a perfect square trinomial on the left.

**d.** Factor the left side.

**e.** Find the square root of both sides. Because $2ax + b$ may be either positive or negative, we obtain two equations.

**f.** Subtract $b$ from both sides of each equation.

**g.** Divide by $2a$ in each equation.

✦ ✦ ✦ ✦ ✦ ✦ ✦ ✦ ✦ ✦ ✦ ✦ ✦ ✦ ✦ ✦ ✦ ✦ ✦ ✦ ✦ ✦ ✦ ✦ ✦ ✦ ✦ ✦ ✦ ✦ ✦ ✦

Two other useful forms of the quadratic formula are

$$x = \frac{-b \pm \sqrt{b^2 - 4ac}}{2a} \quad \text{and} \quad x = \frac{-b}{2a} \pm \frac{\sqrt{b^2 - 4ac}}{2a}$$

In both forms, the plus or minus sign ($\pm$) allows us to write two equations with one statement. In the second form, the fraction is separated into two parts over the common denominator, $2a$. The separation prevents many simplification errors. The second form also contains graphical information which is explored in Exercises 45 and 46, the Vertex project and the Quadratic Formula and Graphs project.

EXAMPLE  8

Use the quadratic formula to solve these equations.

**a.** $2x^2 + 7x + 3 = 0$     **b.** $4x^2 = 5x + 2$     **c.** $2x^2 + 2x = -1$

SOLUTION     **a.** In $2x^2 + 7x + 3 = 0$, $a = 2$, $b = 7$, and $c = 3$.

$$x = \frac{-b \pm \sqrt{b^2 - 4ac}}{2a} = \frac{-7 \pm \sqrt{7^2 - 4(2)(3)}}{2(2)}$$

$$= \frac{-7 \pm \sqrt{49 - 24}}{4} = \frac{-7 \pm \sqrt{25}}{4}$$

The solution is

$$x = \frac{-7 + 5}{4} = \frac{-2}{4} = \frac{-1}{2} \quad \text{or} \quad x = \frac{-7 - 5}{4} = \frac{-12}{4} = -3.$$

**b.** The equation $4x^2 = 5x + 2$ must be rearranged to equal zero. In $4x^2 - 5x - 2 = 0$, $a = 4$, $b = -5$, and $c = -2$.

$$x = \frac{-b \pm \sqrt{b^2 - 4ac}}{2a} = \frac{-(-5) \pm \sqrt{(-5)^2 - 4(4)(-2)}}{2(4)}$$

$$= \frac{-(-5) \pm \sqrt{25 + 32}}{8} = \frac{5 \pm \sqrt{57}}{8}$$

The solution is

$$x = \frac{5 + \sqrt{57}}{8} \approx 1.569 \quad \text{or} \quad x = \frac{5 - \sqrt{57}}{8} \approx -0.319.$$

**c.** The equation $2x^2 + 2x = -1$ must be rearranged to equal zero. In $2x^2 + 2x + 1 = 0$, $a = 2$, $b = 2$, and $c = 1$.

$$x = \frac{-b \pm \sqrt{b^2 - 4ac}}{2a} = \frac{-2 \pm \sqrt{2^2 - 4(2)(1)}}{2(2)}$$

$$= \frac{-2 \pm \sqrt{4 - 8}}{4} = \frac{-2 \pm \sqrt{-4}}{4}$$

There are no real number solutions because the square root of a negative number is not defined in the real numbers. Thus we say "no real solution."

♦ ♦ ♦ ♦ ♦ ♦ ♦ ♦ ♦ ♦ ♦ ♦ ♦ ♦ ♦ ♦ ♦ ♦ ♦ ♦ ♦ ♦ ♦ ♦ ♦ ♦ ♦ ♦ ♦ ♦ ♦ ♦ ♦ ♦ ♦ ♦ ♦ ♦ ♦

Each part of Example 8 has a different form of answer. Part a has rational solutions; we are able to find the exact square root of 25. The square root of 57 in part b cannot be expressed exactly by a decimal number, and so the solutions to the quadratic equation are irrational (not rational) and are approximated with decimals rounded to the nearest thousandth. As mentioned, the square root of $-4$ in part c is not a real number. In later courses you will learn other types of numbers and solve problems such as the one in part c.

We now examine the graphs related to the equations in Example 8.

EXAMPLE  9

Use the graphs of the following equations to identify the solutions to $y = 0$.

**a.** $y = 2x^2 + 7x + 3$     **b.** $y = 4x^2 - 5x - 2$     **c.** $y = 2x^2 + 2x + 1$

SOLUTION     The solutions to $y = 0$ are the $x$-intercepts.

**a.** The graph in Figure 11 confirms that the solutions to $2x^2 + 7x + 3 = 0$ are at $x = -\frac{1}{2}$ and $x = -3$.

**b.** The graph in Figure 12 confirms that one solution to $4x^2 - 5x - 2 = 0$ is near $x = 1\frac{1}{2}$ and the other is between $x = -\frac{1}{2}$ and $x = 0$.

**c.** The graph in Figure 13 does not cross the $x$-axis, so it has no $x$-intercepts. This confirms that the equation $2x^2 + 2x + 1 = 0$ has no real number solutions.

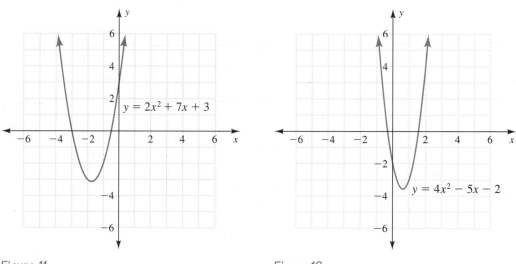

Figure 11                                      Figure 12

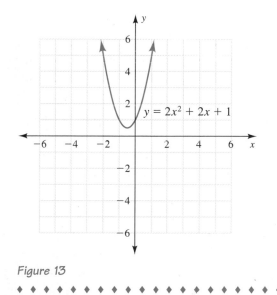

Figure 13

♦ ♦ ♦ ♦ ♦ ♦ ♦ ♦ ♦ ♦ ♦ ♦ ♦ ♦ ♦ ♦ ♦ ♦ ♦ ♦ ♦ ♦ ♦ ♦ ♦ ♦ ♦ ♦ ♦ ♦ ♦

## PATTERNS

**Pythagorean triples** *are sets of three numbers that make the Pythagorean theorem true.* The set of numbers {3, 4, 5} satisfies the Pythagorean theorem: $3^2 + 4^2 = 5^2$.

Before calculators, students and instructors saved considerable time by learning selected sets of Pythagorean triples. Engineering textbooks still make frequent use of triples, as do nationally standardized exams for entry into college or graduate school. The triple {3, 4, 5} is the basis for a rope surveying instrument used by the Egyptians (see the Historical Note).

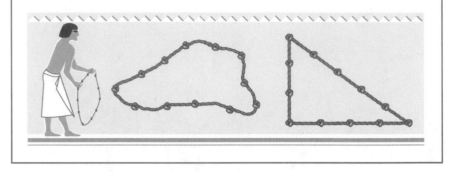

Because {3, 4, 5} is a Pythagorean triple, we might be tempted to think that {4, 5, 6} or {5, 6, 7} or some other set of three consecutive integers would also be a triple. We close this section with Example 10, a proof that the set {3, 4, 5} is the *only set of consecutive positive integers* that satisfies the Pythagorean theorem. We will also use this example to review several ways to solve quadratic equations.

EXAMPLE 10    What sets of three consecutive integers satisfy $a^2 + b^2 = c^2$?

**a.** Let $x$, $x + 1$, and $x + 2$ be the three consecutive integers. Write the Pythagorean theorem using consecutive integer notation, and simplify.

**b.** Find the solutions to the resulting equation by graphing. Use a graphing calculator if one is available.

**c.** Find the solutions to the resulting equation by factoring.

**d.** Find the solutions to the resulting equation with the quadratic formula.

**e.** Substitute the solutions into the expressions in part a and state your conclusions.

SOLUTION    **a.** The consecutive integers are placed on a right triangle in Figure 14. Substitute $x$, $x + 1$, and $x + 2$ into the Pythagorean theorem:

$$x^2 + (x + 1)^2 = (x + 2)^2$$
$$x^2 + x^2 + 2x + 1 = x^2 + 4x + 4 \quad \text{Square the binomials.}$$
$$2x^2 + 2x + 1 - x^2 - 4x - 4 = 0 \quad \text{Subtract } x^2 + 4x + 4 \text{ from both sides.}$$
$$x^2 - 2x - 3 = 0 \quad \text{Add like terms.}$$

**b.** We solve $x^2 - 2x - 3 = 0$ by completing Table 3 for $y = x^2 - 2x - 3$. Then we graph the data (see Figure 15) together with $y = 0$, which is the $x$-axis. The $x$-intercepts, $x = -1$ and $x = 3$, are the solutions to $x^2 - 2x - 3 = 0$. The table also shows $y = 0$ corresponding with the inputs $x = -1$ and $x = 3$.

**c.** We may solve $x^2 - 2x - 3 = 0$ by factoring.

$$x^2 - 2x - 3 = 0$$
$$(x - 3)(x + 1) = 0 \qquad \text{Factor.}$$
Either $(x - 3) = 0$ or $(x + 1) = 0$    Apply the zero product rule.
Either $\qquad x = 3$ or $\qquad x = -1$    Solve the factor equations.

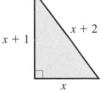

*Figure 14*

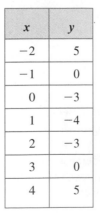

| $x$ | $y$ |
|-----|-----|
| $-2$ | 5 |
| $-1$ | 0 |
| 0 | $-3$ |
| 1 | $-4$ |
| 2 | $-3$ |
| 3 | 0 |
| 4 | 5 |

*Table 3* $y = x^2 - 2x - 3$

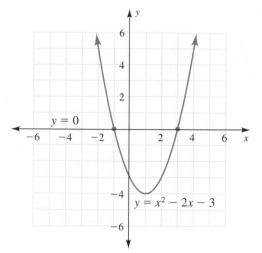

Figure 15

**d.** We may solve $x^2 - 2x - 3 = 0$ with the quadratic formula.

$$x = \frac{-b \pm \sqrt{b^2 - 4ac}}{2a} = \frac{-(-2) \pm \sqrt{(-2)^2 - 4(1)(-3)}}{2(1)}$$

$$= \frac{2 \pm \sqrt{4 + 12}}{2} = \frac{2 \pm \sqrt{16}}{2}$$

The solution is

$$x = \frac{2 + 4}{2} = \frac{6}{2} = 3 \quad \text{or} \quad x = \frac{2 - 4}{2} = \frac{-2}{2} = -1$$

**e.** Substituting the $x$-intercepts, or solutions, into the consecutive integer expressions gives

$$x = -1, \quad x + 1 = 0, \quad \text{and} \quad x + 2 = 1$$
$$x = 3, \quad x + 1 = 4, \quad \text{and} \quad x + 2 = 5$$

Both sets of numbers, $\{-1, 0, 1\}$ and $\{3, 4, 5\}$, satisfy $a^2 + b^2 = c^2$. However, there is no triangle with sides of lengths $-1$, $0$, and $1$, so this solution must be disregarded. Thus the only consecutive integer solution set to the Pythagorean theorem is $\{3, 4, 5\}$.

◆ ◆ ◆ ◆ ◆ ◆ ◆ ◆ ◆ ◆ ◆ ◆ ◆ ◆ ◆ ◆ ◆ ◆ ◆ ◆ ◆ ◆ ◆ ◆ ◆ ◆ ◆ ◆ ◆ ◆ ◆ ◆ ◆ ◆ ◆ ◆ ◆

## EXERCISES 7.3

**In Exercises 1 to 14 multiply the binomials. Identify each answer as a perfect square trinomial (pst) or as a difference of squares (ds).**

1. $(x + 2)(x + 2)$

2. $(x - 2)(x - 2)$

3. $(x + 2)(x - 2)$

4. $(y - 3)(y - 3)$

5. $(y - 3)(y + 3)$

6. $(a - b)(a + b)$

7. $(2x - 1)(2x + 1)$

8. $(2x - 1)(2x - 1)$

9. $(2x + 1)(2x + 1)$

10. $(x + 5)(x - 5)$

11. $(x - 5)(x - 5)$

12. $(3x + 1)(3x + 1)$

13. $(3x - 1)(3x - 1)$

14. $(3x + 1)(3x - 1)$

**Identify each expression in Exercises 15 to 26 as a perfect square trinomial (pst) or a difference of squares (ds) or neither. Change the perfect square trinomials and differences of squares to factored form.**

15. $x^2 - 6x + 9$

16. $x^2 - 16$

17. $x^2 + 25$

18. $9x^2 - 16$

19. $x^2 - 8x + 16$

20. $x^2 - 10x + 10$

21. $y^2 + 12y + 36$

22. $9x^2 + 25$

23. $y^2 + 8y + 8$

24. $x^2 - 4x - 4$

25. $4x^2 - 1$

26. $4y^2 + 4y + 1$

**Which of the sets of numbers in Exercises 27 and 28 are Pythagorean triples? The numbers might not be listed in the $a$, $b$, $c$ order. Which sets satisfy $a^2 + b^2 = c^2$ and yet cannot satisfy the Pythagorean theorem? Explain why.**

27. a. $\{6, 8, 10\}$    b. $\{15, 17, 8\}$    c. $\{4, 7.5, 8.5\}$

d. $\{2.1, 2.9, 2\}$    e. $\{2, \sqrt{3}, 1\}$    f. $\{1, 2, 3\}$

g. $\{-1, 0, 1\}$    h. $\{-5, -4, -3\}$

28. a. $\{12, 13, 5\}$    b. $\left\{\frac{1}{3}, \frac{1}{4}, \frac{1}{5}\right\}$    c. $\{65, 72, 97\}$

d. $\{1.8, 8, 8.2\}$    e. $\{7, 6, \sqrt{13}\}$    f. $\{0, 1, 2\}$

g. $\{-2, -1, 0\}$    h. $\{2, 3, 4\}$

**Solve the equations in Exercises 29 to 38 with the quadratic formula. The equations in 35 to 38 are of the form $-16t^2 + bt + c = 0$. The equations determine the time $t$, in seconds, required for an object to reach the ground ($y = 0$) when it is tossed straight up from height $c$, in feet, with initial velocity $b$, in feet per second. Why is only one time an acceptable solution?**

29. $4x^2 + 3x - 1 = 0$

30. $2x^2 + x - 6 = 0$

31. $7x^2 = 5x + 2$

32. $9x^2 = 2 - 3x$

33. $x = 2 - 10x^2$

34. $x = 2 - 6x^2$

35. $-16t^2 + 30t + 150 = 0$

36. $-16t^2 + 20t + 50 = 0$

37. $-16t^2 + 15t + 150 = 0$

38. $-16t^2 + 25t + 100 = 0$

**Find the missing sides of the similar triangles shown in Exercises 39 and 40. Confirm that they are both right triangles.**

39.

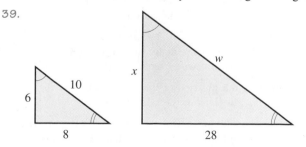

40.

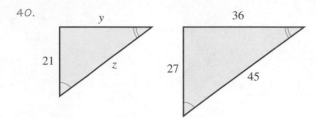

For Exercises 41 and 42, complete the tables, which show enlargements or reductions of {3, 4, 5} right triangles. Each row represents the sides of a different triangle.

41.

| Leg | Leg | Hypotenuse |
|-----|-----|------------|
| 3 | 4 | 5 |
| 6 | | |
| | | 30 |
| | 12 | |
| 1 | | |

42.

| Leg | Leg | Hypotenuse |
|-----|-----|------------|
| 3 | 4 | 5 |
| 9 | | |
| | | 35 |
| | 20 | |
| | | 1 |

Exercises 43 and 44 are modifications of the equations in Exercises 35 to 38. Solve the equations and explain the meaning of the solutions. (*Hint:* If a graphing calculator is available, graph $y$ = the left side of each equation, trace, and compare outputs with the number on the right side of each equation.)

43. a. $-16t^2 + 30t + 150 = 200$

   b. $-16t^2 + 15t + 150 = -50$

44. a. $-16t^2 + 20t + 50 = -40$

   b. $-16t^2 + 25t + 100 = 150$

*PROJECTS*

45. **Vertex.** The highest or lowest point on the graph of a quadratic equation, $y = ax^2 + bx + c$, is called the *vertex*. Return to the graph in Figure 15, Example 10.

   a. Determine the coordinates of the vertex (lowest point) from the graph.

   b. Find the corresponding coordinates in the table for Example 10.

   c. What do you observe about the output values for inputs 1, 2, and 3 units away from the vertex?

   d. Compare the position of the $x$-coordinate of the vertex with the positions of the $x$-intercepts.

   e. Make a table and graph for $y = x^2 - 4x + 3$. Answer parts a to d above for this new quadratic equation.

f. Summarize your conclusions. Describe how to obtain the $x$-coordinate of the vertex if we know the $x$-intercepts. Describe how the vertex is related to the line of symmetry of the quadratic graph. Describe how to obtain the $y$-coordinate of the vertex once we know the $x$-coordinate.

46. **Quadratic Formula and Graphs.** The highest or lowest point on the graph of a quadratic equation, $y = ax^2 + bx + c$, is called the *vertex*. The quadratic formula in the form

$$x = \frac{-b}{2a} \pm \frac{\sqrt{b^2 - 4ac}}{2a}$$

shows why the vertex is symmetrically placed between the $x$-intercepts. The distance between the axis of symmetry and each intercept is

$$\frac{\sqrt{b^2 - 4ac}}{2a}$$

as shown in the figure. The axis of symmetry, which passes through the vertex $V$, has equation

$$x = \frac{-b}{2a}$$

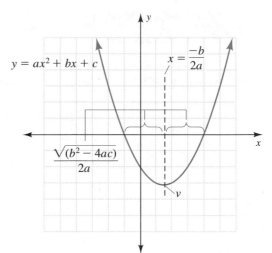

Using either the $x$-intercepts or the axis of symmetry, find the coordinates of the vertex for each quadratic equation:

a. $y = x^2 + 6x - 16$

b. $y = 2x^2 + 3x - 4$

c. $y = -16t^2 + 30t + 150$

d. $y = -16t^2 + 20t + 50$

e. If the graphs for parts c and d represent the height of a ball at time $t$, what is the meaning of the vertex?

f. Find the coordinates of the vertex for each part in Example 9. Check that each vertex agrees with the graphs.

47. **Number Pattern 1.** Use a calculator to evaluate these expressions.

a. $17^2 - 14^2 - (16^2 - 13^2)$    b. $16^2 - 13^2 - (15^2 - 12^2)$

c. $15^2 - 12^2 - (14^2 - 11^2)$    d. $14^2 - 11^2 - (13^2 - 10^2)$

One of the most useful applications of polynomials is in writing descriptions of number patterns and verifying that the patterns are true for all inputs. Write a symbolic description for the number pattern in part a. Let $x$ be the first number in the line. To verify that your description for part a applies to parts b, c, and d, substitute the first number of each for $x$. Show the steps in simplifying your expression (multiply the squared terms and combine like terms).

48. ***Number Pattern 2.*** Write a description of the following pattern in symbols, and simplify to prove that it works.

$$1 \cdot 3 - 2 \cdot 2 = -1$$

$$2 \cdot 4 - 3 \cdot 3 = -1$$

$$3 \cdot 5 - 4 \cdot 4 = -1$$

$$4 \cdot 6 - 5 \cdot 5 = -1$$

$$5 \cdot 7 - 6 \cdot 6 = -1$$
$$\vdots$$
$$49 \cdot 51 - 50 \cdot 50 = -1$$
$$\vdots$$

(*Hint*: Let $x$ be the first number.)

49. ***Garfield Proof.*** James A. Garfield, 20th president of the United States, is credited with a proof of the Pythagorean theorem, based on the area of a trapezoid (see the figure). The area of a trapezoid is half the product of the height of the trapezoid and the sum of the parallel sides:

$$A = \tfrac{1}{2}(\text{height})(\text{top} + \text{bottom})$$

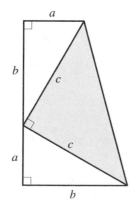

a. Find the area of the trapezoid.

b. Find the area of the trapezoid by adding the areas of the three triangles.

c. Because parts a and b describe the same total area, set the areas equal and show that $a^2 + b^2 = c^2$.

50. ***Pythagorean Theorem.*** The illustrations show two pictorial interpretations of the Pythagorean theorem. Trace and cut out the illustrations. Given that the perpendicular sides of the inner right triangle are $a$ and $b$ and the hypotenuse is $c$, rearrange the pieces to show that $a^2 + b^2 = c^2$. Summarize with numbers in part a and variables in part b to "prove" the Pythagorean theorem.

a.

b.

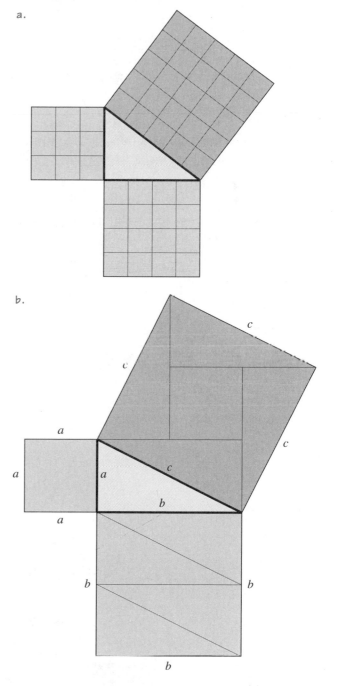

## MID-CHAPTER 7 TEST

1. Which of these sets of numbers represent the sides of a right triangle?

   a. {4, 6, 8}      b. {10, 15, 20}      c. {15, 20, 25}

2. Simplify these exponent and radical expressions.

   a. $\sqrt{2} \cdot \sqrt{18}$      b. $(3\sqrt{2})^2$      c. $\sqrt{18} \cdot \sqrt{8}$

   d. $(2\sqrt{3})^2$      e. $(3\sqrt{3})^2$

3. Find each product. Identify each as a difference of squares (ds), a perfect square trinomial (pst), or neither.

   a. $(x + 4)(x + 4)$      b. $(2x - 5)(2x - 5)$

   c. $(x + 1)(x^2 - x + 1)$      d. $\left(x + \frac{1}{3}\right)\left(x - \frac{1}{3}\right)$

4. Factor each expression. Identify each as a difference of squares (ds), a perfect square trinomial (pst), or neither.

   a. $4x^2 - 25$      b. $x^2 - 20x + 36$

   c. $4x^2 + 8$      d. $x^2 - 12x + 36$

5. Use the graph of $y = x^2 - 2x - 8$ in the figure to solve these equations.

   a. $x^2 - 2x - 8 = 8$      b. $x^2 - 2x - 8 = 0$

6. Solve for $x$.

   a. $x^2 = \frac{16}{144}$      b. $2x^2 - 18 = 0$

   c. $x^2 - 5x + 4 = 0$      d. $x^2 - 4x - 5 = 0$

   e. $2x^2 + 5x - 6 = 0$      f. $2x^2 + 5x + 1 = 0$

7. Solve these formulas for the indicated variable.

   a. $S = 4\pi r^2$ for $r$      b. $h = \frac{v^2}{2g}$ for $v$

8. Each shape has an area of 50 square feet. What is the length of $x$ in each case? The area of an equilateral triangle is

   $$A = \frac{x^2\sqrt{3}}{4}$$

   a. Square, side $x$:

   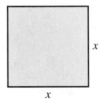

   b. Circle, diameter $x$:

   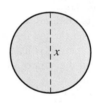

   c. Equilateral triangle:

   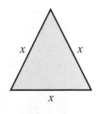

9. What is the distance around each shape in Exercise 8? If each were a small garden plot, which shape would have the lowest cost for fencing material?

---

**7.4**   DISTANCE FORMULA AND PERPENDICULAR LINES

OBJECTIVES    Recognize words for the dimensions of two- and three-dimensional figures formed by straight lines.  ♦  Determine the distance between coordinate points.  ♦  Find the distance between two points when the points are given in variables.  ♦  Identify perpendicular lines by slope.  ♦  Determine whether figures are right triangles.

**WARM-UP**  Calculate or simplify. Assume that the variables are positive. Use a calculator as needed. Explain why some of the answers are equal.

1. $\sqrt{50}$      2. $5\sqrt{2}$      3. $\sqrt{200}$

4. $10\sqrt{2}$      5. $2\sqrt{3}$      6. $\sqrt{12}$

Simplify these expressions.

7. $\dfrac{0 - 1}{3 - 1}$      8. $\sqrt{(3 - 1)^2 + (0 - 1)^2}$      9. $\dfrac{4 - 0}{5 - 3}$

10. $\sqrt{(5 - 3)^2 + (4 - 0)^2}$                    11. $\sqrt{(5 - 1)^2 + (4 - 1)^2}$     ◆

The main focus in this section is on the distance formula, the slopes of perpendicular lines, and further application of algebraic tools to geometric figures.

### DISTANCE AND THE DISTANCE FORMULA

The ordinary English meaning of *distance* and its meaning in the distance formula are slightly different. We ordinarily think of distance in terms of long units such as miles or kilometers. In contrast, the distance formula is used to find the dimensions of figures described by coordinate positions, or how far it is between two points. Thus in this section *distance* describes the lengths of line segments.

To help you recognize words describing length, we summarize the vocabulary of dimension here and use several of the terms in the examples and exercises.

**Dimensions** describe the size of a figure. Dimensions are used to calculate perimeters, areas, and volumes. A rectangular picture or playing field requires two dimensions for a complete description. Boxes and rooms require three dimensions.

The names given to dimensions vary according to the position of the object described. The dimensions of a picture on a wall or of a doorway may be *height* and *width*. A playing field or a hallway rug lies flat and may be described by *length* and *breadth* or *length* and *width*. A triangle has a *base* and *height*. In many instances these words are used somewhat interchangeably. Rectangular-sided boxes are generally described by *length*, *width*, and *height*, but *length*, *breadth*, and *width* are also used.

Height is also used to measure vertical distance. The height of an airplane is its *altitude*. The height of a mountain is its *elevation*. Height in these contexts is relative to sea level.

The distance formula permits us to find the length, or distance, between two coordinate points. Points $(x_1, y_1)$ and $(x_2, y_2)$ are shown in Figure 16.

The distance formula is derived from the Pythagorean theorem. As in finding slope, we find the change in $x$, $\Delta x$, to be $(x_2 - x_1)$ and the change in $y$, $\Delta y$, to be $(y_2 - y_1)$. The angle at $(x_2, y_1)$ (see Figure 16) is a right angle, so the Pythagorean theorem holds:

$$d^2 = (\Delta x)^2 + (\Delta y)^2$$
$$d^2 = (x_2 - x_1)^2 + (y_2 - y_1)^2$$
$$d = \sqrt{(x_2 - x_1)^2 + (y_2 - y_1)^2}$$

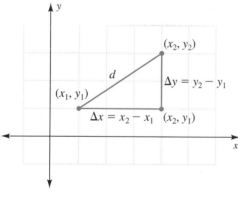

Figure 16

**Distance Formula**

$$d = \sqrt{(x_2 - x_1)^2 + (y_2 - y_1)^2}$$

As in calculating slope, the choice of which point is $(x_1, y_1)$ and which is $(x_2, y_2)$ is arbitrary.

EXAMPLE **1**

Find the three distances between these coordinates: (1, 1), (3, 0), and (5, 4). Determine whether the three points form a right triangle. Make a sketch of the points and lines on coordinate axes. Identify the location of the right angle, if one exists.

SOLUTION     Figure 17 shows the triangle.

Between (1, 1), and (3, 0),

$$d = \sqrt{(3 - 1)^2 + (0 - 1)^2} = \sqrt{2^2 + 1^2} = \sqrt{5}$$

Between (3, 0), and (5, 4),

$$d = \sqrt{(5 - 3)^2 + (4 - 0)^2} = \sqrt{2^2 + 4^2} = \sqrt{20}$$

Between (1, 1) and (5, 4),

$$d = \sqrt{(5 - 1)^2 + (4 - 1)^2} = \sqrt{4^2 + 3^2} = \sqrt{25} = 5$$

By the Pythagorean theorem,

$$\sqrt{5}^2 + \sqrt{20}^2 = 25 = 5^2$$

The right angle is at (3, 0).

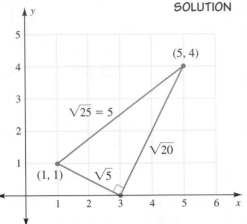

*Figure 17*

♦ ♦ ♦ ♦ ♦ ♦ ♦ ♦ ♦ ♦ ♦ ♦ ♦ ♦ ♦ ♦ ♦ ♦ ♦ ♦ ♦ ♦ ♦ ♦ ♦ ♦ ♦ ♦ ♦ ♦ ♦ ♦ ♦

### PERPENDICULAR LINES

The three coordinates in Example 1 formed the corners of a right triangle. Because the legs of a right triangle are perpendicular, it may be informative to examine their slopes and equations.

EXAMPLE **2**

Find the slope and equation of the line segment forming each leg of the right triangle in Example 1.

**a.** (1, 1), (3, 0)     **b.** (3, 0), (5, 4)

SOLUTION     **a.** *Slope*: $m = \dfrac{\Delta y}{\Delta x} = \dfrac{0 - 1}{3 - 1} = \dfrac{-1}{2}$

*Equation*: $y = -\frac{1}{2}x + b$, where $0 = -\frac{1}{2}(3) + b$ and $b = \frac{3}{2}$
Thus

$$y = -\tfrac{1}{2}x + \tfrac{3}{2}$$

**b.** *Slope*: $m = \dfrac{\Delta y}{\Delta x} = \dfrac{4 - 0}{5 - 3} = \dfrac{4}{2} = 2$

*Equation*: $y = 2x + b$, where $0 = 2(3) + b$ and $b = -6$
Thus

$$y = 2x - 6$$

♦ ♦ ♦ ♦ ♦ ♦ ♦ ♦ ♦ ♦ ♦ ♦ ♦ ♦ ♦ ♦ ♦ ♦ ♦ ♦ ♦ ♦ ♦ ♦ ♦ ♦ ♦ ♦ ♦ ♦ ♦ ♦ ♦

The lines in Example 2 are perpendicular. Their slopes are $-\frac{1}{2}$ and 2, which are negative reciprocals. We are already aware that a number and its reciprocal multiply to 1. Thus the following results seem reasonable:

> Two lines are perpendicular if
>
> **1.** their slopes multiply to $-1$,
>
> **2.** their slopes are negative reciprocals, or
>
> **3.** their slopes are the same as the *x*- and *y*-axes.

In the following two examples and the related exercises we practice using the distance formula and slopes in finding relationships among parts of geometric figures and in solving problems. Convenient placement of geometric shapes onto the coordinate axes will aid in identifying corners and relationships.

EXAMPLE  **3**

A square is placed on the coordinate axes (see Figure 18), with one corner at the origin and one side along the positive *x*-axis. The length of each side is *n*, as indicated by the coordinates at points *A*, *B*, *C*, and *D*.

**a.** Use the distance formula to verify that the diagonals of the square are the same length.

**b.** Use slope to show that the diagonals are perpendicular.

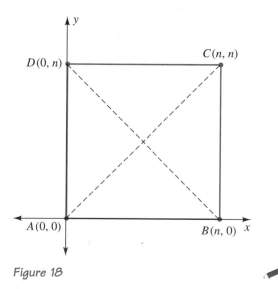

Figure 18

SOLUTION   **a.** *AC* and *BD* are the diagonals.

$$AC: \quad d = \sqrt{(n - 0)^2 + (n - 0)^2} = \sqrt{n^2 + n^2} = \sqrt{2n^2} = |n|\sqrt{2} = n\sqrt{2}$$
$$BD: \quad d = \sqrt{(0 - n)^2 + (n - 0)^2} = \sqrt{n^2 + n^2} = \sqrt{2n^2} = |n|\sqrt{2} = n\sqrt{2}$$

We are able to drop the absolute value signs on *n* because *n* is the side of the square and is positive. *Both diagonals of the square have length* $n\sqrt{2}$ *and are equal.* Because *n* is a variable, we have shown that the diagonals of any square are equal.

**b.**   slope of $AC = \dfrac{\Delta y}{\Delta x} = \dfrac{n - 0}{n - 0} = \dfrac{n}{n} = 1$

slope of $BD = \dfrac{n - 0}{0 - n} = \dfrac{n}{-n} = -1$

The slopes are negative reciprocals and also multiply to $-1$. Because *n* is a variable, the diagonals of any square are perpendicular.

♦ ♦ ♦ ♦ ♦ ♦ ♦ ♦ ♦ ♦ ♦ ♦ ♦ ♦ ♦ ♦ ♦ ♦ ♦ ♦ ♦ ♦ ♦ ♦ ♦ ♦ ♦ ♦ ♦ ♦ ♦ ♦ ♦ ♦ ♦

Example 3 illustrates placing a geometric figure with straight sides onto axes. We generally place one side along the positive *x*-axis. Because one corner is at the origin, the vertex on the *x*-axis is at a distance from the origin equal to the length of the side. This makes it easy to determine the coordinates of other vertices.

In Example 4 we use the distance formula to identify the shape given in Figure 19.

EXAMPLE   **4**    What is the shape in Figure 19?

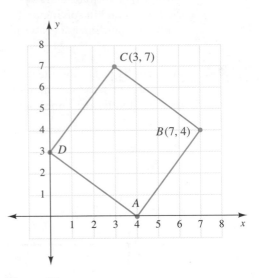

*Figure 19*

**a.** Use the distance formula to find the lengths of sides *AB*, *AD*, *BC*, and *CD* in Figure 19.

**b.** Use the distance formula to find the lengths of the diagonals.

SOLUTION    **a.** Each of the sides of *ABCD* is of length 5.

$$AB: \quad d = \sqrt{(7 - 4)^2 + (4 - 0)^2} = \sqrt{9 + 16} = \sqrt{25} = 5$$
$$BC: \quad d = \sqrt{(3 - 7)^2 + (7 - 4)^2} = \sqrt{16 + 9} = 5$$
$$CD: \quad d = \sqrt{(0 - 3)^2 + (3 - 7)^2} = \sqrt{9 + 16} = 5$$
$$AD: \quad d = \sqrt{(0 - 4)^2 + (3 - 0)^2} = \sqrt{16 + 9} = 5$$

**b.** Each of the diagonals, *AC* and *BD*, is of the same length.

$$AC: \quad d = \sqrt{(3 - 4)^2 + (7 - 0)^2} = \sqrt{1 + 49} = \sqrt{50} \approx 7.07$$
$$BD: \quad d = \sqrt{(0 - 7)^2 + (3 - 4)^2} = \sqrt{49 + 1} = \sqrt{50} \approx 7.07$$

In Example 3 we showed that a square has diagonals of equal length. This figure, *ABCD*, has equal sides and equal diagonals, so it is a square.

♦ ♦ ♦ ♦ ♦ ♦ ♦ ♦ ♦ ♦ ♦ ♦ ♦ ♦ ♦ ♦ ♦ ♦ ♦ ♦ ♦ ♦ ♦ ♦ ♦ ♦ ♦ ♦ ♦ ♦ ♦ ♦ ♦

How might we use slopes to show that Figure 19 is a square? Is the slope calculation enough, or do we also need the lengths of the sides?

## EXERCISES 7.4

**For each pair of points in Exercises 1 to 8**

   a. **find the slope of the line segment connecting the two points, and**

   b. **find the distance between the two points.**

1. (2, 3) and (4, 9)

2. (3, 2) and (9, 4)

3. (2, 2) and (5, −1)

4. (5, 3) and (6, −2)

5. (−3, 3) and (4, 2)

6. (−3, 3) and (2, 4)

7. (−3, −1) and (3, −3)

8. (−2, −1) and (2, −4)

9. Find the equation of the line passing through the points in Exercise 1. Use $y = mx + b$.

10. Find the equation of the line passing through the points in Exercise 2. Use $y = mx + b$.

11. Find the equation of the line passing through the points in Exercise 3.

12. Find the equation of the line passing through the points in Exercise 4.

13. Which line segments from Exercises 1, 3, 5, and 7 are perpendicular? Why?

14. Which line segments from Exercises 2, 4, 6, and 8 are perpendicular? Why?

In Exercises 15 to 20 sets of three coordinates forming the vertices of a triangle are given. Use the distance formula to determine what kind of triangle each is. State the length of each side as well as your conclusion. It is not necessary to simplify radicals. A sketch graph may be helpful. *Note:* An isosceles triangle has two equal sides. An equilateral triangle has three equal sides. A right triangle satisfies the Pythagorean theorem. An isosceles right triangle may also be an option.

15. (3, 4), (0, 1), (6, 1)      16. (3, 1), (0, 4), (−3, 1)

17. (6, 5), (4, 2), (8, 2)      18. (3, 6), (6, 4), (3, 2)

19. (4, 8), (1, 6), (5, 0)      20. (5, 8), (0, 7), (2, −3)

21. Refer to Example 4. The following steps suggest another way to prove that *ABCD* is a square.

  a. Copy the graph containing *ABCD* from Figure 19. We know from the distance formula that the sides of *ABCD* are of length 5, but we also need to know that the corners are right angles (otherwise the figure could be a diamond or a rhombus—a squashed square!).

  b. Draw the square (0, 0), (7, 0), (7, 7), (0, 7). This square surrounds *ABCD*.

  c. What kind of triangles are formed around *ABCD* by the square of side 7? How do we know this?

  d. What do we know about the sum of the two smaller angles in a right triangle?

  e. What can we conclude about the angles forming the corners of *ABCD*?

22. Set up coordinate axes with compass directions: east on the *x*-axis, north on the *y*-axis. Two cars start at the origin at the same time. Car 1 travels east at 60 miles per hour. Car 2 travels north at 30 miles per hour.

  a. Complete the table to find the position of each car, in miles from the origin, at the end of each hour.

  b. Mark the hourly positions on the coordinate axes. Draw a line connecting the positions of the two cars at the end of each hour.

  c. Use the distance formula to find how far apart the cars are at 1 hour, 2 hours, and 3 hours.

  d. How far apart are the cars at time *t*?

| Hours Traveled | Distance Traveled by Car 1 (along the x-axis) | Distance Traveled by Car 2 (along the y-axis) |
|---|---|---|
| 1 hr | | |
| 2 hr | | |
| 3 hr | | |
| *t* hr | | |

23. Refer to Exercise 22. This time, though, Car 1 travels east at 40 miles per hour, and Car 2 travels north at 20 miles per hour. Rework parts a, b, c, and d, with the table below.

| Hours Traveled | Distance Traveled by Car 1 (along the x-axis) | Distance Traveled by Car 2 (along the y-axis) |
|---|---|---|
| 1 hr | | |
| 2 hr | | |
| 3 hr | | |
| *t* hr | | |

24. A parallelogram has height 3, base 6, and slant sides of length 5. Place the parallelogram on coordinate axes (first quadrant) so that one vertex is at the origin and a second is at (6, 0). Label the coordinates of the other two vertices.

25. Is a triangle with sides {9, 16, 25} similar to one with sides {3, 4, 5}? Why or why not? Do sides {9, 16, 25} make a right triangle?

26. *Parallel Lines and Intersecting Segments.* Trace the figure shown. Cut it out, and rearrange the pieces to answer these questions.

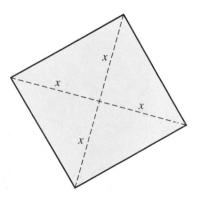

  a. What is the area of the square in terms of *x*?

  b. What is the length of a side of the square in terms of *x*?

PROJECT

27. Refer to the graph.

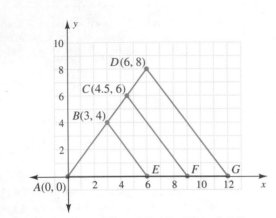

a. Find the slopes of lines *BE*, *AD*, *CF*, and *DG*.

b. How are the lines related?

c. Find these lengths: *BE*, *CF*, *DG*, *AB*, *AC*, and *AD*.

d. Write out and compare these ratios:

*AB* to *AC* with *BE* to *CF*
*AC* to *AD* with *CF* to *DG*

e. How are the triangles within the figure related? Why?

f. Find these lengths: *AB*, *BC*, *CD*, *AE*, *EF*, and *FG*.

g. Write out and compare these ratios:

*AB* to *AE*
*BC* to *EF*
*CD* to *FG*

h. Describe the effect of parallel lines passing through line segments in terms of the ratios found in part g.

---

### 7.5    SOLVING RADICAL EQUATIONS AND FORMULAS

**OBJECTIVES**    Use properties of radicals to simplify expressions. ◆ Use absolute value in simplifying radicands containing exponents. ◆ Solve equations by squaring both sides of an equation once. ◆ Given a graph of a square root expression, solve an equation. ◆ Use formulas involving square root expressions to solve problems.

**WARM-UP**    1. Sketch a triangle with corners *M*, *N*, and *P* at $M(1, 1)$, $N(5, 2)$, and $P(3, 10)$.

2. Determine the lengths of the three sides.

3. Which two sides, if any, are perpendicular?    ◆

I n this section we apply absolute value in simplifying radical expressions, solve equations containing radicals, and work applications with formulas containing radicals.

#### ADDITIONAL PROPERTIES OF RADICALS

For much of our work with roots we assumed that the expression under the radical—the radicand—was positive. If the expression under the radical contains a variable to the first power, we must find the inputs that make it negative and exclude those inputs. Note that the variable *x* may represent expressions in the following statement.

---

The square root of a negative number is not defined in the real number system. The expression $\sqrt{x}$ is defined in the real number system only if *x* is zero or positive.

---

EXAMPLE **1**

For what inputs, $x$, is $\sqrt{x+2}$ defined?

SOLUTION   $\sqrt{x+2}$ is defined if $x+2$ is zero or positive.

$$x + 2 \geq 0 \qquad \text{Determine when the radicand is zero or positive.}$$
$$x + 2 - 2 \geq 0 - 2 \qquad \text{Subtract 2 from each side.}$$
$$x \geq -2$$

Thus $\sqrt{x+2}$ is defined when $x \geq -2$.

♦ ♦ ♦ ♦ ♦ ♦ ♦ ♦ ♦ ♦ ♦ ♦ ♦ ♦ ♦ ♦ ♦ ♦ ♦ ♦ ♦ ♦ ♦ ♦ ♦ ♦ ♦ ♦ ♦ ♦ ♦ ♦ ♦ ♦

If the expression under the radical contains an even power, there is a possibility of a negative output. However, the output from a square root is defined as positive, so if there is any possibility of a negative output we use absolute value around the output. The use of absolute value is illustrated in this summary.

Summary

---

♦ Roots of variables with exponents if $x$ is a positive number:

$$\sqrt{x^2} = x$$
$$\sqrt{x^4} = x^2$$
$$\sqrt{x^6} = x^3$$
$$\sqrt{x^8} = x^4$$

♦ Roots of variables with exponents if $x$ is any real number:

$$\sqrt{x^2} = |x| \qquad \text{Because x could be negative}$$
$$\sqrt{x^4} = x^2 \qquad \text{Because an even power is positive}$$
$$\sqrt{x^6} = |x^3| \qquad \text{Because x could be negative}$$
$$\sqrt{x^8} = x^4 \qquad \text{Because an even power is positive}$$

---

EXAMPLE **2**

Simplify. Let $x$ be any real number.

a. $\sqrt{4x^2}$    b. $\sqrt{(-2)^2}$    c. $\sqrt{x^6 y^4}$    d. $\sqrt{(x-1)^2}$

SOLUTION   a. $\sqrt{4x^2} = 2|x|$
$\sqrt{4} = 2$, but $x$ could be negative.
b. $\sqrt{(-2)^2} = |-2| = 2$
We must have a positive principal square root.
c. $\sqrt{x^6 y^4} = \sqrt{x^6}\sqrt{y^4} = |x^3|y^2$
Recall that $x^3 \cdot x^3 = x^6$. The $x^3$ may be negative, but the $y^2$ is positive.
d. $\sqrt{(x-1)^2} = |x-1|$
The $x - 1$ might be negative.

♦ ♦ ♦ ♦ ♦ ♦ ♦ ♦ ♦ ♦ ♦ ♦ ♦ ♦ ♦ ♦ ♦ ♦ ♦ ♦ ♦ ♦ ♦ ♦ ♦ ♦ ♦ ♦ ♦ ♦ ♦ ♦ ♦ ♦

## SOLVING EQUATIONS INVOLVING SQUARE ROOTS

We have used several techniques to solve an equation. We have added, subtracted, multiplied, and divided expressions on both sides. We have also taken the square root of both sides of an equation. Finally, we now square both sides of an equation.

---

To solve an equation containing a square root, we can square both sides.

---

The following example shows why squaring both sides is a reasonable operation.

EXAMPLE **3**

In Example 1 we determined that $\sqrt{x + 2}$ is defined if $x \geq -2$. Solve $\sqrt{x + 2} = 4$ by

**a.** using symbols     **b.** examining the graph

SOLUTION    **a.** Squaring both sides of an equation is a modification of multiplying both sides by the same number.

$$\sqrt{x + 2} = 4$$
$$4 \cdot \sqrt{x + 2} = 4 \cdot 4 \qquad \text{Multiply both sides by 4.}$$
$$\sqrt{x + 2} \cdot \sqrt{x + 2} = 4 \cdot 4 \qquad \begin{array}{l}\text{Because } \sqrt{x + 2} = 4, \text{ we can replace}\\ \text{the 4 on the left by } \sqrt{x + 2}; \text{ the result}\\ \text{is the same as squaring both sides.}\end{array}$$
$$x + 2 = 16 \qquad \text{Recall that x + 2 is positive.}$$
$$x = 14$$

*Check*: Check the solution in the original equation. Make sure that the answer does not make the equation false or undefined because of a negative radicand. $\sqrt{14 + 2} \overset{?}{=} 4$ ✔

**b.** The intersection of $y = \sqrt{x + 2}$ with $y = 4$ in Figure 20 gives the solution to $\sqrt{x + 2} = 4$. This point of intersection is $(14, 4)$, so $x = 14$. The graph also shows the condition $x \geq -2$ in that there are no points on the square root graph to the left of $x = -2$.

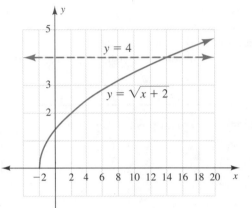

Figure 20

♦ ♦ ♦ ♦ ♦ ♦ ♦ ♦ ♦ ♦ ♦ ♦ ♦ ♦ ♦ ♦ ♦ ♦ ♦ ♦ ♦ ♦ ♦ ♦ ♦ ♦ ♦ ♦ ♦ ♦ ♦ ♦ ♦ ♦ ♦

EXAMPLE **4**

After determining inputs for which the equation $y = \sqrt{x - 5}$ is defined, solve $\sqrt{x - 5} = 3$ by

**a.** using symbols     **b.** examining the graph

SOLUTION    The expression $x - 5$ must be positive, because the square root of a negative is not defined in the real numbers. The statement that $x - 5$ is positive is written $x - 5 \geq 0$. By adding 5 to both sides, we find that this inequality has the solution $x \geq 5$. The equation is defined for all $x$ larger than or equal to 5.

**a.** We solve the equation by squaring both sides.

$$\sqrt{x - 5} = 3 \qquad \text{Square both sides.}$$
$$(\sqrt{x - 5})^2 = 3^2$$
$$x - 5 = 9 \qquad (\sqrt{x-5})^2 = x - 5 \text{ because x - 5 is positive.}$$
$$x = 14$$

*Check*: $\sqrt{14 - 5} \overset{?}{=} 3$ ✔

**b.** The intersection of $y = 3$ with $y = \sqrt{x - 5}$ is at $(14, 3)$ in Figure 21. Thus $x = 14$ is the solution to $\sqrt{x - 5} = 3$. There are no points on the square root graph to the left of 5 because the expression $\sqrt{x - 5}$ is undefined for $x < 5$.

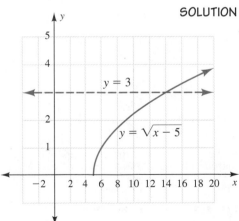

Figure 21

♦ ♦ ♦ ♦ ♦ ♦ ♦ ♦ ♦ ♦ ♦ ♦ ♦ ♦ ♦ ♦ ♦ ♦ ♦ ♦ ♦ ♦ ♦ ♦ ♦ ♦ ♦ ♦ ♦ ♦ ♦ ♦ ♦ ♦ ♦

Example 5 shows the importance of checking the symbolic solutions.

EXAMPLE **5**

After determining the inputs for which $y = \sqrt{3 - x}$ is defined, solve $\sqrt{3 - x} = x - 1$ by

**a.** using symbols     **b.** examining the graph

SOLUTION   The equation is defined for positive radicands, $3 - x \geq 0$. The solution to this inequality is $3 \geq x$, or $x \leq 3$.

**a.** Solve $\sqrt{3 - x} = x - 1$ by squaring both sides.

$$\sqrt{3 - x} = x - 1$$
$$(\sqrt{3 - x})^2 = (x - 1)^2$$
$$3 - x = x^2 - 2x + 1$$
$$0 = x^2 - x - 2$$
$$0 = (x + 1)(x - 2)$$

Either   $(x + 1) = 0$   or   $(x - 2) = 0$

Either   $x = -1$   or   $x = 2$

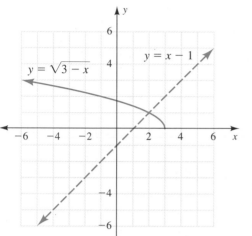

*Figure 22*

*Check*:   $\sqrt{3 - (-1)} \stackrel{?}{=} (-1) - 1$

$\sqrt{3 - (2)} \stackrel{?}{=} 2 - 1$

The input $x = 2$ gives a true statement. The input $x = -1$ gives $2 = -2$, which is false. The solution $x = -1$ must be discarded. ✔

**b.** Figure 22 shows the intersection of $y = \sqrt{3 - x}$ and $y = x - 1$ at $x = 2$, but it shows no intersection at $x = -1$. This confirms the single solution found symbolically.

♦ ♦ ♦ ♦ ♦ ♦ ♦ ♦ ♦ ♦ ♦ ♦ ♦ ♦ ♦ ♦ ♦ ♦ ♦ ♦ ♦ ♦ ♦ ♦ ♦ ♦ ♦ ♦ ♦ ♦ ♦ ♦

---

Solutions that give false statements are called *extraneous roots*.

---

APPLICATIONS

We close this section with two applications of square roots.

EXAMPLE   **6**

**Washington Monument**   As a child I dropped a paper-wrapped sugar cube from the open window of the Washington Monument in our nation's capital. Fortunately, it was a cold winter day in the 1950s, and no one was at the base of the 555-foot tower.

The time required for a dropped object to hit the ground is given by

$$t = \sqrt{\frac{2d}{g}}$$

where $d$ is the distance traveled and $g$ is the acceleration due to gravity, assumed to be 32.2 feet per second squared. The downward speed, or velocity $v$, such an object will be traveling is given by $v = gt$.

**a.** How long did it take for the cube to hit the ground?

**b.** What speed in feet per second was the cube traveling when it hit the ground?

**c.** Use unit analysis to change the speed to miles per hour.

SOLUTION   **a.** Assuming the distance from the window (just below the top of the tower) to the ground is 550 feet, we have

$$t = \sqrt{\frac{2d}{g}} = \sqrt{\frac{2 \cdot 550 \text{ ft}}{32.2 \frac{\text{ft}}{\text{sec}^2}}} \approx \sqrt{34.1615 \text{ sec}^2} \approx 5.8 \text{ sec}$$

**b.** Speed is $v = gt$.

$$v = gt = \frac{32.2 \text{ ft}}{\sec^2} \cdot 5.8 \text{ sec} \approx 187 \text{ ft per sec}$$

**c.** Using unit analysis to change feet per second to miles per hour, we have

$$\frac{187 \text{ ft}}{\sec} \cdot \frac{1 \text{ mi}}{5280 \text{ ft}} \cdot \frac{60 \text{ sec}}{1 \text{ min}} \cdot \frac{60 \text{ min}}{1 \text{ hr}} \approx 128 \text{ mi per hr}$$

**P.S.**   My parents were furious! The Washington Monument windows are now sealed.

◆ ◆ ◆ ◆ ◆ ◆ ◆ ◆ ◆ ◆ ◆ ◆ ◆ ◆ ◆ ◆ ◆ ◆ ◆ ◆ ◆ ◆ ◆ ◆ ◆ ◆ ◆ ◆ ◆ ◆ ◆ ◆ ◆ ◆

E X A M P L E    **7**    *Distance to the horizon at the seashore*    On earth the distance to the horizon is

$$d \approx \sqrt{\frac{3h}{2}}$$

where $d$ is in miles to the horizon and $h$ is in feet above sea level.

**a.** Find the distance seen from a height of 5 feet.

**b.** Find the distance seen from a height of 10 feet.

**c.** Find the distance seen from a height of 24 feet.

**d.** How high would we have to be to see 12 miles?

SOLUTION    **a.** $d \approx \sqrt{\dfrac{3h}{2}} \approx \sqrt{\dfrac{3 \cdot 5}{2}} \approx 2.7 \text{ mi}$    **b.** $d \approx \sqrt{\dfrac{3 \cdot 10}{2}} \approx 3.9 \text{ mi}$

**c.** $d \approx \sqrt{\dfrac{3 \cdot 24}{2}} \approx 6 \text{ mi}$    **d.**   $12 \approx \sqrt{\dfrac{3h}{2}}$

$$144 \approx \frac{3h}{2}$$
$$288 \approx 3h$$
$$h \approx 96 \text{ ft}$$

◆ ◆ ◆ ◆ ◆ ◆ ◆ ◆ ◆ ◆ ◆ ◆ ◆ ◆ ◆ ◆ ◆ ◆ ◆ ◆ ◆ ◆ ◆ ◆ ◆ ◆ ◆ ◆ ◆ ◆ ◆ ◆ ◆ ◆

## EXERCISES 7.5

**Simplify each expression in Exercises 1 to 6. The variables may represent any real number, so absolute values may be needed.**

1. $\sqrt{ab^2}$        2. $\sqrt{a^2b}$        3. $\sqrt{a^2b^2}$

4. $\sqrt{a^2b^4}$        5. $\sqrt{a^4b^2}$        6. $\sqrt{a^6b^4}$

7. Explain why $\sqrt{x^2y} = |x|\sqrt{y}$ needs absolute value, but $\sqrt{x^4y} = x^2\sqrt{y}$ does not.

8. Explain why $\sqrt{x^6y} = |x^3|\sqrt{y}$ needs absolute value, but $\sqrt{x^8y} = x^4\sqrt{y}$ does not.

**In Exercises 9 to 16 simplify the expressions, taking square roots where possible. Assume variables represent positive numbers so no absolute values are needed.**

9. $\sqrt{49x^2}$        10. $\sqrt{121y^2}$

11. $\sqrt{x^2y}$        12. $\sqrt{y^2x}$

13. $\sqrt{p^1p^2}$        14. $\sqrt{p^4}$

15. $\sqrt{p^8}$        16. $\sqrt{p^1p^8}$

**Solve the equations in Exercises 17 to 24. Assume the radicands (expressions within the radical sign) are positive. For what inputs, $x$, are the radical expressions defined?**

17. $\sqrt{x-2} = 3$        18. $\sqrt{x-3} = 2$

19. $\sqrt{2x+2} = 4$        20. $\sqrt{3x+1} = 5$

21. $\sqrt{3x-3} = 6$        22. $\sqrt{2x-1} = 7$

23. $\sqrt{4-x} = x-2$        24. $\sqrt{2-x} = x-1$

For Exercises 25 to 28 use the graphs to solve the equations. Why are there no outputs for some inputs, $x$?

25. $\sqrt{x-2} = 2$

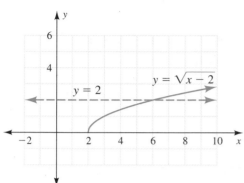

26. $\sqrt{x+2} = 2$

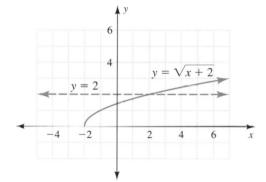

27. $\sqrt{x+1} = 1$

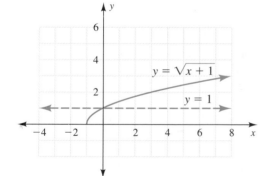

28. $\sqrt{x-1} = 2$

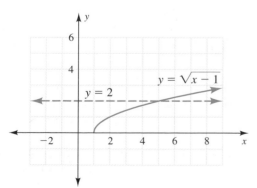

On the moon the distance seen in miles from a height of $h$ feet is given by

$$d \approx \sqrt{\frac{3h}{8}}$$

For Exercises 29 to 32 calculate how far can be seen from the given heights. Why might the distance be different from that on earth?

29. 24 feet

30. 5 feet

31. 96 feet

32. 10 feet

Assume a clear day and flat terrain in the nearby region. How far could you see, in miles, from the top of each of the buildings whose heights are given in Exercises 33 to 36? See text examples for formulas.

33. Sears Tower, Chicago, 1454 feet

34. Bank of China, Hong Kong, 1209 feet

35. Texas Commerce Tower, Houston, 1002 feet

36. Transamerica Pyramid, San Francisco, 853 feet

37. How high a structure is needed to see 30 miles on earth?

38. How high a structure is needed to see 30 miles on the moon? Explain the difference between this and the answer to Exercise 37. A sketch may help.

39. To see the same distance on the moon as on earth, how many times higher do we need to be on the moon?

40. To see twice as far on the moon as on earth, how many times higher do we need to be?

How long would it take an object to fall from the top of each of the places whose heights are given in Exercises 41 to 44? How fast would the object be traveling in feet per second when it hit the ground? See text examples for formulas.

41. C.N. Tower, Toronto, 1821 feet

42. World Trade Center, New York City, 1368 feet

43. C&C Plaza, Atlanta, 1063 feet

44. Eiffel Tower, Paris, 984 feet

45. How tall a building is required for an object to take 12 seconds to fall to the ground?

46. How tall a building is required for an object to take 10 seconds to fall to the ground?

47. How long would it take an object to fall from one mile up? (Assume no air resistance, wind, or other complicating factors.)

48. If a piece of ice falls off an airplane at 35,000 feet, how long until it will hit the ground? (Assume the ice does not melt.)

PROJECTS

49. *Graphing with a Graphing Calculator*

a. Complete the table for the equation $x^2 + y^2 = 25$. There are two columns of coordinates. If there is more than one answer for a particular $x$ or $y$, use $\pm$ or expand the table.

| $x$ | $y$ | $x$ | $y$ |
|-----|-----|-----|-----|
| 0   |     |     | −5  |
|     | 0   | −3  |     |
| 3   |     | −4  |     |
|     | 4   |     | −3  |
| 4   |     | −5  |     |
|     | 3   |     | −4  |

b. Plot the coordinates.

c. What do you observe about the points? Why? Identify the graph.

d. Do you observe any relation to one of the Pythagorean triples?

e. Solve $x^2 + y^2 = 25$ first for $y^2$ and then for $y$. (*Hint*: When taking the square root, remember that $y$ can be either positive or negative, so use a $\pm$ sign in the equation. The $\pm$ creates two equations.) Enter both equations on the graphing calculator. The graph may look oval. Experiment with different window settings to make the graph appear as in your plot. Record the dimensions of the viewing window that makes the best graph. (The upper and lower part of the graph might not connect. Again, a different window setting may improve the shape, but it may not completely connect the two halves.)

50. *Pendulum Swing.* A complete swing of a pendulum is the time it takes for the pendulum to swing away and return approximately to its original position.

For this project you need fishing line or strong, thin string; a lead fishing weight or some other small heavy object; a watch with a second hand or a stop watch; a meter stick; and a balcony with no one standing directly below it.

a. How long, $t$, will it take a lead weight to make one complete swing for each of the pendulum lengths given in the table? Start the swing about 45° off perpendicular to the ground, as shown in the figure. Time the swing from position $A$ to $B$ to $A$ (see the figure). If you measure in feet rather than meters, list from 2 to 12 feet in your table.

| Length, $l$ (m) | Time, $t$ (sec) |
|-----------------|-----------------|
| 0.5             |                 |
| 1               |                 |
| 2               |                 |
| 3               |                 |
| 4               |                 |

One complete swing: $A$ to $B$ to $A$

b. Graph the data from the table. What is the shape of the graph?

c. Why is a linear graph not reasonable?

d. Use your graph to determine what length pendulum, $l$, is needed for a 1-second swing and for a 3-second swing.

e. Does the release position make any difference in the time, $t$? Try a 2-meter pendulum at three different release points. *Note*: A release near or above the level of the support for the pendulum will cause erratic motion and ruin the experiment. Keep the angle of swing less than 45° away from perpendicular to the ground.

---

## 7.6    RANGE AND STANDARD DEVIATION

**OBJECTIVES**    Determine the range from a set of data.  ♦  Calculate the median and quartiles of a set of data, and summarize with a box and whisker plot.  ♦  Calculate the mean absolute variation within a set of data.  ♦  Calculate the standard deviation for a set of data.

**WARM-UP**    Problems 1 to 4 each contain a list of measurements of students drawings of a 3-inch line. Calculate the mean and median of each set. As needed, review the calculation of averages from Section 4.5.

**1.** 1.25, 1.25, 2.25, 2.75, 7.5

**2.** 6, 3.50, 0.75, 1.75, 2.50, 3.50

**3.** 4, 2, 3, 2.25, 3.75

**4.** 3.25, 3.25, 3.25, 2, 3.25

Calculate the distance between these points.

**5.** (3, 7) and (−2, −5)

**6.** (−3, 5) and (5, −10)

Without any rulers or coins, draw the following.

**7.** A line that you estimate to be 3 inches long

**8.** A line that you estimate to be 4 centimeters long

**9.** A circle that you estimate to be the size of a dime

**10.** A circle that you estimate to be the size of a quarter

When everyone has finished drawing, take out a ruler and coins and measure the lengths of the lines and the diameters of the circles. Measure lines to the nearest quarter inch and circles to the nearest $\frac{1}{8}$ inch.                                    ♦

S ection 4.5 introduced the mean, median, and mode. These calculations are called *measures of central tendency* because they indicate the average, middle, or center of a set of data. What do the measures of central tendency indicate about our warm-up drawings? Do they indicate the accuracy of our drawings?

This section will introduce ways you might determine how well you have drawn your lines and circles compared with others in the class. We will examine measures of variation or dispersion — that is, measures that describe how close to the middle or how scattered a set of data is. In studying measures of dispersion we will consider range, box and whisker plots, mean absolute variation, and standard deviation. You may want to use techniques to summarize how your class performed on the drawing tasks. You might also think about why the 3-inch line is likely to be more accurate than the 4-centimeter line and how could we verify that it is.

### RANGE

The problems in the Warm-up describe results when students drew 3-inch lines. The four sets have the same mean; that is, when we added the numbers and divided by the number of numbers, we obtained 3. The middle number, or median, in each set is also close to 3. The mean and median indicate little about the data set itself. If we looked only at the mean and median, we might conclude that the students did very well at drawing a 3-inch line. These measures of central tendency may be deceptive, so it is important to consider other measures as well in summarizing data.

Finding the range is a first step in determining how sets of numbers differ. The **range** is calculated by subtracting the lowest number from the highest number.

*Range*

> The range is the difference between the largest and the smallest number in a data set:
>
> $$\text{range} = \text{maximum number} - \text{minimum number}$$

EXAMPLE **1**  As we saw in the Warm-up, the mean for each set of measurements is 3. Find the range of each set.

**a.** 1.25, 1.25, 2.25, 2.75, 7.50    **b.** 6, 3.50, 0.75, 1.75, 2.50, 3.50

**c.** 4, 2, 3, 2.25, 3.75    **d.** 3.25, 3.25, 3.25, 2, 3.25

SOLUTION  **a.** The range is $7.50 - 1.25 = 6.25$ in.

**b.** The range is $6 - 0.75 = 5.25$ in.

**c.** The range is $4 - 2 = 2$ in.

**d.** The range is $3.25 - 2 = 1.25$ in.

♦ ♦ ♦ ♦ ♦ ♦ ♦ ♦ ♦ ♦ ♦ ♦ ♦ ♦ ♦ ♦ ♦ ♦ ♦ ♦ ♦ ♦ ♦ ♦ ♦ ♦ ♦ ♦ ♦ ♦ ♦ ♦ ♦

Although the sets all have the same mean, their ranges differ considerably. The range indicates the spread of the data, which is considerably greater in parts a and b than in parts c and d. Because the mean of the sets is 3, the desired length of the line, the student measures in parts c and d are more accurate.

### BOX AND WHISKER PLOTS

The median, as found in the Warm-up, may be used with the range to draw a **box and whisker plot**, which gives a visual summary of the data.

**Drawing a Box and Whisker Plot**

1. Find the median—the middle number when data are arranged in smallest to largest order. If there is no single middle number, average the two middle numbers. Locate the median on a horizontal line with a scale appropriate to the data set. (See Figure 23.)

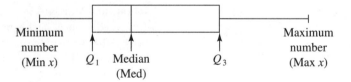

*Figure 23*

2. Find the **quartiles**, $Q_1$ and $Q_3$, as follows: $Q_1$ is *the middle number of the numbers below the median. $Q_3$ is the middle number of the numbers above the median.* If there is no single middle number, average the two middle numbers. Draw a box from $Q_1$ to $Q_3$, with length to scale.

3. Draw a line from the left end of the box to the minimum number. Draw another line from the right end of the box to the maximum number. These lines form the whiskers.

EXAMPLE **2**  Calculate $Q_1$ and $Q_3$ for each data set in Example 1. Draw a box and whisker plot for each set.

SOLUTION  **a.** In the set {1.25, 1.25, 2.25, 2.75, 7.50}, the median is 2.25. $Q_1$ is 1.25, halfway between the numbers smaller than the median: 1.25 and 1.25. $Q_3$ is 5.125, halfway between the numbers larger than the median: 2.75 and 7.5.

The box and whisker plot is in Figure 24. $Q_1$ and the minimum are at the same point, so there is no whisker on the left.

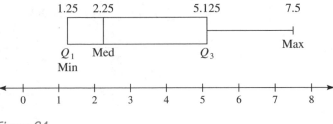

Figure 24

**b.** In the set {0.75, 1.75, 2.50, 3.50, 3.50, 6}, the median is 3, halfway between 2.50 and 3.50. $Q_1$ is 1.75, the middle number of the three below the median. $Q_3$ is 3.50, the middle number of the three above the median. The box and whisker plot is in Figure 25.

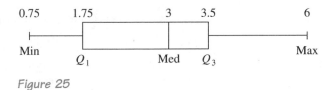

Figure 25

**c.** In the set {2, 2.25, 3, 3.75, 4}, the median is 3. $Q_1$ is 2.125, halfway between the numbers smaller than the median: 2 and 2.25. $Q_3$ is 3.875, halfway between the numbers larger than the median: 3.75 and 4. The box and whisker plot is in Figure 26.

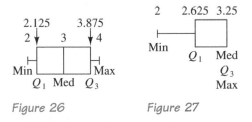

Figure 26                                 Figure 27

**d.** In the set {2, 3.25, 3.25, 3.25, 3.25}, the median is 3.25. $Q_1$ is 2.625, halfway between the numbers smaller than the median: 2 and 3.25. $Q_3$ is 3.25, halfway between the numbers larger than the median: 3.25 and 3.25. This time the median, $Q_3$, and the maximum are all the same, which creates a short box plot with no whisker showing on the right. The box and whisker plot is in Figure 27.

◆ ◆ ◆ ◆ ◆ ◆ ◆ ◆ ◆ ◆ ◆ ◆ ◆ ◆ ◆ ◆ ◆ ◆ ◆ ◆ ◆ ◆ ◆ ◆ ◆ ◆ ◆ ◆ ◆ ◆ ◆ ◆ ◆

## MEAN ABSOLUTE DEVIATION

The box and whisker plot indicates visually where the data are located. A disadvantage of the box and whisker plot is that it does not give us a numerical description. One way to obtain a numerical measure of the relative location and spread of the data might be to add the differences between each data point and the mean. *The difference between each number and the mean is called the* **deviation**.

EXAMPLE 3    Find the sum of the deviations between the data points, {1.25, 1.25, 2.25, 2.75, 7.50}, and the mean, 3.

SOLUTION

| Data Point, $x$ | Deviation, $x$ − mean |
|---|---|
| 1.25 | $1.25 - 3 = -1.75$ |
| 1.25 | $1.25 - 3 = -1.75$ |
| 2.25 | $2.25 - 3 = -0.75$ |
| 2.75 | $2.75 - 3 = -0.25$ |
| 7.50 | $7.50 - 3 = 4.50$ |

The sum of the differences between each number and the mean is

$$-1.75 + (-1.75) + (-0.75) + (-0.25) + (4.50) = -4.5 + 4.5 = 0$$

◆ ◆ ◆ ◆ ◆ ◆ ◆ ◆ ◆ ◆ ◆ ◆ ◆ ◆ ◆ ◆ ◆ ◆ ◆ ◆ ◆ ◆ ◆ ◆ ◆ ◆ ◆ ◆ ◆ ◆ ◆ ◆ ◆ ◆ ◆

A zero result is not very descriptive, so let's try another set.

EXAMPLE 4    Calculate the sum of the deviations between the data points, {0.75, 1.75, 2.50, 3.50, 3.50, 6}, and the mean, 3.

SOLUTION

| Data Point, $x$ | Deviation, $x$ − mean |
|---|---|
| 0.75 | $0.75 - 3 = -2.25$ |
| 1.75 | $1.75 - 3 = -1.25$ |
| 2.50 | $2.50 - 3 = -0.50$ |
| 3.50 | $3.50 - 3 = 0.50$ |
| 3.50 | $3.50 - 3 = 0.50$ |
| 6.00 | $6.00 - 3 = 3$ |

The sum of the deviations is

$$-2.25 + (-1.25) + (-0.50) + (0.50) + (0.50) + 3 = -4 + 4 = 0$$

◆ ◆ ◆ ◆ ◆ ◆ ◆ ◆ ◆ ◆ ◆ ◆ ◆ ◆ ◆ ◆ ◆ ◆ ◆ ◆ ◆ ◆ ◆ ◆ ◆ ◆ ◆ ◆ ◆ ◆ ◆ ◆ ◆ ◆ ◆

In both Example 3 and Example 4 the deviations added to zero, and yet the sets of numbers were quite different. The deviations added to zero because the total amount of deviation above the mean was exactly the same as the total amount below the mean. The proof that the sum of the deviations is always zero is left as an exercise.

The above method did not solve our problem of finding a number to describe the variation within the set. We might solve the problem if we could find a way to eliminate the zero sum. If the deviations were all positively valued, they would no longer add to zero. One way to make them positive is by applying the absolute value.

EXAMPLE 5    Calculate the sum of the absolute values of the deviations between the data points, {2, 2.25, 3, 3.75, 4}, and the mean, 3.

SOLUTION

| Data Points, $x$ | Deviation, $x -$ mean | Absolute Deviation |
|---|---|---|
| 2 | $2 - 3 = -1$ | $\lvert -1 \rvert = 1$ |
| 2.25 | $2.25 - 3 = -0.75$ | $\lvert -0.75 \rvert = 0.75$ |
| 3 | $3 - 3 = \phantom{-}0$ | $\lvert 0 \rvert = 0$ |
| 3.75 | $3.75 - 3 = \phantom{-}0.75$ | $\lvert 0.75 \rvert = 0.75$ |
| 4 | $4 - 3 = \phantom{-}1$ | $\lvert 1 \rvert = 1$ |

The total deviation (the sum of the absolute values of the deviations) is

$$\lvert -1 \rvert + \lvert -0.75 \rvert + \lvert 0 \rvert + \lvert 0.75 \rvert + \lvert 1 \rvert = 3.50$$

◆ ◆ ◆ ◆ ◆ ◆ ◆ ◆ ◆ ◆ ◆ ◆ ◆ ◆ ◆ ◆ ◆ ◆ ◆ ◆ ◆ ◆ ◆ ◆ ◆ ◆ ◆ ◆ ◆ ◆ ◆

The total deviation varies according to the number of pieces of data in the set, so we can improve the information by dividing by the number of pieces of data. When we *divide the total deviation by the number of data points*, we obtain the **mean absolute deviation**. The mean absolute deviation tells us the average amount by which the numbers in the set vary from the mean. The mean absolute deviation for the data in Example 5 is $3.5 \div 5 = 0.7$. The measures for this group vary an average of less than an inch from the mean.

**Finding the Absolute Mean Deviation**

1. Find the mean, and calculate the deviation of each data point from the mean.
2. Find the absolute value of each deviation.
3. Find the mean of the absolute values of the deviations.

Before calculators and computers, the mean absolute deviation was somewhat awkward to work with because of the absolute values. As a result, statisticians (people working with data) adopted another way to make the deviations positive.

## STANDARD DEVIATION

In developing the formula for what British mathematician Karl Pearson was to call "standard deviation" in 1894, statisticians may have thought about how we find the distance between two points.

The distance between $(x_1, y_1)$ and $(x_2, y_2)$ is given by

$$d = \sqrt{(x_2 - x_1)^2 + (y_2 - y_1)^2}$$

The distance formula finds the deviation between the *x*-coordinates and between the *y*-coordinates. To assure that the distances are positive, we square the deviations and then add. To compensate for squaring the deviations, we then take the square root of the sum to obtain the distance. Observe the similarity between the distance formula and the formula for the **sample standard deviation**.

**Sample Standard Deviation**

For a set of $n$ numbers $x_1, x_2, x_3, \ldots, x_n$ with mean $m$, drawn randomly from a population, the standard deviation is

$$s_x = \sqrt{\frac{(x_1 - m)^2 + (x_2 - m)^2 + (x_3 - m)^2 + \cdots + (x_n - m)^2}{n - 1}}$$

The division by $n - 1$ in the sample standard deviation formula is similar to the division in the mean absolute deviation process.

EXAMPLE   6

Find the sample standard deviation for the set of measures in Example 3, {1.25, 1.25, 2.25, 2.75, 7.5}.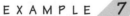

SOLUTION   standard deviation, $s_x$

$$= \sqrt{\frac{(1.25 - 3)^2 + (1.25 - 3)^2 + (2.25 - 3)^2 + (2.75 - 3)^2 + (7.5 - 3)^2}{5 - 1}}$$

$$= \sqrt{\frac{(-1.75)^2 + (-1.75)^2 + (-0.75)^2 + (-0.25)^2 + (4.5)^2}{4}}$$

$$= \sqrt{\frac{27}{4}} \approx 2.6$$

◆ ◆ ◆ ◆ ◆ ◆ ◆ ◆ ◆ ◆ ◆ ◆ ◆ ◆ ◆ ◆ ◆ ◆ ◆ ◆ ◆ ◆ ◆ ◆ ◆ ◆ ◆ ◆ ◆ ◆ ◆ ◆

There is another form of the standard deviation: the **population standard deviation**, $\sigma_x$. The Greek letter sigma ($\sigma$) is used in the population standard deviation. In the $\sigma_x$ formula, $n$ replaces the $n - 1$ used in the earlier standard deviation formula. The population standard deviation is used when the data represent the entire population, such as the set of all ages of the presidents of the United States. The sample standard deviation, $s_x$, is more commonly used because our data set is usually not the *population* (the set of all possible outcomes available). Example 7 illustrates the population standard deviation.

NOTE   Unless $\sigma_x$ is specifically requested in a problem, use the sample standard deviation, $s_x$.

EXAMPLE   7

Find the population standard deviation, $\sigma_x$, for the set of measures in Example 6.

SOLUTION   For the population standard deviation, we use $n$ instead of $n - 1$ in the formula and divide by 5 instead of 4.

population standard deviation, $\sigma_x$

$$= \sqrt{\frac{(x_1 - m)^2 + (x_2 - m)^2 + (x_3 - m)^2 + \cdots + (x_n - m)^2}{n}}$$

$$= \sqrt{\frac{(1.25 - 3)^2 + (1.25 - 3)^2 + (2.25 - 3)^2 + (2.75 - 3)^2 + (7.5 - 3)^2}{5}}$$

$$= \sqrt{\frac{(-1.75)^2 + (-1.75)^2 + (-0.75)^2 + (-0.25)^2 + (4.5)^2}{5}}$$

$$= \sqrt{\frac{27}{5}} \approx 2.32$$

◆ ◆ ◆ ◆ ◆ ◆ ◆ ◆ ◆ ◆ ◆ ◆ ◆ ◆ ◆ ◆ ◆ ◆ ◆ ◆ ◆ ◆ ◆ ◆ ◆ ◆ ◆ ◆ ◆ ◆ ◆ ◆ ◆

Our purpose here is not to perform tedious calculations with numbers, but rather to understand the logic behind the formulas and how the results are found. The standard deviations for Problems 1 to 4 in the Warm-up are given in Example 8 so that you can practice finding them. Graphing calculators and some scientific calculators will calculate the standard deviation.

Graphing Calculator Technique
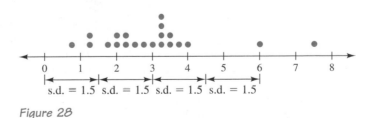

> The mean and the standard deviation may be calculated with the statistical function on a graphing calculator.
>
> For the TI 82, enter the set of data into the first list, $L_1$, obtained with $\boxed{\text{STAT}}$ 1. If there are already data in $L_1$, clear them with $\boxed{\text{STAT}}$ 4 $\boxed{\text{2nd}}$ $L_1$.
>
> To calculate the mean and the standard deviation, press $\boxed{\text{STAT}}$ $\boxed{\blacktriangleright}$ 1 to get the calculation of one-variable statistics. The mean is an x with a line over it, $\bar{x}$, and the standard deviations are listed as $\sigma_x$ and $s_x$.
>
> $Q_1$, the median, and $Q_3$ for the box and whisker plot are listed after the standard deviations.

EXAMPLE  8

Use a calculator to find the population standard deviation and the sample standard deviation for each of the four sets of measures in the Warm-up.

SOLUTION  **a.** The population standard deviation for $\{1.25, 1.25, 2.25, 2.75, 7.50\}$ is $\sigma_x \approx 2.32$. The sample standard deviation is $s_x \approx 2.60$, as described in Examples 6 and 7.

**b.** For $\{0.75, 1.75, 2.50, 3.50, 3.50, 6\}$, the population standard deviation is $\sigma_x \approx 1.65$. The sample standard deviation is $s_x \approx 1.81$.

**c.** For $\{2, 2.25, 3, 3.75, 4\}$, $\sigma_x \approx 0.79$ and $s_x \approx 0.88$.

**d.** For $\{2, 3.25, 3.25, 3.25, 3.25\}$, $\sigma_x \approx 0.5$ and $s_x \approx 0.56$.

Note that the standard deviation is larger for widely scattered data than for data that are close together.

♦ ♦ ♦ ♦ ♦ ♦ ♦ ♦ ♦ ♦ ♦ ♦ ♦ ♦ ♦ ♦ ♦ ♦ ♦ ♦ ♦ ♦ ♦ ♦ ♦ ♦ ♦ ♦ ♦ ♦ ♦ ♦ ♦ ♦ ♦ ♦ ♦ ♦ ♦

Because we divide by a smaller number ($n - 1$ instead of $n$), the sample standard deviation gives a larger number than the population standard deviation. The subtle distinctions that determine when to use each form of the standard deviation are presented in a full statistics course; they will not be discussed here.

PROCESS CONTROL

The standard deviation from the mean is an important tool in manufacturing process control. Figure 28 shows a plot of the four sets of measures in the Warm-up. The data are indicated by dots above the line. The standard deviation for the combined data, s.d. $\approx 1.5$, is shown below the line.

Figure 28

Observe that all data points but one are within two standard deviations of the mean, 3. It is common for data to fall within two standard deviations on each side of the mean; thus control processes are often designed to determine whether sample measurements fit within two standard deviations of the mean.

EXAMPLE 9

**Tennis balls**   Suppose that a manufacturer is setting up a machine to produce professional-quality tennis balls, which must be between $2\frac{1}{2}$ and $2\frac{5}{8}$ inches in diameter. The first task of the manufacturer's quality controller is to calculate the desired mean and standard deviation for the diameter of the balls produced. Then, to see whether the machine is achieving the desired quality, the quality controller will plot data points for a sample of balls and see whether they fall within two standard deviations on each side of the mean. Calculate the desired mean and standard deviation for the diameter of the balls.

SOLUTION   The desired mean, halfway between $2\frac{1}{2}$ (2.5) and $2\frac{5}{8}$ (2.625), is

$$\frac{2.5 + 2.625}{2} = 2.5625$$

A distance of 0.0625 is available on each side of the mean to contain two standard deviations. Thus half of 0.0625, or 0.03125, is the desired standard deviation. (See Figure 29.)

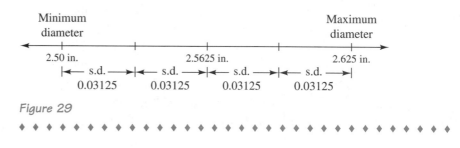

Figure 29

◆ ◆ ◆ ◆ ◆ ◆ ◆ ◆ ◆ ◆ ◆ ◆ ◆ ◆ ◆ ◆ ◆ ◆ ◆ ◆ ◆ ◆ ◆ ◆ ◆ ◆ ◆ ◆ ◆ ◆ ◆ ◆ ◆ ◆ ◆ ◆ ◆ ◆

## EXERCISES 7.6

**Calculate the mean and the median of each set of incomes in Exercises 1 to 4.**

1. $8000, $10,000, $12,000, $13,000, $13,000, $100,000

2. $4000, $8000, $9000, $9000, $100,000

3. $20,000, $27,500, $27,500, $27,500, $27,500

4. $23,000, $25,000, $26,000, $27,000, $29,000

**In Exercises 5 to 8 find $Q_1$ and $Q_3$ for each set and draw a box and whisker plot.**

5. Data from Exercise 1

6. Data from Exercise 2

7. Data from Exercise 3

8. Data from Exercise 4

9. Find the mean absolute deviation for the data in Exercise 1.

10. Find the mean absolute deviation for the data in Exercise 2.

**In Exercises 11 and 12 the numbers represent thousands of dollars.**

11. Find the standard deviation for the data in Exercise 1, using the following equation:

$$s_x = \sqrt{\frac{(8-26)^2 + (10-26)^2 + (12-26)^2 + (13-26)^2 + (13-26)^2 + (100-26)^2}{6-1}}$$

12. Find the standard deviation for the data in Exercise 2, using the following equation:

$$s_x = \sqrt{\frac{(4-26)^2 + (8-26)^2 + (9-26)^2 + (9-26)^2 + (100-26)^2}{5-1}}$$

13. Using a calculator, find the mean, population standard deviation ($\sigma_x$), and sample standard deviation ($s_x$) for the data in Exercise 3.

14. Using a calculator, find the mean, population standard deviation ($\sigma_x$), and sample standard deviation ($s_x$) for the data in Exercise 4.

**The average price of a home in one city in February 1995 was $107,761. This average was calculated from sales of 265 homes. The median home price was $96,500. Use this information in Exercises 15 to 18.**

15. Why might the average be larger than the median?

16. What was the total value of homes sold?

17. Are we able to calculate the standard deviation from the given information?

18. How many homes sold for less than $96,500?

19. What might account for the variability in students' drawings of 3-inch lines in the Warm-up?

20. Why might some students have drawn a 4-centimeter line more accurately than a 3-inch line in the Warm-up?

21. Describe an experiment to determine whether drawings would improve if students were given a second opportunity to draw a 3-inch line in the Warm-up.

22. Describe an experiment to determine whether students are better able to draw a circle the size of a dime if they are given a nickel for reference. What results would you expect?

**In Exercises 23 to 27 suppose the process is in normal operation if the current output or reading is within two standard deviations. For the mean and standard deviation given, calculate the outputs for which the process is normal.**

23. Ceramic furnace temperature: mean = 3500°F, s.d. = 15°F

24. Room heating control: mean = 20°C, s.d. = 0.75°C

25. Milling machine control: mean diameter = 6.075 cm, s.d. = 0.013 cm

26. Pressure control: mean = 84.9 psi, s.d. = 0.28 psi

27. Waiting line time: mean = 2.5 min, s.d. = 0.75 min

**Some controls may be set for normal operation at other than two standard deviations from the mean. Exercises 28 to 31 give the mean, standard deviation, and condition for normal operation for several processes. Give the outputs for which the process is normal.**

28. Steam flow: mean = 3.58 lb/sec, s.d. = 0.04 lb/sec, condition = ±1 s.d.

29. Air pollution: mean = 0.3 ppm (parts per million), s.d. = 0.06 ppm, condition = ±1.5 s.d.

30. Concrete strength: mean = 3000 psi, s.d. = 28 psi, condition = ±2.5 s.d.

31. Gasoline flow: mean = 48.4 gal/min, s.d. = 0.7 gal/min, condition = ±0.5 s.d.

32. Prove that the sum of the differences between the numbers in a set and the mean of the set is zero. (Parts a and b provide hints for part c.)

    a. Suppose that there are just two numbers in the set, $a$ and $b$. The mean is $\dfrac{a + b}{2}$. The sum of the differences is

    $$a - \frac{a + b}{2} + b - \frac{a + b}{2}$$

    Simplify the sum.

    b. Suppose that there are three numbers in the set, $a$, $b$, and $c$. The mean is $\dfrac{a + b + c}{3}$. The sum of the differences is

    $$a - \frac{a + b + c}{3} + b - \frac{a + b + c}{3} + c - \frac{a + b + c}{3}$$

    Simplify the sum.

    c. Suppose that there are $n$ numbers in the set. Set up an expression for the mean and use it to write an expression for the sum of the differences. Show that the sum simplifies to zero.

## PROJECTS

33. a. *Penny Plot.* Plot the data given below, with date on the horizontal axis. (The first number is the date; the second is the weight of the penny in grams.)

    1983D, 2.501; 1994D, 2.510; 1969S, 3.161; 1982D, 2.518; 1972D, 3.107; 1964, 3.070; 1974, 3.130; 1967, 3.135; 1994D, 2.497; 1968D, 3.085; 1960D, 3.111; 1966, 3.100; 1977D, 3.084; 1963, 3.078; 1981D, 3.051; 1985D, 2.515; 1984D, 2.515; 1988D, 2.548; 1984D, 2.628; 1989D, 2.440; 1962, 3.037; 1973D, 3.134; 1979, 3.055; 1970, 3.140; 1991D, 2.538; 1978D, 3.100; 1980D, 3.119

    b. What do you observe from your graph? Find a way to use the mean and standard deviation to justify your observation.

34. *Drawing to Measure.* Do the experiment you described in Exercise 21 or 22. See Exercise 35 for suggestions on writing up your experiment.

35. *Learning a Skill.* Design an experiment based on improving how someone learns a skill. Write out a plan for your experiment and discuss it with your instructor. In your plan, state your idea about the outcome and describe your experiment. Carry out the experiment. Calculate appropriate measures. Summarize the results with graphs or charts. State any conclusions.

---

## CHAPTER 7 SUMMARY

### ◤ Vocabulary

*For definitions and page references, see the Glossary/Index.*

box and whisker plot

converse of the Pythagorean theorem

deviation

difference of squares

dimensions

distance formula

extraneous roots

irrational numbers

mean absolute deviation

parameters

perfect square trinomial

population standard deviation

principal square root

proof

Pythagorean theorem

Pythagorean triples

quadratic formula

quartiles

radicand

range

sample standard deviation

zero product rule

### Concepts

The Pythagorean theorem may be used to determine the lengths of sides of a right triangle, the diagonal of a square or rectangle, the height of an equilateral triangle, and the coordinates of points on some geometric figures placed on coordinate axes.

Two Pythagorean triples are {3, 4, 5} and {5, 12, 13}.

The square root of a negative is undefined in the set of real numbers.

The **principal square root** of a number is the positive root.

A negative root or both positive and negative roots of a number are given only when specified.

The quadratic formula, $x = \dfrac{-b \pm \sqrt{b^2 - 4ac}}{2a}$, solves

$$ax^2 + bx + c = 0$$

The radical property for products is given by

$$\sqrt{a \cdot b} = \sqrt{a} \cdot \sqrt{b}$$

The radical property for quotients is given by

$$\sqrt{\frac{a}{b}} = \frac{\sqrt{a}}{\sqrt{b}}$$

Use absolute value to assure a positive root:

$$\sqrt{a^2} = |a|, \qquad \sqrt{a^6} = |a^3|$$

## CHAPTER 7 REVIEW EXERCISES

1. Which of these sets of numbers represent the sides of a right triangle?

    a. {2, 3, 4}      b. {5, 17, 18}      c. {8, 15, 17}

2. Find the length of each side marked with an $x$.

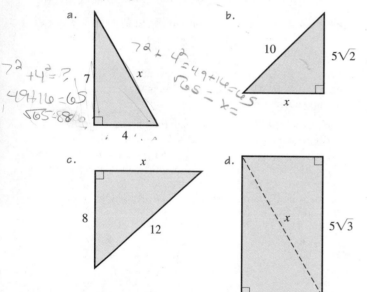

3. An extension ladder is to reach 24 feet up a wall. The safe ladder position for the base is 6 feet from the wall. How long a ladder is needed?

4. Simplify these exponent and radical expressions.

    a. $\sqrt{6} \cdot \sqrt{24}$      b. $(5\sqrt{2})^2$      c. $\sqrt{72} \cdot \sqrt{2}$

    d. $(2\sqrt{5})^2$      e. $\sqrt{3} \cdot \sqrt{12}$

5. Which of the three expressions in each set has the same value as the radical given first?

    a. $\sqrt{60}$      $\{6\sqrt{10},\ 10\sqrt{6},\ 2\sqrt{15}\}$

    b. $\sqrt{63}$      $\{7\sqrt{3},\ 9\sqrt{7},\ 3\sqrt{7}\}$

    c. $\sqrt{54}$      $\{6\sqrt{3},\ 3\sqrt{6},\ 9\sqrt{6}\}$

6. Use the Pythagorean triple shown to find the missing sides of these triangles.

    a. $\{1, 1, \sqrt{2}\}$; 5, 5, $x$

    b. $\{1, 1, \sqrt{2}\}$; $x$, $x$, 8

    c. $\{1, \sqrt{3}, 2\}$; 5, $x$, 10

    d. $\{1, \sqrt{3}, 2\}$; $x$, 8, $y$

7. Find each product. Identify it as a difference of squares (ds), a perfect square trinomial (pst), or neither.

    a. $(x - 4)(x + 4)$

    b. $(x - 2)(x^2 + 2x + 4)$

    c. $(3x + 2)(3x - 2)$

    d. $\left(x - \frac{1}{2}\right)\left(x - \frac{1}{2}\right)$

8. Factor each expression. Identify it as a difference of squares (ds), a perfect square trinomial (pst), or neither.

    a. $x^2 - 49$

    b. $x^2 - 3x + 2$

    c. $4x^2 - 4x + 1$

    d. $x^2 + 6x + 8$

9. Use the distance formula to find the lengths of the diagonals of the four-sided shapes below. (These figures are not drawn to scale.) If the diagonals are equal, then the shape is a rectangle (or square). Which are rectangles?

    a.

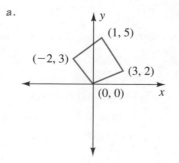

b.

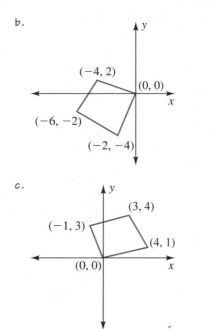

c.

10. The sides of a square or rectangle are perpendicular. Determine the slope of two consecutive sides of each shape in Exercise 9. Verify that the two rectangles do have two consecutive perpendicular sides and the other shape does not.

11. Use the distance formula and then properties of similar triangles to determine coordinate $a$ and the lengths of sides $b$ to $f$ in the triangles in the figures.

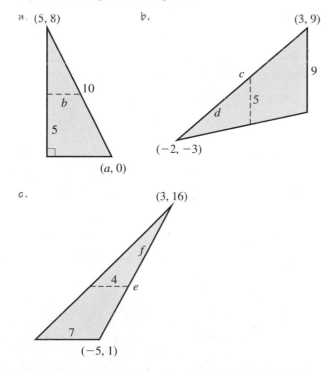

12. Simplify these expressions. Assume that the variables represent only positive numbers.

a. $\sqrt{25x^2y^4}$         b. $\sqrt{169x^6y^2}$

c. $\sqrt{2.25a^3}$         d. $\sqrt{0.64b^5}$

13. How would each answer in Exercise 12 change if the variables represented any real number?

14. Solve these equations. Assume the radicands are positive. Indicate with an inequality or interval inputs that are defined.

a. $\sqrt{7x - 3} = 5$       b. $\sqrt{2 - x} = x - 2$

c. $\sqrt{4x - 3} = 7$

15. Solve for all inputs that make these statements true. The variable $x$ may represent any real number.

a. $x^2 = \frac{25}{49}$         b. $x^2 = \frac{81}{16}$

16. Solve the equations from the graphs. How can we determine which inputs make the radical expressions undefined?

a. $\sqrt{5x + 1} = 11$

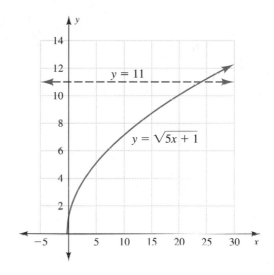

b. $\sqrt{5x - 1} = 7$

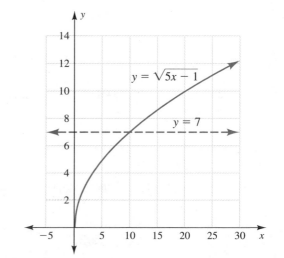

17. Solve for $x$.

a. $x^2 + 3x - 18 = 0$     b. $x^2 - 4x - 21 = 0$

c. $3x^2 = 4x + 7$         d. $4x - 3 + 4x^2 = 0$

18. On the moon the distance seen in miles from a height of $h$ feet is given by

$$d \approx \sqrt{\frac{3h}{8}}$$

   a. How far can be seen from a height of 20 feet?

   b. How high would an astronaut need to climb to see 4 miles?

19. The height of an equilateral triangle passes through the midpoint of the base of the triangle. Show how the Pythagorean theorem is used to find the height of an equilateral triangle with side length $x$. Use your equation for height to find the formula for the area of an equilateral triangle with side $x$.

20. The f-stops on a camera lens are the numbers 1.4, 2, 2.8, 4, 5.6, 8, 11, and 16. Complete the table, and compare the results with the f-stops.

| $n$ | 1 | 2 | 3 | 4 | 5 | 6 | 7 | 8 |
|-----|---|---|---|---|---|---|---|---|
| $(\sqrt{2})^n$ | | | | | | | | |

21. Solve these formulas for the indicated variable.

   a. $A = \pi r^2$ for $r$      b. $p = \frac{1}{2}dv^2$ for $v$

   c. $n = \frac{1}{2l}\sqrt{\frac{T}{m}}$ for $T$      d. $n = \frac{1}{2rl}\sqrt{\frac{T}{\pi d}}$ for $T$

22. In traffic accident investigations, tire skid tests are used to find the coefficient of friction between tires and the road surface near an accident scene. An investigator measures the following tire skid marks, in feet, for a skid at 30 miles per hour:

   *Test 1:*   Left front, 50; right front, 49; left rear, 47; right rear, 48

   *Test 2:*   Left front, 47; right front, 50; left rear, 48; right rear, 51

   a. Find the mean, range, and sample standard deviation ($s_x$) for each of the two tests.

   b. Find the coefficient of friction for each test with $f = \dfrac{S^2}{30D}$, where $f$ is the coefficient of friction, $S$ is the speed of the car making the tests, and $D$ is the mean skid distance of the four tires.

23. The following data give the number of children of each of the presidents of the United States:

   0, 5, 6, 0, 2, 4, 0, 4, 10, 15, 0, 6, 2, 0, 3, 4, 5, 4, 8, 5, 3, 5, 3, 2, 6, 3, 3, 0, 0, 2, 2, 5, 1, 1, 2, 2, 2, 4, 4, 2, 6, 1

   a. Find the median of the data.

   b. Make a box and whisker plot.

   c. If a graphing calculator is available, find the mean and population standard deviation ($\sigma_x$) for the number of children.

### STRETCHERS

24. What is the longest umbrella that can be carried inside a suitcase with interior dimensions, in inches, of 19.5 by 12 by 5?

25. Plot a triangle on coordinate axes. Place the base of the triangle on the positive $x$-axis. Label the endpoints of the base $(0, 0)$ and $(a, 0)$. Choose an arbitrary point $(b, c)$ as the top of the triangle.

   a. Find the midpoints of the sides (not the base) and connect them.

   b. Compare the line connecting the midpoints with the base. Discuss both slope and length.

26. When listing integer factor pairs of a number, how do we know when to stop? (*Hint:* Finish these lists of factor pairs. What relation does the last pair of numbers in the list have to the original number?)

| **48** | **64** | **60** |
|--------|--------|--------|
| $1 \cdot 48$ | $1 \cdot 64$ | $1 \cdot 60$ |
| $2 \cdot 24$ | $2 \cdot 32$ | $2 \cdot 30$ |

   When listing all the numbers that divide evenly into a given number, what is the largest number we need to try?

27. The area of an equilateral triangle is given by

$$A = \frac{x^2\sqrt{3}}{4}$$

   If we double the side, $x$ to $2x$, how does the area change? (*Hint:* Replace $x$ with $2x$ in the area formula. Write and simplify the ratio of the new area to the original area.)

## CHAPTER 7 TEST

1. Which of these sets of numbers represent sides of a right triangle?

   a. $\{3, 4, 5\}$      b. $\{1, 2, 3\}$      c. $\{1, \sqrt{3}, 2\}$

   d. $\{\sqrt{2}, \sqrt{2}, 2\}$      e. $\{\sqrt{3}, \sqrt{4}, \sqrt{5}\}$

2. Which of the three expressions in each set has the same value as the radical given first?

   a. $\sqrt{45}$      $\{9\sqrt{5}, 5\sqrt{9}, 3\sqrt{5}\}$

   b. $\sqrt{44}$      $\{4\sqrt{11}, 2\sqrt{11}, 11\sqrt{4}\}$

3. Romeo plans to use an extension ladder to reach Juliet's balcony, 20 feet above the ground. He will set the base of the ladder 5 feet away from the wall below the edge of the balcony. To preserve this safe ladder position, how long a ladder does he need?

4. Simplify these expressions. Assume that the variables represent positive numbers.

   a. $\sqrt{5} \cdot \sqrt{20}$      b. $\sqrt{3} \cdot \sqrt{27}$      c. $(3\sqrt{6})^2$

   d. $(2\sqrt{7})^2$      e. $\sqrt{36x^2y}$      f. $\sqrt{0.81x^4y^3}$

5. Find each product. Identify each product as a difference of squares (ds), a perfect square trinomial (pst), or neither.

   a. $(x - 5)(x - 5)$      b. $(3x + 8)(3x - 8)$

   c. $(x + 4)^2$

6. Factor each expression, if possible.

   a. $x^2 - 6x + 9$        b. $4x^2 - 25$

7. Solve for all inputs, $x$, that make these equations true. Use an inequality or an interval to indicate inputs that are defined.

   a. $x^2 = \frac{36}{121}$              b. $\sqrt{3 - x} = x + 3$

   c. $x^2 + x - 2 = 0$            d. $\sqrt{5x - 6} = 12$

   e. $2x^2 = 8 - 15x$          f. $8x^2 + 5x = 4$

8. Solve these formulas for the indicated variable.

   a. $E = \frac{1}{2}mv^2$ for $v$        b. $E = \frac{kH^2}{8\pi}$ for $H$

9. Hydroplaning occurs when a tire slides on the surface of the water on a pavement instead of gripping the pavement's surface. The *Advanced Pilot's Flight Manual* gives the relationship between the minimum hydroplaning speed $s$, in miles per hour, and tire pressure $t$, in pounds per square inch, as

$$s = 8.6\sqrt{t}$$

The implication of this formula may not be obvious. Perhaps these thoughts will help: The softer the tire, the greater the surface area and the greater the tendency to slide along the surface, or hydroplane. A harder tire tends to cut through the water to the paved surface.

   a. If the tire pressure is 36 pounds per square inch, what is the speed at which the tire will hydroplane?

   b. If the tire pressure is 100 pounds per square inch, what is the speed at which the tire will hydroplane?

   c. If a plane lands at 120 miles per hour on wet pavement, what is the tire pressure at which hydroplaning will occur?

10. Find the missing sides in the similar right triangles below. Use the marks in the angles to determine which sides are proportional.

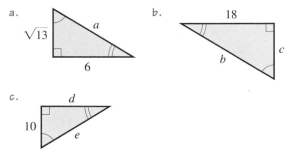

a.   b.   c.

11. The coordinates (3, 5), (6, 8), and (3, 11) form a triangle.

   a. Sketch them on coordinate axes.

   b. What is the length of each side of the triangle?

   c. Use slope to determine whether any two sides are perpendicular.

   d. What kind of a triangle is this triangle? Explain why.

12. The table below gives food energy and sodium content for a variety of dry cereals.

| Dry Cereal | Food Energy (calories) | Sodium Content (milligrams) |
|---|---|---|
| Cap'n Crunch® | 120 | 145 |
| Froot Loops® | 110 | 145 |
| Super Sugar Crisp® | 105 | 25 |
| Sugar Frosted Flakes® | 110 | 230 |
| Sugar Smacks® | 105 | 75 |
| Trix® | 110 | 181 |

   a. Find the mean and sample standard deviation ($s_x$) for food energy.

   b. Find the median and make a box and whisker plot for sodium content.

## Rational

## Expressions

*Figure 1*

**IN SHERMAN COUNTY STRONG WINDS** break open the ripe wheat heads and the grain falls to the ground. Occasionally, the crop is beaten into the ground by hail storms. Thus it is imperative that the wheat be harvested as rapidly as possible after ripening. Although using more than one harvester will speed up the process and save the crop, the equipment is expensive. Individual farmers cannot afford more than one machine.

(continued)

Fortunately, the elevation change in the county is such that the wheat crops in the north end of the county are ready to harvest a month before those in the south end. Several farmers from opposite ends of the county have begun to share harvest crews and equipment.

Farmer Lee's equipment could cut her north-end crop in 12 days. Farmer Terry of the south end of the county estimates he could cut the same acreage in 18 days. If the two farmers work together, how many days will it take to cut Lee's crop? If the farmers also cut Terry's crop, but in a different total time, how will they determine the amount of money needed to make the exchange come out even? We will examine these problems in Section 8.5.

## 8.1  RATIONAL EXPRESSIONS AND APPLICATIONS

**OBJECTIVES**    Identify rational expressions. ♦ Identify inputs that create zero denominators. ♦ Explore graphical behavior near inputs that create undefined expressions. ♦ Use graphs to investigate applications of rational expressions.

**WARM-UP**
1. Solve for $y$: $2x + 2y = 38$.
2. Solve for $y$: $x \cdot y = 48$.
3. Solve for $t$: $r \cdot t = 10$.
4. Solve for $b$: $A = \frac{1}{2}bh$.
5. Solve for $b$: $A = \frac{h}{2}(a + b)$.
6. Complete Table 1 using $y = \frac{x}{x - 5}$.
7. Complete Table 2 using $y = \frac{x + 3}{(x + 2)(x - 3)}$.

| $x$ | $y$ |
|-----|-----|
| $-2$ | |
| $-1$ | |
| $0$ | |
| $1$ | |
| $2$ | |
| $3$ | |
| $4$ | |
| $5$ | |

Table 1

| $x$ | $y$ |
|-----|-----|
| $-2$ | |
| $-1$ | |
| $0$ | |
| $1$ | |
| $2$ | |
| $3$ | |
| $4$ | |
| $5$ | |

Table 2

♦

Т he purpose of this chapter is to provide an overview of operations with
expressions and equations containing fractions. From the student view-
point the material is helpful in three ways.

First, there are occasions when more than one form of an answer is correct, as in solving $A = \dfrac{h}{2}(a + b)$ for $b$. In Sections 8.3 and 8.4 we compare
several solutions and discuss why one might be preferable to another.

Second, many applications involve units of measure containing fractions
or rates. In problem solving we sometimes start with the units given in the answer and work backwards from this answer to select formulas or to understand
processes. Sections 8.2 and 8.3 will review operations with units of measure
that involve fractions or rates.

Third, there are times when the textbook answer indicates that there is *no
real number solution* to a problem. We have seen *undefined* or *no real number
solution* in two situations: the square root of a negative number (Section 7.1)
and division by zero (Sections 2.2, 3.6, and 4.2). In this section we return to
division by zero to look at its graphical implications.

## DIVISION BY ZERO

Because **rational numbers** *are the set of numbers containing fractions*, we
define **rational expressions** as *the set of symbolic expressions containing
fractions*. Similarly, we define **rational equations** as *equations containing
fractions*. These definitions are informal. A more formal definition of rational
expressions would exclude fractions that have only whole numbers in the denominator. We will include expressions containing whole numbers in the denominators because the symbolic technique is the same.

One of the simplest rational expressions is $\dfrac{1}{x}$. The variable $x$ is in the denominator. In some situations we might want to substitute a zero for $x$. However, we cannot have a denominator equal to zero. A zero denominator would
imply division by zero, which is an undefined operation. Thus we must identify
inputs that create a zero denominator and exclude them from our set of inputs.

**E X A M P L E   1**

For what inputs will each fractional expression be undefined?

**a.** $\dfrac{1}{x}$   **b.** $\dfrac{x}{x - 5}$   **c.** $\dfrac{x + 3}{(x + 2)(x - 3)}$

**SOLUTION**   Inputs that give zero denominators lead to undefined expressions.

**a.** When $x = 0$,

$$\frac{1}{x} = \frac{1}{0}$$

**b.** When $x = 5$,

$$\frac{x}{x - 5} = \frac{5}{5 - 5} = \frac{5}{0}$$

**c.** When $x = -2$,

$$\frac{x + 3}{(x + 2)(x - 3)} = \frac{-2 + 3}{(-2 + 2)(-2 - 3)} = \frac{1}{(0)(-5)} = \frac{1}{0}$$

or when $x = 3$,

$$\frac{x + 3}{(x + 2)(x - 3)} = \frac{3 + 3}{(3 + 2)(3 - 3)} = \frac{6}{(5)(0)} = \frac{6}{0}$$

◆ ◆ ◆ ◆ ◆ ◆ ◆ ◆ ◆ ◆ ◆ ◆ ◆ ◆ ◆ ◆ ◆ ◆ ◆ ◆ ◆ ◆ ◆ ◆ ◆ ◆ ◆ ◆ ◆ ◆ ◆ ◆ ◆ ◆ ◆

GRAPHS OF RATIONAL EXPRESSIONS

If an expression is undefined for a certain input, there is no point on the expression's graph for that input. Furthermore, the graph near such a point has unusual features. The equation we examine in Example 2 was presented in the discussion of proportions in Section 4.2.

**EXAMPLE 2**    What happens to the graph of $y = \dfrac{2}{x}$ as $x$ gets close to zero?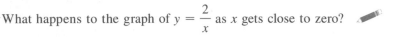

SOLUTION    Tables 3 and 4 show outputs for $y = \dfrac{2}{x}$ as $x$ gets close to zero. The data in each table form a curve when graphed (see Figure 2). If we trace the third-quadrant curve from left to right, we find that as the inputs, $x$, approach zero, the curve turns downward. Although the curve gets close to the $y$-axis, it never touches the $y$-axis. We describe this behavior by saying *y approaches negative infinity as x approaches zero*.

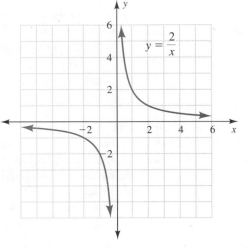

Figure 2

| Input: $x$ | Output: $y = 2/x$ |
|:---:|:---:|
| 5 | $2/5 = 0.4$ |
| 4 | $2/4 = 0.5$ |
| 2 | $2/2 = 1.0$ |
| 1 | $2/1 = 2.0$ |
| 0.5 | $2/0.5 = 4.0$ |
| 0.25 | $2/0.25 = 8.0$ |
| 0.1 | $2/0.1 = 20.0$ |

**Table 3** $y = 2/x$ *as x approaches zero from the right*

| Input: $x$ | Output: $y = 2/x$ |
|:---:|:---:|
| $-5$ | $2/-5 = -0.4$ |
| $-4$ | $2/-4 = -0.5$ |
| $-2$ | $2/-2 = -1.0$ |
| $-1$ | $2/-1 = -2.0$ |
| $-0.5$ | $2/-0.5 = -4.0$ |
| $-0.25$ | $2/-0.25 = -8.0$ |
| $-0.1$ | $2/-0.1 = -20.0$ |

**Table 4** $y = 2/x$ *as x approaches zero from the left*

If we trace the first-quadrant graph from right to left, we see that as the inputs, $x$, approach zero, the curve rises. As the curve rises, it gets closer to the $y$-axis, but, like the third-quadrant curve, it never touches the $y$-axis. We describe this behavior by saying *y approaches positive infinity as x approaches zero*.

Because the equation $y = \dfrac{2}{x}$ has no output at $x = 0$, we say *the equation is undefined at x = 0*.

◆ ◆ ◆ ◆ ◆ ◆ ◆ ◆ ◆ ◆ ◆ ◆ ◆ ◆ ◆ ◆ ◆ ◆ ◆ ◆ ◆ ◆ ◆ ◆ ◆ ◆ ◆ ◆ ◆ ◆

> The graph of a simplified rational expression approaches infinity in a vertical direction whenever the denominator approaches zero.

Together the two curves in Figure 2 form a *rectangular hyperbola*. The name comes from the fact that the curve approaches but does not intersect the rectangular coordinate axes. Look for these curves and changes in their position in the remainder of this chapter.

EXAMPLE  **3**    Make a table and graph for $y = \dfrac{-2}{x + 4}$. For what inputs is the expression undefined? What do you observe about the behavior of the graph near that input?

SOLUTION    The graph in Figure 3 becomes nearly vertical as we approach $x = -4$ from either the left (Table 5) or the right (Table 6).

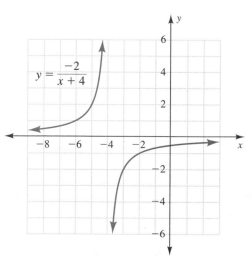

$$y = \frac{-2}{x + 4}$$

Figure 3

| $x$ | $y$ |
| --- | --- |
| $-9$ | 0.4 |
| $-8$ | 0.5 |
| $-7$ | 0.67 |
| $-6$ | 1.0 |
| $-5$ | 2.0 |
| $-4$ | undefined |

Table 5

| $x$ | $y$ |
| --- | --- |
| $-3$ | $-2$ |
| $-2$ | $-1$ |
| $-1$ | $-0.67$ |
| 0 | $-0.5$ |
| 1 | $-0.4$ |
| 2 | $-0.33$ |

Table 6

◆ ◆ ◆ ◆ ◆ ◆ ◆ ◆ ◆ ◆ ◆ ◆ ◆ ◆ ◆ ◆ ◆ ◆ ◆ ◆ ◆ ◆ ◆ ◆ ◆ ◆ ◆ ◆ ◆

Before we examine applications with rational expressions, two facts should be noted about calculators and zero denominators. First, as we observed earlier, all calculators are designed to reject an attempt to divide by zero. When we trace along the graph of a rational expression such as the one in Example 2 or Example 3, the output will be blank each time the input creates a zero denominator. It often takes considerable adjustment of the viewing window to locate such an input, however.

Second, in certain viewing window settings the calculator draws an almost vertical line on the graph at an undefined point. This line is an error made by the calculator. The calculator evaluates inputs to the left and right of the undefined point and connects the outputs. The line will disappear if the calculator is set on dot mode rather than connected mode or if the viewing window is adjusted to force the calculator to evaluate the input that makes the expression undefined.

Although you need not be concerned about it at this point in your study, there is a line associated with an undefined input for rational expressions. When the number $n$ makes the denominator in a simplified rational expression zero, we say the line with equation $x = n$ is a *vertical asymptote*. *Asymptote* is the name given to the line that a graph approaches. We will focus on the numbers that create zero denominators rather than on the asymptotes.

### APPLICATIONS

In our applications of rational equations we return to two settings—distance and geometry (area and perimeter)—and introduce two new settings—resource management and light intensity.

In Section 3.1 we examined two linear equations based on $D = rt$. The first linear equation had a rate of 55 miles per hour: $D = 55t$. The second linear equation had a time of 3 hours: $D = 3r$. In these linear equations we identified distance as output and either rate or time as input. We cannot assume that the relationship among the variables $D$, $r$, and $t$ will always be linear. Suppose we let distance be a fixed number (or, as we usually say, *hold distance constant*) and examine the behavior of the two variables, rate and time.

EXAMPLE   **4**      *Rate of exercise*    The rates in Table 7, of 1 to 10 miles per hour, range from a slow walk to a leisurely bicycling speed.

**a.** Complete Table 7.

| Rate, $r$ mph | Time, $t$ hours | Distance, $D = r \cdot t$ |
|:---:|:---:|:---:|
| 1 | 10 | |
| 2 | 5 | |
| 5 | 2 | |
| 10 | 1 | |

*Table 7*  *Variable rate and time*

**b.** Set up a graph, with rate along the $x$-axis and time along the $y$-axis. Label both axes from 0 to 12. Graph the points $(r, t)$ from the table. Connect the points from left to right.

**c.** What do you observe about the graph?

**d.** Change $r \cdot t = 10$ to a rational equation that uses $r$ as input and $t$ as output.

SOLUTION   **a.** The third column is 10 miles for all entries.

**b.** The graph is shown in Figure 4.

**c.** The graph of the points $(r, t)$ in Figure 4 does not make a straight line. Thus the relationship between rate and time is not linear when the product (distance) is constant.

**d.** The third column in Table 7 is $D = r \cdot t$. The relationship between $r$ and $t$ is $r \cdot t = 10$. If we solve for $t$ as output with $r$ as input, the equation becomes

$$t = \frac{10}{r}$$

The expression $\dfrac{10}{r}$ is undefined for $r = 0$.

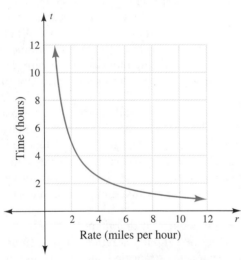

Figure 4

◆ ◆ ◆ ◆ ◆ ◆ ◆ ◆ ◆ ◆ ◆ ◆ ◆ ◆ ◆ ◆ ◆ ◆ ◆ ◆ ◆ ◆ ◆ ◆ ◆ ◆ ◆ ◆ ◆ ◆ ◆ ◆ ◆

In the next example we examine the application of rational expressions in resource management. Many resources, such as minerals, are limited in quantity. Strategic planners estimate the total quantity available and attempt to plan for future shortages.

EXAMPLE   **5**      *Resource management*    Silver is used to produce electrical and electronic products and photographic film, as well as flatware and jewelry. The estimated world reserves of silver were 420,000 metric tons in 1990.

**a.** If the rate of use of silver reserves is $x$ metric tons per day, write an equation for how many days the reserves will last. Write an equation for how many years the reserves will last.

**b.** If the 1990 world production of silver was 40 metric tons per day, how many years will the 1990 reserves last?

**c.** Graph the equation, letting $x$ be use in metric tons per day and $y$ be supply in years.

SOLUTION   **a.** The reserves will last $y$ days, as shown by the equation

$$y = \frac{420{,}000}{x}$$

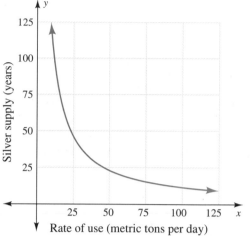

Figure 5

and $y$ years, as shown by the equation

$$y = \frac{420{,}000}{365x}$$

**b.** If the world silver reserves are used up at the rate of 40 metric tons per day, the 1990 reserves will last $y$ years:

$$y = \frac{420{,}000 \text{ metric tons}}{\dfrac{40 \text{ metric tons}}{1 \text{ day}} \cdot \dfrac{365 \text{ days}}{1 \text{ yr}}}$$

$$y = 28.8 \text{ yr}$$

**c.** The graph of $y = \dfrac{420{,}000}{365x}$ is shown in Figure 5.

◆ ◆ ◆ ◆ ◆ ◆ ◆ ◆ ◆ ◆ ◆ ◆ ◆ ◆ ◆ ◆ ◆ ◆ ◆ ◆ ◆ ◆ ◆ ◆ ◆ ◆ ◆ ◆ ◆ ◆ ◆ ◆ ◆

In Sections 8.2 and 8.3 we will practice working with expressions containing units, such as those in part b of Example 5. For now we limit explanation of the simplification in Example 5 to observing that *metric tons* and *days* both cancel and only *years* remain.

In Section 5.2 we determined the length and width of a rectangle with area 48 square centimeters and perimeter 38 centimeters. We repeat the example here and investigate the graphical solution of the resulting system of equations.

**E X A M P L E   6**

Find the length, $x$, and width, $y$, of a rectangle with area 48 square centimeters and perimeter 38 centimeters. Write an equation for the area and another for the perimeter. Use the graphs of the equations to find the length and width that satisfy both equations.

**SOLUTION**   The area is $xy = 48$, or $y = \dfrac{48}{x}$. The perimeter is $2x + 2y = 38$, or $y = 19 - x$. The graphs are shown in Figure 6. There are two intersections: $(16, 3)$ and $(3, 16)$.

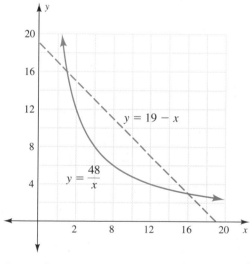

Figure 6

***Check:*** Area $= 16(3) = 3(16) = 48 \text{ cm}^2$
              Perimeter $= 2(16) + 2(3) = 38 \text{ cm}$

Thus the length and width are 16 centimeters and 3 centimeters, respectively. The mathematical results do not specify that length be the larger number, as is customary in everyday use.   ✔

◆ ◆ ◆ ◆ ◆ ◆ ◆ ◆ ◆ ◆ ◆ ◆ ◆ ◆ ◆ ◆ ◆ ◆ ◆ ◆ ◆ ◆ ◆ ◆ ◆ ◆ ◆ ◆ ◆ ◆ ◆ ◆ ◆ ◆ ◆

## INVERSE VARIATION

Equations such as $y = \dfrac{a}{x}$ or $y = \dfrac{a}{x^2}$ illustrate **inverse variation**, or **inverse proportions**. *Two variables vary inversely if one gets smaller as the other gets larger.* In Example 4, for a fixed distance, the rate and time vary inversely. In Example 5, for a fixed amount of silver resources, the amount of daily use and the number of years the supply will last vary inversely. In Example 6, for a fixed area of 48 square centimeters, the length and width of the rectangle vary inversely. In the next example the intensity of light and its distance from the source vary inversely. The farther we are from a light source, the less intense is the light received from that source.

**EXAMPLE 7**   **Light intensity**   Light intensity, $I$, varies inversely with the square of the distance, $d$, from the source of the light. The formula is

$$I = \frac{k}{d^2}$$

where $k$ is a constant that changes the units from square feet to those needed for light intensity. For simplicity we use $k = 1$ here. Make a table and graph for light intensity in terms of distance from the light source.

**SOLUTION**   The values for light intensity are shown in Table 8. The graph in Figure 7 is somewhat the same shape as earlier graphs, but because the distance variable is squared, this curve is not as symmetrically placed in the first quadrant.

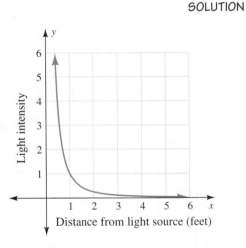

Figure 7

| Distance in Feet, $d$ | Light Intensity, $I$ |
|:---:|:---:|
| 1 | 1 |
| 2 | $\frac{1}{4}$ |
| 3 | $\frac{1}{9}$ |
| 4 | $\frac{1}{16}$ |
| 5 | $\frac{1}{25}$ |

Table 8

◆ ◆ ◆ ◆ ◆ ◆ ◆ ◆ ◆ ◆ ◆ ◆ ◆ ◆ ◆ ◆ ◆ ◆ ◆ ◆ ◆ ◆ ◆ ◆ ◆ ◆ ◆ ◆ ◆ ◆ ◆ ◆ ◆ ◆

Project 31 describes a light intensity activity related to Example 7. You can experiment with light intensity without any measuring device. In a darkened room with only one light source, hold a book 1 foot from the bulb and read. Then move to 2 feet away and read. Move to 3 feet away and read. Observe how quickly it becomes difficult to read the book.

## EXERCISES 8.1

For what inputs will each expression in Exercises 1 to 4 be undefined?

1. $\dfrac{2}{x + 1}$

2. $\dfrac{3}{x - 1}$

3. $\dfrac{x}{2x - 1}$

4. $\dfrac{x}{2x + 3}$

Using a calculator to build a table and a graph or using a graphing calculator, describe the behavior of the graph of each equation in Exercises 5 to 8 as $x$ approaches the indicated number.

5. $y = \dfrac{x}{x - 5}$, as $x$ approaches 5

6. $y = \dfrac{-x}{x+3}$, as $x$ approaches $-3$

7. $y = \dfrac{x+3}{(x+2)(x-3)}$, as $x$ approaches $-2$

8. $y = \dfrac{x+3}{(x+2)(x-3)}$, as $x$ approaches $3$

9. In Chapter 1 we explored the product of positive and negative numbers with these exercises.

   a. Fill in the missing numbers:

   $-3 \cdot \underline{\hspace{1cm}} = -12$      $-4 \cdot \underline{\hspace{1cm}} = -12$
   $-4 \cdot \underline{\hspace{1cm}} = 12$       $3 \cdot \underline{\hspace{1cm}} = -12$
   $4 \cdot \underline{\hspace{1cm}} = -12$       $-3 \cdot \underline{\hspace{1cm}} = 12$
   $-2 \cdot \underline{\hspace{1cm}} = -12$      $-6 \cdot \underline{\hspace{1cm}} = 12$
   $2 \cdot \underline{\hspace{1cm}} = -12$       $-6 \cdot \underline{\hspace{1cm}} = -12$
   $-2 \cdot \underline{\hspace{1cm}} = 12$       $6 \cdot \underline{\hspace{1cm}} = -12$

   b. Plot coordinates with $x \cdot y = 12$ on one graph.

   c. Plot coordinates with $x \cdot y = -12$ on another graph.

   d. On each graph find and label four additional coordinates that fit the rule $x \cdot y = 12$ or $x \cdot y = -12$. Connect the points from left to right.

   e. Should the graphs be connected across the $y$-axis?

   f. What do you observe about the graphs?

   g. Is there a number that, when multiplied by zero, gives 12 or $-12$? Will either graph have a point on the $x$-axis?

10. a. Complete the table, and graph the points formed by the ordered pairs $(x, y)$.

| $x$ | $y$ | $xy$ |
|-----|-----|------|
| $-20$ | $-1$ | |
| $-10$ | $-2$ | |
| $-5$ | $-4$ | |
| $-4$ | | $20$ |
| | $-10$ | $20$ |
| $-1$ | $-20$ | |

   b. Connect the points from left to right.

   c. Find six ordered pairs of positive real numbers that give $y = \dfrac{20}{x}$. Graph them on the same axes as in part b.

   d. Write a statement describing the shape of the graphs. Are there any $x$- or $y$-intercepts? Why?

11. Estimated world crude oil reserves in 1990 were about 1000 billion barrels. If the estimated world production of crude oil is $x$ barrels per day, what equation describes how long the 1990 oil reserves will last?

12. A potential landfill site contains 1,000,000 cubic yards of space. If the estimated fill per day is $x$ cubic yards, what equation describes how long the landfill site may be used?

13. World production of crude oil uses up oil reserves. If crude oil production in 1990 was 60 million barrels per day, how many years will the oil reserves last? (See Exercise 11.)

14. If a city with a population of 100,000 produces 600 cubic yards of garbage each day, how many years will the landfill in Exercise 12 last?

15. A two-year college has financial aid available for a total of 90 credits. The input, $x$, is the number of credits taken each term.

   a. What is the meaning of the output, $y$, for the equation $y = \dfrac{90}{x}$?

   b. Describe the problem situation if $x = 12$ credits per term.

16. A student earns a total of $2800 over summer vacation. The input, $x$, is dollars spent per month.

   a. What is the meaning of the output, $y$, for the equation $y = \dfrac{2800}{x}$?

   b. Describe the problem situation if $x = \$400$.

17. In Example 4, $r = 0$ makes the equation $t = \dfrac{10}{r}$ undefined. How do we interpret $r = 0$ in the problem setting?

18. In Example 4, Figure 4 shows a first-quadrant graph only. Give five $(r, t)$ coordinates from other quadrants that make the equation $t = \dfrac{10}{r}$ true. Why are these points not relevant to the problem situation?

19. In the equation $t = \dfrac{10}{r}$ is there an input $r$ that makes $t = 0$?

20. As time, $t$, changes from 0.1 to 0.01 hour, what happens to the rate, $r$, in the equation $t = \dfrac{10}{r}$?

21. a. In the table list possible widths and lengths of rectangles with area of 30 square inches. Calculate the perimeter for each rectangle.

| Width | Length | Area | Perimeter |
|-------|--------|------|-----------|
| | | 30 | |
| | | 30 | |
| | | 30 | |
| | | 30 | |
| | | 30 | |
| | | 30 | |
| | | 30 | |

   b. Graph the length ($x$-axis) and the perimeter ($y$-axis).

   c. Where is the location of the point on the graph describing the smallest perimeter?

   d. Find the equation (perimeter in terms of length, $x$) describing the graph.

22. Refer to the table completed in Exercise 21.

a. Graph the length and width pairs on coordinate axes, placing width on the $x$-axis and length on the $y$-axis. What shape is formed?

b. What length and width give a perimeter smaller than 22 for an area of 30?

c. Find the equation (length in terms of width) describing the graph.

The equations $y = \dfrac{1}{x}$ and $y = \dfrac{1}{x^2}$ are graphed in the figure below. Refer to this figure in Exercises 23 to 26.

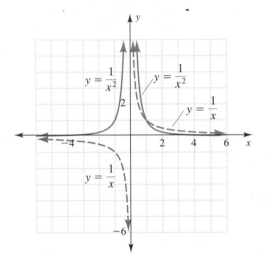

23. Draw the line of symmetry for each graph and write its equation.

24. Why does $y = \dfrac{1}{x^2}$ not have points in the third quadrant?

25. The coordinates $\left(2, \dfrac{1}{2}\right)$ and $\left(\dfrac{1}{2}, 2\right)$ both lie on $y = \dfrac{1}{x}$. Is there such a reversal for all points on the graph? Why may the coordinate be reversed?

26. For inputs $x > 1$, which graph is on top: $y = \dfrac{1}{x}$ or $y = \dfrac{1}{x^2}$? Why?

27. In Example 7 are there other coordinate points that make $I = \dfrac{1}{d^2}$ true? Would these points have any meaning in the light intensity problem situation?

28. Make a graph of the equation $I = \dfrac{1}{d^2}$, from Example 7, for $d = -5$ to $d = 5$. Describe the symmetry in the graph.

29. The force due to gravitational attraction between planets varies inversely with the square of the distance between the planets. Which equation would describe gravitational attraction?

a. $F = km_1m_2d^2$    b. $F = \dfrac{km_1m_2}{d}$    c. $F = \dfrac{km_1m_2}{d^2}$

30. Under certain circumstances the intensity of a magnetic field varies inversely with the cube of the distance from the magnet. Which equation would describe the intensity?

a. $H = \dfrac{M}{r^2}$    b. $H = \dfrac{M}{r^3}$    c. $H = Mr^3$    d. $H = Mr^2$

PROJECT

31. **Measuring Light Intensity.** Light intensity is measured by the formula $I = \dfrac{k}{d^2}$, where $k$ is a constant. Obtain a light meter or other device for measuring light intensity. In a darkened room measure light intensity at several distances from a single light source (such as a bare light bulb).

a. Construct a three-column table. Record your distance, $d$, and intensity, $I$, data in the first two columns.

b. In the third column calculate $k = I \cdot d^2$.

c. Plot your data on a graph, with distance on the horizontal axis and intensity on the vertical axis.

d. Comment on how accurately your experiment reflects the formula. In a perfect experiment, how should the $k$ numbers in the table compare?

e. What factors might change the accuracy of your results?

---

## 8.2    SIMPLIFYING RATIONAL EXPRESSIONS

**OBJECTIVES**    Simplify rational expressions. ♦ Simplify rational expressions containing units of measurement. ♦ Determine when a rational expression simplifies to 1 or to −1. ♦ Add, subtract, multiply, and divide fractions.

**WARM-UP**   Factor these binomials and trinomials by guess-and-check or the table method.

**1.** $x^2 - 4$          **2.** $x^2 + 4x - 5$          **3.** $2x^2 + 7x + 6$

**4.** $x^2 + 2x$          **5.** $x^2 - 5x + 6$          **6.** $2x^2 + 6x$

**7.** $x^2 + 3x + 2$          **8.** $x^2 + 6x + 9$          **9.** $4x^2 + 6x$

**10.** $x^2 - 9$          **11.** $6 - x - x^2$          **12.** $4 - 3x - x^2$          ◆

In this section we simplify rational expressions containing variables and units of measurement. We investigate the simplified result when the numerator is the opposite of the denominator, and we review operations with fractions.

### SIMPLIFICATION OF RATIONAL EXPRESSIONS

As mentioned in Section 2.2, when we simplify a fraction to an equivalent fraction, we are using the **simplification property of fractions**.

**Simplification Property of Fractions**

> For all real numbers, $a$ not zero and $c$ not zero,
>
> $$\frac{ab}{ac} = \frac{a}{a} \cdot \frac{b}{c} = 1 \cdot \frac{b}{c} = \frac{b}{c}$$

When we simplify a fraction we factor the numerator and denominator and remove the common factors, such as $\frac{a}{a}$. If there are no common factors, the fraction cannot be simplified. We now include factoring polynomials within our work in simplifying fractions.

**EXAMPLE   1**   Simplify the following:

**a.** $\dfrac{2a}{a^2}, a \neq 0$          **b.** $\dfrac{6xy}{2y^2}, y \neq 0$

**c.** $\dfrac{x - 3}{(x + 3)(x - 3)}, x \neq -3$ and $x \neq 3$          **d.** $\dfrac{x + x^2}{1 + x}, x \neq -1$

**e.** $\dfrac{x^2 - 4}{x^2 + 3x + 2}, x \neq -2$ and $x \neq -1$          **f.** $\dfrac{x^2 + 4x - 5}{x^2 - 2x + 1}, x \neq 1$ ✎

**SOLUTION**   **a.** $\dfrac{2a}{a^2} = \dfrac{2 \cdot a}{a \cdot a} = \dfrac{2}{a}$

**b.** $\dfrac{6xy}{2y^2} = \dfrac{2 \cdot 3 \cdot x \cdot y}{2 \cdot y \cdot y} = \dfrac{3x}{y}$

**c.** $\dfrac{x - 3}{(x + 3)(x - 3)} = \dfrac{1(x - 3)}{(x + 3)(x - 3)} = \dfrac{1}{(x + 3)}$

**d.** $\dfrac{x + x^2}{1 + x} = \dfrac{x(1 + x)}{(1 + x)} = \dfrac{x}{1} = x$

**e.** $\dfrac{x^2 - 4}{x^2 + 3x + 2} = \dfrac{(x + 2)(x - 2)}{(x + 1)(x + 2)} = \dfrac{(x - 2)}{(x + 1)}$

**f.** $\dfrac{x^2 + 4x - 5}{x^2 - 2x + 1} = \dfrac{(x + 5)(x - 1)}{(x - 1)(x - 1)} = \dfrac{(x + 5)}{(x - 1)}$

◆ ◆ ◆ ◆ ◆ ◆ ◆ ◆ ◆ ◆ ◆ ◆ ◆ ◆ ◆ ◆ ◆ ◆ ◆ ◆ ◆ ◆ ◆ ◆ ◆ ◆ ◆ ◆ ◆ ◆ **383**

Another way to change a fraction to an equivalent fraction is to use the simplification property in reverse. We multiply the numerator and denominator of the fraction by a common factor.

**Equivalent Fraction Property**

> For all real numbers, $a$ not zero and $c$ not zero,
>
> $$\frac{b}{c} = \frac{b}{c} \cdot \frac{a}{a} = \frac{ab}{ac}$$

**EXAMPLE 2**   Change each of the following to an equivalent expression with the indicated numerator or denominator. Indicate which numbers must be excluded from the inputs.

a. $\dfrac{4}{5} = \dfrac{12}{\phantom{xx}}$    b. $\dfrac{a}{2} = \dfrac{\phantom{xx}}{2b}$    c. $\dfrac{3}{x} = \dfrac{3x}{\phantom{xx}}$    d. $\dfrac{2}{x+2} = \dfrac{\phantom{xx}}{2x(x+2)}$

e. $\dfrac{x}{x+2} = \dfrac{\phantom{xx}}{x^2 + 3x + 2}$

**SOLUTION**  a. $\dfrac{4}{5} = \dfrac{4 \cdot 3}{5 \cdot 3} = \dfrac{12}{15}$

b. $\dfrac{a}{2} = \dfrac{a \cdot b}{2 \cdot b} = \dfrac{ab}{2b}, \ b \neq 0$

c. $\dfrac{3}{x} = \dfrac{3 \cdot x}{x \cdot x} = \dfrac{3x}{x^2}, \ x \neq 0$

d. $\dfrac{2}{x+2} = \dfrac{2x \cdot 2}{2x(x+2)} = \dfrac{4x}{2x(x+2)}, \ x \neq 0 \text{ and } x \neq -2$

e. $\dfrac{x}{x+2} = \dfrac{x(x+1)}{(x+2)(x+1)} = \dfrac{x^2+x}{x^2+3x+2}, \ x \neq -2 \text{ and } x \neq -1$

◆ ◆ ◆ ◆ ◆ ◆ ◆ ◆ ◆ ◆ ◆ ◆ ◆ ◆ ◆ ◆ ◆ ◆ ◆ ◆ ◆ ◆ ◆ ◆ ◆ ◆ ◆ ◆ ◆ ◆ ◆ ◆ ◆ ◆

We are not always told whether to simplify a fraction or to change it to an equivalent fraction in unsimplified form. Example 3 uses both operations.

**EXAMPLE 3**   Find the missing number in each pair of fractions. Indicate whether the first fraction was simplified to the second or changed to an equivalent fraction in unsimplified form.

a. $\dfrac{5}{8}, \dfrac{10}{\phantom{xx}}$    b. $\dfrac{8}{6}, \dfrac{4}{\phantom{xx}}$    c. $\dfrac{2}{5}, \dfrac{\phantom{xx}}{15}$    d. $\dfrac{4}{3}, \dfrac{12}{\phantom{xx}}$

e. $\dfrac{5x}{10x}, \dfrac{\phantom{xx}}{2}$    f. $\dfrac{3a}{6a^2}, \dfrac{\phantom{xx}}{12a^3}$

**SOLUTION**  a. $\dfrac{5}{8} = \dfrac{5 \cdot 2}{8 \cdot 2} = \dfrac{10}{16}$; not simplified    b. $\dfrac{8}{6} = \dfrac{2 \cdot 4}{2 \cdot 3} = \dfrac{4}{3}$; simplified

c. $\dfrac{2}{5} = \dfrac{2 \cdot 3}{5 \cdot 3} = \dfrac{6}{15}$; not simplified    d. $\dfrac{4}{3} = \dfrac{4 \cdot 3}{3 \cdot 3} = \dfrac{12}{9}$; not simplified

e. $\dfrac{5x}{10x} = \dfrac{5 \cdot x}{2 \cdot 5 \cdot x} = \dfrac{1}{2}$; simplified

f. $\dfrac{3a}{6a^2} = \dfrac{3a \cdot 2a}{6a^2 \cdot 2a} = \dfrac{6a^2}{12a^3}$; not simplified

◆ ◆ ◆ ◆ ◆ ◆ ◆ ◆ ◆ ◆ ◆ ◆ ◆ ◆ ◆ ◆ ◆ ◆ ◆ ◆ ◆ ◆ ◆ ◆ ◆ ◆ ◆ ◆ ◆ ◆ ◆ ◆ ◆ ◆

As noted earlier, many expressions involve units of measurement. The simplification property indicates that fractions containing units, such as

$$\frac{\text{meters}}{\text{meters}}, \quad \frac{\text{inches}}{\text{inches}}, \quad \frac{\text{gallons}}{\text{gallons}}, \quad \frac{\text{miles}}{\text{miles}}, \quad \text{or} \quad \frac{\text{hours}}{\text{hours}}$$

all simplify to 1.

EXAMPLE  4

Simplify the following:

a. $\dfrac{48 \text{ cm}^3}{16 \text{ cm}}$   b. $\dfrac{5000 \text{ foot-pounds}}{10 \text{ feet}}$   c. $\dfrac{24 \text{ degree days}}{6 \text{ days}}$

SOLUTION

a. $\dfrac{48 \text{ cm}^3}{16 \text{ cm}} = \dfrac{16 \cdot 3 \text{ cm} \cdot \text{cm} \cdot \text{cm}}{16 \text{ cm}} = \dfrac{3 \text{ cm}^2}{1}$

b. $\dfrac{5000 \text{ foot-pounds}}{10 \text{ feet}} = \dfrac{500 \cdot 10 \text{ foot-pounds}}{10 \text{ feet}} = \dfrac{500 \text{ pounds}}{1}$

c. $\dfrac{24 \text{ degree days}}{6 \text{ days}} = \dfrac{6 \cdot 4 \text{ degree days}}{6 \text{ days}} = \dfrac{4 \text{ degrees}}{1}$

♦ ♦ ♦ ♦ ♦ ♦ ♦ ♦ ♦ ♦ ♦ ♦ ♦ ♦ ♦ ♦ ♦ ♦ ♦ ♦ ♦ ♦ ♦ ♦ ♦ ♦ ♦ ♦ ♦ ♦ ♦

The concept of opposites is central to the next two examples, so we restate the definition here.

Definition

> Two expressions are **opposites** if they add to zero.

We use opposites as we look at some special forms of simplifying. The examples provide a shortcut and a caution.

EXAMPLE  5

Simplify the following:

a. $\dfrac{3-4}{4-3}$   b. $\dfrac{5-(-2)}{-2-5}$

SOLUTION

The numerator and denominator in each fraction are opposites. Both fractions simplify to $-1$.

a. $\dfrac{3-4}{4-3} = \dfrac{-1}{1} = -1$   b. $\dfrac{5-(-2)}{-2-5} = \dfrac{7}{-7} = -1$

♦ ♦ ♦ ♦ ♦ ♦ ♦ ♦ ♦ ♦ ♦ ♦ ♦ ♦ ♦ ♦ ♦ ♦ ♦ ♦ ♦ ♦ ♦ ♦ ♦ ♦ ♦ ♦ ♦ ♦

CAUTION   Some expressions, such as $a - b$ and $a + b$, appear to be opposites. However, the sum of $a - b$ and $a + b$ is $2a$, not zero, so the expressions are not opposites. The opposite of $a - b$ is $-a + b$ or $b - a$.  ♦

> ♦ When the numerator and denominator of a fraction are the same, the fraction simplifies to 1.
> ♦ When the numerator and denominator of a fraction are opposites, the fraction simplifies to $-1$.

Example 6 demonstrates why rational expressions containing opposite numerators and denominators simplify to $-1$.

EXAMPLE **6**    Show that $\dfrac{a-b}{b-a} = -1$, $a \neq b$.

SOLUTION    At least three methods are possible.

***Method 1***: Multiply the numerator and denominator of the fraction by $-1$:

$$\frac{a-b}{b-a} = \frac{(-1)(a-b)}{(-1)(b-a)} = \frac{(-1)(a-b)}{(a-b)} = -1$$

***Method 2***: Multiply either the numerator or the denominator by $(-1)(-1)$, which equals 1 and will not change the fraction:

$$\frac{a-b}{b-a} = \frac{(-1)(-1)(a-b)}{(b-a)} = \frac{(-1)(b-a)}{(b-a)} = -1$$

***Method 3***: Factor $-1$ from either the numerator or the denominator:

$$\frac{a-b}{b-a} = \frac{(-1)(b-a)}{(b-a)} = -1$$

With the first two methods we multiplied out only one of the $-1$ factors. This changed one of the expressions to its opposite and thus permitted simplification.

♦ ♦ ♦ ♦ ♦ ♦ ♦ ♦ ♦ ♦ ♦ ♦ ♦ ♦ ♦ ♦ ♦ ♦ ♦ ♦ ♦ ♦ ♦ ♦ ♦ ♦ ♦ ♦ ♦ ♦ ♦ ♦

It does not matter which method is used to simplify fractions containing opposites to $-1$. Choose a method that makes sense and consistently gives you the correct result.

## OPERATIONS WITH FRACTIONS (REVIEW)

In the next two examples we review operations with fractions. Look carefully at the steps because these same steps are repeated when we work with fractions containing expressions. If the operations look too easy, focus on what is the same and what is different from one operation to the next.

EXAMPLE **7**    Add, subtract, multiply, and divide $\frac{1}{18}$ and $\frac{1}{25}$.

SOLUTION    We find a common denominator in addition and subtraction.

$$\frac{1}{18} + \frac{1}{25} = \frac{25}{18 \cdot 25} + \frac{18}{18 \cdot 25} = \frac{25 + 18}{18 \cdot 25} = \frac{43}{450}$$

$$\frac{1}{18} - \frac{1}{25} = \frac{25}{18 \cdot 25} - \frac{18}{18 \cdot 25} = \frac{25 - 18}{18 \cdot 25} = \frac{7}{450}$$

No common denominator is needed for multiplication.

$$\frac{1}{18} \cdot \frac{1}{25} = \frac{1}{18 \cdot 25} = \frac{1}{450}$$

To divide we change the division to multiplication by the reciprocal of the second fraction.

$$\frac{1}{18} \div \frac{1}{25} = \frac{1}{18} \cdot \frac{25}{1} = \frac{25}{18}$$

♦ ♦ ♦ ♦ ♦ ♦ ♦ ♦ ♦ ♦ ♦ ♦ ♦ ♦ ♦ ♦ ♦ ♦ ♦ ♦ ♦ ♦ ♦ ♦ ♦ ♦ ♦ ♦ ♦ ♦ ♦ ♦

The multiplication and division of fractions follow these rules:

$$\frac{a}{b} \cdot \frac{c}{d} = \frac{ac}{bd}, \quad b \neq 0, \, d \neq 0$$

$$\frac{a}{b} \div \frac{c}{d} = \frac{a}{b} \cdot \frac{d}{c}, \quad b \neq 0, \, c \neq 0, \, d \neq 0$$

## EXERCISES 8.2

Add, subtract, multiply, and divide each pair of fractions in Exercises 1 to 8. Describe how the operations with each pair are the same as and different from the operations with the pair of fractions in the preceding exercise.

1. $\frac{1}{3}$ and $\frac{1}{4}$       2. $\frac{1}{2}$ and $\frac{1}{5}$       3. $\frac{1}{4}$ and $\frac{1}{12}$

4. $\frac{1}{8}$ and $\frac{1}{12}$       5. $\frac{3}{4}$ and $\frac{1}{6}$       6. $\frac{2}{3}$ and $\frac{1}{6}$

7. $\frac{a}{b}$ and $\frac{c}{d}$       8. $\frac{w}{x}$ and $\frac{y}{z}$

In Exercises 9 to 12 identify the equal expressions in each set. How are the other two expressions related?

9. $4, \frac{1}{4}, -4, 4^{-1}$              10. $-3, 3, \frac{1}{3}, 3^{-1}$

11. $\frac{1}{-a}, a, a^{-1}, -a$            12. $b^{-1}, \frac{1}{b}, b, -b$

What is the opposite of each expression in Exercises 13 to 18?

13. $a - b$          14. $a + b$          15. $-a + b$

16. $b - a$          17. $b + a$          18. $-a - b$

19. List the expressions in Exercises 13 to 18 that equal $a - b$.

20. List the expressions in Exercises 13 to 18 that equal $b - a$.

21. Evaluate the expressions in Exercises 13 to 18 if $a = 5$ and $b = -4$. How many different outcomes are listed?

22. Evaluate the expressions in Exercises 13 to 18 if $a = 3$ and $b = -5$. How many different outcomes are listed?

Simplify, if possible, each expression in the pairs in Exercises 23 to 28. Which pairs are equal? Why?

23. $\frac{-2}{8}, \frac{2}{-8}$                    24. $\frac{5}{-15}, -\frac{5}{15}$

25. $\frac{-8}{-4}, -\frac{8}{4}$                    26. $\frac{6}{-4}, \frac{6}{4}$

27. $\frac{-a}{-b}, \frac{a}{b}, b \neq 0$          28. $\frac{-a}{b}, \frac{a}{-b}, b \neq 0$

Simplify the fractions in Exercises 29 to 32. What do you observe? Why?

29. $\frac{7 - 4}{4 - 7}$                        30. $\frac{15 - 9}{9 - 15}$

31. $\frac{-3 - 4}{4 - (-3)}$                    32. $\frac{6 - (-2)}{-2 - 6}$

In Exercises 33 to 40, what numerator or denominator is needed in the equation to make a true statement? State any restrictions on the denominator needed to eliminate undefined expressions.

33. $\dfrac{x + 2}{\phantom{xxx}} = 1$     34. $\dfrac{x + 2}{\phantom{xxx}} = -1$     35. $\dfrac{x - 2}{\phantom{xxx}} = -1$

36. $\dfrac{x - 2}{\phantom{xxx}} = 1$     37. $\dfrac{\phantom{xxx}}{a - b} = -1$     38. $\dfrac{b - a}{\phantom{xxx}} = 1$

39. $\dfrac{3 - x}{\phantom{xxx}} = 1$     40. $\dfrac{3 - x}{\phantom{xxx}} = -1$

In Exercises 41 to 44 use both of the following statements to test whether the two expressions are opposites:

a. Opposites add to zero.

b. If we multiply an expression by $-1$, we get its opposite.

41. $n - m$ and $n + m$          42. $n - m$ and $m - n$

43. $x - 2$ and $2 - x$          44. $x - 2$ and $x + 2$

In Exercises 45 and 46 indicate whether the statement is true or false. If true, explain why; if false, give an example that shows why it is false.

45. A rational expression must be factorable in order to be simplified by canceling.

46. A rational expression can have the same variables in the numerator and denominator and still not be simplified.

State any restrictions on the expressions in Exercises 47 to 66, then simplify.

47. $\dfrac{2ab}{6d^2b}$                        48. $\dfrac{3ac}{15ac^2}$

49. $\dfrac{12cd^2}{28c^2d}$                     50. $\dfrac{15b^2c^3}{10b^3c}$

51. $\dfrac{(x - 2)(x + 3)}{x + 3}$             52. $\dfrac{(x - 2)(x + 3)}{(x + 3)(x + 2)}$

53. $\dfrac{2 - x}{(x + 2)(x - 2)}$             54. $\dfrac{1 - x}{(x - 1)(x + 2)}$

55. $\dfrac{3ab + 3ac}{5b^2 + 5bc}$              56. $\dfrac{2ac + 4bc}{4ad + 8bd}$

57. $\dfrac{4x^2 + 8x}{2x^2 - 4x}$              58. $\dfrac{3x^2 - 6x}{6x^2 + 12x}$

59. $\dfrac{x^2 - 4}{x^2 + 5x + 6}$

60. $\dfrac{x^2 - 1}{x^2 + 4x + 3}$

61. $\dfrac{x^2 + x - 6}{x^2 - 2x}$

62. $\dfrac{x^2 + x - 2}{2x + 4}$

63. $\dfrac{x - 3}{6 - 2x}$

64. $\dfrac{x - 4}{12 - 3x}$

65. $\dfrac{x^2 - 5x + 6}{x^2 - 9}$

66. $\dfrac{x^2 - 3x - 4}{x^2 - 16}$

**In Exercises 67 to 72 simplify.**

67. $\dfrac{108 \text{ m}^2}{6 \text{ m}}$

68. $\dfrac{125 \text{ in.}^3}{5 \text{ in.}}$

69. $\dfrac{144 \text{ in.}^2}{1728 \text{ in.}^3}$

70. $\dfrac{27 \text{ yd}^3}{9 \text{ yd}^2}$

71. $\dfrac{2060 \text{ degree gallons}}{103 \text{ degrees}}$

72. $\dfrac{1200 \text{ foot-pounds}}{200 \text{ pounds}}$

**In Exercises 73 and 74 replace the variables with numbers to show that these expressions may *sometimes*, *always*, or *never* be simplified. Explain your conclusions. What do you observe about the numbers that lead to a *sometimes* response?**

73. a. $\dfrac{b - a}{a - b}$   b. $\dfrac{a + b}{a}$   c. $\dfrac{a \cdot b}{a}$   d. $\dfrac{b}{a \cdot b}$

74. a. $\dfrac{x - 2}{x + 2}$   b. $\dfrac{2x}{x + 2}$   c. $\dfrac{x - 3}{3 - x}$   d. $\dfrac{x}{x - 2}$

**Find the missing number in each pair of fractions in Exercises 75 to 78. Indicate whether the first fraction was simplified to the second or changed to an equivalent fraction in unsimplified form.**

75. $\dfrac{6}{9}, \dfrac{}{45}$   76. $\dfrac{15}{10}, \dfrac{3}{}$   77. $\dfrac{24x}{3x^2}, \dfrac{8}{}$   78. $\dfrac{24x}{3x^2}, \dfrac{72x^3}{}$

79. If $\dfrac{a}{a} = 1$, list three fractions containing positive or negative $a$'s that would simplify to $-1$.

80. What happens when we divide opposites?

81. Describe the role of factoring in simplifying fractions.

82. What may be concluded about $-\dfrac{a}{b}, \dfrac{-a}{b},$ and $\dfrac{a}{-b}$?

*GRAPHING CALCULATOR EXPLORATION*

83. a. Graph the expressions in Exercises 51, 53, 57, and 59 before and after simplifying them.

   b. Is there any major difference between the graphs before and after simplifying?

   c. Which give straight lines?

   d. What is the slope of the lines?

   e. Are there any holes in the lines? If so, where? Why?

   f. Are there any nearly vertical parts? If so, where? Why?

84. a. Graph the expressions in Exercises 52, 54, 58, and 62 before and after simplifying them.

   b. Is there any major difference in the graphs before and after simplifying?

   c. Which give straight lines?

   d. What is the slope of the lines?

   e. Are there any holes in the lines? If so, where? Why?

   f. Are there any nearly vertical parts? If so, where? Why?

*PROJECTS*

85. *Fraction Pattern*

   a. Simplify:
   $$\dfrac{1 + 2 + 3}{4 + 5 + 6} \qquad \dfrac{7 + 8 + 9}{10 + 11 + 12} \qquad \dfrac{13 + 14 + 15}{16 + 17 + 18}$$

   b. Predict the values of these fractions:
   $$\dfrac{50 + 51 + 52}{53 + 54 + 55} \qquad \dfrac{100 + 101 + 102}{103 + 104 + 105}$$

   c. Show that your pattern always works by building a rational expression (a fraction) with $x$ as the first number, $x + 1$ as the second number, and so on. Simplify the expression.

86. *Wheat Harvest.* Refer to the wheat harvest problem at the beginning of the chapter. Suppose each large rectangle in the following figure represents the entire wheat harvest. If Terry harvests the wheat in 18 days, then each day he harvests $\frac{1}{18}$ of the crop. The shading in the first rectangle represents the amount Terry harvests in one day. The shading in the second rectangle is the amount Lee harvests in one day: $\frac{1}{12}$. Together they harvest the amount shown in the third rectangle. The second day of harvest is shown by the second row of rectangles.

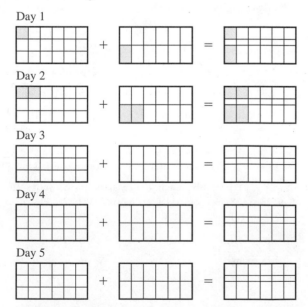

   a. Shade the rectangles for each subsequent day's accumulative harvest. Add more rectangles as needed.

   b. How will the rectangle look when the harvest is complete?

   c. Estimate the total number of days needed to complete the harvest.

   d. What is a better way to show the fractional parts $\frac{1}{12}$ and $\frac{1}{18}$ so that we can improve our estimate of the total number of days needed to complete the harvest?

## 8.3  RATIONAL EXPRESSIONS: MULTIPLICATION AND DIVISION

**OBJECTIVES**   Factor, simplify, and multiply rational expressions. ♦ Change division problems to multiplication and complete the multiplication. ♦ Simplify complex rational expressions by changing to division. ♦ Apply multiplication and division principles to expressions containing units of measurement. ♦ Apply the distributive property to multiplication of rational expressions.

**WARM-UP**   Translate each phrase into symbols, and perform the indicated operation.

1. The product of 3 and $\frac{1}{2}$
2. The quotient of 3 and $\frac{1}{2}$
3. The quotient of $\frac{1}{2}$ and 3
4. The quotient of $\frac{1}{2}$ and $\frac{1}{3}$
5. The product of $\frac{1}{3}$ and $\frac{3}{4}$

Factor these numbers or expressions.

6. 35
7. 32
8. $cx + c$
9. $x^2 - 5x - 6$
10. $x^2 + 3x + 2$
11. $x^2 - 3x - 4$
12. $x^2 - 16$     ♦

In this section we apply principles of multiplication and division to rational expressions, complex rational expressions, and expressions containing units of measurement. We close by applying the distributive property to rational expressions.

### THE WARM-UP PROBLEM

We now examine answers to a problem posed in the Warm-up to Section 8.1: solving $A = \dfrac{h}{2}(a + b)$ for $b$. Before you read on, try solving for $b$.

Suppose four students, Jo, James, Jacques, and Jincy, solved the equation $A = \dfrac{h}{2}(a + b)$ for $b$ as follows:

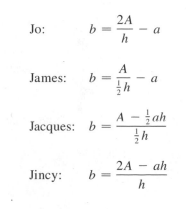

$$\text{Jo:} \qquad b = \frac{2A}{h} - a$$

$$\text{James:} \qquad b = \frac{A}{\frac{1}{2}h} - a$$

$$\text{Jacques:} \qquad b = \frac{A - \frac{1}{2}ah}{\frac{1}{2}h}$$

$$\text{Jincy:} \qquad b = \frac{2A - ah}{h}$$

Does your answer agree with any of these four? We could check each solution in the original equation, but instead we verify one solution and then use several rational expression techniques to compare it with the others.

EXAMPLE 1    Verify Jo's answer.

SOLUTION    Jo's solution was to reverse the order of operations on $b$.

$$A = \frac{h}{2}(a + b) \qquad \text{Solve for } b.$$

$$2A = h(a + b) \qquad \text{Multiply by 2.}$$

$$\frac{2A}{h} = a + b \qquad \text{Divide by } h.$$

$$\frac{2A}{h} - a = b \qquad \text{Subtract } a.$$

◆ ◆ ◆ ◆ ◆ ◆ ◆ ◆ ◆ ◆ ◆ ◆ ◆ ◆ ◆ ◆ ◆ ◆ ◆ ◆ ◆ ◆ ◆ ◆ ◆ ◆ ◆ ◆ ◆ ◆ ◆ ◆ ◆ ◆ ◆ ◆ ◆ ◆ ◆

Jo and James have two terms in their answers, but their first terms are slightly different. Before reading on, pause to recall what the simplification of fractions indicates about the first terms. Is there a way to change James's answer into Jo's?

EXAMPLE 2    Show that James's and Jo's solutions are equal.

SOLUTION

$$b = \frac{A}{\frac{1}{2}h} - a \qquad \text{James's solution}$$

$$b = \frac{2 \cdot A}{2 \cdot \frac{1}{2}h} - a \qquad \text{Multiply the fraction by } \frac{2}{2}.$$

$$b = \frac{2A}{h} - a \qquad \text{Jo's solution}$$

◆ ◆ ◆ ◆ ◆ ◆ ◆ ◆ ◆ ◆ ◆ ◆ ◆ ◆ ◆ ◆ ◆ ◆ ◆ ◆ ◆ ◆ ◆ ◆ ◆ ◆ ◆ ◆ ◆ ◆ ◆ ◆ ◆ ◆ ◆ ◆ ◆ ◆ ◆

Jacques and Jincy have a single fraction on the right side of their equations. Again, pause to consider whether we can use the same technique as in Example 2 to show that their answers are equal.

EXAMPLE 3    Show that Jacques's and Jincy's solutions are equal.

SOLUTION

$$b = \frac{A - \frac{1}{2}ha}{\frac{1}{2}h} \qquad \text{Jacques's solution}$$

$$b = \frac{2\left(A - \frac{1}{2}ha\right)}{2\left(\frac{1}{2}h\right)} \qquad \text{Multiply the fraction by } \frac{2}{2}.$$

$$b = \frac{2A - ha}{h} \qquad \text{Jincy's solution}$$

◆ ◆ ◆ ◆ ◆ ◆ ◆ ◆ ◆ ◆ ◆ ◆ ◆ ◆ ◆ ◆ ◆ ◆ ◆ ◆ ◆ ◆ ◆ ◆ ◆ ◆ ◆ ◆ ◆ ◆ ◆ ◆ ◆ ◆ ◆ ◆ ◆ ◆ ◆

If we now find a way to change Jincy's answer into Jo's answer, then all four solutions will be equivalent and thus correct. Before reading on, pause to think about how they might be made the same.

EXAMPLE 4    Show that Jincy's and Jo's solutions are equal.

SOLUTION

$$b = \frac{2A - ha}{h} \qquad \text{Jincy's solution}$$

$$b = \frac{2A}{h} - \frac{ha}{h} \qquad \text{Divide numerator terms by } h.$$

$$b = \frac{2A}{h} - a \qquad \text{Jo's solution}$$

◆ ◆ ◆ ◆ ◆ ◆ ◆ ◆ ◆ ◆ ◆ ◆ ◆ ◆ ◆ ◆ ◆ ◆ ◆ ◆ ◆ ◆ ◆ ◆ ◆ ◆ ◆ ◆ ◆ ◆ ◆ ◆ ◆ ◆ ◆ ◆ ◆ ◆ ◆

MULTIPLICATION OF RATIONAL EXPRESSIONS

In Examples 2 to 4 the fractions were changed by using the fact that $\dfrac{a}{a} = 1$. The same fact was used to factor and simplify in Section 8.2. In the next three examples observe the role of $\dfrac{a}{a} = 1$ in simplifying the multiplication and division of rational expressions.

In each case the solution is found most easily by simplifying expressions before doing the final multiplication. In all solutions we use the multiplication property of fractions.

**Multiplication Property of Fractions**

If $b \neq 0$ and $d \neq 0$, then

$$\frac{a}{b} \cdot \frac{c}{d} = \frac{a \cdot c}{b \cdot d}$$

**EXAMPLE  5**    Multiply $\frac{8}{35} \cdot \frac{25}{32} \cdot \frac{14}{9}$.

SOLUTION

$$\frac{8}{35} \cdot \frac{25}{32} \cdot \frac{14}{9} = \frac{8 \cdot 25 \cdot 14}{35 \cdot 32 \cdot 9} = \frac{2 \cdot 2 \cdot 2 \cdot 5 \cdot 5 \cdot 2 \cdot 7}{5 \cdot 7 \cdot 2 \cdot 2 \cdot 2 \cdot 2 \cdot 2 \cdot 3 \cdot 3} = \frac{5}{2 \cdot 3 \cdot 3} = \frac{5}{18}$$

In this solution we use the multiplication property to change the three fractions to a single fraction with one numerator and one denominator. We simplify and then multiply the remaining factors in the numerator and denominator. Changing to all primes may not be necessary.

♦ ♦ ♦ ♦ ♦ ♦ ♦ ♦ ♦ ♦ ♦ ♦ ♦ ♦ ♦ ♦ ♦ ♦ ♦ ♦ ♦ ♦ ♦ ♦ ♦ ♦ ♦ ♦ ♦ ♦ ♦ ♦ ♦ ♦ ♦ ♦ ♦

**EXAMPLE  6**    Multiply $\dfrac{ax^2}{cx} \cdot \dfrac{cx + c}{a^2}$, $a \neq 0$, $c \neq 0$, $x \neq 0$.

SOLUTION

$$\frac{ax^2}{cx} \cdot \frac{cx + c}{a^2} = \frac{ax^2(cx + c)}{cx \cdot a^2} = \frac{a \cdot x \cdot x \cdot c(x + 1)}{c \cdot x \cdot a \cdot a} = \frac{x(x + 1)}{a}$$

In this solution we again write the numerators and denominators as products and factor them completely. Simplification eliminates any further need for multiplication.

♦ ♦ ♦ ♦ ♦ ♦ ♦ ♦ ♦ ♦ ♦ ♦ ♦ ♦ ♦ ♦ ♦ ♦ ♦ ♦ ♦ ♦ ♦ ♦ ♦ ♦ ♦ ♦ ♦ ♦ ♦ ♦ ♦ ♦ ♦ ♦ ♦

**EXAMPLE  7**    Multiply $\dfrac{x + 2}{x + 6} \cdot \dfrac{x^2 - 5x - 6}{x^2 + 3x + 2}$, $x \neq -6$, $x \neq -2$, and $x \neq -1$.

SOLUTION

$$\frac{x + 2}{x + 6} \cdot \frac{x^2 - 5x - 6}{x^2 + 3x + 2} = \frac{(x + 2)(x^2 - 5x - 6)}{(x + 6)(x^2 + 3x + 2)}$$

$$= \frac{(x + 2)(x + 1)(x - 6)}{(x + 6)(x + 1)(x + 2)} = \frac{x - 6}{x + 6}$$

We apply the multiplication property of fractions by writing the two fractions as one fraction. We then factor the numerator and denominator expressions and simplify the fraction.

♦ ♦ ♦ ♦ ♦ ♦ ♦ ♦ ♦ ♦ ♦ ♦ ♦ ♦ ♦ ♦ ♦ ♦ ♦ ♦ ♦ ♦ ♦ ♦ ♦ ♦ ♦ ♦ ♦ ♦ ♦ ♦ ♦ ♦ ♦ ♦ ♦

CAUTION   If we first multiply the fractions in Example 7, the resulting expression is not easily simplified. It requires factoring techniques beyond the level of this or the next mathematics course.

$$\frac{x+2}{x+6} \cdot \frac{x^2 - 5x - 6}{x^2 + 3x + 2} = \frac{(x+2)(x^2 - 5x - 6)}{(x+6)(x^2 + 3x + 2)} = \frac{x^3 - 3x^2 - 16x - 12}{x^3 + 9x^2 + 20x + 12} = \ ?$$

If your homework solutions contain similar expressions, go back to the original problem, factor, and simplify before multiplying.   ♦

In all multiplication problems keep in mind that we are canceling factors, not terms. Any units are treated the same way as factors. We used this idea earlier in unit analysis.

E X A M P L E   **8**

**Water flow**   A shower head permits a flow of 5 gallons per minute. How many gallons of water are used in a $3\frac{1}{2}$-minute shower?

SOLUTION

$$\frac{5 \text{ gal}}{1 \text{ min}} \cdot 3.5 \text{ min} = \frac{5(3.5)}{1} \frac{\text{gal} \cdot \text{min}}{\text{min}} = 17.5 \text{ gal}$$

♦ ♦ ♦ ♦ ♦ ♦ ♦ ♦ ♦ ♦ ♦ ♦ ♦ ♦ ♦ ♦ ♦ ♦ ♦ ♦ ♦ ♦ ♦ ♦ ♦ ♦ ♦ ♦ ♦ ♦ ♦ ♦ ♦ ♦ ♦ ♦ ♦ ♦ ♦ ♦ ♦ ♦ ♦

DIVISION OF RATIONAL EXPRESSIONS

Division of rational expressions is based on the same property as division of fractions.

**Division of Fractions**

> To divide fractions, multiply the first fraction by the reciprocal of the second.

E X A M P L E   **9**

Divide $\dfrac{x^2 - 3x - 4}{x - 3} \div \dfrac{x^2 - 16}{x^2 - 9}$, $x \neq -4, -3, 3,$ and $4$.

SOLUTION

$$\frac{x^2 - 3x - 4}{x - 3} \div \frac{x^2 - 16}{x^2 - 9} = \frac{x^2 - 3x - 4}{x - 3} \cdot \frac{x^2 - 9}{x^2 - 16}$$

$$= \frac{(x - 4)(x + 1)(x - 3)(x + 3)}{(x - 3)(x - 4)(x + 4)}$$

$$= \frac{(x + 1)(x + 3)}{(x + 4)}$$

After changing the problem to a multiplication problem, we factor the expression and simplify. No further simplification is possible because there are no common factors in the numerator and denominator.

♦ ♦ ♦ ♦ ♦ ♦ ♦ ♦ ♦ ♦ ♦ ♦ ♦ ♦ ♦ ♦ ♦ ♦ ♦ ♦ ♦ ♦ ♦ ♦ ♦ ♦ ♦ ♦ ♦ ♦ ♦ ♦ ♦ ♦ ♦ ♦ ♦ ♦ ♦ ♦ ♦ ♦ ♦

E X A M P L E   **10**

Divide 450 miles by 60 miles per hour.

SOLUTION

$$450 \text{ mi} \div \frac{60 \text{ mi}}{1 \text{ hr}} = 450 \text{ mi} \cdot \frac{1 \text{ hr}}{60 \text{ mi}} = 7.5 \text{ hr}$$

♦ ♦ ♦ ♦ ♦ ♦ ♦ ♦ ♦ ♦ ♦ ♦ ♦ ♦ ♦ ♦ ♦ ♦ ♦ ♦ ♦ ♦ ♦ ♦ ♦ ♦ ♦ ♦ ♦ ♦ ♦ ♦ ♦ ♦ ♦ ♦ ♦ ♦ ♦ ♦ ♦ ♦ ♦

The technique of changing division to multiplication by a reciprocal may be applied to more complicated forms of rational expressions—the complex rational expressions.

## COMPLEX RATIONAL EXPRESSIONS

*Rational expressions that contain fractions in either the numerator or the denominator are called* **complex fractions**. When the numerator, the denominator, or both are single fractions, recall that the fraction bar means division and change the complex fraction to a division problem.

**E X A M P L E   11**   Simplify the complex fraction $\dfrac{\dfrac{a}{b}}{\dfrac{c}{d}}$.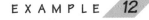

**SOLUTION**

$$\frac{\dfrac{a}{b}}{\dfrac{c}{d}} = \frac{a}{b} \div \frac{c}{d} = \frac{a}{b} \cdot \frac{d}{c} = \frac{ad}{bc}$$

We simplify the complex fraction by writing it as a division problem, with the longer fraction bar replaced by a division sign. The division is then changed to multiplication by a reciprocal.

♦ ♦ ♦ ♦ ♦ ♦ ♦ ♦ ♦ ♦ ♦ ♦ ♦ ♦ ♦ ♦ ♦ ♦ ♦ ♦ ♦ ♦ ♦ ♦ ♦ ♦ ♦ ♦ ♦ ♦ ♦ ♦ ♦ ♦ ♦ ♦ ♦

We may include units in simplifying complex fractions.

**E X A M P L E   12**   Answer the question by writing a complex fraction and simplifying.

**a.** How many half-dollars are in $5.00?

**b.** How many fourths are in $\frac{5}{8}$?

**c.** If a 19-passenger jet flying at 459 nautical miles (nm) per hour consumes 397 gallons of fuel per hour, how many nautical miles does it get per gallon?

**SOLUTION**   In parts a and b *how many* implies division.

**a.** $\dfrac{5.00 \text{ dollars}}{\frac{1}{2} \text{ dollar}} = 5 \div \dfrac{1}{2} = 5 \cdot \dfrac{2}{1} = 10$

**b.** $\dfrac{\frac{5}{8}}{\frac{1}{4}} = \dfrac{5}{8} \div \dfrac{1}{4} = \dfrac{5}{8} \cdot \dfrac{4}{1} = \dfrac{20}{8} = \dfrac{5}{2} = 2\frac{1}{2}$

In part c we are looking for nautical miles per gallon, so the expression containing nautical miles should be placed in the numerator.

**c.** $\dfrac{\dfrac{459 \text{ nm}}{\text{hr}}}{\dfrac{397 \text{ gal}}{\text{hr}}} = \dfrac{459 \text{ nm}}{\text{hr}} \div \dfrac{397 \text{ gal}}{\text{hr}} = \dfrac{459 \text{ nm}}{\text{hr}} \cdot \dfrac{\text{hr}}{397 \text{ gal}} \approx 1.16 \dfrac{\text{nm}}{\text{gal}}$

Unit analysis can also be used to solve this problem. We start with 459 nm per hour and multiply by an expression containing hours in the numerator in order to cancel the hours:

$$\frac{459 \text{ nm}}{\text{hr}} \cdot \frac{\text{hr}}{397 \text{ gal}} \approx 1.16 \frac{\text{nm}}{\text{gal}}$$

♦ ♦ ♦ ♦ ♦ ♦ ♦ ♦ ♦ ♦ ♦ ♦ ♦ ♦ ♦ ♦ ♦ ♦ ♦ ♦ ♦ ♦ ♦ ♦ ♦ ♦ ♦ ♦ ♦ ♦ ♦ ♦ ♦ ♦ ♦ ♦

We close this section by applying the distributive property of multiplication to rational expressions.

RATIONAL EXPRESSIONS AND THE DISTRIBUTIVE PROPERTY

We return to the table form of multiplication because tables remind us how important it is to multiply both terms in parentheses by the expression in front. Recall that the distributive property is

$$a(b + c) = ab + ac$$

Application of the distributive property to $3x(x + 1)$ appears in area form in Figure 8. In the table in Example 13, the corresponding area is the part enclosed by the double lines.

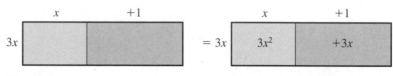

Figure 8

EXAMPLE  13    Complete these tables. Summarize by writing the solution as a simplified product.

a.

| Multiply | $\dfrac{1}{4}$ | $+\dfrac{1}{3x}$ |
|---|---|---|
| $12x$ | | |

b.

| Multiply | $\dfrac{1}{x - 3}$ | $+\dfrac{5}{x - 2}$ |
|---|---|---|
| $(x - 2)(x - 3)$ | | |

SOLUTION   a.

| Multiply | $\dfrac{1}{4}$ | $+\dfrac{1}{3x}$ |
|---|---|---|
| $12x$ | $\dfrac{12x}{4}$ | $+\dfrac{12x}{3x}$ |

*Summary:*

$$12x\left(\frac{1}{4} + \frac{1}{3x}\right) = 12x\left(\frac{1}{4}\right) + 12x\left(\frac{1}{3x}\right)$$

$$= \frac{12x}{4} + \frac{12x}{3x} = 3x + 4$$

b.

| Multiply | $\dfrac{1}{x - 3}$ | $+\dfrac{5}{x - 2}$ |
|---|---|---|
| $(x - 2)(x - 3)$ | $x - 2$ | $+5(x - 3)$ |

*Summary:*

$$(x - 2)(x - 3)\left(\frac{1}{x - 3} + \frac{5}{x - 2}\right) = \frac{(x - 2)(x - 3)}{x - 3} + \frac{5(x - 2)(x - 3)}{x - 2}$$

$$= (x - 2) + 5(x - 3) = 6x - 17$$

◆ ◆ ◆ ◆ ◆ ◆ ◆ ◆ ◆ ◆ ◆ ◆ ◆ ◆ ◆ ◆ ◆ ◆ ◆ ◆ ◆ ◆ ◆ ◆ ◆ ◆ ◆ ◆ ◆ ◆ ◆ ◆ ◆ ◆ ◆ ◆ ◆ ◆ ◆ ◆ ◆ ◆

### EXERCISES 8.3

1. Calculate these fractional expressions. What may be observed about each pair of problems? What do they tell us about multiplication and division?

   a. $100 \div \frac{4}{1}$ and $100 \cdot \frac{1}{4}$   b. $100 \div \frac{1}{5}$ and $100 \cdot \frac{5}{1}$

2. In part b of Example 12 we divided $\frac{5}{8}$ by $\frac{1}{4}$, and the answer was $2\frac{1}{2}$. The figure below shows $\frac{5}{8}$ of one rectangle shaded and $\frac{1}{4}$ of an identical rectangle shaded. Trace these rectangles and show why there are $2\frac{1}{2}$ fourths in $\frac{5}{8}$.

In Exercises 3 and 4 translate each expression into symbols, then perform the indicated operation. Assume that neither $a$ nor $b$ is zero.

3. a. The product of $\frac{1}{a}$ and $\frac{1}{b}$

   b. The quotient of $a$ and $\frac{1}{a}$

   c. The product of $\frac{1}{b}$ and $a$

   d. The quotient of $\frac{1}{b}$ and $\frac{1}{a}$

   e. The product of $\frac{a}{b}$ and $\frac{1}{b}$

   f. The quotient of $\frac{a}{b}$ and $\frac{1}{a}$

4. a. The quotient of $\frac{1}{a}$ and $\frac{1}{b}$

   b. The product of $b$ and $\frac{1}{b}$

   c. The quotient of $\frac{1}{b}$ and $a$

   d. The quotient of $\frac{1}{b}$ and $b$

   e. The quotient of $\frac{a}{b}$ and $\frac{1}{b}$

   f. The product of $\frac{1}{b}$ and $\frac{1}{a}$

In Exercises 5 and 6 multiply or divide as indicated. Assume that none of the variables, $a$, $b$, or $x$, is zero.

5. a. $\frac{1}{x} \cdot \frac{x^2}{1}$   b. $\frac{1}{a} \div \frac{a^2 b^2}{1}$

   c. $\frac{a}{b} \cdot \frac{b^2}{a^2}$   d. $\frac{a}{b} \div \frac{a^2}{b^3}$

6. a. $\frac{1}{x} \cdot \frac{x^3}{1}$   b. $\frac{1}{b} \div \frac{a^2 b^2}{1}$

   c. $\frac{b}{a} \div \frac{a^2}{b^2}$   d. $\frac{a^2}{b^3} \div \frac{a}{b}$

For what inputs will the fractions in Exercises 7 and 8 be undefined? Multiply or divide, as indicated. Simplify.

7. a. $\dfrac{x^2 + 2x + 1}{x + 1} \cdot \dfrac{x}{x^2 + x}$

   b. $\dfrac{x^2 - 4}{x + 2} \cdot \dfrac{1}{x^2 - x}$

   c. $\dfrac{x + 2}{x^2 - 4x + 4} \div \dfrac{x^2 + 2x}{x - 2}$

   d. $\dfrac{x^2 - 5x}{x^2 + 5x} \div \dfrac{x^2 - 10x + 25}{x}$

   e. $\dfrac{x^2 - 6x + 9}{x^2 + 3x} \div \dfrac{x^2 - 9}{x}$

   f. $\dfrac{x^2 - x}{x^2 - 3x + 2} \div \dfrac{1 - x^2}{x^2 - 2x + 1}$

8. a. $\dfrac{x^2 - 7x + 12}{x^2 - 4} \cdot \dfrac{x^2 + 2x}{x - 3}$

   b. $\dfrac{x^2 - 2x}{x} \cdot \dfrac{x^2}{x^2 - 3x + 2}$

   c. $\dfrac{x - 3}{x^2 - 4x + 3} \div \dfrac{x^2 + x}{x - 1}$

   d. $\dfrac{x^2 - 6x + 9}{x^2 + 3x} \cdot \dfrac{x + 3}{x - 3}$

   e. $\dfrac{x^2 + 3x}{x} \cdot \dfrac{x^2 - x - 6}{x^2 - 9}$

   f. $\dfrac{x^2 + 4x + 4}{x^2 - 4} \div \dfrac{x^2 + 2x}{2 - x}$

9. The current, $I$, in an electrical circuit is found by dividing the voltage, $V$, by the resistance, $R$. Suppose the voltage and resistance vary with time as in the equations

$$V = \frac{t^2 - 4}{2t^2 - 3t - 2} \quad \text{and} \quad R = \frac{t + 2}{t^2}$$

Find a formula in terms of $t$ that gives the current, $I$.

10. For many people the price determines the brand of soft drink purchased. These people are said to exhibit "constant product of price and demand behavior." Find the product of the price and the demand for the given expressions, where $x$ is the number of units of soft drink.

$$\text{price} = 3x + 6 \quad \text{and} \quad \text{demand} = \frac{800}{x^2 + 2x}$$

11. A student familiar with simplifying fractions such as $\frac{12}{15}$ is puzzled by the canceling of threes in $\frac{3}{5} \cdot \frac{4}{3}$. Explain why the product can be simplified when the threes are in different fractions.

12. Explain why

$$\frac{x(x+1)(x-3)}{x(x-3)} = x + 1$$

is correct and

$$\frac{x^2 + x + 1}{x^2} = x + 1$$

is not correct.

For Exercises 13 to 16 complete the tables. Write the expressions described by the tables, and simplify the answers.

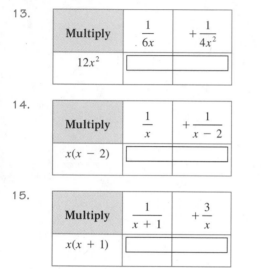

13.

| Multiply | $\frac{1}{6x}$ | $+\frac{1}{4x^2}$ |
|---|---|---|
| $12x^2$ | | |

14.

| Multiply | $\frac{1}{x}$ | $+\frac{1}{x-2}$ |
|---|---|---|
| $x(x-2)$ | | |

15.

| Multiply | $\frac{1}{x+1}$ | $+\frac{3}{x}$ |
|---|---|---|
| $x(x+1)$ | | |

16.

| Multiply | $\frac{1}{2}$ | $+\frac{1}{12x}$ | $+\frac{1}{x^2}$ |
|---|---|---|---|
| $12x^2$ | | | |

**Multiply the expressions in Exercises 17 to 22, and simplify the results.**

17. $12x\left(\frac{1}{12} + \frac{2}{3x}\right)$

18. $8x\left(\frac{1}{4x} + \frac{1}{2}\right)$

19. $4x^2\left(\frac{1}{2x} + \frac{3}{x^2}\right)$

20. $6x^2\left(\frac{2}{3x^2} + \frac{1}{6}\right)$

21. $x(x-1)\left(\frac{1}{x-1} + \frac{1}{x}\right)$

22. $x(x+2)\left(\frac{1}{x} + \frac{3}{x+2}\right)$

23. Why do the denominators disappear when we multiply the expressions in Exercises 13 to 22?

24. Why is it that the product of $a \cdot b$ and $c$ and is the same as $a$ times the product of $b$ and $c$?

25. When we multiply $a \cdot b \cdot c$, why do we not multiply $a \cdot b$ and $a \cdot c$?

26. What expression would eliminate the denominator when multiplied times $\left(\frac{1}{3x} + \frac{1}{5x^2} + \frac{4}{x}\right)$?

**In Exercises 27 to 40 use properties of fractions to simplify these expressions containing units of measurement. The word *per* means division and may be replaced by a fraction bar.**

27. $\dfrac{\dfrac{\text{miles}}{\text{miles}}}{\text{hours}}$

28. $\dfrac{\dfrac{\text{kilometers}}{\text{kilometers}}}{\text{minute}}$

29. $\dfrac{93{,}000{,}000 \text{ miles}}{186{,}000 \text{ miles per second}}$

30. $\dfrac{5280 \text{ feet}}{1130 \text{ feet per second}}$

31. $\dfrac{300 \text{ miles per hour}}{100 \text{ gallons per hour}}$

32. $\dfrac{60 \text{ miles per hour}}{25 \text{ miles per gallon}}$

33. $\dfrac{100 \text{ aspirin per bottle}}{2 \text{ aspirin per 12 hours}}$

34. $\dfrac{12 \text{ cookies per dozen}}{\$2.98 \text{ per dozen}}$

35. $\dfrac{\dfrac{12 \text{ stitches per inch}}{1 \text{ yard}}}{36 \text{ inches}}$

36. $\dfrac{130 \text{ heartbeats per minute}}{1 \text{ mile per 4 minutes}}$

37. $\dfrac{85 \text{ words per minute}}{300 \text{ words per page}}$

38. $\dfrac{880 \text{ cycles per second}}{344 \text{ meters per second}}$

39. $\dfrac{40 \text{ moles}}{12 \text{ moles per liter}}$

40. $\dfrac{186 \text{ days}}{5 \text{ days per week}}$

41. Choose one of the expressions from Exercises 27 to 40, and give a situation in which it would make sense.

42. Make up a division problem containing units of measure, and explain what the answer means.

*GRAPHING CALCULATOR EXPLORATION*

43. Graph Exercises 17 and 19 before and after multiplication. Is there any difference between the graphs? How could we use graphs to check our work?

*PROJECT*

44. ***Exiting a Theater.*** A movie theater has two exits. One door, by itself, can empty the theater in 5 minutes. The second door, by itself, can empty the theater in 8 minutes. Suppose we wish to answer this question: If both doors are available, how many minutes will it take to empty the theater?

Each large rectangle in the figure below represents a full theater. During each minute the first door permits $\frac{1}{5}$ of the theater to empty. During each minute the second door permits $\frac{1}{8}$ of the theater to empty.

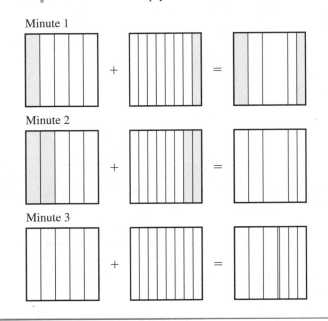

Minute 1

Minute 2

Minute 3

a. What does the shading in the first rectangle for Minute 1 represent?

b. What does the shading in the second rectangle for Minute 1 represent?

c. What does the shading in the third rectangle for Minute 1 represent?

d. Shade the rectangles for each subsequent minute's departures. Add more rectangles as needed.

e. How will the last rectangle look when the theater is empty?

f. Estimate the total number of minutes needed to empty the theater.

g. What is a better way to draw the $\frac{1}{5}$ and $\frac{1}{8}$ fractions to improve the addition?

## MID-CHAPTER 8 TEST

1. For what inputs, $x$, will the expression $\dfrac{1}{(x+2)(x-1)}$ be undefined?

For Exercises 2 and 3 assume that the budget for a credit union's annual meeting is $4800. The budget covers food, chair set-up charges, and a souvenir gift for each member attending.

2. Make a table and graph for the possible spending per person for zero to 1200 members. Use number of members attending as input and spending per member as output.

3. What equation describes the relationship between the number of members attending the meeting and the spending per member?

**Describe the behavior of the graph in the figure for the situations specified in Exercises 4 to 6.**

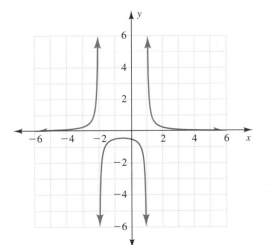

4. As $x$ approaches $-2$

5. As $x$ approaches 1

6. As $x$ approaches 0

**In Exercises 7 to 9, simplify these expressions containing units of measure.**

7. $\dfrac{63 \text{ days}}{7 \text{ days per week}}$

8. $\dfrac{16 \text{ stitches per second}}{8 \text{ stitches per inch}}$

9. From dosage computation: $\dfrac{\dfrac{10 \text{ mL}}{100 \text{ mL}} \cdot 400 \text{ mL}}{\dfrac{50 \text{ mL}}{100 \text{ mL}}}$

**For what inputs are the expressions in Exercises 10 to 13 undefined? Simplify. If the expression does not simplify, explain why.**

10. $\dfrac{24ac}{28a^2}$

11. $\dfrac{a+2}{a-2}$

12. $\dfrac{x+2}{x^2-4}$

13. $\dfrac{x^2-2x+1}{x^2-3x+2}$

**Perform the indicated operations in Exercises 14 to 20. Simplify the answers.**

14. $\dfrac{3x}{y^2} \div \dfrac{x^2}{y}$

15. $\dfrac{2x}{y} \div \dfrac{x^2}{y}$

16. $\dfrac{x^2 - 5x - 6}{x + 2} \cdot \dfrac{x^2 - 4}{x - 1}$

17. $\dfrac{x^2 - 3x}{x^2 - 16} \div \dfrac{x - 3}{x + 4}$

18. $6x\left(\dfrac{3}{2x} + \dfrac{4}{3x}\right)$

19. $x(x + 1)\left(\dfrac{2}{x + 1} + \dfrac{4}{x}\right)$

20. $\dfrac{\dfrac{2}{3x}}{\dfrac{x^2}{6}}$

---

## 8.4   RATIONAL EXPRESSIONS: COMMON DENOMINATOR, ADDITION, AND SUBTRACTION

**OBJECTIVES**   Determine the common denominator for two or more rational expressions. ◆ Add and subtract rational expressions with like denominators. ◆ Convert rational expressions to like denominators, and complete the addition or subtraction. ◆ Simplify complex fractions, using multiplication by a common denominator.

**WARM-UP**   Add or subtract, as indicated.

**1.** $\frac{4}{5} + \frac{5}{6}$    **2.** $\frac{1}{3} + \frac{1}{9}$    **3.** $\frac{1}{4} - \frac{1}{6}$    **4.** $\frac{1}{18} + \frac{1}{12}$    **5.** $\frac{2}{18} + \frac{2}{12}$    **6.** $\frac{3}{18} + \frac{3}{12}$    ◆

I n this section we use a common denominator to add and subtract fractions and to simplify complex fractions.

### THE WARM-UP PROBLEM

We return to the recurring theme of the variety of answers obtained when $A = \dfrac{h}{2}(a + b)$ is solved for $b$. In the last section we determined that all four student solutions were correct by showing them to be equivalent to one verified solution. We may show the answers to be equivalent in more than one way. In Example 1 we add together the terms on the right side of Jo's answer to obtain Jincy's answer.

**EXAMPLE  1**   Use common denominators and addition or subtraction of fractions to show that Jo's answer, $b = \dfrac{2A}{h} - a$, is equivalent to Jincy's, $b = \dfrac{2A - ha}{h}$.

**SOLUTION**
$b = \dfrac{2A}{h} - a$    Jo's answer

$b = \dfrac{2A}{h} - \dfrac{a}{1}$    Change $a$ to fraction form.

$b = \dfrac{2A}{h} - \dfrac{h}{h} \cdot a$    Build an equivalent fraction with a common denominator.

$b = \dfrac{2A - ha}{h}$    Jincy's answer

◆ ◆ ◆ ◆ ◆ ◆ ◆ ◆ ◆ ◆ ◆ ◆ ◆ ◆ ◆ ◆ ◆ ◆ ◆ ◆ ◆ ◆ ◆ ◆ ◆ ◆ ◆ ◆ ◆ ◆ ◆ ◆ ◆ ◆

In Example 1 we found a common denominator for the terms on the right side and combined them. A common denominator is needed to add or subtract rational expressions.

**Addition and Subtraction of Rational Numbers**

1. Convert to a common denominator, if necessary.
2. Add or subtract numerators, and place the result over the common denominator.
3. Factor and simplify the answer, if necessary.

**EXAMPLE 2**

Add or subtract these rational expressions containing like denominators.

**a.** $\dfrac{8}{y} - \dfrac{1}{y}, y \neq 0$  **b.** $\dfrac{2x}{x+2} + \dfrac{5}{x+2}, x \neq -2$

**SOLUTION**

**a.** $\dfrac{8}{y} - \dfrac{1}{y} = \dfrac{8-1}{y} = \dfrac{7}{y}$  **b.** $\dfrac{2x}{x+2} + \dfrac{5}{x+2} = \dfrac{2x+5}{x+2}$

In both answers the numerators and denominators contain no common factors, and therefore the fractions are simplified.

♦ ♦ ♦ ♦ ♦ ♦ ♦ ♦ ♦ ♦ ♦ ♦ ♦ ♦ ♦ ♦ ♦ ♦ ♦ ♦ ♦ ♦ ♦ ♦ ♦ ♦ ♦ ♦ ♦ ♦

### LEAST COMMON DENOMINATOR

Factoring plays an important role in finding the **least common denominator** (**LCD**), *the smallest number into which both denominators divide evenly.*

**EXAMPLE 3**

Add $\frac{3}{4} + \frac{5}{6}$.

**SOLUTION**

To find the least common denominator, we look at the factors of each denominator:

$$4 = 2 \cdot 2$$
$$6 = 2 \cdot 3$$

The least common denominator needs to be divisible by both denominators and needs two factors of 2 and one factor of 3. The product of these factors, $2 \cdot 2 \cdot 3$, gives the least common denominator: 12.

$$\dfrac{3}{4} = \dfrac{3 \cdot 3}{4 \cdot 3} = \dfrac{9}{12}$$

$$+ \dfrac{5}{6} = \dfrac{5 \cdot 2}{6 \cdot 2} = \dfrac{10}{12}$$

$$\dfrac{19}{12}$$

♦ ♦ ♦ ♦ ♦ ♦ ♦ ♦ ♦ ♦ ♦ ♦ ♦ ♦ ♦ ♦ ♦ ♦ ♦ ♦ ♦ ♦ ♦ ♦ ♦ ♦ ♦ ♦ ♦ ♦

The solutions in Example 2 were written horizontally. In Example 3 the solution was written vertically. Try one or two problems each way, and then use whichever form makes sense to you.

Although any common denominator can be used to add or subtract fractions, there are two advantages of using the least common denominator. First, the expressions are simpler. Second, the answer is less likely to need simplifying. In the next example we extend the common denominator to rational expressions.

EXAMPLE 4    Add $\dfrac{3}{ab^2} + \dfrac{5}{abc}$, $a$, $b$, and $c \neq 0$.

SOLUTION    To find the least common denominator, we factor each denominator:

$$ab^2 = a \cdot b \cdot b$$

$$abc = a \cdot b \cdot c$$

The least common denominator needs to be divisible by both denominators and needs two factors of $b$ as well as one each of $a$ and $c$. The product of these factors, $a \cdot b \cdot b \cdot c$, gives the least common denominator: $ab^2c$.

$$\frac{3}{ab^2} + \frac{5}{abc} = \frac{3 \cdot c}{ab^2 \cdot c} + \frac{5 \cdot b}{abc \cdot b}$$

$$= \frac{3c}{ab^2c} + \frac{5b}{ab^2c} = \frac{3c + 5b}{ab^2c}$$

◆ ◆ ◆ ◆ ◆ ◆ ◆ ◆ ◆ ◆ ◆ ◆ ◆ ◆ ◆ ◆ ◆ ◆ ◆ ◆ ◆ ◆ ◆ ◆ ◆ ◆ ◆ ◆ ◆ ◆ ◆ ◆ ◆ ◆ ◆

We also use factors to find the least common denominator for rational expressions containing binomials. Again, by using the least common denominator rather than any other common denominator we produce answers that are less likely to need simplification.

EXAMPLE 5    Subtract $\dfrac{2}{x + 1} - \dfrac{x}{(x + 1)^2}$, $x \neq -1$.

SOLUTION    $\dfrac{2}{x + 1} - \dfrac{x}{(x + 1)^2} = \dfrac{2(x + 1)}{(x + 1)(x + 1)} - \dfrac{x}{(x + 1)^2}$   Set up the common denominator.

$$= \frac{2(x + 1) - x}{(x + 1)^2}$$   Combine numerators.

$$= \frac{2x + 2 - x}{(x + 1)^2}$$   Distributive property

$$= \frac{x + 2}{(x + 1)^2}$$   Add like terms.

◆ ◆ ◆ ◆ ◆ ◆ ◆ ◆ ◆ ◆ ◆ ◆ ◆ ◆ ◆ ◆ ◆ ◆ ◆ ◆ ◆ ◆ ◆ ◆ ◆ ◆ ◆ ◆ ◆ ◆ ◆ ◆ ◆ ◆ ◆

Subtraction problems must be worked carefully because there may be a sign change when numerators are subtracted. The next example illustrates both finding the common denominator and changing signs with subtraction.

EXAMPLE 6    Subtract $\dfrac{x}{x^2 + 3x + 2} - \dfrac{3}{2x + 2}$, $x \neq -1$, $x \neq -2$.

SOLUTION    We factor the denominators:

$$x^2 + 3x + 2 = (x + 1)(x + 2)$$

$$2x + 2 = 2(x + 1)$$

The least common denominator will be the product of the three factors, $(x + 1)$, $(x + 2)$, and 2.

$$\frac{x}{x^2 + 3x + 2} - \frac{3}{2x + 2} = \frac{x}{(x + 1)(x + 2)} - \frac{3}{2(x + 1)}$$

Factor the denominators.

$$= \frac{2 \cdot x}{2(x + 1)(x + 2)} - \frac{3(x + 2)}{2(x + 1)(x + 2)}$$

Set up the LCD.

$$= \frac{2x - 3(x + 2)}{2(x + 1)(x + 2)}$$

Combine numerators.

$$= \frac{2x - 3x - 6}{2(x + 1)(x + 2)}$$

Distributive property

$$= \frac{-x - 6}{2(x + 1)(x + 2)}$$

Add like terms.

$$= \frac{-1(x + 6)}{2(x + 1)(x + 2)}$$

Factor the numerator.

Each time we subtract rational expressions we must watch for sign changes.

♦ ♦ ♦ ♦ ♦ ♦ ♦ ♦ ♦ ♦ ♦ ♦ ♦ ♦ ♦ ♦ ♦ ♦ ♦ ♦ ♦ ♦ ♦ ♦ ♦ ♦ ♦ ♦ ♦ ♦ ♦ ♦ ♦ ♦ ♦ ♦ ♦ ♦

**Finding the Least Common Denominator (LCD)**

1. If the denominators have no common factors, the LCD is the product of the denominators.

2. If the denominators have common factors, the LCD is the product formed by including each factor the highest number of times it appears in any one denominator.

## APPLICATIONS

Fractions and rational expressions are found in many applications of mathematics.

**EXAMPLE 7**

*The van der Waal equation*   In physics the van der Waal equation for gases is

$$P = \frac{RT}{v - b} - \frac{a}{v^2}$$

Change the expression on the right to a single fraction using a common denominator.

**SOLUTION**

$$P = \frac{RT}{v - b} - \frac{a}{v^2}$$

$$= \frac{v^2 RT}{v^2(v - b)} - \frac{a(v - b)}{v^2(v - b)}$$

$$= \frac{RTv^2 - av + ab}{v^2(v - b)}$$

♦ ♦ ♦ ♦ ♦ ♦ ♦ ♦ ♦ ♦ ♦ ♦ ♦ ♦ ♦ ♦ ♦ ♦ ♦ ♦ ♦ ♦ ♦ ♦ ♦ ♦ ♦ ♦ ♦ ♦ ♦ ♦ ♦ ♦ ♦ ♦ ♦ ♦

Recall that subscripts are small numbers placed on variables to distinguish different items in a set of similar objects. In the next example subscripts are used to distinguish three different doors in a theater.

EXAMPLE 8

**Theater exits**    In architecture, the rate of flow of traffic through doors is important. A movie theater has three exit doors of slightly different sizes. The first can empty the theater in $t_1$ minutes; the second, in $t_2$ minutes; and the third, in $t_3$ minutes. The formula to determine the number of minutes, $t$, required to empty the theater if all three doors are functioning is

$$\frac{1}{t_1} + \frac{1}{t_2} + \frac{1}{t_3} = \frac{1}{t}$$

Add the three fractions on the left side of the equation.

SOLUTION

$$\frac{1}{t_1} + \frac{1}{t_2} + \frac{1}{t_3} = \frac{1}{t}$$

$$\frac{1 \cdot t_2 t_3}{t_1 \cdot t_2 t_3} + \frac{1 \cdot t_1 t_3}{t_2 \cdot t_1 t_3} + \frac{1 \cdot t_1 t_2}{t_3 \cdot t_1 t_2} = \frac{1}{t}$$

$$\frac{t_2 t_3 + t_1 t_3 + t_1 t_2}{t_1 t_2 t_3} = \frac{1}{t}$$

◆ ◆ ◆ ◆ ◆ ◆ ◆ ◆ ◆ ◆ ◆ ◆ ◆ ◆ ◆ ◆ ◆ ◆ ◆ ◆ ◆ ◆ ◆ ◆ ◆ ◆ ◆ ◆ ◆ ◆ ◆ ◆

If we solve for $t$ by cross multiplication or by taking the reciprocal of both sides, we obtain

$$t = \frac{t_1 t_2 t_3}{t_2 t_3 + t_1 t_3 + t_1 t_2}$$

Thus the time required to empty the theater is not a simple sum of the individual door times.

## COMMON DENOMINATORS AND COMPLEX FRACTIONS

We may use common denominators to simplify complex fractions.

EXAMPLE 9

Determine the slope of the line connecting $\left(\dfrac{a}{2}, \dfrac{b}{2}\right)$ with $(a, b)$.

SOLUTION

A sketch of the coordinates is shown in Figure 9. We substitute the coordinates directly into the slope formula.

$$\text{slope} = m = \frac{y_2 - y_1}{x_2 - x_1} = \frac{b - \dfrac{b}{2}}{a - \dfrac{a}{2}} = \frac{\dfrac{b}{2}}{\dfrac{a}{2}}$$

We have two ways to simplify the resulting complex fraction. We can change the fraction bar into division, as in Section 8.3, or we can multiply top and bottom (numerator and denominator) by the common denominator of the two fractions.

*Method 1:*

$$\frac{\dfrac{b}{2}}{\dfrac{a}{2}} = \frac{b}{2} \div \frac{a}{2} = \frac{b}{2} \cdot \frac{2}{a} = \frac{b}{a}$$

*Method 2:*

$$\frac{\dfrac{b}{2}}{\dfrac{a}{2}} \cdot \frac{2}{2} = \frac{\dfrac{b}{2} \cdot \dfrac{2}{1}}{\dfrac{a}{2} \cdot \dfrac{2}{1}} = \frac{b}{a}$$

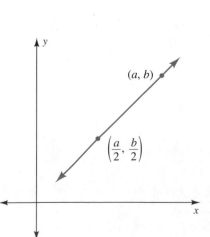

Figure 9

Thus the slope of the line connecting $\left(\dfrac{a}{2}, \dfrac{b}{2}\right)$ with $(a, b)$ is $\dfrac{b}{a}$.

◆ ◆ ◆ ◆ ◆ ◆ ◆ ◆ ◆ ◆ ◆ ◆ ◆ ◆ ◆ ◆ ◆ ◆ ◆ ◆ ◆ ◆ ◆ ◆ ◆ ◆ ◆ ◆ ◆ ◆ ◆ ◆ ◆ ◆

In this section we examined the techniques needed to add or subtract two or more fractions or rational expressions. In the next section we work with equations containing rational expressions.

## EXERCISES 8.4

**Add or subtract the fractions or rational expressions in Exercises 1 and 2.**

1. a. $\dfrac{11}{7} - \dfrac{4}{7}$   b. $\dfrac{2}{3} + \dfrac{x}{3}$   c. $\dfrac{3}{2x} - \dfrac{5}{2x}$

   d. $\dfrac{4}{x^2 + 1} - \dfrac{x^2}{x^2 + 1}$   e. $\dfrac{2}{x - 1} - \dfrac{x + 1}{x - 1}$

2. a. $\dfrac{13}{6} - \dfrac{5}{6}$   b. $\dfrac{2}{5} - \dfrac{x}{5}$   c. $\dfrac{5}{3x} - \dfrac{8}{3x}$

   d. $\dfrac{4}{x^2 + 2} + \dfrac{x^2}{x^2 + 2}$   e. $\dfrac{2}{x + 1} - \dfrac{x - 1}{x + 1}$

**What is the common denominator for each set of fractions or rational expressions in Exercises 3 and 4?**

3. a. $\dfrac{5}{12} , \dfrac{7}{20}$   b. $\dfrac{2}{x} , \dfrac{5}{2x}$   c. $\dfrac{8}{y} , \dfrac{1}{y^2}$

   d. $\dfrac{3}{b} + \dfrac{2}{a} + \dfrac{5}{b} - \dfrac{3}{a}$   e. $\dfrac{A(x+2)}{x - 3} , \dfrac{2}{x^2 - 9}$

   f. $\dfrac{4}{x^2 + 5x + 6} - \dfrac{2}{x^2 - 9}$

4. a. $\dfrac{1}{2} + \dfrac{3}{8}$   b. $\dfrac{5}{8} + \dfrac{7}{18}$   c. $\dfrac{2}{3} - \dfrac{3}{a}$

   d. $\dfrac{b}{a} - \dfrac{c}{a^2}$   e. $\dfrac{2}{x^2 + 2x} + \dfrac{5}{x^2 - 4}$

   f. $\dfrac{4}{x^2 - 2x + 1} - \dfrac{3}{x^2 - 1}$

5. Add or subtract, as indicated, the expressions in Exercise 3.

6. Add or subtract, as indicated, the expressions in Exercise 4.

**In Exercises 7 to 20 add or subtract the rational expressions, as indicated.**

7. $\dfrac{3}{2b} + \dfrac{3}{4a}$

8. $\dfrac{5}{3a} + \dfrac{7}{6b}$

9. $\dfrac{1}{x - 1} + \dfrac{1}{x}$

10. $\dfrac{1}{x + 1} + \dfrac{1}{x - 1}$

11. $\dfrac{8}{x + 1} - \dfrac{3}{x}$

12. $\dfrac{5}{x + 1} - \dfrac{8}{x}$

13. $\dfrac{2}{x + 3} + \dfrac{3}{(x + 3)^2}$

14. $\dfrac{4}{x - 2} + \dfrac{2}{(x - 2)^2}$

15. $\dfrac{1}{a} - \dfrac{1}{a^2}$

16. $\dfrac{2}{b^2} - \dfrac{3}{b}$

17. $\dfrac{2}{ab} - \dfrac{3}{2b}$

18. $\dfrac{3}{ac} - \dfrac{5}{2a}$

19. $\dfrac{3}{x^2 - x} + \dfrac{x}{x^2 - 3x + 2}$

20. $\dfrac{x}{x^2 - 6x + 9} + \dfrac{5}{x^2 + 3x}$

**Add the fractions on the right side of each equation in Exercises 21 to 32 to obtain a single fraction for the application formula. Assume nonzero variables.**

21. Temperature change: $\Delta T = \dfrac{T_0}{T} - 1$

22. Ventilation fans in an attic: $\dfrac{1}{t} = \dfrac{1}{t_1} + \dfrac{1}{t_2}$

23. Resistors in parallel in an electrical circuit:
    $\dfrac{1}{R} = \dfrac{1}{R_1} + \dfrac{1}{R_2}$

24. Condensers in series in an electrical circuit:
    $\dfrac{1}{C} = \dfrac{1}{C_1} + \dfrac{1}{C_2}$

25. Days to complete a wheat harvest with two machines:
    $\dfrac{1}{D} = \dfrac{1}{D_1} + \dfrac{1}{D_2}$

26. Total time for a round trip: $t = \dfrac{D}{r_1} + \dfrac{D}{r_2}$

27. Traffic flow through parallel doors: $\dfrac{1}{m} = \dfrac{1}{m_1} + \dfrac{1}{m_2}$

**28.** Focal distance for a lens, in optics: $\dfrac{1}{F} = \dfrac{1}{f_1} + \dfrac{1}{f_2}$

**29.** Traffic accident analysis, preliminary to calculating vehicle speed: $R = \dfrac{C^2}{8M} + \dfrac{M}{2}$

**30.** Radius of curvature of a surface: $F = \dfrac{L^2}{6d} + \dfrac{d}{2}$

**31.** Approximating an exponential function:

$$e^x \approx 1 + x + \dfrac{x^2}{2} + \dfrac{x^3}{6} + \dfrac{x^4}{24}$$

**32.** Approximating a trigonometric function:

$$\sin x \approx x - \dfrac{x^3}{6} + \dfrac{x^5}{120} - \dfrac{x^7}{5040}$$

**For Exercises 33 to 38 determine the slope of the line connecting the points. Simplify to eliminate fractions from the numerator and the denominator.**

**33.**

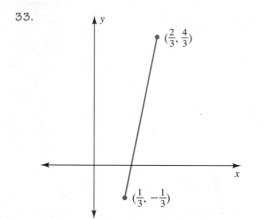

**34.**

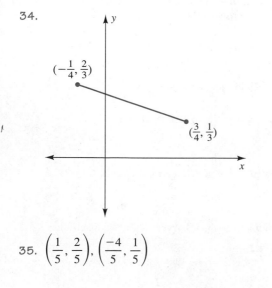

**35.** $\left(\dfrac{1}{5}, \dfrac{2}{5}\right), \left(\dfrac{-4}{5}, \dfrac{1}{5}\right)$

**36.** $\left(\dfrac{2}{3}, \dfrac{1}{4}\right), \left(\dfrac{-2}{3}, \dfrac{3}{4}\right)$

**37.**

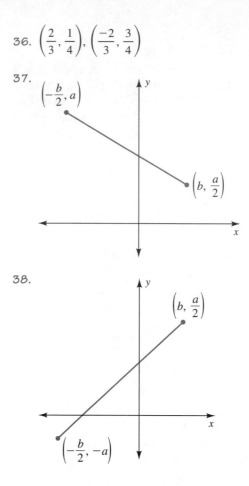

**38.**

**In Exercises 39 and 40 determine the slopes of the line segments connecting the indicated points in each figure. What do you observe about the slopes?**

**39.** Line segment $(a, b)$ to $(c, 0)$

Line segment $\left(\dfrac{a}{2}, \dfrac{b}{2}\right)$ to $\left(\dfrac{c}{2}, 0\right)$

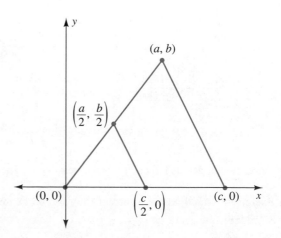

**40.** Line segment $(0, 0)$ to $(c, 0)$

Line segment $\left(\dfrac{a}{2}, \dfrac{b}{2}\right)$ to $\left(\dfrac{a + c}{2}, \dfrac{b}{2}\right)$

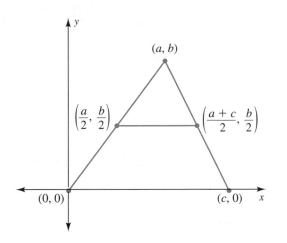

**Simplify the complex fractions in Exercises 41 to 48 to eliminate fractions from the numerators and denominators.**

**41.** $\dfrac{5}{\frac{1}{5} + 1}$

**42.** $\dfrac{3}{1 + \frac{1}{3}}$

**43.** $h = \dfrac{A}{\frac{1}{2}b}$

**44.** $h = \dfrac{V}{\frac{1}{3}\pi r^2}$

**45.** $\dfrac{x + \dfrac{x}{2}}{2 - \dfrac{x}{3}}$

**46.** $\dfrac{\dfrac{x}{3} + x}{3 - \dfrac{x}{2}}$

**47.** Refrigeration cycle: $\dfrac{1}{\dfrac{Q_H}{Q_L} - 1}$

**48.** Heat transfer: $\dfrac{1}{1 - \dfrac{Q_L}{Q_H}}$

**49.** Use the distance formula to find the lengths of the indicated line segments in Exercise 39. What do you observe about the segments?

**50.** Use the distance formula to find the lengths of the indicated line segments in Exercise 40. What do you observe about the segments?

**51.** Refer to Figure 9 in Example 9. Does the line containing $(a, b)$ and $\left(\dfrac{a}{2}, \dfrac{b}{2}\right)$ pass through the origin? Why or why not?

**52.** Does the line containing $\left(a, \dfrac{a}{2}\right)$ and $\left(b, \dfrac{b}{2}\right)$ pass through the origin?

GRAPHING CALCULATOR EXPLORATION

**53. a.** Graph the equation in Exercise 45, before and after simplifying.

**b.** Compare the graphs before and after simplifying.

**c.** Where is the graph nearly vertical? Why?

**54. a.** Graph the equation in Exercise 46 before and after simplifying.

**b.** Compare the graphs before and after simplifying.

**c.** Where is the graph nearly vertical? Why?

*PROBLEM SOLVING*

**Identify the missing operation symbol ($+$, $-$, $\cdot$, or $\div$) in Exercises 55 to 58.**

**55. a.** $\dfrac{3}{4} \; \square \; \dfrac{2}{5} = \dfrac{6}{20} = \dfrac{3}{10}$

**b.** $\dfrac{3}{4} \; \square \; \dfrac{2}{5} = \dfrac{15}{8}$

**c.** $\dfrac{3}{4} \; \square \; \dfrac{2}{5} = \dfrac{23}{20}$

**56. a.** $\dfrac{2}{5} \; \square \; \dfrac{3}{4} = \dfrac{8}{15}$

**b.** $\dfrac{3}{4} \; \square \; \dfrac{2}{5} = \dfrac{7}{20}$

**c.** $\dfrac{5}{4} \; \square \; \dfrac{2}{3} = \dfrac{23}{12}$

**57. a.** $\dfrac{1}{a} \; \square \; \dfrac{1}{b} = \dfrac{a + b}{ab}$

**b.** $\dfrac{1}{a} \; \square \; \dfrac{1}{b} = \dfrac{1}{ab}$

**58. a.** $\dfrac{1}{a} \; \square \; \dfrac{1}{b} = \dfrac{b - a}{ab}$

**b.** $\dfrac{1}{b} \; \square \; \dfrac{1}{a} = \dfrac{a}{b}$

*PROJECTS*

**59. *Fraction Pattern***

**a.** Add these two sets of fractions.

$$\tfrac{1}{6} + \tfrac{1}{9} + \tfrac{1}{18} \qquad \tfrac{1}{10} + \tfrac{1}{15} + \tfrac{1}{30}$$

**b.** Look for a shortcut for finding the sum, and use it to add the next two sets of fractions.

$$\tfrac{1}{22} + \tfrac{1}{33} + \tfrac{1}{66} \qquad \tfrac{1}{18} + \tfrac{1}{27} + \tfrac{1}{54}$$

**c.** Write a formula for the pattern, and show why the pattern always works.

**60.** ***Exiting a Theater.*** Return to the movie theater project in Exercise 44 of Section 8.3. The movie theater has two exits. One door, by itself, can empty the theater in 5 minutes. The second door, by itself, can empty the theater in 8 minutes. Suppose we wish to answer this question: If both doors are available, how many minutes will it take to empty the theater?

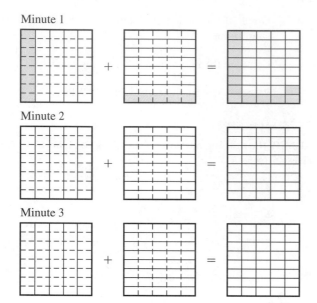

Minute 1

Minute 2

Minute 3

a. How are the rectangles in the figure the same as those in the figure in Exercise 44 of Section 8.3?

b. How are the rectangles different?

c. How many small pieces are there in each large rectangle? Why is this number important to adding the pieces together?

d. What does the shading in the first rectangle in the first row represent?

e. What does the shading in the second rectangle in the first row represent?

f. What does the shading in the third rectangle in the first row represent?

g. Shade the rectangles for each subsequent minute's departures. Add more rectangles as needed.

h. How will the rectangle look when the theater is empty?

i. Estimate the total number of minutes needed to empty the theater.

j. Why might the rectangles in this figure be a better way to represent the fractions $\frac{1}{5}$ and $\frac{1}{8}$ than those in the figure in Exercise 44 of Section 8.3?

k. Make up a problem of your own involving three doors. See Example 8.

---

## 8.5  SOLVING RATIONAL EQUATIONS

**OBJECTIVES**   Find the least common denominator of rational expressions in an equation. ♦ Multiply by the least common denominator to eliminate the denominators in an equation. ♦ Solve equations containing fractions or rational expressions. ♦ Solve application problems related to $\dfrac{1}{a} + \dfrac{1}{b} = \dfrac{1}{c}$. ♦ Solve application problems related to ratios.

**WARM-UP**   Multiply these expressions.

**1.** $\dfrac{12}{1}\left(\dfrac{x}{4} + \dfrac{x}{6}\right)$       **2.** $\dfrac{2x}{1}\left(\dfrac{2}{x} + \dfrac{3x}{2}\right)$       **3.** $x(x + 1)\left(\dfrac{2}{x + 1} + \dfrac{3}{x}\right)$

**4.** $x^2\left(\dfrac{2}{x^2} - \dfrac{3}{x} + 1\right)$       **5.** $36d\left(\dfrac{1}{12} + \dfrac{1}{18}\right)$

Find the least common denominator for each pair of fractions.

**6.** $\dfrac{1}{12}, \dfrac{1}{18}$       **7.** $\dfrac{1}{9}, \dfrac{1}{15}$                                                  ♦

I n this section we solve equations containing fractions and rational expressions. We also examine one aspect of rational equations that has particular appeal to the mathematician: a common formula to describe patterns that occur in a variety of applications.

THE HARVEST PROBLEM

In Example 1 we return to the wheat harvest problem.

E X A M P L E    *1*

*Wheat harvest*    Farmers Lee and Terry can separately harvest Lee's wheat in 12 days and 18 days, respectively. How might we go about calculating the number of days needed to harvest the crop if they work together?

SOLUTION    Consider the following proposed methods.

*Proposed method 1*: Adding the 12 days and 18 days gives 30 days, which is not reasonable. If the farmers work together, the task should take fewer than the 12 days Lee requires alone.

*Proposed method 2*: Suppose we average the number of days:

$$\text{average} = \frac{12 \text{ days} + 18 \text{ days}}{2} = 15 \text{ days}$$

This result is not reasonable, as the two farmers should not take longer than Lee working by herself.

*Proposed method 3*: If we subtract 12 days from the 18 days, we get 6 days. Although subtraction might give a reasonable answer in this setting, consider subtraction in the case where both farmers harvest in 18 days. Subtraction would give zero days working together, whereas a reasonable answer might be 9 days.

*Proposed method 4*: Finding how much the farmers harvest each day is another approach. On the first day, Lee harvests $\frac{1}{12}$ of the total while Terry harvests $\frac{1}{18}$. Together they harvest $\frac{1}{12} + \frac{1}{18}$. Each day they repeat these fractions of the total job. The following summarizes the fraction of the total crop harvested by the end of each day, for 8 days:

Day 1:  $\frac{1}{12} + \frac{1}{18} = \frac{3}{36} + \frac{2}{36} = \frac{5}{36}$

Day 2:  $2\left(\frac{1}{12} + \frac{1}{18}\right) = 2\left(\frac{5}{36}\right) = \frac{10}{36}$

Day 3:  $3\left(\frac{1}{12} + \frac{1}{18}\right) = 3\left(\frac{5}{36}\right) = \frac{15}{36}$

Day 4:  $\qquad\qquad 4\left(\frac{5}{36}\right) = \frac{20}{36}$

Day 5:  $\qquad\qquad 5\left(\frac{5}{36}\right) = \frac{25}{36}$

Day 6:  $\qquad\qquad 6\left(\frac{5}{36}\right) = \frac{30}{36}$

Day 7:  $\qquad\qquad 7\left(\frac{5}{36}\right) = \frac{35}{36}$

Day 8:  $\qquad\qquad 8\left(\frac{5}{36}\right) = \frac{40}{36}$

The harvest is complete when the fraction reaches $\frac{36}{36} = 1$. The harvest is nearly complete at the end of day 7. It is completely finished during day 8, when the fraction exceeds 1. The exact time to finish is $d$ days, where

$$d\left(\frac{1}{12} + \frac{1}{18}\right) = 1$$

Traditionally, this equation is divided on both sides by $d$ and written

$$\frac{1}{12} + \frac{1}{18} = \frac{1}{d}$$

❖ ❖ ❖ ❖ ❖ ❖ ❖ ❖ ❖ ❖ ❖ ❖ ❖ ❖ ❖ ❖ ❖ ❖ ❖ ❖ ❖ ❖ ❖ ❖ ❖ ❖ ❖ ❖ ❖ ❖

SOLVING RATIONAL EQUATIONS

Our first step in solving the harvest equation will be to eliminate the denominators. (Yes, we just divided by $d$ and put it in the denominator. Now we are going to take it out again. Keep in mind that the traditional form of the equation has the variable in the denominator.)

In Section 8.3 and this section's Warm-up we applied the distributive property to rational expressions. When we multiplied by the least common denominator, we eliminated the denominators. We use this approach to eliminate denominators in equations.

> To eliminate fractions from an equation, multiply both sides of the equation by the least common denominator.

**EXAMPLE  2**    Solve $\dfrac{1}{12} + \dfrac{1}{18} = \dfrac{1}{d}$ for $d$.

**SOLUTION**    The least common denominator for all fractions in the equation is $36d$.

$$\frac{1}{12} + \frac{1}{18} = \frac{1}{d}$$

$$36d\left(\frac{1}{12} + \frac{1}{18}\right) = 36d\left(\frac{1}{d}\right)$$

$$\frac{36d \cdot 1}{12} + \frac{36d \cdot 1}{18} = \frac{36d \cdot 1}{d}$$

$$3d + 2d = 36$$

$$5d = 36$$

$$d = \tfrac{36}{5} = 7.2 \text{ days}$$

*Calculator check*: $\frac{1}{12} + \frac{1}{18} \approx 0.0833 + 0.0556 \approx 0.1389 \approx \frac{1}{7.2}$ ✔

◆ ◆ ◆ ◆ ◆ ◆ ◆ ◆ ◆ ◆ ◆ ◆ ◆ ◆ ◆ ◆ ◆ ◆ ◆ ◆ ◆ ◆ ◆ ◆ ◆ ◆ ◆ ◆ ◆ ◆ ◆ ◆ ◆

In Example 2 we could have added the fractions on the left side and created a proportion. The equation could then have been solved by cross multiplication. We leave this method as an exercise.

**EXAMPLE  3**    *Wheat harvest*    When it is time to harvest Terry's wheat, Lee moves her equipment to the south end of the county. Terry normally harvests his wheat in 15 days. Lee's equipment could do the work in 9 days. If they work together, how long will it take to harvest Terry's wheat?

**SOLUTION**    We have the same situation as before but with different numbers of days of work. The equation is

$$\frac{1}{9} + \frac{1}{15} = \frac{1}{d}$$

The least common denominator of the fractions within the equation is $45d$.

$$\frac{1}{9} + \frac{1}{15} = \frac{1}{d}$$

$$45d\left(\frac{1}{9} + \frac{1}{15}\right) = 45d\left(\frac{1}{d}\right)$$

$$\frac{45d \cdot 1}{9} + \frac{45d \cdot 1}{15} = \frac{45d \cdot 1}{d}$$

$$5d + 3d = 45$$

$$8d = 45$$

$$d = \tfrac{45}{8} = 5.625 \text{ days}$$

*Calculator check*: $\frac{1}{9} + \frac{1}{15} \approx 0.1111 + 0.0667 \approx 0.1778 \approx \frac{1}{5.625}$ ✔

◆ ◆ ◆ ◆ ◆ ◆ ◆ ◆ ◆ ◆ ◆ ◆ ◆ ◆ ◆ ◆ ◆ ◆ ◆ ◆ ◆ ◆ ◆ ◆ ◆ ◆ ◆ ◆ ◆ ◆ ◆ ◆ ◆

The individual times to harvest the wheat may be described as $d_1$ and $d_2$. Working individually, the farmers could harvest the whole crop at a rate of $\frac{1}{d_1}$ and $\frac{1}{d_2}$, respectively, each day. Because both are working to finish 1 job, we add the rates for the two farmers to obtain the rate for working together, $\frac{1}{d}$:

$$\frac{1 \text{ wheat crop}}{d_1 \text{ days}} + \frac{1 \text{ wheat crop}}{d_2 \text{ days}} = \frac{1 \text{ wheat crop}}{d \text{ days together}}$$

$$\frac{1}{d_1} + \frac{1}{d_2} = \frac{1}{d}$$

We now return to an application from the projects in Sections 8.3 and 8.4. This application has a surprisingly similar structure to the harvest application.

**EXAMPLE 4**

*Exiting a theater*    Suppose a movie theater has two exits. One is a single door that, by itself, can empty the theater in 8 minutes. The other is a double-wide door that, by itself, can empty the theater in 5 minutes. Build an equation to calculate the time required to empty the theater when both doors are open.

SOLUTION    Converting exit time to a rate for emptying the theater, we say that during each minute the doors permit $\frac{1}{8}$ and $\frac{1}{5}$ of the theater, respectively, to empty. The figure in Project 60 of Section 8.5 shows a related drawing. The following expressions represent the portion of the theater that has been emptied by the end of each minute:

First minute:    $\frac{1}{8} + \frac{1}{5} = \frac{5}{40} + \frac{8}{40} = \frac{13}{40}$

Second minute:  $2\left(\frac{1}{8} + \frac{1}{5}\right) = 2\left(\frac{13}{40}\right) = \frac{26}{40}$

Third minute:   $3\left(\frac{13}{40}\right) = \frac{39}{40}$

Fourth minute:  $4\left(\frac{13}{40}\right) = \frac{52}{40}$

The theater will be entirely empty during the fourth minute, when the fraction $\frac{40}{40} = 1$ is reached. To find the exact time, $x$, needed to empty the theater, we solve the equation

$$x\left(\frac{1}{8} + \frac{1}{5}\right) = 1$$

Again, the traditional form of this equation is

$$\frac{1}{8} + \frac{1}{5} = \frac{1}{x}$$

♦ ♦ ♦ ♦ ♦ ♦ ♦ ♦ ♦ ♦ ♦ ♦ ♦ ♦ ♦ ♦ ♦ ♦ ♦ ♦ ♦ ♦ ♦ ♦ ♦ ♦ ♦ ♦ ♦ ♦ ♦ ♦ ♦ ♦ ♦ ♦ ♦ ♦

**EXAMPLE 5**    Solve $\frac{1}{8} + \frac{1}{5} = \frac{1}{x}$ for $x$, where $x \neq 0$.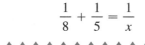

SOLUTION    The least common denominator for the equation is $40x$.

$$\frac{1}{8} + \frac{1}{5} = \frac{1}{x}$$

$$40x\left(\frac{1}{8} + \frac{1}{5}\right) = 40x\left(\frac{1}{x}\right)$$

$$5x + 8x = 40$$

$$13x = 40$$

$$x = \frac{40}{13} \approx 3.1 \text{ minutes}$$

*Calculator check*: $\frac{1}{8} + \frac{1}{5} = 0.125 + 0.200 = 0.325 \approx \frac{1}{3.1}$ ✔

The two-exit system permits the theater to be cleared rapidly.

♦ ♦ ♦ ♦ ♦ ♦ ♦ ♦ ♦ ♦ ♦ ♦ ♦ ♦ ♦ ♦ ♦ ♦ ♦ ♦ ♦ ♦ ♦ ♦ ♦ ♦ ♦ ♦ ♦ ♦ ♦ ♦ ♦ ♦ ♦ ♦ ♦ ♦ ♦

We have now used equations of the form $\frac{1}{a} + \frac{1}{b} = \frac{1}{c}$ in two applications—first in harvesting a crop and then in emptying a theater. This form of equation may also be used to represent filling containers. If $a$ and $b$ are the times needed to fill a child's wading pool with two separate hoses, then $c$ is the time needed to fill it with the two hoses together. What may be more surprising is that the same formula occurs in applications that seem to have nothing to do with harvesting, emptying, or filling. The formula appears in optics and at least twice in basic electronics. The formula to determine the focal length for lenses is

$$\frac{1}{p} + \frac{1}{q} = \frac{1}{f}$$

The formula to calculate resistance for parallel resistors is

$$\frac{1}{R_1} + \frac{1}{R_2} = \frac{1}{R}$$

and the formula to calculate the capacitance of condensers in series is

$$\frac{1}{C_1} + \frac{1}{C_2} = \frac{1}{C}$$

In Section 8.4 we added three rates of emptying a theater:

$$\frac{1}{t_1} + \frac{1}{t_2} + \frac{1}{t_3} = \frac{1}{t}$$

The formula may apply to three doors, pieces of harvesting equipment, hoses, lenses, resistors, or condensers. Each additional door, harvester, hose, lens face, resistor, or condenser adds another fraction.

Keep in mind that the objective in working with these applications is developing skill in using the formula, rather than mastering the applications themselves. Examples 6, 7, and 8 provide more illustrations of solving equations containing rational expressions. The first two review solving equations by factoring.

**EXAMPLE 6**    Solve $\frac{2}{x} + \frac{3x}{2} = 4$ for $x$, where $x \neq 0$.

SOLUTION

$$\frac{2}{x} + \frac{3x}{2} = 4$$

$$2x\left(\frac{2}{x} + \frac{3x}{2}\right) = 2x \cdot 4 \qquad \text{Multiply by the LCD.}$$

$$\frac{2x \cdot 2}{x} + \frac{2x \cdot 3x}{2} = 2x \cdot 4 \qquad \text{Distribute the LCD.}$$

$$4 + 3x^2 = 8x \qquad \text{Reduce the fractions.}$$

$$3x^2 - 8x + 4 = 0 \qquad \text{Solve for the zero form.}$$

$$(3x - 2)(x - 2) = 0 \qquad \text{Factor.}$$

Either    $3x - 2 = 0$    or    $x - 2 = 0$

Either    $x = \frac{2}{3}$    or    $x = 2$

We check by substituting each $x$ into the original equation. The check is left as an exercise.

◆ ◆ ◆ ◆ ◆ ◆ ◆ ◆ ◆ ◆ ◆ ◆ ◆ ◆ ◆ ◆ ◆ ◆ ◆ ◆ ◆ ◆ ◆ ◆ ◆ ◆ ◆ ◆ ◆ ◆ ◆ ◆ ◆

E X A M P L E  **7**   Solve $\dfrac{2}{x+1} + \dfrac{3}{x} = -2$ for $x$, where $x \neq -1$ and $x \neq 0$.

SOLUTION

$$\frac{2}{x+1} + \frac{3}{x} = -2$$

$$x(x+1)\left(\frac{2}{x+1} + \frac{3}{x}\right) = -2 \cdot x(x+1) \qquad \text{Multiply by the LCD.}$$

$$\frac{x(x+1)\cdot 2}{x+1} + \frac{x(x+1)\cdot 3}{x} = -2x(x+1) \qquad \text{Distribute the LCD.}$$

$$2x + 3(x+1) = -2x^2 - 2x \qquad \text{Reduce the fractions.}$$

$$2x + 3x + 3 + 2x^2 + 2x = 0 \qquad \text{Solve for the zero form.}$$

$$2x^2 + 7x + 3 = 0 \qquad \text{Combine like terms.}$$

$$(2x+1)(x+3) = 0 \qquad \text{Factor.}$$

Either $\quad x = -\tfrac{1}{2} \quad$ or $\quad x = -3$

The check by substitution is left as an exercise.

$\blacklozenge \blacklozenge \blacklozenge \blacklozenge \blacklozenge \blacklozenge \blacklozenge \blacklozenge \blacklozenge \blacklozenge \blacklozenge \blacklozenge \blacklozenge \blacklozenge \blacklozenge \blacklozenge \blacklozenge \blacklozenge \blacklozenge \blacklozenge \blacklozenge \blacklozenge \blacklozenge \blacklozenge \blacklozenge \blacklozenge \blacklozenge \blacklozenge \blacklozenge \blacklozenge \blacklozenge \blacklozenge$

E X A M P L E  **8**   Solve $\dfrac{x+1}{x-3} = \dfrac{3}{x-3}$ for $x$, where $x \neq 3$, first by multiplying both sides by the common denominator and then by cross multiplication.

SOLUTION   When we multiply both sides by $x-3$, the solution is $x=2$.

$$\frac{x+1}{x-3} = \frac{3}{x-3}$$

$$(x-3) \cdot \frac{(x+1)}{(x-3)} = (x-3) \cdot \frac{3}{(x-3)}$$

$$x + 1 = 3$$

$$x = 2$$

When we cross multiply, we find a quadratic equation with two solutions.

$$\frac{x+1}{x-3} = \frac{3}{x-3}$$

$$(x-3)(x+1) = 3(x-3)$$

$$x^2 - 2x - 3 = 3x - 9$$

$$x^2 - 5x + 6 = 0$$

$$(x-2)(x-3) = 0$$

Either $\quad x - 2 = 0 \quad$ or $\quad x - 3 = 0$

Either $\qquad x = 2 \quad$ or $\qquad x = 3$

***Check***: In checking our solutions, we find that $x = 2$ satisfies the equation:

$$\frac{2+1}{2-3} \overset{?}{=} \frac{3}{2-3}$$

However, $x = 3$ gives a zero denominator:

$$\frac{3+1}{3-3} \overset{?}{=} \frac{3}{3-3}$$

As noted in the original problem, $x = 3$ has been excluded from the set of possible inputs. The solution $x = 3$ is an extraneous root. ✔

$\blacklozenge \blacklozenge \blacklozenge \blacklozenge \blacklozenge \blacklozenge \blacklozenge \blacklozenge \blacklozenge \blacklozenge \blacklozenge \blacklozenge \blacklozenge \blacklozenge \blacklozenge \blacklozenge \blacklozenge \blacklozenge \blacklozenge \blacklozenge \blacklozenge \blacklozenge \blacklozenge \blacklozenge \blacklozenge \blacklozenge \blacklozenge \blacklozenge \blacklozenge \blacklozenge \blacklozenge \blacklozenge$

## EXERCISES 8.5

**Add the fractions on the left side of each equation in Exercises 1 to 4, and use cross multiplication on the resulting proportion to solve for the indicated variable.**

1. Solve $\dfrac{1}{12} + \dfrac{1}{18} = \dfrac{1}{d}$ for $d$.

2. Solve $\dfrac{1}{9} + \dfrac{1}{15} = \dfrac{1}{d}$ for $d$.

3. Solve $\dfrac{1}{8} + \dfrac{1}{5} = \dfrac{1}{x}$ for $x$.

4. Solve $\dfrac{1}{5} + \dfrac{1}{6} = \dfrac{1}{x}$ for $x$.

**Recall that the reciprocal key, $\boxed{x^{-1}}$ or $\boxed{1/x}$, provides a way to obtain decimals for fractions with 1 in the numerator. The order of keystrokes varies with the brand of calculator, so identify your calculator's order as you work the following exercises. In Exercises 5 to 8 use the reciprocal key to solve the given equation; list your keystrokes.**

5. The equation in Exercise 1

6. The equation in Exercise 2

7. The equation in Exercise 3

8. The equation in Exercise 4

9. Substitute $x = 2$ into the equation in Example 6, and verify that it is a solution.

10. Substitute $x = \frac{2}{3}$ into the equation in Example 6, and verify that it is a solution. Use a calculator as needed.

11. Substitute $x = -\frac{1}{2}$ into the equation in Example 7, and verify that it is a solution. Use a calculator as needed.

12. Substitute $x = -3$ into the equation in Example 7, and verify that it is a solution.

**In Exercises 13 to 28 solve for $x$, using whichever method (proportion, calculator, multiplication by the LCD) seems appropriate. List any inputs that make the fractions undefined.**

13. $\dfrac{x}{4} + \dfrac{x}{6} = 28$

14. $\dfrac{x}{14} + \dfrac{x}{8} = 11$

15. $\dfrac{3}{4} + \dfrac{1}{5} = \dfrac{1}{x}$

16. $\dfrac{2}{3} + \dfrac{2}{5} = \dfrac{1}{x}$

17. $\dfrac{1}{8} + \dfrac{1}{x} = \dfrac{1}{2}$

18. $\dfrac{1}{10} + \dfrac{1}{x} = \dfrac{1}{6}$

19. $\dfrac{1}{x} = \dfrac{1}{3x} + \dfrac{1}{3}$

20. $\dfrac{4}{x} + \dfrac{2}{x} = \dfrac{3}{x}$

21. $\dfrac{3}{x} - \dfrac{2}{x} = \dfrac{4}{x}$

22. $\dfrac{1}{x} = \dfrac{1}{2} - \dfrac{1}{2x}$

23. $\dfrac{1}{x-1} = \dfrac{2}{x+3}$

24. $\dfrac{2}{x-1} = \dfrac{1}{x-4}$

25. $\dfrac{2}{x^2} - \dfrac{3}{x} + 1 = 0$

26. $1 + \dfrac{5}{x} - \dfrac{14}{x^2} = 0$

27. $\dfrac{1}{x-3} - 3 = \dfrac{4-x}{x-3}$

28. $\dfrac{1}{x-4} = \dfrac{5-x}{x-4} + 4$

29. Compare the roles of multiplying by the least common denominator in simplifying complex fractions and in solving equations.

30. What was done wrong in the following solution?

$$12x\left(\dfrac{2}{3x} + \dfrac{1}{4}\right) = \dfrac{4 \cdot 2}{1} + \dfrac{4}{4} = 8 + 1 = 9$$

31. One farmer harvests his barley in 14 days. A second farmer harvests the same crop in 12 days. How many days will it take them working together?

32. One farmer bales her hay in 6 days. A second farmer does it in 8 days. How many days will it take them working together?

33. One ventilation fan changes the air in a house in 4 hours. A second fan vents the same volume of air in 5 hours. How long will it take to vent the house if both fans are working? If building code requires a complete change of air every 3 hours, will the two fans be sufficient?

34. A $\frac{5}{8}$-inch garden hose fills a child's pool in 1 hour. A $\frac{1}{2}$-inch garden hose fills a child's pool in 1.5 hours. How long will it take to fill the pool if both hoses are used at once?

35. One ventilation fan changes the air in a house in 5 hours. A second fan is to be installed. In order to satisfy code, both fans working together must change the air in 3 hours. Describe the fan needed to satisfy code.

36. A $1\frac{1}{2}$-inch pipe fills a swimming pool directly from a farm well in 5 days. A $\frac{5}{8}$-inch hose feeding water from the cistern is added, and the pool fills in 4 days. How long would the hose take to fill the pool by itself?

37. One door can clear a theater in 9 minutes. A second door can clear the theater in 6 minutes. How fast can the theater be emptied if both doors are available?

38. A large theater has three exits. Operating individually, each door can clear the theater in 6 minutes. How fast can the theater be emptied if all three doors are available?

**In Exercises 39 to 42 solve for the indicated letter. Assume none of the variables is zero.**

39. $\dfrac{1}{a} + \dfrac{1}{b} = \dfrac{1}{c}$ for $b$

40. $\dfrac{1}{a} + \dfrac{1}{b} = \dfrac{1}{c}$ for $a$

41. $\dfrac{1}{a} + \dfrac{1}{b} = \dfrac{1}{c}$ for $c$

42. $\dfrac{1}{R} = \dfrac{1}{R_1} + \dfrac{1}{R_2}$ for $R_1$

PROJECTS

43. **Harvest Problem Scale Drawing.** A graphical solution to the types of problems in this section has been described by J. E. Thompson.* We can apply this graphical model to our harvest problem: If Lee is able to harvest the crop in 12 days and Terry is able to harvest the crop in 18 days, how many days will it take both working together to harvest the crop?

On a horizontal line, we draw two perpendicular lines to scale (say, $AC = 12$ centimeters and $BD = 18$ centimeters). As shown in the figure, these lines are parallel; they may be any distance, $AB$, apart. Each centimeter in the vertical direction represents 1 day. We then draw diagonals $AD$ and $BC$. The point of intersection of $AD$ and $BC$ is labeled $E$. The number of days required for the farmers to harvest the crop together is the vertical distance from the line $AB$ to the point $E$. To determine the length of $EF$, we measure carefully. Use the scale to convert back to days.

a. Repeat Example 3, using this technique.

b. Repeat Example 5, using this technique. Let each centimeter represent 1 minute.

44. **Wheat Harvest.** Return to Examples 1 to 3. After the harvest the two farmers wish to equalize the costs. Which farmer should pay the other and for how many days' work? There may be no simple answer. Consider

harvest time working alone,

actual harvest time working together,

how many days of harvest time each saved the other, and

how many more days one worked on the other's harvest.

How might your cost analysis change if a local campground were the source of a fire every summer that threatened Lee's wheat crop?

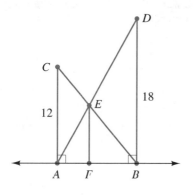

---

# CHAPTER 8 SUMMARY

## Vocabulary

*For definitions and page references, see the Glossary/Index.*

complex fractions

inverse variation

inverse proportion

least common denominator

opposites

rational equations

rational expressions

rational numbers

simplification property of fractions

## Concepts

A vertically infinite graph results when the denominator of a simplified rational expression approaches zero.

A rational expression must be factored in order to simplify by canceling.

Advice from Angie Cowles, student at Lane Community College: "Common factors will be more apparent if you leave the final expressions in factored form."

When the numerator and denominator of a fraction are the same, the fraction equals 1.

When the numerator and denominator of a fraction are opposites, the fraction equals $-1$.

To divide rational expressions, multiply the first expression by the reciprocal of the second expression.

To find the **least common denominator**, form a product by including each factor the highest number of times it appears in any denominator.

To eliminate denominators in an equation, multiply each side by the least common denominator.

When a rational equation is in the form of a proportion, $\dfrac{a}{b} = \dfrac{c}{d}$, the equation may be cross multiplied to $a \cdot d = b \cdot c$ and then solved.

---

*Arithmetic for the Practical Man* (D. Van Nostrand Co., 1946), p. 237.

## CHAPTER 8 REVIEW EXERCISES

1. Explain why $\dfrac{x-6}{x+6}$ will not simplify.

2. For what input, $x$, is the expression $\dfrac{2-x}{x+3}$ undefined?

3. Find ten coordinate points that satisfy $y = \dfrac{6}{x+3}$, and sketch a graph. Is there an input that gives $y = 0$? Why or why not?

4. World reserves of natural gas were estimated at 4000 trillion cubic feet in 1990. Write an equation that describes the number of years the gas reserves will last if $x$ cubic feet are used per day.

5. Simplify each expression. Which simplify to fractions that are equivalent to $\frac{6}{8}$? Why?

   a. $\dfrac{6 \cdot 2}{8 \cdot 2}$    b. $\dfrac{6 \div 2}{8 \div 2}$    c. $\dfrac{6-2}{8-2}$    d. $\dfrac{6+2}{8+2}$

6. Is the pair of fractions $\dfrac{16}{9}$ and $\dfrac{\sqrt{16}}{\sqrt{9}}$ equal or not equal? Explain why, in terms of factors.

**Simplify the rational expressions in Exercises 7 to 12, if possible. If not, explain why. Assume no zero denominators.**

7. $\dfrac{2xy}{x^2}$

8. $\dfrac{x+y}{xy}$

9. $\dfrac{a^2-b^2}{a+b}$

10. $\dfrac{ab}{a^2+b}$

11. $\dfrac{ab}{a^2+b^2}$

12. $\dfrac{ab}{a^2+ab}$

**Simplify the expressions in Exercises 13 and 14. Indicate any inputs that make the expression undefined.**

13. $\dfrac{1-a}{a-1}$

14. $\dfrac{x^2+4x+4}{x^2+3x+2}$

**In Exercises 15 to 20, what numerator or denominator is needed to make a true statement?**

15. $\dfrac{x-3}{\phantom{xxx}} = -1$    16. $\dfrac{\phantom{xxx}}{2-x} = -1$    17. $\dfrac{\phantom{xxx}}{a-b} = 1$

18. $\dfrac{\phantom{xxx}}{a+b} = 1$    19. $\dfrac{\phantom{xxx}}{4-x} = -1$    20. $\dfrac{b-5}{\phantom{xxx}} = -1$

21. a. Calculate the fractional expressions $300 \div \frac{3}{1}$ and $300 \cdot \frac{1}{3}$.

   b. What may be observed about the pair of problems?

   c. What do they tell us about multiplication and division?

**Add or subtract the expressions in Exercises 22 to 26, as indicated.**

22. $\dfrac{1}{2a} + \dfrac{5}{6ab}$

23. $\dfrac{x}{x+2} - \dfrac{2}{x^2-4}$

24. $\dfrac{x}{x+1} - \dfrac{2}{(x+1)^2}$

25. $\dfrac{a}{a+b} - \dfrac{b}{a-b}$

26. $\dfrac{4}{5} - \dfrac{3}{x} + \dfrac{2}{x^2}$

**In Exercises 27 to 30 multiply or divide, as indicated. Factor and simplify.**

27. $\dfrac{x}{x-1} \cdot \dfrac{x^2-1}{x^2}$

28. $\dfrac{1-x}{x+1} \div \dfrac{x^2-1}{x^2}$

29. $\dfrac{x^2-9}{3-x} \div \dfrac{x^2}{x+3}$

30. $\dfrac{n-2}{n(n-1)} \cdot \dfrac{(n+1)n(n-1)}{n-2}$

**Multiply the expressions in Exercises 31 and 32, and simplify the result.**

31. $x^2 \left( \dfrac{1}{x} + \dfrac{2}{x^2} \right)$

32. $(x+2)(x-2) \left( \dfrac{2}{x-2} + \dfrac{1}{x+2} \right)$

**Identify the word clues that indicate the necessary operations (addition, subtraction, multiplication, or division), then answer the question.**

33. John ate $\frac{1}{3}$ of the pie; Sue ate $\frac{1}{4}$ of the pie. Altogether they ate what portion of the pie?

34. How many servings are there in 4 large pizzas if each person eats one-third of a pizza?

35. Sred stitches $\frac{2}{3}$ of the hem. Evelyn rips out $\frac{1}{2}$ of Sred's work. What fraction remains to be finished?

36. Sally ran $\frac{3}{4}$ mile. Jim ran half as far. How far did Jim run?

37. A box of corn flakes contains 18 ounces. A serving is $1\frac{1}{10}$ ounces. How many servings are in the box?

38. Half an animal shelter's funding comes from cat and dog licensing fees. A third comes from property taxes. What fraction remains to be raised from private donations?

**Use the order of operations to simplify the expressions in Exercises 39 and 40.**

39. $\dfrac{1}{a} + \dfrac{2a}{3} \cdot \dfrac{6}{a} \div \dfrac{1}{3} - \dfrac{a}{3}$

40. $\dfrac{3}{c} - \dfrac{4}{3c} \div \dfrac{2}{3} + \dfrac{1}{4} \cdot \dfrac{8}{c}$

**In Exercises 41 and 42 simplify the expression or equation containing units of measurement.**

41. $\dfrac{4 \text{ buttons per card}}{12 \text{ buttons per shirt}}$

42. $d = -\dfrac{1}{2} \left( \dfrac{9.81 \text{ m}}{\text{sec}^2} \right) (5 \text{ sec})^2 + \left( \dfrac{8 \text{ m}}{\text{sec}} \right) (5 \text{ sec}) + 50 \text{ m}$

**In Exercises 43 to 45 simplify the expressions from dosage computation.**

43. $\text{gr} \, \dfrac{1}{2} \cdot \dfrac{1 \text{ tab}}{\text{gr} \frac{1}{6}}$

44. $\dfrac{\dfrac{1 \text{ mL}}{25 \text{ mL}} \cdot 400 \text{ mL}}{\dfrac{1 \text{ mL}}{4 \text{ mL}}}$

45. $\text{gr} \, \dfrac{1}{150} \cdot \dfrac{1 \text{ mL}}{\text{gr} \frac{1}{750}}$

In Exercises 46 and 47 describe the output behavior of the graph in the figure in the given situations.

46.

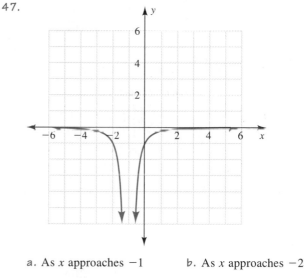

a. As $x$ approaches $-1$    b. As $x$ approaches 3

c. As $x$ approaches 0

47.

a. As $x$ approaches $-1$    b. As $x$ approaches $-2$

c. As $x$ approaches 0

In Exercises 48 and 49 write the expression on the right side of the equation as a single fraction that approximates the trigonometric function on the left.

48. $\tan x \approx x + \dfrac{x^3}{3} + \dfrac{2x^5}{15} + \dfrac{17x^7}{315} + \dfrac{62x^9}{2835}$

49. $\cos x \approx 1 - \dfrac{x^2}{2} + \dfrac{x^4}{24} - \dfrac{x^6}{720}$

50. Simplify the expression on the right side of the equation so that it contains no fractions in the numerator or denominator:

$$t^2 = \frac{d}{\frac{1}{2}g}$$

**Solve the equations in Exercises 51 to 54. For what inputs are the fractions not defined?**

51. $\dfrac{1}{5} = \dfrac{1}{9} + \dfrac{1}{x}$

52. $\dfrac{1}{5} + \dfrac{1}{x} = \dfrac{8}{5x}$

53. $\dfrac{3}{x} + \dfrac{1}{x-1} = \dfrac{17}{4x}$

54. $\dfrac{1}{2x} = \dfrac{2}{x} - \dfrac{x}{24}$

55. Find the slope of the line connecting the points $\left(\dfrac{a}{2}, b\right)$ and $\left(\dfrac{b}{2}, \dfrac{a}{3}\right)$.

56. The hot water line fills the clothes washer in 3 minutes. The cold water line fills the washer in 2 minutes. How long will it take to fill the washer on the "warm" setting, using both water lines?

57. One clerk is able to issue all advance mail-order tickets for an event in 30 days. An additional clerk is hired. The two clerks together get the job finished in 20 days. In how many days could the second clerk have done the job alone?

58. Explain the role of factoring in the addition or subtraction of rational expressions.

*GRAPHING CALCULATOR*

59. a. Enter $y_1 = \dfrac{2}{x} + \dfrac{1}{x-2}$ and $y_2 = \dfrac{2(x-2)+x}{x(x-2)}$.

b. Use the table function to compare the outputs. Try TblSet with TblMin $= -3$, $\Delta$Tbl $= 1$.

c. Why are there no outputs at $x = 0$ and $x = 2$?

d. Graph the equations.

e. What may be concluded about the two equations? Why?

---

## CHAPTER 8 TEST

1. For what input is the expression $\dfrac{4-x}{x-4}$ undefined?

2. Explain why the expression $\dfrac{4-x}{x-4}$ may or may not be simplified.

3. Evaluate $y = \dfrac{4}{x-2}$ for $x$ as integers between $-5$ and 5. Sketch a graph.

4. Simplify each expression. Which simplify to fractions equivalent to $\frac{6}{9}$? Why?

a. $\dfrac{6 \div 3}{9 \div 3}$    b. $\dfrac{6 - 3}{9 - 3}$    c. $\dfrac{6 + 3}{9 + 3}$    d. $\dfrac{6 \cdot 3}{9 \cdot 3}$

5. Using factors, show whether the fractions $\dfrac{4}{25}$ and $\dfrac{\sqrt{4}}{\sqrt{25}}$ are or are not equal.

**Simplify the expressions in Exercises 6 to 9, if possible. If not, explain why.**

6. $\dfrac{b^2}{3ab}$

7. $\dfrac{a - b}{ab}$

8. $\dfrac{a^2 - b^2}{a - b}$

9. $\dfrac{xy}{xy + y}$

**Multiply or divide the expressions in Exercises 10 to 13, as indicated. Factor and simplify.**

10. $\dfrac{1 - x}{x + 1} \cdot \dfrac{x^2 - 1}{x^2}$

11. $\dfrac{x - 1}{1 - x^2} \div \dfrac{x^2 - 2x + 1}{1 + x}$

12. $\dfrac{n + 1}{n} \div \dfrac{n(n - 1)}{n^2}$

13. $\dfrac{x^2 - 2x}{x^2 - 4} \cdot \dfrac{x - 2}{x}$

**Multiply this expression, and simplify.**

14. $(x + 1)(x - 1)\left(\dfrac{1}{x + 1} + \dfrac{2}{x - 1}\right)$

15. One door can empty a meeting room in 16 minutes. A different width door is to be added. Both doors working together must be able to empty the room in 5 minutes. How fast must the new door empty the room?

**Simplify each expression or equation in Exercises 16 to 18.**

16. 4 yd per shirt · $8.98 per yd

17. $\dfrac{\dfrac{\$2.50}{1 \text{ gal}}}{\dfrac{25 \text{ mi}}{1 \text{ gal}}}$

18. $3 \text{ mg} \cdot \dfrac{\text{gr } 1}{60 \text{ mg}} \cdot \dfrac{1 \text{ tab}}{\text{gr } \frac{1}{120}}$

19. Is $\dfrac{2\frac{1}{7}}{10} = \dfrac{3}{14}$ a true statement? Explain your answer.

20. Write the expression on the left side of this optics equation as a single fraction:

$$\frac{1}{p} + \frac{1}{q} = \frac{1}{f}$$

21. Simplify the right side:

$$h = \frac{A}{\frac{1}{2}b}$$

**Add or subtract the expressions in Exercises 22 to 25.**

22. $\dfrac{2}{9} + \dfrac{7}{15}$

23. $\dfrac{2}{ab^2} - \dfrac{a}{2b}$

24. $\dfrac{x}{x + 2} + \dfrac{2}{x^2 - 4}$

25. $\dfrac{2}{3} + \dfrac{3}{x} - \dfrac{4}{x^2}$

26. Simplify $\dfrac{a}{2} + \dfrac{2}{3} \cdot \dfrac{a}{2} - \dfrac{3a}{2} \div \dfrac{9}{2} + \dfrac{1}{4}$.

**Solve the equations in Exercises 27 to 30.**

27. $\dfrac{1}{3} + \dfrac{1}{6} = \dfrac{1}{x}$

28. $\dfrac{1}{x} = \dfrac{1}{2x} + \dfrac{1}{2}$

29. $\dfrac{3}{x} = \dfrac{2}{x - 3}$

30. $\dfrac{x - 2}{4} + \dfrac{1}{x} = \dfrac{19}{4x}$

31. Explain the role of factoring in multiplying and dividing rational expressions.

1. a. Make an input–output table for $y = 3x - 4$ for inputs in the interval $[-3, 1]$.

   b. Graph the equation.

   c. Where do we find the slope in a linear equation? Explain how to find the slope from a table. Show the slope on the graph, with rise and run.

   d. Where is the $y$-intercept in the equation? How do we find the $y$-intercept from the table? Indicate the $y$-intercept on the graph.

   e. If we set $y = 0$ and solve $0 = 3x - 4$ for $x$, what point on the graph have we found?

   f. Solve the equation $3x - 4 = -6.4$. Show the steps.

   g. Describe how we might estimate the solution to $3x - 4 = -6.4$ from the graph.

2. A credit card payment schedule is shown in the table.

   | Charge Balance, $x$ | Payment Due, $y$ |
   |---|---|
   | ($0, $30] | Full amount |
   | ($30, $100] | $30 + 20% of amount in excess of $30 |
   | ($100, +∞) | $50 + 50% of amount in excess of $100 |

   a. How much is paid for charge balances of $25, $35, $95, $100, and $105?

   b. Write an equation that describes the payment due in terms of the charge balance for each of the three categories. The equations should use $y$ in terms of $x$.

   c. Explain the difference between the brackets, [ ], and the parentheses, ( ), in the charge balance column.

   d. Does it matter which interval notation—brackets or parentheses—is used with the infinity symbol (∞)?

   e. Write each charge balance interval using an inequality expression.

3. a. The pairs of coordinates, $(x, y)$, use the same linear rule to get from $x$ to $y$. Guess the rule, and give the missing $y$ value. (*Hint:* the rule for the coordinates $(1, 4)$ and $(2, 8)$ is $y = 4x$.)

   $$(3, 3), (5, 5), (6, \underline{\quad})$$

   $$(1, 2), (3, 6), (4, \underline{\quad})$$
   $$(3, 1), (9, 3), (6, \underline{\quad})$$
   $$(2, -1), (8, -4), (6, \underline{\quad})$$
   $$(4, -4), (2, -2), (8, \underline{\quad})$$
   $$(8, 4), (2, 1), (6, \underline{\quad})$$
   $$(4, -8), (6, -12), (3, \underline{\quad})$$
   $$(2, 6), (3, 9), (5, \underline{\quad})$$

   b. What is the slope of the line connecting each pair of coordinates in part a?

   c. The rule describing how we get from $x$ to $y$ is the equation of the line connecting the points. Match the following equations with the sets of coordinates in part a above.

   $$y = \tfrac{1}{2}x \qquad y = -\tfrac{1}{2}x$$
   $$y = x \qquad y = -x$$
   $$y = 3x \qquad y = \tfrac{1}{3}x$$
   $$y = 2x \qquad y = -2x$$

4. The volume of a cube, or box with all sides equal to $x$, is found by the formula $V = x^3$. The surface area of the box is the amount of area covering all six sides. The surface area of a cube is $6x^2$.

   a. Fill in this table.

   | Length of Side of Box, $x$ | Volume of Cube, $x^3$ | Surface Area $6x^2$ |
   |---|---|---|
   | 10 | | |
   | 20 | | |
   | 40 | | |
   | $n$ | | |
   | $2n$ | | |

   b. If we double the length of a side of the box, what happens to the volume?

   c. If we double the length of a side of the box, what happens to the surface area?

   d. Use the Pythagorean theorem to determine the length of the diagonal on the side of a cubical box if the length of the side is 20.

417

5. List all the types of numbers that are appropriate for each of the following problem situations. Choose from the following: real numbers, rational numbers, irrational numbers, positive numbers, negative numbers, repeating decimals, terminating decimals, integers, zero.

  a. Coins in a pocket or purse

  b. Dimensions of geometric figures

  c. Number of people

  d. Checking account balance

  e. Pi, $\sqrt{2}$, $\sqrt{3}$

6. Use the given facts and unit analysis, as needed, to answer the following questions.

The speed of light is 186,000 miles per second.
The minimum distance from Pluto to the sun is 2756.4 × 10⁶ miles.
The maximum distance from Pluto to the sun is 4551.4 × 10⁶ miles.
There are 60 seconds in a minute.
There are 60 minutes in an hour.
The speed limit on a freeway is 65 miles per hour.
There are 5280 feet per mile.
There are 1609 meters in a mile.
There are 1000 meters in a kilometer.

  a. What is the minimum distance from Pluto to the sun, written in the standard form of scientific notation?

  b. How long does it take light to travel the maximum distance from the sun to Pluto?

  c. Mary Meagher set a world record in the butterfly in 1981. She swam 200 meters in 125.96 seconds. What was her average speed in miles per hour?

  d. The orbit velocity of the space shuttle is 28,300 km/hr. How many miles per hour is this?

7. Solve for the indicated variable.

  a. $2x + 3y = 12$ for $y$

  b. $2x - 3y = 7$ for $x$

  c. $ax + by = c$ for $y$

  d. $\dfrac{x}{x+1} = \dfrac{3}{8}$ for $x$

  e. $\dfrac{P_1 V_1}{T_1} = \dfrac{P_2 V_2}{T_2}$ for $P_2$

  f. $3(x + 2) = 7(x - 10)$ for $x$

  g. $4^2 + x^2 = 8^2$ for $x$

  h. $d^2 = [4 - (-3)]^2 + (5 - 2)^2$ for $d$

  i. $6 = \sqrt{3x - 3}$ for $x$

8. A washing machine tub has an 11-inch radius and a 14-inch height. It is shaped like a cylinder.

The volume of a cylinder is $V = \pi r^2 h$.
There are 231 cubic inches in a gallon.
Winter cold water is 35°F.
Summer cold water is 60°F.
Hot water is 140°F.

  a. One winter day you turn the washing machine on and forget to look at the temperature setting. By the time you remember it, there are 5 gallons of cold water in the washtub. You want to wash clothes at 100°. How many gallons of hot water must be added to raise the temperature to 100°F? (A quantity and value table might be helpful.)

  b. Is the volume of the tub large enough for the water in part a?

  c. Is there room in the tub for both the water and the clothes?

  d. How many gallons of each—hot and cold water—are needed to produce 8 gallons of 100° water in the tub during the summer? (A quantity and value table might be helpful.)

9. Set up equations for each problem. If the problem describes a system of equations, solve using substitution or elimination or a graphing calculator.

  a. Three strings of holiday lights and two packages of giftwrap cost $18.90. Two strings of holiday lights and five packages of giftwrap cost $19.86. What is the cost of each item?

  b. Four cups of milk and a cup of cottage cheese contain 685 calories. Three cups of milk and two cups of cottage cheese contain 770 calories. How many calories are in each?

  c. The three angles of a triangle have measures that add to 180°. The largest angle is four times the smallest. The middle angle is 9 less than the largest. Find the measure of each angle.

  d. The equations $y = \frac{3}{5}x$ and $y = -\frac{3}{5}x + 6$ form the diagonals of a rectangle. What is the point of intersection of the two lines? If two sides of the rectangle lie on the coordinate axes, what is the area of the rectangle? What is the length of the diagonal of the rectangle?

  e. A birthday party at McDonald's® costs $29.95 for a party of 10 and $2.50 for each additional child. What equation describes the cost of a party for $x$ people?

  f. A jumbo package contains 20 ounces of potato chips. If the input is the number of ounces eaten by each person, what equation describes the number of people served by the package?

10. Multiply these polynomials.

  a. $3(x^2 - 6x + 4)$      b. $-4x(x^2 - 2x - 3)$

  c. $(x - 4)(x + 4)$      d. $(x - 4)^2$

  e. $(x - 4)(x^2 - 8x + 16)$      f. $(2x - 3)(3x - 2)$

11. Factor these expressions, using the table method as needed.

  a. $12xy + 3xy^2 + 6x^2y^2$      b. $x^2 + x$

  c. $x^2 + x - 20$      d. $4x^2 - 9$

  e. $6x^2 + x - 2$      f. $6x^2 - 11x - 2$

12. Solve these equations.

    a. $x^2 + 3x - 4 = 0$    b. $x^2 - 4x + 3 = 0$

    c. $2x^2 + 8x + 8 = 0$

13. Use symmetry and Pascal's Triangle to finish these expressions.

    a. $(x + y)^2 = x^2 + 2xy +$

    b. $(x + y)^3 = x^3 + 3x^2y +$

    c. $(a + 1)^2 = a^2 + 2a +$

    d. $(1 + b)^3 = 1 + 3b +$

    e. $(x + 1)^4 = x^4 + 4x^3 + 6x^2 +$

    f. $(a - b)^4 = a^4 - 4a^3b +$

14. Multiply out each of these binomial squares.

    a. $(n - 3)^2$

    b. $(n - 1)^2$

    c. $(n - 4)^2$

    d. Simplify $(n - 1)^2 - (n - 4)^2$.

    e. Show that $n^2 - (n - 3)^2 - [(n - 1)^2 - (n - 4)^2] = 6$.

15. Simplify these expressions using properties of exponents and radicals.

    a. $3^0 + 4^2$

    b. $5^2 + 5^{-1}$

    c. $\sqrt{25} + 16^{1/2}$

    d. $\sqrt{225} + 25^{1/2}$

    e. $\sqrt{64x^2}$, assuming $x$ is either positive or negative

    f. $\sqrt{64x^2}$, assuming $x$ is only positive

    g. $0^0$

    h. $\sqrt{x}$, where $x$ is a negative number

16. The expressions in Exercises 15g and 15h are undefined. Two other operations giving undefined values are dividing by zero and calculating the slope of a vertical line. Draw an example that shows why calculating the slope of a vertical line actually results in division by zero.

17. Factor and simplify $\dfrac{4x^2 - 1}{2x^2 + 5x + 2}$.

18. For what inputs, $x$, will $\dfrac{-1}{(x - 3)(x + 1)}$ be undefined?

19. Simplify these expressions containing units of measure.

    a. $\dfrac{\$6.00 \text{ per hour}}{3 \text{ rooms cleaned per hour}}$

    b. $d = -\dfrac{1}{2}\left(\dfrac{32.2 \text{ ft}}{\text{sec}^2}\right)(3 \text{ sec})^2 + \left(\dfrac{22 \text{ ft}}{\text{sec}}\right)(3 \text{ sec}) + 100 \text{ ft}$

20. Simplify these expressions. Exclude any inputs that make the expression undefined.

    a. $\dfrac{2a + 2b}{a^2 - b^2}$

    b. $\dfrac{4 - x}{6x - 24}$

21. Perform the indicated operations. Leave the answers in reduced form.

    a. $\dfrac{2 - x}{xy^2} \cdot \dfrac{x^2}{x - 2}$

    b. $\dfrac{x^2 - 3x - 4}{x^2 - 2x} \cdot \dfrac{x^2 - 4}{4 - x}$

    c. $\dfrac{x - 4}{x + 2} + \dfrac{x^2 - 3x - 4}{x^2 - 4}$

    d. $\dfrac{x^2 + 3x + 2}{x^2 + 6x + 9} \div \dfrac{x + 2}{x + 3}$

    e. $\dfrac{1}{3x} - \dfrac{2}{5x}$

    f. $8x\left(\dfrac{1}{2x} - \dfrac{3}{x}\right)$

    g. $3(x - 2)\left(\dfrac{4}{x - 2} + \dfrac{x}{3}\right)$

22. A vacation savings account contains $1000.

    a. Make a table and graph for the possible per day spending for a 0- to 10-day vacation. Use days as input and spending per day as output.

    b. What equation describes the number of vacation days and the possible spending per day?

**This Answer Section also contains answers for both the even-numbered and the odd-numbered exercises in the Mid-Chapter Tests and the Chapter Tests.**

## EXERCISES 1.1

1. Surgeon is the boy's mother.
3. 52 partitions
7. One possible answer is a double row of triangles; 37 partitions.
9. One possible answer is 20 working days.

## EXERCISES 1.2

1. 2    3. $\frac{5}{6}$    5. $\frac{1}{3}$    7. $\frac{4}{15}$    9. $1\frac{5}{12}$    11. $2\frac{2}{15}$
13. $288    15. 16 credits    17. $+0.00027$ cc
19. $-0.00004$ cc    21. $-0.00067$ cc    25. $45
33. $3\frac{7}{8}$ yards    35. $\frac{5}{8}$ yard    37. 0
39. $\frac{3}{4}$; positive rational numbers
41.

| Input | Output |
|-------|--------|
| 0 | 5 |
| 1 | 2 |
| 2 | 5 |
| 3 | 6 |
| 4 | 5 |
| 5 | 10 |
| 6 | 5 |
| 7 | 14 |
| 8 | 5 |

43. 2; 4    45. 9; 3    47. 12; 1
49.

| $a + b$ | $a - b$ | $a \cdot b$ | $a \div b$ |
|---------|---------|-------------|------------|
| 20 | 10 | 75 | 3 |
| $1\frac{3}{20}$ | $\frac{7}{20}$ | $\frac{3}{10}$ | $1\frac{7}{8}$ |
| 0.42 | 0.30 | 0.0216 | 6 |
| 6.3 | 4.9 | 3.92 | 8 |
| 3.75 | 0.75 | 3.375 | 1.5 |

## MID-CHAPTER 1 TEST

1. 20 pens use 41 partitions
2. a. true    b. true    c. true    d. true    e. true
3. a. 12    b. 22    c. 42    d. 42.25
4. a. $2.45    b. $6.50    c. $4.95

## EXERCISES 1.3

1. a. d    b. input $+ 5$    c. $n + 5$
   d.

| Input | Output |
|-------|--------|
| 5 | 10 |
| 50 | 55 |
| 100 | 105 |
| $n$ | $n + 5$ |

3. a. b    b. input times 4    c. $4x$
   d.

| Input | Output |
|-------|--------|
| 3 | 12 |
| 50 | 200 |
| 100 | 400 |
| $x$ | $4x$ |

5. a. $n \cdot n = n^2$    b. $3n - 2$    c. $2n + 2$
   d. $\frac{7n}{2}$    e. $\frac{n}{8}$    f. $\frac{2}{n} - 3$    g. $n - 4$
7. a. $0.10n$    b. $0.125n$    c. $0.065n$    d. $1.5n$
15. a. 6 stitches/1 inch    b. 10 miles/1 day
    c. 12 eggs/1 dozen    d. $120/1 cord
17.

| Input | Output |
|-------|--------|
| 1 | $0.86 |
| 2 | 1.72 |
| 3 | 2.58 |
| $n$ | $0.86n$ |

19.

| Input | Output |
|-------|--------|
| 0 | $35 |
| 100 | 40 |
| 200 | 45 |
| 300 | 50 |
| 400 | 55 |
| 500 | 60 |
| $m$ | $\$35 + 0.05m$ |

21. $n - 35$    23. $1.065n$    25. 0; 7; 0; 0; 101; 0; $n$

**27.**

| Input | Output |
|---|---|
| −3 | 3 |
| −2 | 2 |
| −1 | 1 |
| 0 | 0 |
| 1 | 1 |
| 2 | 2 |
| 3 | 3 |

**29.**

| Input | Output |
|---|---|
| −3 | 0 |
| −2 | 0 |
| −1 | 0 |
| 0 | 0 |
| 1 | 1 |
| 2 | 2 |
| 3 | 3 |

**31.** 13; $3n + 1$; 61; 151    **33.** 10; $2n + 2$; 42    **35.** 36

**37.** *a.*

| Input: $x$ | Input: $y$ | Output: $xy$ | Output: $x + y$ |
|---|---|---|---|
| 2 | 6 | 12 | 8 |
| 10 | 2 | 20 | 12 |
| 3 | 5 | 15 | 8 |
| 1 | 15 | 15 | 16 |
| 9 | 2 | 18 | 11 |
| 4 | 1 | 4 | 5 |

There may be more than one correct answer for some rows. $xy$ is the product of the inputs; $x + y$ is the sum of the inputs.

## EXERCISES 1.4

**1.** $A(2, 4)$; $B(0, 2)$; $C(-2, 0)$; $D(-6, -2)$; $E(-4, -6)$; $F(0, -6)$; $G(2, -4)$; $H(4, -2)$; $I(5, 0)$

**5.** *a.* 2   *b.* 4   *c.* 3   *d.* 3

**7.** *a.* $y$-axis   *b.* $x$-axis   *c.* $x$-axis

**11.**

| $x$ | $y$ |
|---|---|
| 0 | 1 |
| $-\frac{1}{2}$ | 0 |
| 2 | 5 |
| 3 | 7 |
| 1 | 3 |
| −2 | −3 |
| $\frac{1}{2}$ | 2 |

**13–19.**

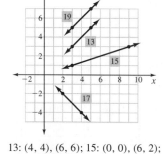

13: (4, 4), (6, 6); 15: (0, 0), (6, 2); 17: (3, −3), (5, −5), 19: (4, 6), (6, 8)

**21.** *a.* #18   *b.* #13   *c.* #20

**31.** *a.*

| Pounds | Dollars |
|---|---|
| 0 | 0 |
| 1 | 1.29 |
| 2 | 2.58 |
| 3 | 3.87 |
| 4 | 5.16 |

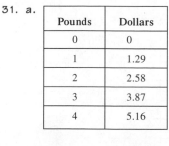
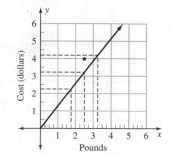

*b.* yes; yes; the cost is higher per pound

*c.* Answer within $0.10 of these: $2\frac{1}{2}$ lb = $3.23; $1\frac{3}{4}$ lb = $2.26; $3\frac{1}{4}$ lb = $4.19

*d.* bulk candy

**33.** *a.*

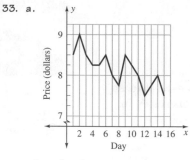

*b.* Stock prices are falling.

**37.** $10^2, 100^2, 40^2$   **39.** none exist   **41.** 1, 4, 5, 6, 9, 0

**43.**

| Square Root | Value | Type |
|---|---|---|
| $\sqrt{36}$ | 6 | rational |
| $\sqrt{15}$ | 3.873 | irrational |
| $\sqrt{25}$ | 5 | rational |
| $\sqrt{35}$ | 5.916 | irrational |
| $\sqrt{12.25}$ | 3.5 | rational |
| $\sqrt{2.25}$ | 1.5 | rational |
| $\sqrt{6}$ | 2.449 | irrational |
| $\sqrt{16}$ | 4 | rational |
| $\sqrt{26}$ | 5.099 | irrational |

**45.** $A(0, -2)$; $B(-3, 0)$   **47.** $A(6, 1)$; $B(4, -2)$

## CHAPTER 1 REVIEW EXERCISES

**1.** $3n + 1$

**3.** *a.* $3.25   *b.* $7.25   *c.* $15   *d.* $50   *e.* $50   *f.* $56

**5.** *a.* real, rational   *b.* real, irrational
*c.* real, rational, integer, whole   *d.* real, rational
*e.* real, rational   *f.* real, rational, integer

**9.**

| Pounds | Cost |
|---|---|
| $x$ | $0.37x |
| 1 | 0.37 |
| 2 | 0.74 |
| 3 | 1.11 |
| 4 | 1.48 |

rule: $y = \$0.37x$

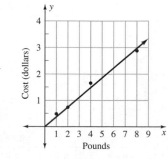

**13.** $y = -\frac{1}{2}x$

**15.**

| Hours: $h$ | Cost: $0.04h + 5$ |
|---|---|
| 100 | $9 |
| 200 | 13 |
| 400 | 21 |
| 500 | 25 |

**17.** *a.* 42   *b.* Answers will vary; $(6.5)^2 = 42.25$ largest

**19.** *a.* D   *b.* B   *c.* G   *d.* C

**21.** $A(7, 3)$; $B(2, 0)$

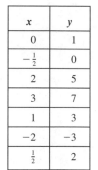
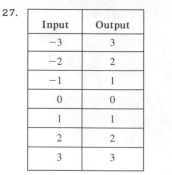

# CHAPTER 1 TEST

1. difference  2. set

3. a. origin  b. y-axis  c. x-axis  d. coordinate
   e. quadrant

4.

| $x$ | $2x + 3$ |
|-----|----------|
| 0 | 3 |
| 1 | 5 |
| 2 | 7 |

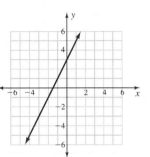

5. a. miles driven; cost; miles on $x$ and cost on $y$

c.

| Miles: $x$ | Cost: $0.10x + 150$ |
|------------|---------------------|
| 50 | $155 |
| 100 | 160 |
| 150 | 165 |
| 200 | 170 |

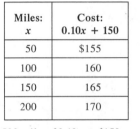

d. 500 miles; $0.10x + $150

6. Answers will vary: 1, 18; 3, 16; 6, 13; 8, 11; 9, 10

   a. For the above: 18; 48; 78; 88; 90  b. 90  c. $9.5^2 = 90.25$

7. a. 1; rational  b. 3.162  c. 10; rational  d. 31.623
   e. 100; rational  f. 1000; rational

8. $A(-5, 4); B(-2, 5)$

9.

| Input | Output |
|-------|--------|
| 0 | 0 |
| 1 | 2 |
| 2 | 1 |
| 3 | 6 |
| 4 | 2 |
| 5 | 10 |
| 6 | 3 |
| 7 | 14 |

10. Option 1 is worth $1023. Option 2 is worth $255. Option 1 is better.

11. Output is total number of footprints after each hop. Final entry in table is 20; $4x$

# EXERCISES 2.1

1. $+3 + (-2) = +1$; net charge $= +1$

3. $+4 + (-8) = -4$; net charge $= -4$

5. $+6 + (-6) = 0$; net charge $= 0$  7. $-1$  9. $-2$

11. $1.27  13. $-$0.74  15. $-14$  17. $-1$

19. a. 4  b. 5  c. 5  d. $-7$

21. a. 270 mi  b. 252 mi  c. 329 mi  23. $40

27. a. 4  b. 3  c. $-3$  d. $-4$  e. $-3$  f. $-4$

29. $3 \cdot (4) + 4(-3) = 0$  31. $-19$  33. $-17$

35. a. $-1$  b. 1  c. $-1$  d. 1  37. line d  39. line a

41. All are false. Subtraction and division are not commutative.

43. a. 6  b. 0  c. 8  d. 2
    Subtraction and division are not associative.

45.

| Input | Output |
|-------|--------|
| $-3$ | 3 |
| $-2$ | 2 |
| $-1$ | 1 |
| 0 | 0 |
| 1 | 1 |
| 2 | 2 |
| 3 | 3 |

47.

| Input: $x$ | Output: $|x + 3|$ |
|-----------|------------------|
| $-4$ | 1 |
| $-3$ | 0 |
| $-2$ | 1 |
| $-1$ | 2 |
| 0 | 3 |
| 1 | 4 |
| 2 | 5 |
| 3 | 6 |
| 4 | 7 |

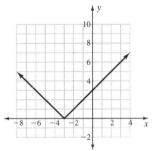

49.

| Input: $x$ | Output: $|x| - 2$ |
|-----------|------------------|
| $-4$ | 2 |
| $-3$ | 1 |
| $-2$ | 0 |
| $-1$ | $-1$ |
| 0 | $-2$ |
| 1 | $-1$ |
| 2 | 0 |
| 3 | 1 |
| 4 | 2 |

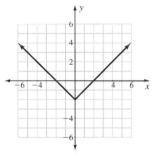

51.

| Input: $x$ | Input: $y$ | Output: $xy$ | Output: $x + y$ |
|-----------|-----------|-------------|----------------|
| 3 | 4 | 12 | 7 |
| $-2$ | 3 | $-6$ | 1 |
| $-7$ | $-3$ | 21 | $-10$ |
| $-3$ | 5 | $-15$ | 2 |
| 2 | $-3$ | $-6$ | $-1$ |
| 4 | $-3$ | $-12$ | 1 |
| $-1$ | $-2$ | 2 | $-3$ |
| 3 | $-1$ | $-3$ | 2 |

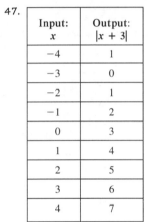

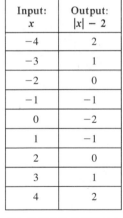

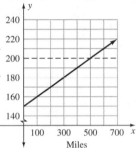

## EXERCISES 2.2

1. a. $-5$   b. $\frac{1}{2}$   c. $-0.4$   d. $-x$   e. $2x$

3. a. $\frac{1}{4}$   b. $-\frac{1}{2}$   c. $2$   d. $-1\frac{1}{3}$   e. $2$

5. a. $\frac{3}{10}$   b. $\frac{2}{13}$   c. $\frac{1}{x}, x \neq 0$   d. $\frac{b}{a}, a \neq 0$   e. $\frac{1}{-x}, x \neq 0$

7. a. $\frac{a}{c}$   b. $\frac{b}{c}$   c. $\frac{a}{b}$

9. a. One answer: 2, 4, 6    b. One answer: $3, \frac{3}{2}, 6$

11. a. $-3$   b. $3$   c. $-19$   d. $-5$   e. $-21$

13. a. $0$   b. $4$   c. $5$   d. $-1$   e. $-9$

15. $+1 - (+2) = -1$    17. $-2 - (-4) = +2$    19. $-1$

21. $-5$    23. $-214$    25. a. $\frac{1}{3}$   b. $-\frac{2}{3}$   c. $\frac{2}{5yz}$

27.

| $x$ | $y$ | $x \cdot y$ | $x + y$ |
|---|---|---|---|
| 2 | $-2$ | $-4$ | 0 |
| 3 | $-2$ | $-6$ | 1 |
| $-4$ | 3 | $-12$ | $-1$ |
| $-3$ | 3 | $-9$ | 0 |
| 2 | $-3$ | $-6$ | $-1$ |

29.

| $x$ | $\dfrac{3-x}{x-3}$ | $\dfrac{x-2}{2-x}$ |
|---|---|---|
| 6 | $-1$ | $-1$ |
| 4 | $-1$ | $-1$ |
| 0 | $-1$ | $-1$ |
| $-3$ | $-1$ | $-1$ |
| $-5$ | $-1$ | $-1$ |

31. a. $45$   b. $\frac{5}{6}$   c. $-\frac{15}{16}$   d. $-3$   e. $-\frac{5}{12}$   f. $-1$

35. a. $-\frac{x}{y}$   b. $\frac{x}{-y}$   c. $\frac{-x}{y}$    37. 2330 m    39. 6052 m

## EXERCISES 2.3

1. a. Base is $x$; $3 \cdot x \cdot x$    b. Base is $x$; $-3 \cdot x \cdot x$

    c. Base is $(-3x)$; $(-3x)(-3x)$    d. Base is $x$; $a \cdot x \cdot x$

    e. Base is $x$; $-1 \cdot x \cdot x$    f. Base is $(-x)$; $(-x)(-x)$

3. a. $243$   b. $64$   c. $4$   d. $-27$   e. $\frac{4}{9}$   f. $\frac{1}{4}$   g. $-\frac{1}{27}$

5. $n = 3$

7. a. $m^3 n^5$ is already simplified.    b. $n^8$   c. $a^8$   d. $a^8$

9. 1, 11; 2, 10; 3, 9; 4, 8; 5, 7; 6, 6

11. a. $512a^9$   b. $108a^2b^3$   c. $-108a^3$   d. $-12a$   e. $240x^4y$

13. a. $x^3$   b. $a^3$   c. $\frac{x^2}{y^2}$   d. $\frac{x^2}{y^4}$

15. a. $x^4$   b. $8x^6$   c. $\frac{a^4}{16}$   d. $10x^8y^2$   e. $80x^4y^6$

17. a. $4a^2$   b. $9a^2$   c. $8x^3$   d. $27x^3$    19. $\frac{4}{9}; \frac{8}{27}$

21. 3 cm    23. 4,096,000,000    25. a. 1   b. 15

27. a. 74   b. 20   c. 28

29. a. 0.293   b. 1.816   c. $\sqrt{6} \approx 2.449$

31. a. $-2$   b. $6$    33. a. 4   b. 0    35. associative

37. no    39. Student #3 is right.

41. a.

| $a$ | $b$ | $a^2 + b^2$ | $\sqrt{a^2 + b^2}$ | $a + b$ |
|---|---|---|---|---|
| 3 | 4 | $9 + 16 = 25$ | 5 | 7 |
| 9 | 12 | $81 + 144 = 225$ | 15 | 21 |
| 5 | 12 | $25 + 144 = 169$ | 13 | 17 |
| 6 | 8 | $36 + 48 = 100$ | 10 | 14 |
| 9 | 40 | $81 + 1600 = 1681$ | 41 | 49 |

   b. $\sqrt{a^2 + b^2} \neq \sqrt{a^2} + \sqrt{b^2}$

## MID-CHAPTER 2 TEST

1. a. $-3$   b. 2   c. 2   d. 7   e. $-7$   f. 3.5

2. $+6 - (+4) = (+2)$; net charge = 2

3. $0 - (-3) = 3$; net charge = 3    4. 4    5. $-3$

6. \$0.72    7. \$3.72    8. 135    9. 3    10. 780    11. 350

12.

| Input: $x$ | Output: $y = x^2$ |
|---|---|
| $-2$ | 4 |
| $-1$ | 1 |
| 0 | 0 |
| 1 | 1 |
| 2 | 4 |
| 3 | 9 |

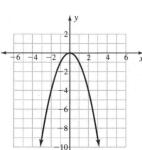

13.

| Input: $x$ | Output: $y = -x^2$ |
|---|---|
| $-2$ | $-4$ |
| $-1$ | $-1$ |
| 0 | 0 |
| 1 | $-1$ |
| 2 | $-4$ |
| 3 | $-9$ |

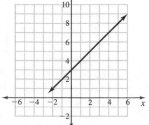

14.

| Input: $x$ | Output: $y$ |
|---|---|
| $-2$ | 1 |
| $-1$ | 2 |
| 0 | 3 |
| 1 | 4 |
| 2 | 5 |
| 3 | 6 |

*Note:* On a more complete graph, as $x$ values decreased beyond $-3$, $y$ values would increase.

15.

| Input: $x$ | Output: $y$ |
|---|---|
| $-2$ | 5 |
| $-1$ | 4 |
| 0 | 3 |
| 1 | 4 |
| 2 | 5 |
| 3 | 6 |

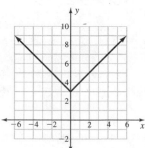

**16.** 9250 m    **17.** 6280 m    **18.** Mauna Kea is 1082 ft "taller."

**19. a.** 81    **b.** $\frac{1}{16}$    **c.** $243m^2n^3$    **d.** $810xy^4$    **e.** $81x^8$

**f.** $\frac{9}{16} \cdot b^4$    **g.** $\frac{4}{9x^2}$    **h.** $-\frac{0.008x^3}{y^9}$

**20.** 10; multiplication is done before subtraction.

**21.** $-2$; multiplication is done before subtraction.

**22.** $1\frac{1}{2}$    **23.** 0.293    **24.** $-1$    **25.** 8

## EXERCISES 2.4

**1.** $5.77/hr    **3.** $38.46/hr

**5.** Rental option #2 is better; it's cheaper.    **7.** 44 ft/sec

**9.** $117\frac{1}{3}$ ft/sec    **11.** 60 mph    **13.** 150 mph    **15.** 352 ft

**17.** 5.83 cal/cracker    **19.** 1219.2 m

**21. a.** $P = 9.9$ cm; $A = 4.8$ cm$^2$    **b.** $P = 12.6$ cm; $A = 6.8$ cm$^2$
   **c.** $P = 12.56$ ft; $A = 12.56$ ft$^2$    **d.** $P = 9.6$ cm; $A = 3.99$ cm$^2$

**23. a.** $P = 104$ mm; $A = 549$ mm$^2$    **b.** $P = 94$ mm; $A = 285$ mm$^2$
   **c.** $P = 88$ mm; $A = 333$ mm$^2$    **d.** $P = 75$ mm; $A = 452$ mm$^2$

**27.** $P = 21.6$ in.    **29. a.** 1.94 cm$^2$    **b.** 6.28 cm$^2$    **c.** 4.86 cm$^2$

**31.** 2.25 times larger    **33.** 425.25 in.$^3$; $\approx 246$ ft$^3$

**35. a.** SA $= 50.24$ cm$^2$; $V = 25.12$ cm$^3$
   **b.** SA $= 12.56$ cm$^2$; $V = 3.14$ cm$^3$
   **c.** SA $= 50.24$ cm$^2$; $V = 33.49$ cm$^3$
   **d.** SA $= 12.56$ cm$^2$; $V = 4.19$ cm$^3$

**37. a.** 3.0 ft per side    **b.** 3.375 ft$^3$    **c.** 27 ft$^3$    **d.** 240 lb

**39. a.**

| Radius of Sphere | Volume of Sphere |
|---|---|
| 1 | 4 in.$^3$ |
| 2 | 33 in.$^3$ |
| 3 | 113 in.$^3$ |
| 4 | 268 in.$^3$ |
| 5 | 523 in.$^3$ |

**b. and c.**

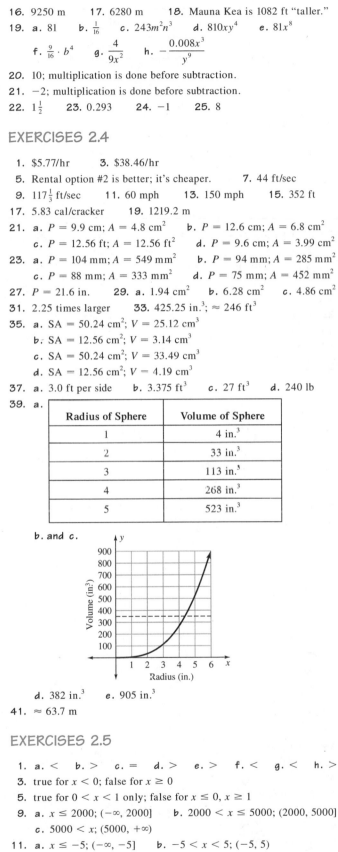

**d.** 382 in.$^3$    **e.** 905 in.$^3$

**41.** $\approx 63.7$ m

## EXERCISES 2.5

**1. a.** $<$    **b.** $>$    **c.** $=$    **d.** $>$    **e.** $>$    **f.** $<$    **g.** $<$    **h.** $>$

**3.** true for $x < 0$; false for $x \geq 0$

**5.** true for $0 < x < 1$ only; false for $x \leq 0$, $x \geq 1$

**9. a.** $x \leq 2000$; $(-\infty, 2000]$    **b.** $2000 < x \leq 5000$; $(2000, 5000]$
   **c.** $5000 < x$; $(5000, +\infty)$

**11. a.** $x \leq -5$; $(-\infty, -5]$    **b.** $-5 < x < 5$; $(-5, 5)$
   **c.** $5 \leq x$; $[5, +\infty)$

**13. a.** $x < 5$; $(-\infty, 5)$    **b.** $5 \leq x \leq 50$; $[5, 50]$
   **c.** $50 < x$; $(50, +\infty)$

**15.** f; s    **17.** e; w    **19.** b, r

**21.**

| $x$ | $y$ |
|---|---|
| $-5$ | 8 |
| $-4$ | 7 |
| $-3$ | 6 |
| $-2$ | 5 |
| $-1$ | 4 |
| 0 | 3 |
| 1 | 2 |
| 2 | 1 |
| 3 | 0 |
| 4 | 1 |
| 5 | 2 |

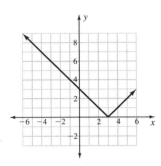

**23.**

| $x$ | $y$ |
|---|---|
| $-5$ | $-3$ |
| $-4$ | $-2$ |
| $-3$ | $-1$ |
| $-2$ | 0 |
| $-1$ | $-1$ |
| 0 | $-2$ |
| 1 | $-3$ |
| 2 | $-4$ |
| 3 | $-5$ |

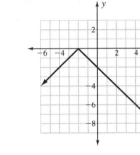

**25.** $y - |x - 3|$    **27.** $y = -|x + 2|$

**29. a.**

| Income | Tax |
|---|---|
| $ 0 | $ 0 |
| 1500 | 75 |
| 2000 | 100 |
| 3500 | 205 |
| 5000 | 310 |
| 7000 | 490 |
| 8000 | 580 |

**b.**

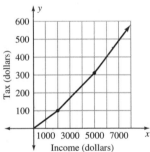

**c.** $0 \leq x \leq $2000$; [0, $2000]
$2000 < x \leq $5000$; ($2000, $5000]
$5000 < x$; ($5000, +\infty)

## CHAPTER 2 REVIEW EXERCISES

**1. a.**

| $x$ | $-x - 2$ |
|---|---|
| $-2$ | 0 |
| $-1$ | $-1$ |
| 0 | $-2$ |
| 1 | $-3$ |
| 2 | $-4$ |

**b.**

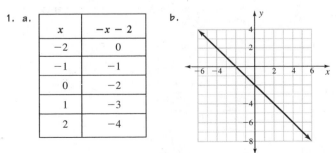

**3.** a. $-54$   b. $54$   c. $54$   d. $-54$   e. $56$   f. $-56$   g. $-56$
   h. $56$   i. $2$   j. $-4$   k. $-8$   l. $8$   m. $16$   n. $-16$

**5.** a. $-4$   b. $10$   c. $12$   d. $-9$   e. $5$   f. $10$   g. $15$
   h. $a^2b^2$   i. $2.5$   j. $4a^2b^4$   k. $a^2b^2$   l. $9$   m. $4a^2b^4$
   n. $8$   o. $\dfrac{16x^2}{y^2}$   p. $\dfrac{x^2}{9y^2}$   q. $m^9$   r. $m^9$   s. $m^3$   t. $m^3$

**7.** a. $15$   b. $110$   c. $-30$   d. $60$

**9.** associative property

**13.** a.

| $r$ | $V$ |
|---|---|
| $1$ | $\frac{4}{3}\pi$ |
| $2$ | $\frac{32}{3}\pi$ |
| $4$ | $\frac{256}{3}\pi$ |
| $8$ | $\frac{2048}{3}\pi$ |

b.

| $r$ | $V$ |
|---|---|
| $1$ | $3\pi$ |
| $2$ | $12\pi$ |
| $4$ | $48\pi$ |
| $8$ | $192\pi$ |

**19.** $0 \le x < \$50$; $[0, \$50)$
   $\$50 \le x \le \$500$; $[\$50, \$500]$
   $\$500 < x$; $(\$500, +\infty)$

**21.** $A = \frac{1}{2}hb$

## CHAPTER 2 TEST

**1.**

| $x$ | $|x - 2|$ |
|---|---|
| $-2$ | $4$ |
| $0$ | $2$ |
| $2$ | $0$ |
| $4$ | $2$ |

**2.**

| $x$ | $-2x + 3$ |
|---|---|
| $-2$ | $7$ |
| $-1$ | $5$ |
| $0$ | $3$ |
| $1$ | $1$ |
| $2$ | $-1$ |

**3.** commutative property   **4.** associative property

**5.** a. $4$   b. $-2.5$   c. $-1$   d. $60$   e. $-1$   f. $10$
   g. $m^{11}$   h. $m^4$   i. $37$

**6.** $x^2 \ne -x^2$. The negative on $x^2$ is placed after $x$ is squared. We want
   the opposite of $x^2$, not $(-x)^2$.

**7.** $100$ times larger   **8.** $1.646$ yd$^3$   **9.** $3$ trays/can

**10.** a. $12\pi$   b. circle

**11.**

| Inequality | Interval |
|---|---|
| $0 \le x \le 20$ | $[0, 20]$ |
| $20 < x < 50$ | $(20, 50)$ |
| $50 \le x$ | $[50, +\infty)$ |

**12.**

| $x$ | $y$ | $x + y$ | $x - y$ | $x \cdot y$ | $x \div y$ |
|---|---|---|---|---|---|
| $-8$ | $4$ | $-4$ | $-12$ | $-32$ | $-2$ |
| $-6$ | $-2$ | $-8$ | $-4$ | $12$ | $3$ |
| $6$ | $-3$ | $3$ | $9$ | $-18$ | $-2$ |

## EXERCISES 3.1

**1.** $(-4, 1)$; $(-6, 0)$; $(-10, -2)$   **3.** $(1.7, 4)$; $(0.7, 0)$; $(-1.2, -8)$

**5.** $(7, 3)$; $(10, 0)$; $(12, -2)$   **7.** $(-2, 9)$; $(1, 0)$; $\left(2\frac{1}{3}, -4\right)$

**9.** $x = -4$ is a solution   **11.** $x = \frac{1}{2}$ is a solution

**13.** $x = -2$ is not a solution   **15.** $x = 0.8$ is a solution

**17.** a. any $y > 4.6$   b. solution not possible

**19.** a. any $y > 2$ or $y < 1$   b. $y = -4$ or $y = 0$

**21.** a. any $1 < y < 2$   b. solution not possible

**23.** a. $x = \pm 2$   b. $x \approx \pm 1.6$   c. $x = 0$

**25.** a. $D = 35x$   b. $D = 200 + 35x$

**27.** a. $T = 0.08x$   b. $C = 1.08x$

**29.** a. $137.5$ mi   b. $7.3$ hr   c. $10$ hr   d. $20$ mph

**33.** a ladder or cliff   **35.** wheelchair access ramp

**37.** $y = x$   **39.** b   **41.** c   **43.** a, b, d   **45.** b, c, d

## EXERCISES 3.2

**1.** f   **3.** c   **5.** b

**9.** 3 minus 4 divided by $-2$, $x = \frac{1}{2}$; 3 plus 4 multiplied by 2, $x = 14$;
   3 plus 2 multiplied by 4, $x = 20$

**11.** $x = 12$   **13.** $x = 7$   **15.** $x = 1$   **17.** $x = -2$

**19.** $x = -10$   **21.** $x = -6$   **23.** $x = -9$   **25.** $x = -1.5$

**27.** $x = 2.5$   **29.** $C = 100$

**31.** Equations are equivalent; multiplication; $\frac{1}{2}$

**33.** Equations are equivalent; addition; $-5$   **35.** not equivalent

**37.** Equations are equivalent; addition; $-4$

**39.** Equations are equivalent; addition; 4   **41.** not equivalent

**43.** $t \approx 3.64$   **45.** $r = 66\frac{2}{3}$   **47.** $p = \$1.67$   **49.** $x \approx 1884.25$

**51.** $h = 10$   **53.** a. $(2, 3)$   **55.** a. $(-1, -8)$

## EXERCISES 3.3

**1.** Change $\$4.97$ to $\$5.00 - 0.03$ and then multiply by 4; $\$19.88$

**3.** Change $\$10.98$ to $\$11.00 - 0.02$ and then multiply by 3; $\$32.94$

**5.** a. $6x + 12$   b. $-3x + 9$   c. $-6x - 24$   d. $-2 + x$
   e. $x^2 - 3x$   f. $-3x - 3y + 15$

**7.** $x^2 + xy + xz$   **9.** $24$ in.$^2$   **11.** $24 - 16x + 24y$

**13.** $2a^2 - 2ab + 2ac$   **15.** $-18a^2 + 12ab - 9ac$   **17.** $2a$

**19.** 7 yd; 3yd; 7yd   **21.** $5(a + b)$   **23.** $2(x - 2y)$   **25.** $11a$

**27.** $13.5$ m   **29.** $\$7.11$   **31.** a. $-4$   b. $1$   c. $-1$

**33.** $y = 150 - (x + 20)$, $x > 0$

**35.** $y = 30$, $0 < x \le 4$; $y = 30 + 10(x - 4)$, $x > 4$

**37.** $y = 25(x + 3)$, $x > 0$

**39.** Student subtracted $7 - 4$; correct answer is $19 - 4x$.

**41.** commutative   **43.** commutative   **45.** associative

**47.** 3 factors   **49.** 4 factors   **51.** 5 factors   **53.** 4 terms

**55.** 1 term   **57.** 3 terms   **59.** no

## EXERCISES 3.4

**1.** $x = 4$   **3.** $x = 1$   **5.** $x = -2$   **7.** $x = 1$

**9. a.** $x = 0$   **b.** $x = 2$     **11. a.** $x = 0$   **b.** $x = -2$

**13.**

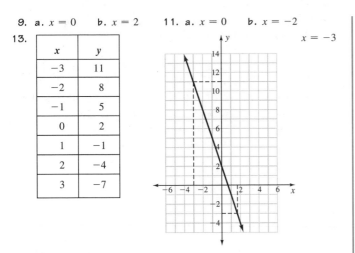

| $x$ | $y$ |
|-----|-----|
| $-3$ | $11$ |
| $-2$ | $8$ |
| $-1$ | $5$ |
| $0$ | $2$ |
| $1$ | $-1$ |
| $2$ | $-4$ |
| $3$ | $-7$ |

$x = -3$

**15.** See graph in 13; $x \approx 1.7$     **17.** $3x^2 + 3x + 6 = 3(x^2 + x + 2)$

**19. a.** $3x$   **b.** $-11y^2$   **c.** $5a + 15$   **d.** $x + 1$   **e.** $4 - x$

**21. a.** 0 sec   **b.** 7 in.   **c.** 12 m   **d.** 8 in.

**23.** 4 yd or 12 ft or 144 in.     **25.** $1\frac{3}{8}$     **27.** $x = 3$

**29.** $x = -3\frac{1}{2}$     **31.** $x = \frac{1}{3}$     **33.** $x = 1\frac{1}{2}$     **35.** $x = \frac{2}{3}$

**37.** $x = 4$     **45.** $150 - (x + 20) = 98.75, x > 0; x = \$31.25$

**47.** $30 + 10(x - 4) = 70, x > 4; x = 8$ hr

**49.** $25(x + 3) = 1200, x > 0; x = 45$ seats

## MID-CHAPTER 3 TEST

**1. a.** $4x - 12$   **b.** $-2x + 2y - 6$   **c.** $5 - x$   **d.** $12 - 2x$
   **e.** $3x^3 - 6x^2 + 3x$

**2. a.** $4(x^2 - 2x + 3)$   **b.** $3(2x + 4y - 5)$   **c.** 3.75 in.
   **d.** 5 m   **e.** $2xy(2x - 4xy + y)$

**3. a.** $-4x - 2y$   **b.** $x^3 - x^2 - 7x + 3$   **c.** $8x - 7$
   **d.** $5x + 10$

**4. a.** 370 cm   **b.** 255 mm

**5. a.** $x = 2$   **b.** $x = 8$   **c.** $x = 0$   **d.** $x = 5$

**6. a.** $x = 2$   **b.** $x = -4$   **c.** $x = -1$

**7. a.** $x = -4, x = 3$   **b.** $x = -3, x = 2$   **c.** $x = -1, x = 0$

**8. a.** $(1, -2)$   **b.** $-2 = -1 - 1; -2 = 2(1) - 4$   **c.** $x = 1$
   **d.** $x = 1$ is a solution.

**9.** $x = 1$     **10.** $x = 4\frac{1}{2}$     **11.** $x = 3$     **12.** $x = 2$

**13.** $b = -2$     **14.** $b = 10$     **15.** $F = 98.6$     **16.** $b = 9$

## EXERCISES 3.5

**7.** $y = \dfrac{-4}{x}$     **9.** $y = 3x - 10$     **11.** $y = \dfrac{x + 5}{2}$

**13.** $x = 3y - 6$     **15.** $x = \dfrac{7}{y}$     **17.** $x = y - 5$

**19.** $l = \dfrac{A}{w}$     **21.** $C = 2\pi r$     **23.** $h = \dfrac{A}{b}$     **25.** $t = \dfrac{I}{pr}$

**27.** $d = \dfrac{C}{\pi}$     **29.** $R = P + C$     **31.** $n = \dfrac{PV}{RT}$     **33.** $V_1 = \dfrac{C_2 V_2}{C_1}$

**35.** $C = P - a - b$     **37.** $h = \dfrac{2A}{a + b}$

**39. a.** $r = \dfrac{A - P}{Pt}$   **b.** $r = 5\frac{1}{4}\% = 0.0525$

**41. a.** $F = \frac{9}{5}C + 32$   **b.** $F = 98.6°$

**43. a.** $b = -2$   **b.** $b = 10$     **45.** 182 points

**47. a.** $t = \dfrac{v - v_0}{g}$   **b.** $t = 2.05$ sec   **c.** $t = 1.37$ sec

## EXERCISES 3.6

**5.** 1890s through 1920s; 1940s     **7.** 1930s; 1950s

**9.** 1890s through 1910s; 1940s     **11.** 1960s through 1990s

**13. a.** 3   **b.** $\frac{2}{3}$     **15. a.** $-2$   **b.** $\frac{3}{2}$

**17.** slope = \$0.32/lb     **19.** linear; slope = \$1.50/gal

**21.** linear; slope = \$9.00/hr     **23.** linear; slope = $-\$0.25$/copy

**25.** not linear     **27. b.** slope = $\frac{1}{4}$     **29. b.** slope = $-\frac{7}{2}$

**31. b.** slope undefined     **33. b.** slope = 0

**35.** $m = \dfrac{b - d}{a - c}$ and $m = \dfrac{d - b}{c - a}$     **37.** $m = -\dfrac{b}{a}$

**39.** yes; signs will be opposite in both numerator and denominator and expressions will simplify to the same number.

**43.** 1.35; \$/gal     **45.** 6.25; \$/hr

**47.** positive slope at $(-2, 0)$, $(2, 4)$, $(6, 8)$; negative slope at $(0, 2)$, $(4, 6)$, $(8, 10)$; slope is never zero; slope is undefined at $-2, 0, 2, 4, 6, 8, 10$

**49.** slope = $\frac{9}{100}$; rise = 950 ft     **51. a.** $\frac{5}{7}$   **b.** $\frac{27}{41}$   **c.** $\frac{29}{43}$   **d.** $\frac{5}{12}$

**55.** $\frac{1}{4} = \frac{25}{100} = 25\%$     **57.** slope = $\frac{1}{2}$; pitch = $\frac{1}{4}$

**59.** slope = $\frac{2}{3}$; pitch = $\frac{1}{3}$

**61. a.** $6°$   **b.** $11°$   **c.** $17°$   **d.** $37°$   **e.** $3°$

**63. a.** $0.18$   **b.** $0.37$   **c.** $0.57$

**65. a.** $35.5°$   **b.** $33.4°$   **c.** $34°$   **d.** $22.6°$

## EXERCISES 3.7

**9.** Celsius temperature when Fahrenheit temperature is zero. $x = -17.78°C$

**11.** $22.2°C$     **13.** $x = -5$; no meaning     **15.** $x = 100; x = 200$

**17.** $AD$: slope = 2, $y = 2x$; $BC$: slope = 2, $y = 2x - 20$

**19.** $m = 55; b = 0$     **21.** $m = 2\pi; b = 8$

**23.** $m = 2.98; b = 0.50$     **25.** $m = -0.29; b = 50$

**27.** $y = -0.4x + 3.8$     **29.** $y = -8x - 34$     **31.** $y = 2x - 20$

**33–37.**

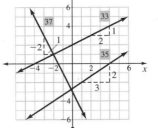

**39.** $y = -\$2x + \$50$     **41.** $y = \$42x + \$28$

**45.** $y = \$0.78x + \$0.55$

**47. a.** $y = 700x + 2400$

   **b.** No heat is produced below the $x$-intercept.

   **c.** $y = 2400$; amount of heat produced when outside temperature is $0°F$

   **d.** $x = -3.4°$; point where no heat is produced

**49.** parallel     **51.** not parallel     **53.** not parallel

**55.** All have the same slope; the coefficient of $x$ is 2; $m = 2$

**57.** All have the same $y$-intercept; the constant term is 3; $y$-intercept; $b = 3$

## CHAPTER 3 REVIEW EXERCISES

1. **a.** $x$, miles   **b.** $y$, cost   **c.** \$45.00   **d.** 275 mi
   **e.** \$35.00; cost/day   **f.** \$0.08; cost/mile
   **g.** $y = \$0.08x + \$35.00$   **h.** $x = -437.5$; no meaning

3. **a.** $3(5x - y + 2z)$   **b.** $2(4x^2 - 2x - 3)$   **c.** 11.4 cm
   **d.** $3\frac{3}{4}$ ft   **e.** $3xy(2y + 1 + 3xy)$

5. **a.** 56.4 mm   **b.** 5.58 m

7. **a.** $x = 1$   **b.** $x = -1$   **c.** $x = -2.5$

9. **a.** $(1, 2)$   **b.** $2 = 1 + 1$; $2 = 3 - 1$   **c.** $x = 1$
   **d.** The solution to the equation is the point of intersection of the two lines.

11. $x = -9$   13. $x = 2$   15. $b = 0$   17. $t = \dfrac{D}{r}$

19. $x = \dfrac{c - by}{a}$   21. $P_2 = \dfrac{P_1 V_1}{V_2}$   23. $a = \dfrac{2A}{h} - b$

25. $m = \dfrac{y_2 - y_1}{x_2 - x_1}$   27. Yes; student 2 needs 115 points

29.

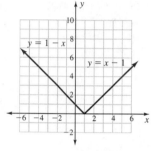

31. not linear   33. linear; slope $= -\frac{5}{2}$   35. slope $= -\frac{2}{1}$

37. slope $= \frac{1}{3}$

39. **a.** (2, \$0.26), (7, \$0.91)   **b.** \$0.13   **c.** cost/min

41. **a.** (6, 2), (12, 3)   **b.** 0.167   **c.** hr/dozen

43–45.

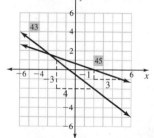

47. positive slope at $(-2, 0)$, $(0, 2)$, $(2, 4)$, $(4, 6)$, $(6, 8)$; slope is never negative; slope is never zero; slope is undefined at $-2, 0, 2, 4, 6, 8$

49. **a.** $y = 0.25x + 9$   **b.** hr/ft   **c.** set-up time

51. not parallel   53. not parallel   55. parallel   57. not parallel

## CHAPTER 3 TEST

1. **a.** $x = $ input, $y = $ output; $y = 0.25x$
   **b.** $x = $ input, $y = $ output; $y = 2x$
   **c.** $C = $ monthly cost, $x = $ no. of checks;
   $C = \$0.50(x - 15) + \$3.00, x \geq 15$; $C = \$3.00, 0 \leq x \leq 15$

2. $x + 12$   3. $xy - 5xz + 8yz$   4. $w = 13$; $l = 2a + 3$

5. $7(2x - y)$   6. $28ab - 14ac$   7. $3d(5a - 12c)$

8. $x = 1, y = 3$   9. 2.5 (halfway between 2 and 3)

10. $x$-intercept   11. $-2$   12. line 2   13. 0

14. 10 mph   15. 50 mph   16. undefined

17. You must multiply before subtracting in the expression $17 - 3(x + 4)$; in that case $5 - 3x \neq 14x + 56$

18. $x = 7$   19. $x = 1$   20. $x = -16$   21. $b = 7$

22. $x = 12$   23. $x = -2$   24. $V_2 = \dfrac{P_1 V_1}{P_2}$   25. $h = \dfrac{2A}{b}$

26. $+1.4$ million/yr   27. $-0.4$ million/yr   28. 48 million

30. **a.** $x$, miles   **b.** $y$, cost   **c.** \$42.50   **d.** 270 mi
   **e.** \$30.00; cost if no miles are driven   **f.** \$0.10; cost/mi
   **g.** $y = \$0.10x + \$30$   **h.** $x = -300$; no meaning

## EXERCISES 4.1

1. 3   3. $-4$   5. $\frac{1}{2}$   7. 2.98   9. $-0.05$

11. **a.** 5; 3:7   **b.** 16; 3:1   **c.** 240; 22:15   **d.** $4x$; $3:x$
   **e.** $2xy$; $1:3x$

13. **a.** $(b + c):1$   **b.** $(b + c):1$   **c.** $b:e$   **d.** $(1 + \sqrt{3}):1$

15. **a.** 6:1   **b.** 3:1   **c.** 2:5   **d.** 4:1   **e.** 1:100
   **f.** 4:25   **g.** 4:3   **h.** $m:60h$

17. no; ratio is the same   19. 3:1   21. \$40/credit

23. 1:16   25. 15 cc/hour   27. **a.** .357   **b.** .260

29. 9:8:4   31. 1:1:3   33. \$400,000; \$400,000; \$200,000

35. Virginia and James, \$1250; Ursula, \$2500

39. 63.64 kg   41. 300 g   43. \$137.50 per day

45. 20.83 microdrops/min   47. All have 2.16 oz alcohol.

49. 42.2 g HCN   51. 2.27 mph   53. 3.33 mi

55. **a.** 360°/hr   **b.** 30°/hr   **c.** 21,600°/hr

## EXERCISES 4.2

1. unsafe; slips   3. unsafe; tips   5. safe

9. $\frac{6}{8} \overset{?}{=} \frac{15}{20}$; $120 = 120$; proportion   11. $\frac{4}{6} \overset{?}{=} \frac{6}{9}$; $36 = 36$; proportion

13. $\frac{9}{21} \overset{?}{=} \frac{21}{35}$; $441 \neq 315$; false statement   15. $x = 11\frac{1}{4}$   17. $x = 3\frac{1}{3}$

19. $x = 2\frac{1}{7}$   21. $x = 9\frac{1}{3}$   23. 1.65 m   25. 24.15 km

27. 27,878,400 ft$^2$   29. $x = 3$   31. $x = -5$   33. $x \leq 0$

35. all inputs are okay; $x = 9$   37. all inputs are okay; $x = 2$

39. $x \neq -3, -7$; $x = 5$   41. $x \neq -\frac{1}{2}, 0$; $x = 2$

43. $5\frac{1}{4}$ ft   45. 36 ft   47. 12.6 ft   49. $\frac{1}{6864}$

51. 133,333.3 units   53. 488 pheasants

55. **a.** $t = \frac{15}{8}$   **b.** $t = \frac{24}{5}$   57. $V_1 T_2 = V_2 T_1$   59. $P_1 V_1 = P_2 V_2$

61. $\dfrac{V_1}{C_2} = \dfrac{V_2}{C_1}$   63. $\dfrac{T_2}{V_2} = \dfrac{T_1}{V_1}$   65. $l = 12$; $w = 6$; $A = 72$ units$^2$

## EXERCISES 4.3

1. 5 cm   3.

5. similar; $\angle B$ and $\angle D$, $\angle T$ and $\angle E$, $\angle A$ and $\angle N$, $BA$ and $DN$, $AT$ and $NE$, $BT$ and $DE$

7. not similar; $\angle S$ and $\angle O$; no others correspond

9. similar   11. similar

15. **a.** 6.7 cm by 8.04 cm   **b.** 15 cm by 18 cm

17. 0.2 in. by 0.56 in.   19. 4.67 ft   21. 40.5 ft

23. $A = (0, -1); B = \left(\frac{2}{3}, -3\right)$
25. $A = (1, 2); B = (4, 0); C = \left(0, \frac{8}{3}\right)$
27. $A = (6, 0); B = (3, 2)$    29. $A = (0, 2); B = (0, 4.8)$
31.
33.
35. $x = 3$    37. $x = 6\frac{2}{3}$    39. $\approx 1.4$; yes    41. 127.3 ft
43. $\approx 0.85$    45. ratios not the same    47. not linear

## MID-CHAPTER 4 TEST

1. $\frac{4x}{5yz}; y \neq 0, z \neq 0$    2. $\frac{x-2}{x}; x \neq -2, 0$    3. $1:4$
4. $4:1$    5. $60x:y$    6. gal per mi to oil a road
7. number of lb per week lost on a diet    8. $x = \frac{1}{100,000}$
9. fat = 300; carbs = 450; protein = 750    10. $45°; 45°; 90°$
11. 48 ft    12. $x = 26.67$    13. $x = 15$    14. $x = 17$
15. $x = 2$    16. $b = \frac{ad}{c}$    17. $x = 11.25$
18. $A = (5, 0); B = (7, 5.6)$    19. 24 ft    20. 400 native trout

## EXERCISES 4.4

1.

| Quantity | Value |
|---|---|
| 100 lb | 0.12 protein |
| 50 lb | 0.15 protein |

3.

| Quantity | Value |
|---|---|
| 5 kg | $8.80 per kg |
| 2 kg | $24.20 per kg |

5.

| Quantity | Value |
|---|---|
| 150 mL | 18 molar |
| 100 mL | 3 molar |

7.

| Quantity | Value |
|---|---|
| 15 dimes | $0.10 |
| 20 quarters | $0.25 |

9.

| Quantity | Value |
|---|---|
| 3 hr | 80 km/hr |
| 2 hr | 30 km/hr |

11. a.

| Item | Quantity | Value ($) | Total ($) |
|---|---|---|---|
| grapes | 3 lb | 0.98 | 2.94 |
| potatoes | 5 lb | 0.49 | 2.45 |
| broccoli | 2 lb | 0.89 | 1.78 |

b. yes; total lb purchased    c. yes; average cost per lb
d. $7.17
13. a.

| Quantity | Value | Total |
|---|---|---|
| 0.2 L | 16 molar | 3.2 moles |
| 0.3 L | 6 molar | 1.8 moles |

b. yes; total L    c. yes; molar value of mixture    d. 10 molar
15. $\approx 10.3$ gal    17. $87.5°$    19. 18.3 gal    21. 200 shares
23. 0.2 L    25. $(300 - x)$ lb    27. $\$15,000 - x$
29. $(16 - x)$ L
31. Colombian = 180 lb; Sumatran = 120 lb; ratio = 3:2
33. 2nd-year student
35. a. cashews = 7.14 kg; peanuts = 42.86 kg; ratio = 1:6
    b. cashews = 21.4 kg; peanuts = 28.6 kg; ratio = 3:4
37. a. $1200    b. $750    c. $975    d. $1050

## EXERCISES 4.5

1. mean = 58; median = 57; mode = 57
3. mean = 58.4; median = 61; no mode
17. a. $(2.5, 3)$    b. $(1, 0)$
19. a. $\left(\frac{a}{2}, \frac{b}{2}\right)$    b. $\left(\frac{a}{2}, \frac{a}{2}\right)$    c. $\left(0, \frac{b}{2}\right)$
21. a. $(0, 3), (4.5, 0), (4.5, 3)$    b. $(0, 0), (-2, 2), (2, 2)$
23.
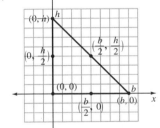
25.
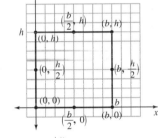
27. Both midpoints are $\left(\frac{b}{2}, \frac{h}{2}\right)$.
29. a. $(0, 2), (1.5, 0), (1.5, 2)$
    b. $y = \left(\frac{4}{3}\right)x; y = \left(-\frac{8}{3}\right)x + 4; y = \left(-\frac{2}{3}\right)x + 2$
35. a. $\left(\frac{-4 + (-4) + 12 + 12}{4}, \frac{0 + 8 + 8 + 0}{4}\right) = (4, 4)$
    b. $\left(\frac{-30 + 0 + (-15)}{3}, \frac{24 + 24 + 0}{3}\right) = (-15, 16)$

## CHAPTER 4 REVIEW EXERCISES

1. $4:x^2; x \neq 0$    3. $10:3$    5. $m:4n$
9. Pepsi = 55,000; Coke = 55,000; other = 11,000

**11.** 450 cal; 50 g     **13.** 109.36 yd     **15.** 20 ft     **17.** 4.5 gal

**19.** $x = 9$     **21.** $x = 8$     **23.** $V_2 = \dfrac{V_1 C_1}{C_2}$

**27.** $A = (0, 1.5)$; $B = (-3, 0)$     **29.** 5000 bats

**31.**

| Quantity (hr) | Value (mph) | Item Total (mi) |
|---|---|---|
| 3.85 | 130 | 500 |
| 7.50 | 200 | 1500 |
| 11.35 | 176 | 2000 |

A 2000-mi trip takes 11.35 hr, for an average of 176.2 mph.

**33.**

| Quantity (gal) | Value (octane) | Item Total |
|---|---|---|
| 6 | 88 | 528 |
| 10 | 92 | 920 |
| 16 | 90.5 | 1448 |

The mixture is 16 gal total, with an average octane of 90.5.

**35.**

| Quantity (coins) | Value ($) | Item Total |
|---|---|---|
| 6 dimes | 0.10 | 0.60 |
| 4 quarters | 0.25 | 1.00 |
| 10 coins | 0.16 | 1.60 |

The ten coins have an average value of $0.16, which is probably not meaningful.

**37.** $15,000

**39.** mean = 31.725 cm; median = 31.75 cm; mode = 31.8 cm

**41.** mean = 44.8 mL; median = 45.0 mL; mode = 45.0 mL

**43.** **a.** $EG = (1, 7)$; $EF = (1, 2)$; $HE = (-3, 7)$; $HF = (1, 7)$
**b.** $HF = -\frac{5}{4}$; $EG = \frac{5}{4}$
**c.** Division by zero is not defined.
**d.** (1, 7)

**45.** (1, 1)     **47.** $\frac{8}{27}$

## CHAPTER 4 TEST

**1.** $\frac{3}{7}$; $\frac{6}{14}$; $\frac{9}{21}$     **2.** $\frac{3}{9}$; $\frac{1}{3}$; $\frac{6}{18}$     **3.** $\dfrac{b}{4a}$     **4.** $\dfrac{1}{a-b}$     **5.** 5:1

**6.** $60h$ to $m$     **7.** 196.85 in.     **8.** 1.4 ft     **9.** $x = 32.5$

**10.** $x = 15$     **11.** 94%     **12.** median = $0.29; mean = $0.59

**13.** median = $0.69; mean = $0.59

**14.** Exercise 13 gives more useful mean; one large price makes Exercise 12 less useful.

**15.** (5, 5); (7.5, 5); (5.5, 2)     **16.** (6, 4)

**17.**

| Quantity | Value | Total |
|---|---|---|
| $17,000 | 0.058 | $986 |
| $2,000 | 0.149 | $298 |

total debt = $19,000; total interest = $1284; avg. interest rate = 6.76%

**18.** $x = 16$; $x = 4.8$     **19.** 12 credits

**20.** $I_1 = 10$ mA; $V$ and $I$ are not proportional

## EXERCISES 5.1

**1.** Skate World: $y = 55 + 6x$; Lane Co. Ice: $y = 85 + 4.75x$

**3.** $A = B = 55°$; $C = 70°$

**5.** **a.** 23 dimes; 45 quarters
**b.** $23 + 45 = 68$; $0.10(23) + 0.25(45) = 13.55$

**21.** **a.** $x + 5 = 10 - x$     **b.** $A + B = 10$; $A + 5 = B$

**23.** **a.** $A = 2x$; $C = 2x + 20$
**b.** $A + B + C = 180$; $A = 2B$; $C = A + 20$

**25.** **a.** $0.05x + 0.25(22 - x) = 2.90$
**b.** $n + q = 22$; $0.05n + 0.25q = 2.90$

**27.** **a.** $0.10x + 0.25(26 - x) = 5.45$
**b.** $d + q = 26$; $0.10d + 0.25q = 5.45$

**29.** **a.** $x + x - 20 + 2x = 620$
**b.** $L + W + C = 620$; $C = L - 20$; $W = 2L$

**31.** **a.** $2x + 2(7x) = 32$     **b.** $2w + 2l = 32$; $l = 7w$

**33.** **a.** $30 + 3.50(x - 8) = 3.00x$, $x > 8$
**b.** $y = 30 + 3.50(x - 8)$, $x > 8$; $y = 3.00x$

**35.** **a.** one     **b.** two     **c.** three

## EXERCISES 5.2

**1.** rate per day     **3.** [0, 3750); $0 \le x < 3750$

**5.** % of sales received as commission

**7.** Slopes of parallel lines are equal.     **9.** (3, 16); $3 < x < 16$

**11.** no $x$-intercept

**13.** The first 10 people are included in the $45 cost.

**15.** parallel lines     **17.** not parallel     **19.** parallel lines

**21.** not coincident     **23.** coincident     **25.** Supply decreases.

**27.** Demand decreases.     **29.** Slope is positive.

**31.** Supply does not change.     **33.** $E$ also changes upward.

**35.** Price decreases.     **37.** Quantity does not change.

**39.** **a.** $525     **b.** $500     **41.** $3.75     **43.** $25 loss

**45.** **a.** $900 cost     **b.** $1000 revenue     **c.** $100 profit

**47.** Revenue is on top.     **49.** $x = 120$

**51.** **a.** $A = (0, 0)$; $B = (4, 16)$     **b.** $x^2 = 4x$     **c.** $y = x^2$; $y = 4x$

**53.** (23, 45)

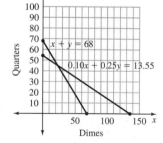

## EXERCISES 5.3

**1.** $W = \dfrac{L}{2}$     **3.** $r = \dfrac{C}{2\pi}$     **5.** $d = \dfrac{C}{\pi}$     **7.** $x = 3$

**9.** $x = 3$     **11.** $y = 4 - 3x$     **13.** $x = 5 + 4y$

**15.** $x = 5y - 9$     **17.** $y = 3x + 2$

**19.** **a.** $y = 5 - 2x$     **b.** $y = 3 - x$
intercept is (2, 1); $1 = 5 - 2(2)$; $1 = 3 - 2$

**21.** $x = 3$; $y = -5$     **23.** $x = 2$; $y = 15$     **25.** $x = 4$; $y = -1$

**27.** $x = 5$; $y = 9$     **29.** $x = 3$; $y = -2$

**31.** $A = 108$; $B = 54$; $C = 18$     **33.** $x = 2$; $y = \frac{8}{3}$

**35.** $\square = 2\heartsuit$; $\odot = 3\heartsuit$     **37.** $\clubsuit = 2\heartsuit$; $\spadesuit = 5\heartsuit$

**39.** $8\frac{1}{2}$ yd; $11\frac{1}{2}$ yd     **41.** 17 quarters; 7 nickels

**43.** 17 dimes; 11 quarters     **45.** 23 people

**47.** width = 8 in.; height = 12 in.

**49.** height = 19 cm; width = 10 cm    **51.** 45°; 45°; 90°

**53.** $x = 4; y = 5; z = -3$    **55.** $A = \dfrac{b^2\sqrt{3}}{4}$    **57.** $A = \dfrac{C^2}{4\pi}$

**59.** $d = \dfrac{C}{\pi}$

## MID-CHAPTER 5 TEST

**1.** $y = 2x - 5000$    **2.** $h = \dfrac{3V}{\pi r^2}$    **3.** $x = 1500; y = 3500$

**4.** $x = 2500; y = 2500$

**5. a.** $y = 6 - \frac{2}{3}x; y = x + 1$    **b.** $(4.5, 5.5)$    **c.** $(4.5, 5.5)$
   **d.** $(12, -2)$

**6.** $t$ = turtle, $s$ = ostrich; $t + s = 195, t = 12s$;
   turtle = 180, ostrich = 15

**7.** $f$ = 5-lb bags, $t$ = 10-lb bags; $5f + 10t = 10,000, f = 3t$;
   400 10-lb bags, 1200 5-lb bags

**8.** $\dfrac{l}{w} = \dfrac{13}{6}, 2l + 2w = 114$; width = 18 ft, length = 39 ft

**9.** $w$ = twins, $t$ = triplets, $q$ = quadruplets; any three of the four
   equations $q + t + w = 27, 4q + 3t + 2w = 69, 4q = w, w - 9 = t$;
   16 sets of twins, 7 sets of triplets, 4 sets of quadruplets

**10.** The intersection is where the same input gives the same output in
   each equation.

## EXERCISES 5.4

**1.** $x = 3; y = -5$    **3.** $n = -4; m = 7$    **5.** $x = -2, y = 3$
**7.** $a = -10; b = 15$    **9.** $x = -3; y = 3$    **11.** $p = 0; q = 3$
**13.** $m - 2; b = -3$    **15.** $x = 2; y - 4$    **17.** 8.5; 16.5
**19.** 4.75; 15.25    **21.** $f - 9; p - 4$    **23.** $y = x + 3$
**25.** $y = \frac{1}{2}x - 2$    **27.** $y = 2$
**29.** $4s + 2g = 296, 3s + 10g = 329$;
   sugar cookie = 67.7 cal, ginger snap = 12.6 cal
**31.** $15c + 22g = 126, 20c + 11g = 113$; cherry = 4 cal, grape = 3 cal
**33.** $8g + 5r = 285, 4g + 10r = 330$;
   green olive = 20 cal, ripe olive = 25 cal
**35.** $6a + 3s = 58.50, 5a + 4s = 54.00$;
   adult = \$8.00; student = \$3.50
**37.** $3s + 2t = 109.95, 4s + t = 119.95$; shirt = \$25.99, tie = \$15.99
**39.** airspeed = 100 mph; wind = 20 mph
**41.** boat speed = 7 mph; current = 3 mph

## EXERCISES 5.5

**1. a.** some possibilities: (0, 4); (1, 3); (2, 2); (3, 2); (4, 1); (5, 1);
   (6, 0); (7, 0)
   **b.**

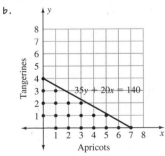

**c.** $35y + 20x \le 140, 0 \le y, 0 \le x$; whole numbers

**3.**

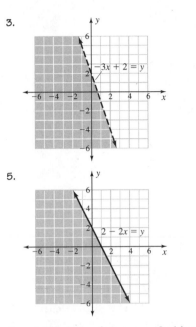

**5.**

**7.** $y < 150, 0 \le y, 0 \le x$    **9.** 4th quadrant, negative $y$-axis
**11.** 2nd quadrant, neither axis    **13.** $x > 0, y < 0$
**15.** $x > 2$    **17.** $x < 1$    **19.** $x > 2$    **21.** $x < 4$
**23.** $x < 1$    **25.** $2 \ge x$    **31.** $2 < x$    **33.** $x < 1$
**35.** $2 < x$    **37.** $4 > x$    **39.** $1 > x$    **41.** $2 \ge x$
**43.** $x < 1$    **45.** $x < 1$

## CHAPTER 5 REVIEW EXERCISES

**1.** $y = 2x + 5000$    **3.** $r = \dfrac{C}{2\pi}$    **5.** neither; $\left(1, -\frac{2}{3}\right)$

**7.** parallel    **9.** $m = -\frac{1}{5}; b = \frac{13}{5} = 2\frac{3}{5}$    **11.** $x = 500; y = 4500$

**13. a.** line from (0, 30) to (30, 0)
   **b.** line from (0, 0) to (10, 20)
   **c.** (10, 20); spending entire budget
   **d.** line from (0, 20) to (20, 0)
   **e.** (15, 30); spending if budget increases to \$45

**15.** $B$ = Boeing 747, $b$ = Boeing 707; $B + b = 721, B = b + 279$;
   Boeing 747 holds 500 persons, Boeing 707 holds 221 persons

**17.** $c + s = 2000, 0.09c + 0.44s = 400$;
   1371.4 lb of corn, 628.6 lb of soybean meal

**19.** $x$ = shorter side, $y$ = longer sides; $x + y + y = 32, y = 2.5 + x$;
   sides are 9 in., 11.5 in., 11.5 in.

**21.** $p + f + c = 37, 4p + 4c + 9f = 198, p = c + 5$;
   16 g protein, 10 g fat, 11 g carbohydrate

**25.** b    **27.** $x > 5$
**29.** solid line is $y = 15 - 2x$; broken line is $y = x - 6$; $x > 7$
**31.** $x \ge 2$    **33.** $2x - 1 > 2 - x$
**35.** $85 + 4.75(x - 10) \le 150; x \le 23$

## CHAPTER 5 TEST

**1.** $y = 3x - 400$    **2.** $b = \dfrac{2A}{h}$    **3.** $T = \dfrac{AH}{I}$

**4.** $x = 13; y = 9$    **5.** $x = 1500; y = 600$    **6.** $x = -25; y = 8$

**7.** when no variables remain and we have a false statement

**8.** $y = 8 - x; y = 4 - \frac{1}{3}x$; the point (6, 2) makes both true

**9.** $m = \frac{3}{4}; b = -\frac{7}{4}$

**10.** $c = 10 + b$, $8b + 6c = 144$;
caterpillar = 16 legs, butterfly = 6 legs

**11.** $16p + 5c = 135$, $20p + 25c = 405$;
peanuts = 4.5 cal, cashews = 12.6 cal

**12.** $135 = 3(d + c)$, $100 = 4(d - c)$;
current = 10 mph; dolphin = 35 mph

**13. a.** $y = x - 1$ is the solid line; $y = 5 - x$ is the broken line
**b.** $(3, 2)$    **c.** $x > 3$

**14.** $x \le 2$

## EXERCISES 6.1

**1. a.** $-5$   **b.** $-5$   **c.** $-13$   **d.** $-10$   **e.** $-4$   **f.** 7

**3. a.** $a + 12b - 8c$   **b.** $3m - n$   **c.** 21 cm
**d.** $x^2 - x - 12$   **e.** $-x^2 + 2x - 3$

**5.** 2 yd + 2 in.    **7.** 1 day + 19 hr + 5 min

**9. a.** $x^2 + 5x + 6$; trinomial   **b.** $3x^2 + 7x + 2$; trinomial
**c.** $17 - 4x$; binomial   **d.** $6x^2 + 5x + 1$; trinomial
**e.** $a^2 - b^2$; binomial

**11. a.** $x^2 - 4xy + 4y^2$; trinomial   **b.** $x^2 - 4y^2$; binomial
**c.** $x^3 + 3x^2 + 3x + 1$; polynomial
**d.** $x^3 + 2x^2 - 4x - 8$; polynomial   **e.** $a^3 + b^3$; binomial

**13.** See answers to exercises 9 and 11 above.

**17.** $x(x^2 + 4x + 4)$    **19.** $b(a^2 + ab + b^2)$    **21.** $2x(3x + 1)$

**23.** $3y(5y - 1)$    **25.** $4(-x - 3)$; $-4(x + 3)$

**27.** $2(-x^2 - 4x - 4)$; $-2(x^2 + 4x + 4)$    **29.** $y(y^2 + 2xy + x^2)$

**31.** $2ab(a - 2b + 3b^2)$    **33.** $3 \cdot 37$    **35.** $7 \cdot 11 \cdot 13$

**37. a.** $\frac{4}{11}$   **b.** $\frac{1}{15}$   **c.** $\frac{5}{27}$   **d.** $\frac{n}{p}$   **e.** $\frac{p}{6m}$

**45. a.** width $2a$; length $1 - 2b$   **b.** width $xy$; length $x + y$

**47. a.** 5   **b.** 10.32   **c.** $73a + 23g$   **d.** $2.5x$   **e.** $3 + \pi$   **f.** 9.5 cm

**49. a.** $3c$   **b.** $2.5y$   **c.** 5 m; 5 m    **51.** 50 foot-pounds

**53.** 20 foot-pounds    **55.** $xy$ foot-pounds    **57.** $5000 + 100y$

**59.** 322 ft 8 in. or $\approx 322.6$ ft    **61.** $\frac{121}{3}x$ ft or $\approx 40.3x$ ft

## EXERCISES 6.2

**1.** $x^2 - 4x + 4$    **3.** $x^2 - 4$    **5.** $a^2 + 10a + 25$

**7.** $b^2 - 25$    **9.** $a^2 - b^2$    **11.** $a^2 - 2ab + b^2$

**13.** $x^2 + 8x + 7$    **15.** $x^2 + 6x - 7$    **17.** $b^2 + 2b + 1$

**19.** $a^2 - 49$    **21.** $x^2 + 2xy + y^2$    **23.** $x^2 - y^2$

**25.** $(x + 1)(x + 1)$    **31.** $x^2 + 11x - 12$    **33.** $x^2 + x - 12$

**35.** $x^2 - 4x - 12$    **37.** $x^3 - 3x^2 + 3x - 1$    **39.** $a^3 - b^3$

**41.** $x^3 + 3x^2y + 3xy^2 + y^3$    **43.** $4x^2 + 12x + 9$

**45.** $4x^2 - 9$    **47.** $9x^2 - 12x + 4$    **49.** $9x^2 - 4$

**53.** $(x + 5)(x + 4)$    **55.** $(x + 2)(x - 10)$    **59.** $(x + 3)(x + 3)$

**61.** $(x + 5)(x + 6)$    **63.** $(x + 15)(x - 2)$    **65.** $(x - 8)(x + 2)$

**67.** $(x - 1)(x + 16)$    **69.** $(x + 5)(x - 5)$    **71.** $(x + 9)(x - 9)$

**73.** $(x + 4)(2x + 3)$    **75.** $(x - 3)(2x + 3)$

**77.** $(n - 1)(2n + 3)$    **79.** $(x + 2)(3x - 1)$

**81.** $(a - 4)(3a + 1)$    **83.** $(3x + 7)(3x - 7)$

**85.** $(4x + 3)(4x - 3)$    **87.** $(3x + 2)(2x - 1)$

**89.** $(2x + 3)(3x - 2)$    **91.** $(n + 5)(2n - 1)$

**93.** $(5x - 6)(5x + 6)$    **95.** $3(x + 3)(x + 1)$

## EXERCISES 6.3

**1.** 10 ways    **3.** 21 ways

**5.** Last row is 1, 10, 45, 120, 210, 252, 210, 120, 45, 10, 1.

**7.** $2^1 = 2$, $2^2 = 4$, $2^3 = 8$, $2^4 = 16$, $2^5 = 32$, $2^6 = 64$, $2^7 = 128$, $2^8 = 256$, $2^9 = 512$, $2^{10} = 1024$; the numbers in a particular row of Pascal's triangle add to the power of 2 that corresponds to that particular row if we begin with $2^0$ corresponding to the first row. The row sums are then 1, 2, 4, 8, 16, 32, 64, 128, 256, 512, 1024

**9.** GGGG, GGGB, GGBB, GBBB, BBBB, GGBG, GBGB, BGBB, GBGG, GBBG, BBGB, BGGG, BGBG, BBBG, BGGB, BBGG

**11.** $1 + 4r + 6r^2 + 4r^3 + r^4$    **13.** $1 - 4a + 6a^2 - 4a^3 + a^4$

**15.** $x^6 + 6x^5y + 15x^4y^2 + 20x^3y^3 + 15x^2y^4 + 6xy^5 + y^6$

**17.** $b^4 - 4b^3 + 6b^2 - 4b + 1$

**19.** $1 + 5z + 10z^2 + 10z^3 + 5z^4 + z^5$

**23.** $+70x^4y^4 + 56x^3y^5 + 28x^2y^6 + 8xy^7 + y^8$

**25.** $+126x^5y^4 + 126x^4y^5 + 84x^3y^6 + 36x^2y^7 + 9xy^8 + y^9$

**27.** 2nd power

**29.** $\overline{A}\ \overline{B}\ \overline{C}\ \overline{D}\ \overline{E}\ \overline{H}$
$\overline{I}\ \overline{M}$    The others have no line of symmetry.

**31. a.**    **b.**

## MID-CHAPTER 6 TEST

**1.** $P = 10x + 2$; $A = 6x^2 + 3x$    **2.** $3x$    **3.** $-2x^2 + 4x - 2$

**4. a.** $2(3x^2 - x + 4)$   **b.** $a(2bc - 3c + 4b)$

**5. a.** $(3x - 2)(5x + 3) = 15x^2 - x - 6$
**b.** $6x^2 - 7x - 20 = (3x + 4)(2x - 5)$

**6. a.** $x^2 - 5x - 14$   **b.** $4x^2 + 8x + 3$

**7. a.** $(x + 2)(x + 2)$   **b.** $(x - 4)(x - 4)$   **c.** $(x - 2)(2x + 1)$
**d.** $2(x - 2)(x + 2)$

**8. a.** $x^3 + 3x^2y + 3xy^2 + y^3$   **b.** $x^4 - 4x^3 + 6x^2 - 4x + 1$

## EXERCISES 6.4

**1. a.** 1   **b.** $\frac{1}{25}$   **c.** 5   **d.** 5

**3. a.** 3   **b.** 1   **c.** 3   **d.** $\frac{1}{9}$

**5. a.** 4   **b.** 1   **c.** $\frac{1}{2}$   **d.** $\frac{1}{2}$

**7. a.** 4   **b.** 0.1   **c.** 2.5   **d.** 0.5   **e.** 50   **f.** 20

**9.** false; $\left(\frac{1}{4}\right)^{-1} = 4$, the reciprocal of a number between 0 and 1 is a larger number

**11. a.** $\frac{1}{x}$   **b.** $\frac{y}{x}$   **c.** $\frac{x}{y}$   **d.** 1   **e.** 1   **f.** $\frac{bc}{a}$

**13. a.** $\frac{1}{x^2}$   **b.** $\frac{1}{y^3}$   **c.** $\frac{x^2}{y^2}$   **d.** $\frac{b^3}{a^3}$   **e.** $\frac{c^2}{16a^4}$   **f.** $\frac{b^6}{a^3}$   **g.** $b^3$

**15. a.** $x$   **b.** $x$   **c.** $x$

**17.**

| $x$ | $10^x$ | $10^x$ |
|---|---|---|
| 0 | 1 | 1 |
| $-1$ | $\frac{1}{10}$ | 0.1 |
| $-2$ | $\frac{1}{100}$ | 0.01 |
| $-3$ | $\frac{1}{1000}$ | 0.001 |
| $-4$ | $\frac{1}{10000}$ | 0.0001 |
| $-5$ | $\frac{1}{100000}$ | 0.00001 |

**19.**

| $x$ | $2^x$ |
|-----|-------|
| 2 | 4 |
| 1 | 2 |
| 0 | 1 |
| −1 | $\frac{1}{2}$ |
| −2 | $\frac{1}{4}$ |
| −3 | $\frac{1}{8}$ |

**21.**

| $x$ | $5^x$ |
|-----|-------|
| 2 | 25 |
| 1 | 5 |
| 0 | 1 |
| −1 | $\frac{1}{5}$ |
| −2 | $\frac{1}{25}$ |
| −3 | $\frac{1}{125}$ |

**23.** $8x^3 + 36x^2 + 54x + 27$

**25.** $81x^4 - 216x^3 + 216x^2 - 96x + 16$   **27.** $x^3 - \frac{3}{2}x^2 + \frac{3}{4}x - \frac{1}{8}$

**29.** $2.7564 \times 10^9$ mi   **31.** $1.8 \times 10^3$ g   **33.** 200,000,000 yr

**35.** 0.000 000 000 000 000 000 000 000 001 675 kg

**37.** electron   **39.** 0.023 4   **41.** 62,800,000

**43.** 456,000,000   **45.** 0.000 634   **49.** $4.8 \times 10^{-7}$

**51.** $7.0 \times 10^1$, or 70   **53.** $4.0 \times 10^{-7}$

**55.** $1.54 \times 10^2$ sec or 2.6 min   **57.** $4.91 \times 10^2$ sec or 8.2 min

## EXERCISES 6.5

**1. a.** 4; 4   **b.** $[(x + 6)3 - 6] \div 3 - x = 4$

**3.** $x - 2, x - 1, x$

**5.** $x - 2, x, x + 2$ for $x = 7$; $x - 4, x - 2, x$ for $x = 9$

**7.** $x = 1, x + 1 = 2, x + 3 = 4$; $x = 2, x + 1 = 3, x + 3 = 5$

**9.** They are the same.   **11.** 1942, 1945; $x, x + 3, x + 6$

**13.** 2012; $x + 4, x + 8, x + 12$

**15.** $x + (x + 1) + (x + 2) = 42$; 13, 14, 15   **17.** 1942

**19.** $x + (x + 2) + (x + 4) = 177$; 57, 59, 61   **21.** 42; $6 \cdot 7$

**23.** 50°, 60°, 70°   **25.** 70°, 70°, 40°   **27.** $46\frac{1}{3}°, 48\frac{1}{3}°, 85\frac{1}{3}°$

**29.** $A = 107°$; $B = 73°$; $C = 77°$; $D = 103°$; $E = 46°$; $F = 44°$; $G = 61°$; $H = 29°$; $I = 77°$; $J = 103°$

## CHAPTER 6 REVIEW EXERCISES

**1. a.** $P = 10x + 2$; $A = 6x^2 + 2x$   **b.** $P = 8x + 4$; $A = 3x^2 + 6x$
**c.** $P = 7x - 3$; $A = 2x^2 + x$

**3. a.** $4x - 20$   **b.** $-7x^2 - 14x + 21$   **c.** $-9x^3 + 6x^2 + 21x$

**5. a.** $3x(2x + 5) = 6x^2 + 15x$   **b.** $3(a^2 - 3) = 3a^2 - 9$

**7. a.** $x^2 + 7x + 12$   **b.** $x^2 + x - 12$   **c.** $4x^2 - 20x + 25$
**d.** $4x^2 - 25$

**11. a.** $90x^2$   **b.** $1x, 90x$; $2x, 45x$; $3x, 30x$; $5x, 18x$; $6x, 15x$; $9x, 10x$
**c.** $2x, 45x$

**13. a.** $(x + 6)(x - 1)$   **b.** $(x - 5)(2x + 7)$   **c.** $(2x - 7)(2x + 5)$
**d.** $4(x - 1)(x - 1)$   **e.** $4(x + 1)(x - 1)$   **f.** $x(x + 2)(x + 2)$

**17. a.**   **b.**

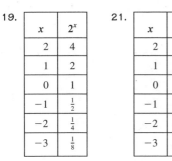

**19. a.** 12   **b.** $\frac{1}{144}$   **c.** 1   **d.** 12

**21. a.** 0.6   **b.** $\frac{25}{9}$   **c.** 0.6   **d.** 1

**23. a.** $\frac{b^2}{a^2}$   **b.** 1   **c.** $\frac{a}{b}$   **d.** $\frac{b^6 c^3}{8a^3}$

**25.** $K^{40} = 1{,}400{,}000{,}000$; $Ca^{41} = 120{,}000$; $Rn^{219} = 0.000\ 000\ 1243$; $Po^{212} = 0.000\ 000\ 3$

**27.** $9.5 \times 10^{-15}$ yr

**29.** Let $x$ = multiple of 5; $x + (x + 5) + (x + 10) = 60$; 15, 20, 25 in.

## CHAPTER 6 TEST

**1.** $P = 12x + 10$; $A = 8x^2 + 20x$   **2.** $3x$

**3. a.** $x^2 + 3x - 28$   **b.** $x^2 - 49$   **c.** $4x^2 - 28x + 49$
**d.** $2x^2 - 16x + 32$   **e.** $x^3 + 14x^2 + 49x$   **f.** $x^3 - 8$

**4. a.** $2y(7x + 3x^2 - 9y)$   **b.** $(x - 4)(x - 4)$
**c.** $6(x - 3)(x + 3)$   **d.** $(3x + 1)(3x + 1)$
**e.** $(5x - 3)(3x - 1)$   **f.** $5(2x - 1)(x + 1)$

**5.** no; $2x - 2$ factors into $2(x - 1)$, and 2 is not a factor of the trinomial.

**6. a.** 9   **b.** 0.4   **c.** 1   **d.** $\frac{625y^4}{81x^4}$

**7. a.** $3{,}600\ 000\ 000\ 000\ 000\ 000\ 000\ 000$; $3.6 \times 10^{21}$
**b.** $0.000\ 000\ 000\ 000\ 000\ 000\ 000\ 008$; $8.0 \times 10^{-21}$

**8.** $a^4 + 4a^3 + 6a^2 + 4a + 1$

**9.** Let $x$ = multiple of 13; $x + (x + 13) + (x + 26) = 156$; 39, 52, 65 in.

**10.** 60°, 120°   **11.**

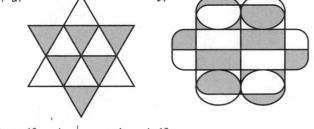

**12.** $a, b, c$ do not indicate the consecutive nature of the numbers; also, one variable is easier to solve for.

**13.** $2158.92

## EXERCISES 7.1

**1.** 316   **3.** 31,622   **5.** age, 44; year, 1936; born, 1892

**7.** not a right triangle   **9.** right triangle   **11.** right triangle

**13.** right triangle   **15.** right triangle

**17. a.** $x = \sqrt{34} \approx 5.83$   **b.** $x = \sqrt{65} \approx 8.06$
**c.** $x = \sqrt{117} \approx 10.82$

**19.** 12.4 ft   **21.** 9.3 ft   **23.** 3.5 ft from base to wall; 14.4 ft

**25.** base, 2.91 ft = 2 ft 11 in.; height, 11.64 ft = 11 ft $7\frac{1}{2}$ in.

**27.** base, 4.37 ft = 4 ft $4\frac{1}{2}$ in.; height, 17.46 ft = 17 ft $5\frac{1}{2}$ in.

**29.** 3 ft $10\frac{1}{2}$ in.; 15 ft 6 in.   **31.** 251.8 mi; 6.8°   **33.** 19.72 ft

**35.** 1028 ft$^2$   **37.** 1232 ft$^2$   **39. a.** 7   **b.** −15   **c.** ±20

**41. a.** $\sqrt{15} \approx 3.87$   **b.** $\sqrt{14} \approx 3.74$   **c.** 45   **d.** 75   **e.** 8

**43. a.** $a$   **b.** $b$   **c.** 11a   **d.** 6x

**45. a.** $\frac{x}{3}$   **b.** $\frac{2}{5}$   **c.** 3   **d.** $\frac{2}{x}$

## EXERCISES 7.2

**1. a.** $x = -4, x = 3$   **b.** $x = -2, x = 1$   **c.** no solution
**d.** $x = -3, x = 2$

**3.** $x = \pm 1$   **5.** $x = \pm \frac{2}{5}$   **7.** $x = \pm \frac{1}{5}$   **9.** $x = \pm \frac{15}{7}$

**11.** $x = \pm \frac{11}{6}$   **13.** $x = \pm \frac{3}{5}$   **15.** $x = -3, x = 2$

**17.** $x = 5, x = -3$   **19.** $x = -2, x = 6$   **21.** $x = -1, x = \frac{3}{2}$

**23.** $x = \pm \frac{5}{2}$   **25.** $t = 0$ sec, $t = 3$ sec   **27.** $t = 1$ sec, $t = 3$ sec

**29.** $t = 0$ sec, $t = 4$ sec   **31.** $a = 2, b = 3, c = 1$

**33.** $a = 1, b = -4, c = 4$   **35.** $a = 1, b = 0, c = -4$

**37.** $a = 4, b = -8, c = 0$   **39.** $a = -0.5g, b = v, c = s$

**41.** $a = \pi, b = 0, c = 0$

**43.**

| $a^2 + b^2$ | $(a + b)^2$ | $\sqrt{(a + b)}$ | $\sqrt{a} + \sqrt{b}$ |
|---|---|---|---|
| 97 | 169 | $\sqrt{13}$ | 5 |
| 10 | 16 | 2 | 2.7 |
| 41 | 81 | 3 | 4.2 |
| 45 | 81 | 3 | 4.2 |

**45.** $x^2 + 5x + 6$   **47.** $x^2 - x - 12$

## EXERCISES 7.3

**1.** $x^2 + 4x + 4$; pst   **3.** $x^2 - 4$; ds   **5.** $y^2 - 9$; ds
**7.** $4x^2 - 1$; ds   **9.** $4x^2 + 4x + 1$; pst   **11.** $x^2 - 10x + 25$; pst
**13.** $9x^2 - 6x + 1$; pst   **15.** pst; $(x - 3)(x - 3)$   **17.** neither
**19.** pst; $(x - 4)(x - 4)$   **21.** pst; $(y + 6)(y + 6)$   **23.** neither
**25.** ds; $(2x - 1)(2x + 1)$
**27.** triples: a to e; g and h are triples but not triangles.
**29.** $x = \frac{1}{4}, x = -1$   **31.** $x = 1, x = -\frac{2}{7}$   **33.** $x = \frac{2}{5}, x = -\frac{1}{2}$
**35.** $t \approx 4.14$ sec   **37.** $t \approx 3.57$ sec
**39.** $x = 21$; $w = 35$; both are right triangles
**41.**

| Leg | Leg | Hypotenuse |
|---|---|---|
| 3 | 4 | 5 |
| 6 | 8 | 10 |
| 18 | 24 | 30 |
| 9 | 12 | 15 |
| 1 | $\frac{4}{3}$ | $\frac{5}{3}$ |

**43. a.** no real solution   **b.** $t \approx 4.04$ sec

## MID-CHAPTER 7 TEST

**1. a.** not a right triangle   **b.** not a right triangle   **c.** right triangle
**2. a.** 6   **b.** 18   **c.** 12   **d.** 12   **e.** 27
**3. a.** $x^2 + 8x + 16$; pst   **b.** $4x^2 - 20x + 25$; pst
**c.** $x^3 + 1$; neither   **d.** $x^2 - \frac{1}{9}$; ds
**4. a.** $(2x + 5)(2x - 5)$; ds   **b.** $(x - 2)(x - 18)$; neither
**c.** $4(x^2 + 2)$; neither   **d.** $(x - 6)(x - 6)$; pst
**5. a.** $x \approx -3.1$; $x \approx 5.1$   **b.** $x = -2$; $x = 4$
**6. a.** $x = \pm\frac{1}{3}$   **b.** $x = \pm3$   **c.** $x = 1$; $x = 4$   **d.** $x = -1$; $x = 5$
**e.** $x \approx 0.886$; $x \approx -3.386$   **f.** $x \approx -0.219$; $x \approx -2.281$
**7. a.** $r = \pm\frac{1}{2}\sqrt{\frac{S}{\pi}}$   **b.** $v = \pm\sqrt{2gh}$
**8. a.** square: $x \approx 7.07$ ft   **b.** circle: $x \approx 7.98$ ft
**c.** triangle: $x \approx 10.75$ ft
**9.** square = 28.28 ft; circle = 25.07 ft; triangle = 32.24 ft; circle costs less

## EXERCISES 7.4

**1.** $m = 3$; $d = 2\sqrt{10} \approx 6.325$   **3.** $m = -1$; $d = 3\sqrt{2} \approx 4.243$
**5.** $m = -\frac{1}{7}$; $d = 5\sqrt{2} \approx 7.071$   **7.** $m = -\frac{1}{3}$; $d = 2\sqrt{10} \approx 6.325$
**9.** $y = 3x - 3$   **11.** $y = -x + 4$
**13.** 1 and 7 are perpendicular because their slopes multiply to $-1$
**15.** $\sqrt{18}, 6, \sqrt{18}$; isosceles right triangle
**17.** $\sqrt{13}, 4, \sqrt{13}$; isosceles triangle
**19.** $\sqrt{13}, \sqrt{52}, \sqrt{65}$; right triangle
**21. c.** forms 3-4-5 right triangles, the legs are on the axis grid

**23. a.**

| Hours | Car 1 | Car 2 |
|---|---|---|
| 1 | 40 | 20 |
| 2 | 80 | 40 |
| 3 | 120 | 60 |
| $t$ | $40t$ | $20t$ |

**c.** 1 hr $\approx 44.7$ mi apart; 2 hr $\approx 89.4$ mi apart; 3 hr $\approx 134.2$ mi apart
**d.** $\approx 44.7t$ mi
**25.** not similar; $9^2 + 16^2 \neq 25^2$; not a right triangle

## EXERCISES 7.5

**1.** $|b|\sqrt{a}, a \geq 0$   **3.** $|ab|$   **5.** $a^2|b|$
**7.** $x^2$ is always positive.   **9.** $7x$   **11.** $x\sqrt{y}$   **13.** $p\sqrt{p}$
**15.** $p^4$   **17.** $x = 11$; $x \geq 2$   **19.** $x = 7$; $x \geq -1$
**21.** $x = 13$; $x \geq 1$   **23.** $x = 3$; $x \leq 4$
**25.** $x = 6$; $\sqrt{x - 2}$ is undefined for $x < 2$
**27.** $x = 0$; $\sqrt{x + 1}$ is undefined for $x < -1$   **29.** 3 mi
**31.** 6 mi   **33.** 46.7 mi   **35.** 38.8 mi   **37.** 600 ft
**39.** 4 times higher   **41.** 10.6 sec; 341 ft/sec
**43.** 8.1 sec; 261 ft/sec   **45.** 2318.4 ft   **47.** 18.1 sec

## EXERCISES 7.6

**1.** $26,000; $12,500   **3.** $26,000; $27,500
**5.** $10,000; $13,000   **7.** $23,750; $27,500

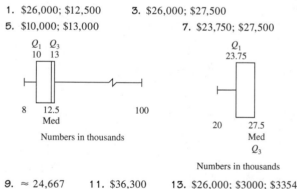

**9.** $\approx 24,667$   **11.** $36,300   **13.** $26,000; $3000; $3354
**15.** A few houses may be sold for large amounts of money. This would raise the average (mean) but not the median.
**17.** No; we need to know individual home prices to calculate the standard deviation.
**19.** Students may not have used measurement with inches recently.
**23.** 3470°F to 3530°F   **25.** 6.049 cm to 6.101 cm
**27.** 1 min to 4 min   **29.** 0.21 ppm to 0.39 ppm
**31.** 48.05 gal/min to 48.75 gal/min

## CHAPTER 7 REVIEW EXERCISES

**1. a.** not a right triangle   **b.** not a right triangle   **c.** right triangle
**3.** 24.7 ft   **5. a.** $2\sqrt{15}$   **b.** $3\sqrt{7}$   **c.** $3\sqrt{6}$
**7. a.** $x^2 - 16$; ds   **b.** $x^3 - 8$; neither   **c.** $9x^2 - 4$; ds
**d.** $x^2 - x + \frac{1}{4}$; pst

9. a. rectangle; $\sqrt{26}$, $\sqrt{26}$   b. rectangle; $\sqrt{40}$, $\sqrt{40}$

   c. $\sqrt{29}$, 5

11. a. coordinate $a = 11$; side $b = 2.25$

   b. side $c = 13$; side $d = 7.22$

   c. side $e = 17$; side $f = 9.71$

15. a. $x = \pm\frac{5}{7}$   b. $x = \pm\frac{9}{4}$

17. a. $x = -6$, $x = 3$   b. $x = 7$, $x = -3$   c. $x = 2\frac{1}{3}$, $x = -1$

   d. $x = \frac{1}{2}$, $x = -\frac{3}{2}$

19. $h = \dfrac{x\sqrt{3}}{2}$; $A = \dfrac{x^2\sqrt{3}}{4}$

21. a. $r = \pm\sqrt{\dfrac{A}{\pi}}$   b. $v = \pm\sqrt{\dfrac{2p}{d}}$   c. $T = 4l^2n^2m$

   d. $T = 4\pi dl^2n^2r^2$

23. a. median = 3   b.

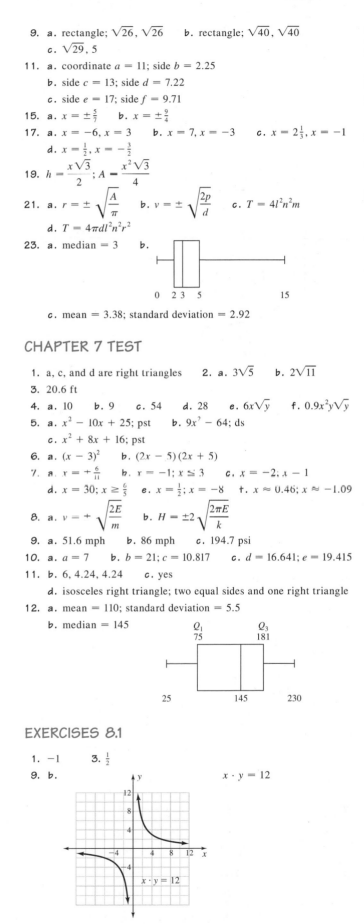

   c. mean = 3.38; standard deviation = 2.92

## CHAPTER 7 TEST

1. a, c, and d are right triangles   2. a. $3\sqrt{5}$   b. $2\sqrt{11}$

3. 20.6 ft

4. a. 10   b. 9   c. 54   d. 28   e. $6x\sqrt{y}$   f. $0.9x^2y\sqrt{y}$

5. a. $x^2 - 10x + 25$; pst   b. $9x^2 - 64$; ds

   c. $x^2 + 8x + 16$; pst

6. a. $(x - 3)^2$   b. $(2x - 5)(2x + 5)$

7. a. $x = +\frac{6}{11}$   b. $x = -1$; $x \leq 3$   c. $x = -2$; $x - 1$

   d. $x = 30$; $x \geq \frac{6}{5}$   e. $x = \frac{1}{2}$; $x = -8$   f. $x \approx 0.46$; $x \approx -1.09$

8. a. $v = +\sqrt{\dfrac{2E}{m}}$   b. $H = \pm 2\sqrt{\dfrac{2\pi E}{k}}$

9. a. 51.6 mph   b. 86 mph   c. 194.7 psi

10. a. $a = 7$   b. $b = 21$; $c = 10.817$   c. $d = 16.641$; $e = 19.415$

11. b. 6, 4.24, 4.24   c. yes

   d. isosceles right triangle; two equal sides and one right triangle

12. a. mean = 110; standard deviation = 5.5

   b. median = 145

## EXERCISES 8.1

1. $-1$   3. $\frac{1}{2}$

9. b.

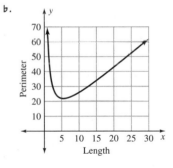

$x \cdot y = 12$

c.

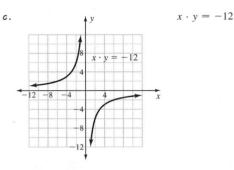

$x \cdot y = -12$

e. no   f. They form rectangular hyperbolas.   g. no; no

11. $\dfrac{1000 \times 10^9}{x} = y$ days   13. 16,700 days or 46 yr

15. a. $y$ is number of terms   b. Available aid will last $7\frac{1}{2}$ terms.

17. You are not moving.   19. no

21. a.

| Width | Length | Area | Perimeter |
|-------|--------|------|-----------|
| 1 | 30 | 30 | 62 |
| 2 | 15 | 30 | 34 |
| 3 | 10 | 30 | 26 |
| 5 | 6 | 30 | 22 |
| 6 | 5 | 30 | 22 |
| 10 | 3 | 30 | 26 |
| 15 | 2 | 30 | 34 |

b.

c. $\approx (5.477, 21.91)$   d. $y = 2x + 2\left(\dfrac{30}{x}\right)$

23. For $y = \dfrac{1}{x}$, the lines of symmetry are $y = x$ and $y = -x$.

   For $y = \dfrac{1}{x^2}$, the line of symmetry is $x = 0$.

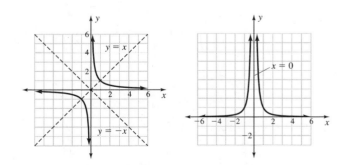

29. equation c

## EXERCISES 8.2

**1.** $\frac{1}{3} + \frac{1}{4} = \frac{7}{12}$   **3.** $\frac{1}{4} + \frac{1}{12} = \frac{1}{3}$   **5.** $\frac{3}{4} + \frac{1}{6} = \frac{11}{12}$
$\frac{1}{3} - \frac{1}{4} = \frac{1}{12}$        $\frac{1}{4} - \frac{1}{12} = \frac{1}{6}$        $\frac{3}{4} - \frac{1}{6} = \frac{7}{12}$
$\frac{1}{3} \cdot \frac{1}{4} = \frac{1}{12}$        $\frac{1}{4} \cdot \frac{1}{12} = \frac{1}{48}$        $\frac{3}{4} \cdot \frac{1}{6} = \frac{1}{8}$
$\frac{1}{3} \div \frac{1}{4} = \frac{4}{3}$        $\frac{1}{4} \div \frac{1}{12} = 3$        $\frac{3}{4} \div \frac{1}{6} = \frac{9}{2}$

**7.** $\frac{a}{b} + \frac{c}{d} = \frac{ad + bc}{bd}$   **9.** $\frac{1}{4} = 4^{-1}$; 4 is the opposite of $-4$
$\frac{a}{b} - \frac{c}{d} = \frac{ad - bc}{bd}$

$\frac{a}{b} \cdot \frac{c}{d} = \frac{ac}{bd}$

$\frac{a}{b} \div \frac{c}{d} = \frac{ad}{bc}$

**11.** $\frac{1}{a} = a^{-1}$; $a$ is the opposite of $-a$   **13.** $-a + b$   **15.** $a - b$

**17.** $-b - a$   **21.** 9; 1; $-9$; $-9$; 1; $-1$; 4 outcomes

**23.** $-\frac{1}{4}$, $-\frac{1}{4}$; equal   **25.** 2, $-2$; not equal   **27.** $\frac{a}{b}$, $\frac{a}{b}$; equal

**29.** $-1$; opposite numerator and denominator simplify to $-1$

**31.** $-1$   **33.** $x + 2$; $x \neq -2$   **35.** $2 - x$; $x \neq 2$

**37.** $b - a$; $a \neq b$   **39.** $3 - x$; $x \neq 3$   **41.** not opposites

**43.** opposites   **45.** true; only factors may be canceled

**47.** $a \neq 0, b \neq 0$; $\frac{1}{3a}$   **49.** $c \neq 0, d \neq 0$; $\frac{3d}{2c}$

**51.** $x \neq -3$; $x - 2$   **53.** $x \neq -2, 2$; $-\frac{1}{x+2}$

**55.** $b \neq 0, -c$; $\frac{3a}{5b}$   **57.** $x \neq 0, 2$; $\frac{2(x+2)}{x-2}$

**59.** $x \neq -3, -2$; $\frac{x-2}{x+3}$   **61.** $x \neq 0, 2$; $\frac{x+3}{x}$   **63.** $x \neq 3$; $-\frac{1}{2}$

**65.** $x \neq 3, -3$; $\frac{x-2}{x+3}$   **67.** 18 m   **69.** $\frac{1}{12 \text{ in.}}$   **71.** 20 gal

**73. a.** always   **b.** sometimes   **c.** always   **d.** always

**75.** 30; not simplified   **77.** $x$; simplified

## EXERCISES 8.3

**1.** Both equal; division by a number is the same as multiplication by its reciprocal.
**a.** 25   **b.** 500

**3. a.** $\frac{1}{ab}$   **b.** $a^2$   **c.** $\frac{a}{b}$   **d.** $\frac{a}{b}$   **e.** $\frac{a}{b^2}$   **f.** $\frac{a^2}{b}$

**5. a.** $x$   **b.** $\frac{1}{a^3b^2}$   **c.** $\frac{b}{a}$   **d.** $\frac{b^2}{a}$

**7. a.** $x \neq -1, 0$; 1   **b.** $x \neq -2, 0, 1$; $\frac{x-2}{x(x-1)}$

**c.** $x \neq -2, 0, 2$; $\frac{1}{x(x-2)}$   **d.** $x \neq -5, 0, 5$; $\frac{x}{(x+5)(x-5)}$

**e.** $x \neq -3, 0, 3$; $\frac{x-3}{(x+3)(x+3)}$

**f.** $x \neq -1, 1, 2$; $-\frac{x(x-1)}{(x-2)(1+x)}$

**9.** $I = \frac{t^2}{2t+1}$   **13.** $2x + 3$   **15.** $4x + 3$   **17.** $x + 8$

**19.** $2x + 12$   **21.** $2x - 1$   **27.** hours   **29.** 500 sec

**31.** 3 mpg   **33.** 600 hr/bottle   **35.** 432 stitches/yd

**37.** 0.283 pg/min   **39.** $\approx 3.33$ L

## MID-CHAPTER 8 TEST

**1.** $-2, 1$

**2.**

| Input | Output | Budget |
|-------|--------|--------|
| 10    | 480    | 4800   |
| 20    | 240    | 4800   |
| 100   | 48     | 4800   |
| 1000  | 4.80   | 4800   |
| 1200  | 4.00   | 4800   |

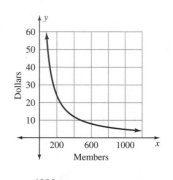

**3.** $y = \frac{4800}{x}$   **4.** approaches $\pm$ infinity

**5.** approaches $\pm$ infinity   **6.** approaches $-\frac{1}{2}$   **7.** 9 weeks

**8.** 2 in./sec   **9.** 80 mL   **10.** $x \neq 0$; $\frac{6c}{7a}$

**11.** $x \neq 2$; no common factors so it does not simplify

**12.** $x \neq -2, 2$; $\frac{1}{x-2}$   **13.** $x \neq 1, 2$; $\frac{x-1}{x-2}$

**14.** $\frac{3}{xy}$, $x \neq 0, y \neq 0$   **15.** $\frac{2}{x}$, $x \neq 0, y \neq 0$

**16.** $\frac{(x-6)(x+1)(x-2)}{x-1}$, $x \neq -2, 1$   **17.** $\frac{x}{x-4}$, $x \neq -4, 3, 4$

**18.** 17, $x \neq 0$   **19.** $6x + 4$, $x \neq -1, 0$   **20.** $\frac{4}{x^3}$, $x \neq 0$

## EXERCISES 8.4

**1. a.** 1   **b.** $\frac{2+x}{3}$   **c.** $-\frac{1}{x}$, $x \neq 0$   **d.** $\frac{4-x^2}{x^2+1}$   **e.** $-1$, $x \neq 1$

**3. a.** 60   **b.** $2x$   **c.** $y^2$   **d.** $ab$   **e.** $(x+3)(x-3)$
**f.** $(x+3)(x+2)(x-3)$

**5. a.** $\frac{23}{30}$   **b.** $\frac{9}{2x}$, $x \neq 0$   **c.** $\frac{8y-1}{y^2}$, $y \neq 0$

**d.** $\frac{8a-b}{ab}$, $a \neq 0, b \neq 0$   **e.** $\frac{4x+10}{(x+3)(x-3)}$, $x \neq -3, 3$

**f.** $\frac{2x-16}{(x+3)(x+2)(x-3)}$, $x \neq -3, -2, 3$

**7.** $\frac{6a+3b}{4ab}$, $a \neq 0, b \neq 0$   **9.** $\frac{2x-1}{x(x-1)}$, $x \neq 0, 1$

**11.** $\frac{5x-3}{x(x+1)}$, $x \neq -1, 0$   **13.** $\frac{2x+9}{(x+3)(x+3)}$, $x \neq -3$

**15.** $\frac{a-1}{a^2}$, $a \neq 0$   **17.** $\frac{4-3a}{2ab}$, $a \neq 0, b \neq 0$

**19.** $\frac{x^2+3x-6}{x(x-1)(x-2)}$, $x \neq 0, 1, 2$   **21.** $\Delta T = \frac{T_0 - T}{T}$

**23.** $\dfrac{1}{R} = \dfrac{R_2 + R_1}{R_1 R_2}$   **25.** $\dfrac{1}{D} = \dfrac{D_2 + D_1}{D_1 D_2}$   **27.** $\dfrac{1}{m} = \dfrac{m_2 + m_1}{m_1 m_2}$

**29.** $R = \dfrac{C^2 + 4m^2}{8M}$   **31.** $e^x \approx \dfrac{24 + 24x + 12x^2 + 4x^3 + x^4}{24}$

**33.** 5   **35.** $\frac{1}{5}$   **37.** $-\dfrac{a}{3b}$   **39.** $\dfrac{b}{a - c}$; slopes are equal

**41.** $\frac{25}{6}$   **43.** $h = \dfrac{2A}{b}$   **45.** $\dfrac{9x}{2(6 - x)}$   **47.** $\dfrac{Q_L}{Q_H - Q_L}$

**49.** One is half as long as the other.

**51.** yes; substituting $(a, b)$ into $y = \dfrac{b}{a}x + B$ gives $B = 0$

## EXERCISES 8.5

**1.** $d = \frac{36}{5}$   **3.** $x = \frac{40}{13}$   **5.** $d = 7.2$   **7.** $x \approx 3.077$

**13.** $x = 67.2$; none   **15.** $x = \frac{20}{19}$; $x \neq 0$   **17.** $x = \frac{8}{3}$; $x \neq 0$

**19.** $x = 2$; $x \neq 0$   **21.** no solution; $x \neq 0$

**23.** $x = 5$; $x \neq -3, 1$   **25.** $x = 1, x = 2$; $x \neq 0$

**27.** no solution; $x \neq 3$   **31.** 6.5 days   **33.** 2.2 hr; yes

**35.** one that vents in $7\frac{1}{2}$ hr   **37.** 3.6 min

**39.** $b = \dfrac{ac}{a - c}$   **41.** $c = \dfrac{ab}{a + b}$

## CHAPTER 8 REVIEW EXERCISES

**3.**

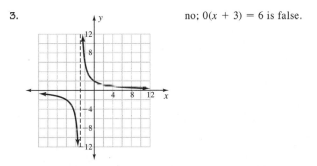

no; $0(x + 3) = 6$ is false.

**5.** Only a and b satisfy the simplification property of fractions.
   **a.** $\frac{3}{4}$; equivalent   **b.** $\frac{3}{4}$; equivalent   **c.** $\frac{2}{3}$   **d.** $\frac{4}{5}$

**7.** $\dfrac{2y}{x}$   **9.** $a - b$   **11.** no common factors   **13.** $-1$; $a \neq 1$

**15.** $3 - x$   **17.** $a - b$   **19.** $x - 4$   **21.** **b.** Both equal 100.

**23.** $\dfrac{x^2 - 2x - 2}{(x + 2)(x - 2)}$   **25.** $\dfrac{a^2 - 2ab - b^2}{(a + b)(a - b)}$   **27.** $\dfrac{x + 1}{x}$

**29.** $-\dfrac{(x + 3)^2}{x^2}$   **31.** $x + 2$   **33.** altogether; $\frac{7}{12}$

**35.** what fraction remains; $\frac{2}{3}$   **37.** how many servings; 16.4

**39.** $\dfrac{3 + 36a - a^2}{3a}$   **41.** $\frac{1}{3}$ shirt/card   **43.** 3 tab   **45.** 5 mL

**47. a.** approaches $-$ infinity   **b.** approaches $-1$   **c.** approaches $-1$

**49.** $\cos x \approx \dfrac{720 - 360x^2 + 30x^4 - x^6}{720}$   **51.** $x = \frac{45}{4}$, $x \neq 0$

**53.** $x = 5$; $x \neq 0, 1$   **55.** $\dfrac{2(3b - a)}{3(a - b)}$   **57.** 60 days

## CHAPTER 8 TEST

**1.** 4   **2.** simplifies to $-1$ because $4 - x$ is the opposite of $x - 4$

**3.**

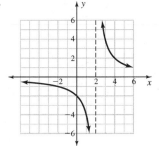

**4.** Only a and d satisfy the simplification property of fractions.
   **a.** $\frac{2}{3}$; equivalent   **b.** $\frac{1}{2}$   **c.** $\frac{3}{4}$   **d.** $\frac{2}{3}$; equivalent

**5.** $\dfrac{\sqrt{4}}{\sqrt{25}} = \dfrac{2}{5}$, not equal   **6.** $\dfrac{b}{3a}$, $a \neq 0, b \neq 0$

**7.** no common factors   **8.** $a + b$, $a \neq b$

**9.** $\dfrac{x}{x + 1}$, $x \neq -1$, $y \neq 0$   **10.** $\dfrac{(1 - x)(x - 1)}{x^2}$, $x \neq 1, 0$

**11.** $\dfrac{1}{(1 - x)(x - 1)}$, $x \neq -1, 1$   **12.** $\dfrac{n + 1}{n - 1}$, $n \neq 0, 1$

**13.** $\dfrac{x - 2}{x + 2}$, $x \neq -2, 0, 2$   **14.** $3x + 1$, $x \neq -1, 1$

**15.** 7.3 min   **16.** \$35.92/shirt   **17.** \$0.10/mi   **18.** 6 tab

**19.** true, by cross multiplication or multiplying the first fraction by $\dfrac{1.4}{1.4}$

**20.** $\dfrac{q + p}{pq}$   **21.** $\dfrac{2A}{b}$   **22.** $\frac{31}{45}$   **23.** $\dfrac{4 - a^2 b}{2ab^2}$, $a \neq 0, b \neq 0$

**24.** $\dfrac{x^2 - 2x + 2}{(x + 2)(x - 2)}$, $x \neq -2, 2$   **25.** $\dfrac{2x^2 + 9x - 12}{3x^2}$, $x \neq 0$

**26.** $\dfrac{2a + 1}{4}$   **27.** $x = 2$, $x \neq 0$   **28.** $x = 1$, $x \neq 0$

**29.** $x = 9$, $x \neq 0, 3$   **30.** $x = 5, x = -3$, $x \neq 0$